Springer INdAM Series

Volume 49

This series will publish textbooks, multi-authors books, thesis and monographs in English language resulting from workshops, conferences, courses, schools, seminars, doctoral thesis, and research activities carried out at INDAM - Istituto Nazionale di Alta Matematica, http://www.altamatematica.it/en. The books in the series will discuss recent results and analyze new trends in mathematics and its applications.

THE SERIES IS INDEXED IN SCOPUS

More information about this series at https://link.springer.com/bookseries/10283

Carla Manni · Hendrik Speleers
Editors

Geometric Challenges in Isogeometric Analysis

Springer

Editors
Carla Manni
Department of Mathematics
University of Rome Tor Vergata
Rome, Italy

Hendrik Speleers
Department of Mathematics
University of Rome Tor Vergata
Rome, Italy

ISSN 2281-518X ISSN 2281-5198 (electronic)
Springer INdAM Series
ISBN 978-3-030-92312-9 ISBN 978-3-030-92313-6 (eBook)
https://doi.org/10.1007/978-3-030-92313-6

This Springer imprint is published by the registered company Springer Nature Switzerland AG
The registered company address is: Gewerbestrasse 11, 6330 Cham, Switzerland

Preface

The INdAM workshop "Geometric Challenges in Isogeometric Analysis" was held in Rome (Italy) from January 27 to 31, 2020. The workshop provided a discussion forum for researchers working in the area of isogeometric analysis and related fields, with a particular focus on the issues pertaining to the discretization of differential problems over complex geometries. The efficient and accurate description of complex geometries with tools that are appropriate for analysis (analysis-suitable) is one of the research fields of greatest interest for—but definitely not limited to—the isogeometric paradigm. Spaces of smooth (spline) functions on unstructured meshes, suitable for both geometric design and numerical simulation, are indispensable for isogeometric analysis to exploit its full potential for the treatment of complex real-life problems.

Thematic survey talks were complemented with talks presenting recent and innovative results on themes central to the scope of the workshop. The addressed topics include

- high-quality spline surfaces on complex and trimmed geometries,
- construction and analysis of smooth spline spaces on unstructured meshes,
- numerical aspects and benchmarking of isogeometric discretizations on unstructured meshes, meshing strategies, and software.

The present volume comprises a selection of fifteen contributions by several workshop participants and their collaborators.

The construction of smooth spline surfaces of any topological genus and of high quality on complex and trimmed geometries is the topic of the papers, "Quadrilateral Orbifold Splines" by Beccari and Prautzsch, "Trimmed Spline Surfaces with Accurate Boundary Control" by Martin and Reif, and "Subdivision and G-Spline Hybrid Constructions for High-Quality Geometric and Analysis-Suitable Surfaces" by Peters and Karčiauskas.

The two papers, "B-Symmetric Univariate Splines and Euler Numbers" by Boafo-Adade et al., and "Simplex-Splines on the Clough-Tocher Split with Arbitrary Smoothness" by Lyche et al., deal with structural properties of specific spline spaces. In particular, B-spline-like bases and Marsden-like identities are proposed for the Clough-Tocher split of any smoothness.

The set of papers, "DPB-Splines: The Decoupled Basis of Patchwork Splines" by Hemelmayr and Jüttler, "Completeness Characterization of Type-I Box Splines" by Villamizar et al., and "THU-Splines: Highly Localized Refinement on Smooth Unstructured Splines" by Wei, deal with certain hierarchical structures and their generalizations, with a special focus on their representation in terms of (truncated) hierachical B-spline bases, their completeness, and appropriate refinement strategies.

The papers, "A Collocation IGA-BEM for 3D Potential Problems on Unbounded Domains" by Falini et al., "Scattered Data Approximation by LR B-Spline Surfaces: A Study on Refinement Strategies for Efficient Approximation" by Skytt and Dokken, and "C^1 Triangular Isogeometric Analysis of the von Karman Equations" by Zareh and Qian, have a more applicative flavor. They present numerical experiments and discuss practical aspects related to applications of suitable spline spaces in different contexts ranging from isogeometric discretizations to approximation methods.

The pair of papers, "Fast Formation and Assembly of Isogeometric Galerkin Matrices for Trimmed Patches" by Marussig, and "A Block ILUT Smoother for Multipatch Geometries in Isogeometric Analysis" by Tielen et al., address the problems of efficient assembly of matrices arising from isogeometric methods and of design of appropriate solvers in the context of trimmed and multipatch geometries.

The interplay between meshing and bases as well as practical aspects of meshing are discussed in the two papers, "Meshing as the Choice of Basis Functions for Finite Element Analysis" by Sabin, and "HexGen and Hex2Spline: Polycube-Based Hexahedral Mesh Generation and Spline Modeling for Isogeometric Analysis Applications in LS-DYNA" by Yu et al.

We express our deepest gratitude to everyone who has contributed to the success of this workshop. We wish to thank all the speakers, as well as all the participants, for the interesting talks, the vivid scientific discussions, and the nice atmosphere during the week in Rome. This volume would not have been possible without the authors of the collected papers, who deserve our sincerest appreciation for their interesting works. Also, special thanks go to the anonymous referees for their precious collaboration.

Finally, we would like to thank INdAM for the principal financial support for the workshop organization, which also included the nice facilities in Rome. Additional support was received from the University of Rome Tor Vergata and its Department of Mathematics (MIUR Excellence Department Project 2018–2022).

Rome, Italy
June 2021

Carla Manni
Hendrik Speleers

Contents

About the Editors

Carla Manni is a Full Professor of Numerical Analysis at the Department of Mathematics, University of Rome Tor Vergata, Italy. She received her Ph.D. in Mathematics from the University of Florence in 1990. Her research interest is primarily in spline functions and their applications, constrained interpolation and approximation, computer aided geometric design and isogeometric analysis. She is the author of more than 100 peer-reviewed research publications.

Hendrik Speleers received his Ph.D. in Engineering (Numerical Analysis and Applied Mathematics) from the University of Leuven, Belgium in 2008. He is currently a Full Professor of Numerical Analysis at the Department of Mathematics, University of Rome Tor Vergata, Italy. His main research interest is in the construction, analysis, and application of multivariate splines. He is the author of more than 75 peer-reviewed scientific papers.

Quadrilateral Orbifold Splines

Carolina Vittoria Beccari and Hartmut Prautzsch

Abstract Orbifold spline surfaces are closed free form surfaces of any topological genus. Their patches have C^k joints along common boundaries after linear rational reparametrizations, where k can be chosen arbitrarily. There are orbifold splines with rational triangular and—with certain restrictions on the patch layout—polynomial quadrilateral patches. Here, we present orbifold splines with rational quadrilateral patches without any restrictions on the patch layout. They are obtained from triangular orbifold splines by rational linear or bi-linear reparametrizations. With linear reparametrizations, we get discontinuously and otherwise continuously parametrized G^k surfaces.

1 Introduction

Constructing smooth free form surfaces necessarily requires so-called G^k constructions, i.e., reparametrizations that typically raise the degree of the polynomial patches of such surfaces. Thus, low-degree reparametrization functions are desirable and, if polynomials are used, their degree can be $k + 1$, where k is the smoothness order [15, 22]. In view of the degree estimates in [18, 21], $k + 1$ seems to be the lowest possible degree in general. Therefore, it is intriguing that linear rational reparametrizations are sufficient if we work with rational patches parametrized over the hyperbolic plane. This then leads to rational orbifold splines introduced to Computer Aided Geometric Design and studied by [2, 3, 19]. The spline manifolds introduced in [7, 8] are related but are built with affine reparametrizations and therefore necessarily have one singular point.

C. V. Beccari
Alma Mater Studiorum - Università di Bologna, Bologna, Italy
e-mail: carolina.beccari2@unibo.it

H. Prautzsch (✉)
Karlsruhe Institute for Technology, Karlsruhe, Germany
e-mail: prautzsch@kit.edu

C. Manni and H. Speleers (eds.), *Geometric Challenges in Isogeometric Analysis*, Springer INdAM Series 49, https://doi.org/10.1007/978-3-030-92313-6_1

The reparametrization functions for G^k constructions are called **transition functions** and linear rational transition functions are projective maps defined by the weighted vertices (see Definition 1) of a triangle and its image. Hence, if we wish to use such functions, it is natural to work with triangular patches as in the references above. Since we may replace any transition function by its k-th degree Taylor polynomial for the G^k constructions, it is also possible to come up with polynomial orbifold splines with linear rational transition functions. The degree, however, goes faster up with k then for rational orbifolds. All this has been derived by J. Peters and coauthors [13, 14, 23] using a purely algebraic approach as opposed to the geometric one employed by the other authors based on hyperbolic geometry.

Since the algebraic approach is quite involved, certain symmetry assumptions helped Peters to get it through with explicit formulae and sophisticated constructions. On the downside of it, this led to certain restrictions on the patch layout.

Although we could give up on Peters' symmetry assumption, a geometric look reveals that there are intrinsic restrictions on the patch layout if we want to work with either polynomial or rational quadrilateral patches and projective, i.e., linear rational transition functions [16] that only reparametrize cross boundary derivatives, but no patch boundaries.

In this paper, we show how to get around these problems by either projective transition functions that lead to reparametrizations of boundary curves and to C^{-1} joints of adjacent patch parametrizations or by bi-projective, i.e., bi-linear, rational transition functions contacting adjacent ones continuously.

The quadrilateral orbifold splines we present have no restrictions on their patch layout and share the same low degree with triangular orbifold splines. However, here it is a bi-degree n, whereas for triangles it is a total degree n. Furthermore, triangular orbifold G^k splines can be built with Powell-Sabin splits, or by enforcing G^{2k} supersmoothness at the vertices, or by solving simply the system of G^k conditions. Our quadrilateral orbifold splines are obtained from any such triangular splines where we pair triangles into quadrilaterals. For each pair, we require G^{2k} contact, but do this only at the two common vertices. Otherwise, we require G^k contact between different pairs.

Since the construction is based on triangular orbifold splines and to keep the paper self-contained, we first introduce the essentials of triangular orbifold spline constructions starting with a review of the hyperbolic plane in Sect. 2, two different Bézier representations for homogenous polynomials in Sect. 3, and simple G^k joints for rational polynomials that can be expressed by C^k joints of homogenous polynomials in Sect. 4. Then in Sect. 5, we explain orbifold splines, come to projective structures in Sect. 6 and finally in Sects. 7 and 8 to the constructions of quadrilateral orbifold splines.

2 The Hyperbolic Plane

In this section, we briefly review three models of the hyperbolic plane $\mathscr{H}^2$ that have been used to introduce and develop orbifold splines in [2, 3, 6, 8, 19, 24].

The Hyperbolic Model

Most important for this paper is the **hyperbolic model** of the hyperbolic plane $\mathscr{H}^2$. In this model, $\mathscr{H}^2$ is formed by the pairs $(\mathbf{x}, -\mathbf{x})$ of points on the two-sheeted hyperboloid $\mathscr{H}$ given by the equation

$$\mathscr{H} : \mathbf{x}^t Q\mathbf{x} = \begin{bmatrix} x & y & z \end{bmatrix} \begin{bmatrix} 1 & & \\ & 1 & \\ & & -1 \end{bmatrix} \begin{bmatrix} x \\ y \\ z \end{bmatrix} = -1. \tag{1}$$

Note that the associated homogenous equation

$$\mathscr{C} : \quad \mathbf{x}^t Q\mathbf{x} = 0$$

defines the asymptotic cone $\mathscr{C}$ of the hyperboloid. We can simplify this model to the upper sheet $\mathscr{U}$ of the hyperboloid by replacing the pairs $(\mathbf{x}, -\mathbf{x})$ by the points $\mathbf{x} \in \mathscr{U}$, see Fig. 1.

A hyperbolic plane has (hyperbolic) lines and any such line is a non-empty intersection of $\mathscr{U}$ with any two-dimensional subspace of $\mathbb{R}^3$. The lines are preserved by the hyperbolic maps, which are the linear maps of $\mathbb{R}^3$ restricted to $\mathscr{U}$ that map

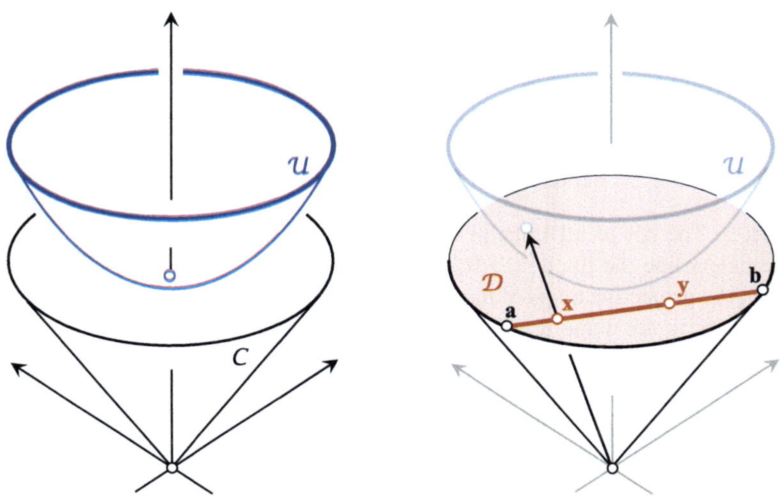

Fig. 1 The hyperbolic (left) and the Klein model (right) $\mathscr{H}$ of the hyperbolic plane

$\mathcal{U}$ into itself. For example, rotations around the z-axis and reflections about planes containing the z-axis represent special hyperbolic maps.

Remark 1 If $\mathcal{U}$ is mapped to itself by a linear map $\mathbf{x} \mapsto A\mathbf{x}$, then the quadratic polynomials $\mathbf{x}^t Q\mathbf{x} + 1$ and $(A\mathbf{x})^t Q(A\mathbf{x}) + 1$ have the same zero sets. This implies that a hyperbolic map is represented by a matrix A for which $A^t QA = Q$. Such a map $\mathbf{x} \mapsto A\mathbf{x}$ also maps the cone $\mathcal{C}$ onto itself. On the other hand, any linear map that maps $\mathcal{C}$ onto itself has a matrix A for which $A^t QA = Q\rho^2$, $\rho \neq 0$, is a multiple of Q or equivalently for which some multiple $A\rho$, $\rho \neq 0$, represents a hyperbolic map.

Remark 2 In incidence geometry, the lines form the structure of a geometry and the hyperbolic plane can be defined by Hilbert's axioms [9], [11, p. 268]. Different from that approach, Felix Klein proposed in his seminal Erlangen program (1872) to classify geometries by the invariants under an automorphism group. The hyperbolic maps form the automorphisms in hyperbolic geometry and their invariants are the lines, angles and distances.

The Klein Model

Orbifold splines satisfy smoothness conditions based on linear rational reparametrizations. Therefore, it is natural to work with rational orbifold splines. The Klein model of the hyperbolic plane shows where the linear rational maps come from.

We obtain the Klein model by the central projection

$$\pi : \quad \mathbb{R}^3 \setminus \{0\} \to \left\{ \begin{bmatrix} \mathbf{x} \\ 1 \end{bmatrix} \middle| \mathbf{x} \in \mathbb{R}^2 \right\}, \quad \begin{bmatrix} \mathbf{x} \\ z \end{bmatrix} \mapsto \begin{bmatrix} \mathbf{x}/z \\ 1 \end{bmatrix}$$

that maps $\mathcal{U}$ onto the open unit disk

$$\mathcal{D} : \quad x^2 + y^2 < 1, \quad z = 1,$$

and the hyperbolic lines into straight line segments, see Fig. 1. Under this projection, the linear maps in $\mathbb{R}^3$ become the linear rational maps of the plane $z = 1$ into itself.

Remark 3 The set of all lines in $\mathbb{R}^3$ through the origin forms a projective plane and we can define the lines $\begin{bmatrix} x & y & 0 \end{bmatrix}^t \mathbb{R}$ to be its ideal points. The linear maps in $\mathbb{R}^3$ represent the projective maps of the projective plane and projective extensions of the rational maps in the plane $z = 1$.

Thus, in the Klein model, the hyperbolic maps are projective maps restricted to $\mathcal{D}$ that map the disk $\mathcal{D}$ onto itself. Note that a projective map is represented by a linear map or a 3×3 matrix A and that any non-zero multiple of A represents the same projective map in accordance with Remark 1.

Projective maps preserve cross ratios. This can be used to define hyperbolic distances that are invariant under hyperbolic maps: Any two hyperbolic points $\mathbf{x}$ and

$\mathbf{y}$ span a hyperbolic line, which is the interior of a line segment $\mathbf{ab}$ in $\mathbb{R}^3$. The non-hyperbolic points $\mathbf{a}, \mathbf{b}$ lie on the boundary of $\mathscr{D}$ and the points on the boundary of $\mathscr{D}$ are the **ideal points** (of $\mathscr{H}^2$), see Fig. 1. If we extend hyperbolic maps to the ideal points, they map them to ideal points. Thus,

$$\text{dist}(\mathbf{x}, \mathbf{y}) := \Big|\log\left(\text{cross ratio}[\mathbf{xy}|\mathbf{ab}]\right)\Big| ,$$

is invariant under hyperbolic maps and is called the **hyperbolic distance** of $\mathbf{x}$ and $\mathbf{y}$ [10, (1) on p. 214].

Remark 4 Any three points $\mathbf{a}, \mathbf{b}, \mathbf{c}$ on $\mathscr{U} \subset \mathbb{R}^3$ define a hyperbolic triangle and there is a unique linear map mapping $\mathbf{a}, \mathbf{b}, \mathbf{c} \in \mathbb{R}^3$ to the vertices in a given order of any other hyperbolic triangle. It represents a projective map in the plane $z = 1$. Note that it is a hyperbolic map only if it maps $\mathscr{D}$ onto itself.

We mention this since, although the domain of an orbifold spline, which is defined in Sect. 5, is the hyperbolic plane, the G^k transition functions in Sect. 4 of an orbifold spline are such projective maps that need not be hyperbolic. Therefore, these transition functions form what is called a projective structure and we come to it in Sect. 6.

Definition 1 We denote the linear span without the zero element of any vectors $\mathbf{v}_1, \ldots, \mathbf{v}_n$ by $\langle \mathbf{v}_1, \ldots, \mathbf{v}_n \rangle$ and the linear span of a matrix A without the zero matrix by $\langle A \rangle$. Hence, for $n = 1$ and $\mathbf{x} = \mathbf{v}_1$, the preimage $\pi^{-1}(\mathbf{x})$ is $\langle \mathbf{x} \rangle$. This span is the class of all (homogeneous) coordinate vectors representing the point $\mathbf{x}$ and we also address it as the point $\langle \mathbf{x} \rangle$. Similarly, $\langle A \rangle$ is the class of all (homogeneous) matrices representing the same projective map and we also call the map $\langle A \rangle$. Further, we call any member $\mathbf{v}$ of $\langle \mathbf{v} \rangle$ a **weighted point**.

The Poincaré Model

For illustrations, such as in Fig. 4, we like to use the Poincaré model of the hyperbolic plane. It is obtained from the Klein model by mapping the disk $\mathscr{D}$ onto itself by the map

$$\begin{bmatrix} \mathbf{x} \\ \mathbf{y} \end{bmatrix} \mapsto \begin{bmatrix} \mathbf{x} \\ \mathbf{y} \end{bmatrix} \Big/ \left(1 + \sqrt{1 - x^2 - y^2}\right),$$

which maps the hyperbolic lines into circular arcs that meet the unit circle orthogonally. This map is a composition of the inverse of the orthogonal projection from the lower half of the sphere with equator $\mathscr{D}$ to $\mathscr{D}$ and the (angle and circle preserving) stereographic projection whose center is the north pole back to the disk $\mathscr{D}$, see Fig. 2.

In the Poincaré model, the hyperbolic maps are the circle relationships that map the disk $\mathscr{D}$ onto itself [12, Theorem 6.1 on p. 286]. Since they preserve angles, the hyperbolic maps do so too and, in this model, the **hyperbolic angles** are just the euclidean ones.

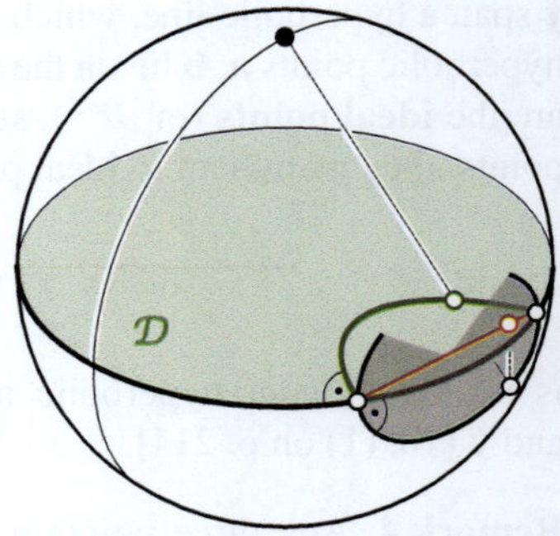

Fig. 2 The Poincaré model (green) of the hyperbolic plane

3 Rational Patches and Their Bézier Forms

We wish to construct piecewise rational splines mapping certain domains in the hyperbolic plane onto smooth closed surfaces. These splines consist of quadrilateral bi-rational patches obtained from pairs of triangular patches.

A triangular patch is defined over some triangle **abc** and first we assume **a**, **b**, **c** to lie in the open disk $\mathscr{D} \subset \mathbb{R}^3$ representing the Klein model of the hyperbolic plane. With the coordinate transformation

$$\mathbf{x} = \begin{bmatrix}\mathbf{a}\ \mathbf{b}\ \mathbf{c}\end{bmatrix}\mathbf{u}, \quad \mathbf{u} = \begin{bmatrix} u \\ v \\ w \end{bmatrix},$$

we write a rational polynomial of degree n as

$$\mathbf{r}(\mathbf{x}) = \frac{\mathbf{p}(\mathbf{u})}{q(\mathbf{u})}, \quad |\mathbf{u}| := u + v + w = 1,$$

or homogeneously as

$$\mathbb{r}(\mathbf{x}) = \begin{bmatrix} \mathbf{p}(\mathbf{u}) \\ q(\mathbf{u}) \end{bmatrix}.$$

With the Bernstein polynomials

$$B_{\mathbf{i}}(\mathbf{u}) = \binom{n}{\mathbf{i}}\mathbf{u}^{\mathbf{i}} := \frac{n!}{i!j!k!}u^i v^j w^k$$

for

$$\mathbf{i} = \begin{bmatrix} i \\ j \\ k \end{bmatrix} \in \Gamma_n^3 := \left\{\mathbf{i} \in \mathbb{Z}_{n+1}^3 \,\middle|\, |\mathbf{i}| = i + j + k = n\right\},$$

we get the **homogeneous Bézier form** or **homogeneous Bézier representation**

$$\mathbb{r}(\mathbf{x}) = \sum_{\mathbf{i} \in \Gamma_n^3} \mathbb{r}_\mathbf{i} B_\mathbf{i}(\mathbf{u})$$

of the rational patch.

Since any nonzero multiple of $\mathbb{r}$ represents the same point $\mathbf{p}/q$, we can omit the restriction $|\mathbf{u}| = 1$ and extend the polynomial $\mathbb{r}$ to a homogeneous polynomial of degree n over $\mathbb{R}^3$, i.e.,

$$\mathbb{r}(\mathbf{x}\rho) = \mathbb{r}(\mathbf{x})\rho^n \quad \text{for any} \quad \rho \in \mathbb{R}.$$

Remark 5 Writing the vectors $\mathbb{r}_\mathbf{i}$ as

$$\mathbb{r}_\mathbf{i} = \begin{bmatrix} \mathbf{r}_\mathbf{i}\rho_\mathbf{i} \\ \varepsilon_\mathbf{i}\rho_\mathbf{i} \end{bmatrix}, \quad \varepsilon_\mathbf{i} \in \{0, 1\},$$

we see the **Bézier points** $\mathbf{r}_\mathbf{i}$ and **weights** $\rho_\mathbf{i}$. If $\varepsilon_\mathbf{i} = 0$, then $\mathbf{r}_\mathbf{i}$ actually is a vector on whose length the weight ρ_i depends.

Remark 6 The homogeneous polynomial $\mathbb{r}(\mathbf{x})$ is completely determined by its values on any plane or surface passing through the cone

$$\begin{bmatrix} \mathbf{a}\ \mathbf{b}\ \mathbf{c} \end{bmatrix} \mathbb{R}^3_{>0}.$$

In particular, we could consider $\mathbb{r}(\mathbf{x})$ over the hyperbolic sheet $\mathscr{U}$ or over any triangle $\mathbf{a}\alpha\ \mathbf{b}\beta\ \mathbf{c}\gamma$ with $\alpha, \beta, \gamma \in \mathbb{R} \setminus \{0\}$ and derive the Bézier form of $\mathbb{r}$ with respect to it: Let $\mathbf{u}$ be the coordinate vector of any $\mathbf{x} \in \mathbb{R}^3$ with respect to the scaled basis $\mathbf{a}\alpha$, $\mathbf{b}\beta$, $\mathbf{c}\gamma$, i.e.,

$$\mathbf{x} = \begin{bmatrix} \mathbf{a}\alpha\ \mathbf{b}\beta\ \mathbf{c}\gamma \end{bmatrix} \mathbf{u} = \begin{bmatrix} \mathbf{a}\ \mathbf{b}\ \mathbf{c} \end{bmatrix} S\mathbf{u}, \quad S = \begin{bmatrix} \alpha & & \\ & \beta & \\ & & \gamma \end{bmatrix}.$$

Then $S\mathbf{u}$ represents $\mathbf{x}$ with respect to the initial basis $\mathbf{a}\ \mathbf{b}\ \mathbf{c}$ and because the Bernstein polynomials are homogeneous in each variable, we obtain from $\mathbb{r}$'s Bézier form

$$\mathbb{r}(\mathbf{x}) = \sum \mathbb{r}_\mathbf{i} B_\mathbf{i}(S\mathbf{u})$$

with respect to $\mathbf{a}\ \mathbf{b}\ \mathbf{c}$ directly the new one w.r.t. $\mathbf{a}\alpha\ \mathbf{b}\beta\ \mathbf{c}\gamma$:

$$\mathbb{r}(\mathbf{x}) = \sum \mathbb{r}_\mathbf{i}\, \alpha^i \beta^j \gamma^k B_\mathbf{i}(\mathbf{u}).$$

Note that the Bézier points $\mathbf{r}_\mathbf{i}$ are still the same. Only the weights $\rho_\mathbf{i}$ changed to $\rho_\mathbf{i}\alpha^i\beta^j\gamma^k$.

Remark 7 Substituting $\mathbf{u}$ by $S\mathbf{u}$ corresponds to a linear reparametrization with the map $\mathbf{u} \mapsto S\mathbf{u}$. If we view $\mathbb{r}$ as a rational polynomial $\mathbf{r}(\mathbf{x})$ over the plane $z = 1$, in which S represents a projective map mapping the triangle **abc** to itself, this reparametrization is a projective reparametrization of $\mathbf{r}(\mathbf{x})$, see [4, p. 244].

The homogeneous polynomial $\mathbb{r}(\mathbf{x})$ also has an **inhomogenous tetrahedral Bézier representation**. Since a homogeneous polynomial of degree n has zero derivatives at the origin $\mathbf{o}$ up to degree $n-1$, we easily get $\mathbb{r}$'s Bézier form

$$\mathbb{r}(\mathbf{x}) = \sum_{\mathbb{i} \in \Gamma_n^4} \mathbb{r}_{\mathbb{i}} B_{\mathbb{i}}(\mathbb{v})$$

over the tetrahedron **oabc** with

$$\mathbf{x} = \begin{bmatrix}\mathbf{o} & \mathbf{a} & \mathbf{b} & \mathbf{c}\end{bmatrix} \mathbb{v}, \qquad B_{\mathbb{i}}(\mathbb{v}) = \binom{n}{\mathbb{i}} \mathbb{v}, \qquad |\mathbb{v}| = 1,$$

and

$$\mathbb{r}_{\mathbb{i}} = \begin{cases} \mathbb{o} & \text{for} \quad \mathbb{i} = \begin{bmatrix} i_0 \\ \mathbf{i} \end{bmatrix} \text{ with } i_0 > 0\,, \\ \mathbb{r}_{\mathbf{i}} & \text{for} \quad \mathbb{i} = \begin{bmatrix} 0 \\ \mathbf{i} \end{bmatrix}. \end{cases}$$

4 Smooth Joints

As follows from the previous section, we can represent any piecewise rational spline over the hyperboloid $\mathscr{U}$, the disk $\mathscr{D}$ or any plane $\mathscr{E} \subset \mathbb{R}^3 \setminus \{0\}$ by a homogeneous spline $\mathbb{r} = \begin{bmatrix} \mathbf{p} \\ q \end{bmatrix}$. If this spline is in C^k, then its restriction to $\mathscr{U}$, $\mathscr{D}$ or $\mathscr{E}$ also is, and vice versa. Further, if the restriction is in C^k, then so is its central projection $\mathbf{r} = \mathbf{p}/q$. Although $\mathbb{r}$ could be less smooth then its projection, we only build rational C^k splines with C^k homogeneous representations since this is sufficient for our purpose and is the simplest.

To build such a homogeneous spline consisting of polynomial segments over pyramidal cones as $\begin{bmatrix}\mathbf{a} & \mathbf{b} & \mathbf{c}\end{bmatrix} \mathbb{R}^3 = \{\begin{bmatrix}\mathbf{a} & \mathbf{b} & \mathbf{c}\end{bmatrix} \mathbf{v} \mid \mathbf{v} \in \mathbb{R}^3\}$, we specify the smoothness conditions between adjacent segments using their homogeneous Bézier forms.

Given two triangles $A = \begin{bmatrix}\mathbf{a} & \mathbf{b} & \mathbf{c}\end{bmatrix}$ and $B = \begin{bmatrix}\mathbf{a} & \mathbf{b} & \mathbf{d}\end{bmatrix}$ in $\mathbb{R}^3$ with homogeneous polynomial pieces

$$\mathbb{r}(A\mathbf{u}) = \sum_{\mathbf{i} \in \Gamma_n^3} \mathbb{r}_{\mathbf{i}} B_{\mathbf{i}}(\mathbf{u}), \quad \mathbf{u} \geq 0,$$

and

$$\mathbb{s}(B\mathbf{u}) = \sum_{\mathbf{i}\in\Gamma_n^3} \mathbb{s}_\mathbf{i} B_\mathbf{i}(\mathbf{u}), \quad \mathbf{u} \geq 0,$$

we use their Bézier forms

$$\sum_{\mathbb{i}\in\Gamma_n^4} \mathbb{r}_\mathbb{i} B_\mathbb{i}(\mathbb{v}) \quad \text{and} \quad \sum_{\mathbb{i}\in\Gamma_n^4} \mathbb{s}_\mathbb{i} B_\mathbb{i}(\mathbb{v})$$

with respect to the tetrahedra $\begin{bmatrix}\mathbf{o}\ \mathbf{a}\ \mathbf{b}\ \mathbf{c}\end{bmatrix}$ and $\begin{bmatrix}\mathbf{o}\ \mathbf{a}\ \mathbf{b}\ \mathbf{d}\end{bmatrix}$, respectively, where

$$\mathbb{r}_\mathbb{i} = \begin{cases} \mathbb{r}_\mathbf{i} & \text{if} \quad \mathbb{i} = \begin{bmatrix} 0 \\ \mathbf{i} \end{bmatrix} \\ \mathbb{0} & \text{else} \end{cases}$$

and

$$\mathbb{s}_\mathbb{i} = \begin{cases} \mathbb{s}_\mathbf{i} & \text{if} \quad \mathbb{i} = \begin{bmatrix} 0 \\ \mathbf{i} \end{bmatrix} \\ \mathbb{0} & \text{else} \end{cases}.$$

We recall from [5] that $\mathbb{r}(\mathbf{x})$ and $\mathbb{s}(\mathbf{x})$ have equal derivatives up to order r over the plane **oab** if and only if, for all $k \leq r$ and all

$$\mathbb{i} = \begin{bmatrix} i_0 \\ i \\ j \\ k \end{bmatrix} = \mathbb{j} + \begin{bmatrix} 0 \\ 0 \\ 0 \\ k \end{bmatrix} \in \Gamma_n^4,$$

we have

$$\mathbb{s}_\mathbb{i} = \sum_{\mathbb{k}\in\Gamma_k^4} \mathbb{r}_{\mathbb{j}+\mathbb{k}} B_\mathbb{k}(\mathbb{d}),$$

where

$$\begin{bmatrix} \mathbf{o}\ \mathbf{a}\ \mathbf{b}\ \mathbf{c} \\ 1\ 1\ 1\ 1 \end{bmatrix} \mathbb{d} = \begin{bmatrix} \mathbf{d} \\ 1 \end{bmatrix}.$$

Since most $\mathbb{r}_\mathbb{i}$ are zero, these conditions simplify and can be written solely in terms of the $\mathbb{r}_\mathbf{i}$ and $\mathbb{s}_\mathbf{i}$. Namely, $\mathbb{r}(\mathbf{x})$ and $\mathbb{s}(\mathbf{x})$ have C^r contact if and only if for all $k \leq r$ and

$$\mathbf{i} = \begin{bmatrix} i \\ j \\ k \end{bmatrix} = \mathbf{j} + \begin{bmatrix} 0 \\ 0 \\ k \end{bmatrix} \in \Gamma_n^3,$$

we have

$$\mathbb{s}_{\mathbf{i}} = \sum_{\mathbf{k}\in\Gamma_k^3} \mathbb{r}_{\mathbf{j}+\mathbf{k}} B_{\mathbf{k}}(A^{-1}\mathbf{d}), \tag{2}$$

which has first been proved in [1], where homogeneous trivariate polynomials are used to define splines over spheres.

Remark 8 Using the local parameters $\mathbf{u} = A^{-1}\mathbf{x}$ and $\mathbf{v} = B^{-1}\mathbf{x}$, we obtain that

$$\mathbb{p}(\mathbf{u}) := \mathbb{r}(A\mathbf{u}) = \mathbb{r}(\mathbf{x}) \quad \text{and} \quad \mathbb{q}(\mathbf{v}) := \mathbb{s}(B\mathbf{v}) = \mathbb{s}(\mathbf{x})$$

have C^k contact after we reparametrize $\mathbb{p}(\mathbf{u})$ by the coordinate transformation

$$\mathbf{v} \mapsto \mathbf{u}(\mathbf{v}) = A^{-1}B\mathbf{v} = [\mathbf{e}_1\mathbf{e}_2A^{-1}\mathbf{d}]\mathbf{v}$$

taking us from system **abd** to **abc**.

Note that the matrix $A^{-1}B$ is completely defined by the vector $A^{-1}\mathbf{d}$ defining the C^r conditions (2). With respect to the first system **abc**, the matrix $A^{-1}B$ represents the linear map mapping **a**, **b**, **c** to **a**, **b**, **d**, respectively.

Remark 9 In the Klein model, the projective map $\langle A^{-1}B\rangle$ maps the points $\langle\mathbf{a}\rangle$, $\langle\mathbf{b}\rangle$, $\langle\mathbf{c}\rangle$ to $\langle\mathbf{a}\rangle$, $\langle\mathbf{b}\rangle$ and $\langle\mathbf{d}\rangle$. It is determined either by the weighted points **a**, **b**, **c** and their weighted images **a**, **b**, **d** or by the (non-weighted) points $\langle\mathbf{a}\rangle$, $\langle\mathbf{b}\rangle$, $\langle\mathbf{c}\rangle$ with the so-called unit point $\langle\mathbf{a}+\mathbf{b}+\mathbf{c}\rangle$ and their images $\langle\mathbf{a}\rangle$, $\langle\mathbf{b}\rangle$, $\langle\mathbf{d}\rangle$, $\langle\mathbf{a}+\mathbf{b}+\mathbf{d}\rangle$. Thus, the rational polynomials $\mathbf{r}(\mathbf{u})$ and $\mathbf{s}(\mathbf{v})$ represented by $\mathbb{r}(\mathbf{u})$ and $\mathbb{s}(\mathbf{v})$, respectively, have a C^r contact after a projective reparametrization. Such a contact is called a $\mathbf{G^r}$ **contact** and the reparametrization map its **transition function**.

Remark 10 Under a projective map $\langle P\rangle$, the triangles A and B are mapped to PA and PB while the matrices

$$(PA)^{-1}PB = A^{-1}B$$

of their transition functions (w.r.t. the systems PA and A, resp.) are the same.

5 Orbifold Splines

Any closed surface S of topological genus $\gamma \geq 2$ can be cut open into an open surface with a boundary. There are many ways of doing that. For example, one could pick a base point on the surface where all cuts start and end. Then any handle could be turned into a single hole by two cuts, and two cuts for every other handle expand this hole. This would turn the surface int a 4γ-gon with all its vertices at the base point, see Fig. 3 for an illustration where a double torus is cut into an octagon on the left. To show that other cuts are possible, this figure also depicts another cut into a dodecagon.

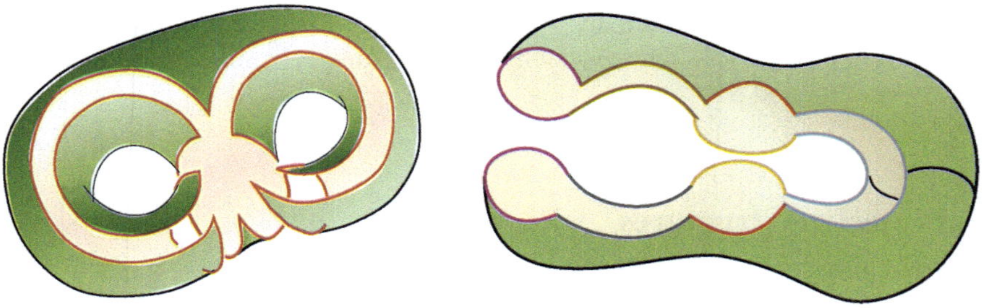

Fig. 3 Cutting a double torus into an octagon and a dodecagon

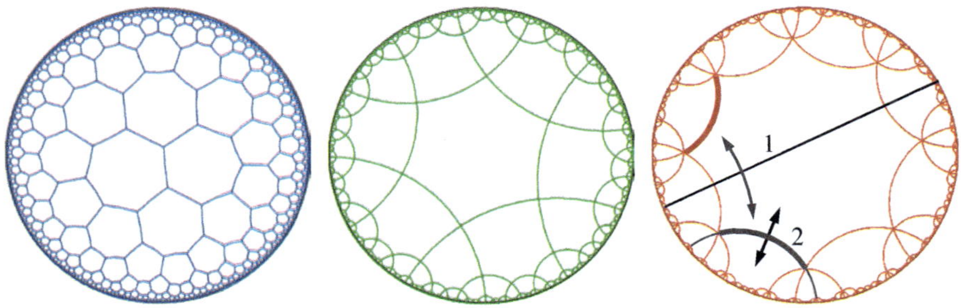

Fig. 4 Tessellations of the hyperbolic plane into hexagons and octagons. The blue and green tessellations on the left side are hyperbolic images of each other

A polygon forming such a closed surface S has pairs of identical edges and (groups of) identical vertices whose angles sum to 2π. For any such surface polygon, there are similar polygons in the hyperbolic plane and we parametrize S over such a polygon F. Here "similar" means that the edges of F with identical images on S are equally long and that the angles at all vertices with the same image on S sum to 2π.

By two hyperbolic reflections, we can first map any edge of F onto its partner and secondly onto itself. These two reflections map F to an adjacent congruent polygon and, iterating this procedure, we obtain a partition of the hyperbolic plane into infinitely many congruent copies of F, see [20] and Fig. 4 for an illustration, where the bold edge of the red tessellation on the right side is first reflected about line 1 and then about line 2.

The polygon F is called a **fundamental domain** and the hyperbolic isometries mapping F to any of the tiles above form a **Fuchsian group** G_F.

A smooth piecewise rational function $\mathbf{s}$ mapping the hyperbolic plane $\mathscr{H}^2$ to some closed surface S is called an **orbifold spline** if, for any member φ of G_F, we have

$$\mathbf{s} \circ \varphi = \mathbf{s}. \tag{3}$$

Hence an orbifold spline $\mathbf{s}$ maps F and any isometric image of it onto the entire surface S; in other words, $\mathbf{s}$ wraps the hyperbolic plane $\mathscr{H}^2$ infinitely often around S or "orbits" around S.

6 Projective Structures

From Sects. 3 and 5, it follows that we can represent any orbifold spline consisting of rational triangular patches by a trivariate homogeneous piecewise polynomial spline, where

- the polynomial segments have Bézier forms with respect to tetrahedra
- that have the origin as a common vertex
- and represent a triangulation, i.e., a tesselation into triangles, of the hyperbolic plane $\mathscr{H}^2$.

The tetrahedra form a bunch of tetrahedra that span cones which, when intersected with the disk $\mathscr{D}$ or the hyperboloid $\mathscr{U}$, give us a triangulation of $\mathscr{H}^2$ in the Klein or hyperboloid model, respectively. Since the triangles of this triangulation are the parameter domains of the patches of an orbifold spline, which satisfies (3), the triangulation is mapped to itself by all members of the corresponding Fuchsian group.

If the homogeneous spline representation of any rational orbifold spline is in C^r, then the rational spline satisfies G^r conditions with projective, i.e., linear rational transition functions determined by the pairs of adjacent tetrahedra as discussed in Sect. 4. Note that, due to Remark 10, any linear image of two adjacent tetrahedra (or any projective image of two adjacent weighted triangles) defines the same G^r conditions.

Definition 2 The set of transition functions above is called a **projective structure**.

Remark 11 As an orbifold spline "wraps" $\mathscr{H}^2$ infinitely often around the surface, it parametrizes any patch of the surface multiply over infinitely many triangles in $\mathscr{H}^2$. For each patch, we can pick one such triangle so as to get a connected domain for one covering of the surface. Thus, this domain is a fundamental domain F and the members of the Fuchsian group G_F map any triangle to all the other triangles that are the domains for the same patch. Hence, G_F induces an equivalence relation among the domain triangles such that any two triangles are equivalent if one of them can be mapped onto the other by some $\varphi \in G_F$.

Remark 12 A projective structure is represented by the tetrahedral bunch above and can easily be changed into other projective structures for the same surface by moving the vertices of the tetrahedra (except for the origin) and splitting or collapsing tetrahedra freely as long as any change within a fundamental domain F is simultaneously transferred to all copies φF, $\varphi \in G$. In addition the interiors of the tetrahedra should stay disjoint to exclude self-overlapping spline surfaces.

Remark 13 In the projective structure above, we can replace any tetrahedron **abc** by a scaled version $\mathbf{a}\alpha\ \mathbf{b}\beta\ \mathbf{c}\gamma$ of it and represent the spline with respect to the scaled tetrahedron, see Remarks 6 and 7. This does not change the surface, but its parametrization and transition functions. A canonical choice is to scale the vertices of a projective structure such that they lie on the hyperboloic sheet $\mathscr{U}$. This guarantees that the hyperbolic maps in the Fuchsian group map the set of all vertices into itself, see Remark 4 and its succeeding paragraph.

Remark 14 The transition functions of tetrahedra $A_1, \ldots, A_n = A_0$ around common edges satisfy

$$\langle A_n A_{n-1}^{-1} \rangle \cdot \langle A_{n-1} A_{n-2}^{-1} \rangle \ldots \langle A_1 A_0^{-1} \rangle = I,$$

which ensures that the derivatives up to order r of n patches around a common vertex end up being the same when we "pass them around" by the G^r conditions based on these transition functions.

Remark 15 Restricting the homogeneous C^r representation of an orbifold spline to any plane not going through the origin gives a polynomial C^r spline over a planar triangulation. It is a bivariate spline where the periodicity, or orbifold condition (3), means that any spline patch equals infinitely many others that are all parametrized differently, as explained in Remark 7 (and Remark 6).

Remark 16 There are several ways of getting a projective structure for an orbifold spline. All we need is the topology of the patch layout or network, i.e., the graph of the patch boundaries. For example, we can take any fundamental domain corresponding to cuts that open the patch network to a topologically planar bounded polygon and triangulate it arbitrarily but consistently with the patch layout and the Fuchsian group. Another possibility is to determine a triangulation of the hyperbolic plane for a given surface from the two conditions that *(i)* its hyperbolic angles around a vertex sum to 2π and *(ii)* that edge lengths, which are used to compute the angles of a triangle, are equal for both triangles they belong to, see [3].

7 Quadrilateral Orbifold Splines with Bi-Projective Structures

Our goal is to construct rational orbifold splines with quadrilateral tensor product patches based on triangular orbifold splines. While it is straightforward to use the tetrahedral Bézier form for rational triangular patches in homogeneous coordinates to build triangular orbifold splines, we cannot easily follow this approach for quadrilateral orbifold splines.

However, we can use pairs of adjacent triangular patches to determine quadrilateral patches with the same four boundary curves and G^r contact along those curves up to any derivative order. In this way it is possible to obtain quadrilateral orbifolds through triangular splines. In the following we present the details:

1. Given a quadrilateral patch layout for an orbifold spline, we split all quadrilaterals into pairs of triangles and compute a projective structure for it.
2. Over the projective structure, i.e., over the bunch of tetrahedra representing the projective structure, we build a piecewise homogeneous polynomial spline such that all pairs of its adjacent segments have C^r joints except for those over any two tetrahedral cones corresponding to a split quadrilateral. Between any two such segments, we require C^{2r} smoothness but only at the common edges of their cones. (Figure 6 shows a quadrilateral cone split into two triangular cones **abc** and **bcd** with common edges $\mathbf{b}\mathbb{R}$ and $\mathbf{c}\mathbb{R}$.)
3. For any initially given quadrilateral, we compute a quadrilateral patch having G^r contact along the four boundary edges with the associated pair of triangular patches.

Step 3 is the essential part and we discuss it in detail: As in Sect. 4, let

$$\mathbb{r}(A\mathbf{u}) = \sum_{\mathbf{i}\in\Gamma_n^3} \mathbb{r}_{\mathbf{i}} B_{\mathbf{i}}(\mathbf{u})$$

and

$$\mathbb{s}(B\mathbf{u}) = \sum_{\mathbf{i}\in\Gamma_n^3} \mathbb{s}_{\mathbf{i}} B_{\mathbf{i}}(\mathbf{u})$$

be two homogeneous polynomials given w.r.t. the triangles $A = [\mathbf{abc}]$ and $B = [\mathbf{bcd}]$. We reparametrize them by the bi-linear map

$$\mathbf{x}(s,t) = \mathbf{a}\, B_{00}^{11}(s,t) + \mathbf{b}\, B_{10}^{11}(s,t) + \mathbf{c}\, B_{01}^{11}(s,t) + \mathbf{d}\, B_{11}^{11}(s,t), \tag{4}$$

where

$$B_{ij}^{nn}(s,t) = \binom{n}{i} s^i (1-s)^{n-i} \binom{n}{j} t^j (1-t)^{n-j}$$

are the tensor product Bernstein polynomials of bi-degree n and where s and t are the affine coordinates w.r.t. the affine system $\mathbf{a}$, $(\mathbf{b}-\mathbf{c})$, $(\mathbf{c}-\mathbf{a})$.

Thus, we get the bi-degree n tensor product polynomials

$$\mathbb{r}(s,t) = \mathbb{r}(\mathbf{x}(s,t)) = \sum \mathbb{r}_{ij} B_{ij}^{nn}(s,t)$$

and

$$\mathbb{s}(s,t) = \mathbb{s}(\mathbf{x}(s,t)) = \sum \mathbb{s}_{ij} B_{ij}^{nn}(s,t)$$

with C^{2r} contacts at **b** and **c** or their parameter values $(1,0)$ and $(0,1)$, respectively. Consequently

$$\mathbb{r}_{ij} = \mathbb{s}_{ij}, \quad \text{for} \quad i = j \pm k \quad \text{and} \quad k \geq n - r.$$

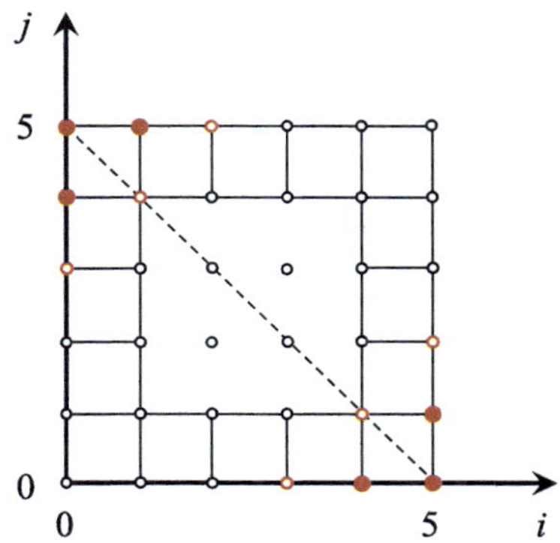

Fig. 5 Indices of the equal Bézier points of two bi-quintic patches obtained by reparametrizing a pair of triangular patches with C^1 (and C^2) contact at the pair's common vertices with bold (and empty) red dots

The indices of the equal Bézier points $\mathbb{r}_{ij}$ and $\mathbb{s}_{ij}$ are marked by red dots in Fig. 5 for $n = 5$ and $r = 1$. The solid red dots show which Bézier points are equal if we required C^1 joints at the vertices only and that then we could not ensure compatible corner twists for $\mathbb{r}$ and $\mathbb{s}$ at $(s, t) = (0, 1)$ and $(1, 0)$.

Hence, any tensor product patch

$$\mathbb{p}(s, t) = \sum_{i,j=0}^{n} \mathbb{p}_{ij} B_{ij}^{nn}(s, t)$$

with

$$\mathbb{p}_{ij} = \begin{cases} \mathbb{r}_{ij}, & i \le r \text{ or } j \le r \\ \mathbb{s}_{ij}, & i \ge n - r \text{ or } j \ge n - r \end{cases}$$

represents a rational polynomial of bi-degree n having G^r contact with the rational triangular patches $\langle \mathbb{r}(\mathbf{x}) \rangle$ and $\langle \mathbb{s}(\mathbf{x}) \rangle$ over the lines $\langle \mathbf{ab} \rangle$ and $\langle \mathbf{ac} \rangle$ or $\langle \mathbf{cd} \rangle$ and $\langle \mathbf{bd} \rangle$, respectively. Note that the transition functions of these G^r contacts are bi-linear rational functions.

Remark 17 The Bézier points $\mathbb{r}_{ij}$ can be obtained from the points $\mathbb{r}_{\mathbf{i}} = \mathbb{r}_{ijk}$ using polar forms. For instance, let

$$\mathbb{r}[\mathbf{x}_1 \dots \mathbf{x}_n]$$

be the polar form of the trivariate homogeneous polynomial $\mathbb{r}(\mathbf{x})$. Then we recall from, e.g., [17, Section 11.10] that $\mathbb{r}(s, t)$ has the tensor product polar form

$$\mathbb{r}[s_1 \dots s_n \,|\, t_1 \dots t_n] = \frac{1}{n!} \sum_{\tau} \mathbb{r}[\mathbf{x}(s_1, \tau_1) \dots \mathbf{x}(s_n, \tau_n)],$$

where the sum is taken over all permutations $\tau = (\tau_1, \dots, \tau_n)$ of $(t_1, \dots, t_n)$, and that

$$\begin{aligned}\mathbb{r}_{ij} &= \mathbb{r}[0 \overset{n-i}{\cdots} 0\; 1 \overset{i}{\cdots} 1 \,|\, 0 \overset{n-j}{\cdots} 0\; 1 \overset{j}{\cdots} 1] \\ &= \sum_{k=0}^{j} \beta_{ijk} \mathbb{r}_{n+k-i-j,i-k,j-k,k}\end{aligned}$$

with

$$\beta_{ijk} = \frac{k!}{n!} \binom{i}{k} \binom{j}{k} \frac{(n-i)!(n-j)!}{(n+k-i-j)!}$$

and the points $\mathbb{r}_{\kappa\lambda\mu\nu}$ obtained after ν steps of de Casteljau's algorithm to compute $\mathbb{r}(\mathbf{d})$ from the Bézier points

$$\mathbb{r}_{ijk0} = \mathbb{r}_{\mathbf{i}}, \quad \mathbf{i} = \begin{bmatrix} i \\ j \\ k \end{bmatrix} \in \Gamma_n^3,$$

which means that

$$\mathbb{r}_{\kappa\lambda\mu\nu} = \alpha\, \mathbb{r}_{\kappa+1,\lambda,\mu,\nu-1} + \beta\, \mathbb{r}_{\kappa,\lambda+1,\mu,\nu-1} + \gamma\, \mathbb{r}_{\kappa,\lambda,\mu+1,\nu-1},$$

where

$$\mathbf{d} = \mathbf{a}\alpha + \mathbf{b}\beta + \mathbf{c}\gamma.$$

8 Quadrilateral Orbifold Splines with Projective Structures

The construction in Sect. 7 results in quadrilateral orbifold splines with rational bi-linear transition functions forming what we could call a **bi-projective structure** and, in general, there is no projective structure consisting of only projective transition functions for G^k quadrilateral orbifold splines that also lie in C^0 [16]. However, there are projective structures for such splines if we give up continuous transitions between the parametrizations of adjacent patches, and in this section we show how to build such C^{-1}/G^k splines. The construction is a special case of the one in Sect. 7, where we replace the bi-linear by linear reparametrizations.

For any quadrilateral **abcd** in $\mathbb{R}^3$ associated to a pair of triangular patches, we determine $\alpha, \beta, \gamma, \delta$ such that $\mathbf{a}\alpha$, $\mathbf{c}\gamma$, $\mathbf{b}\beta$, $\mathbf{d}\delta$ form a parallelogram. Such a parallelogram exists and is parallel to the intersections of each pair of opposite sides of the cone **abdc** with apex **o**, see Fig. 6. Thus,

$$\mathbf{c}\gamma - \mathbf{a}\alpha = \mathbf{d}\beta - \mathbf{b}\delta$$

is a multiple of

$$(\mathbf{a} \times \mathbf{c}) \times (\mathbf{b} \times \mathbf{d}).$$

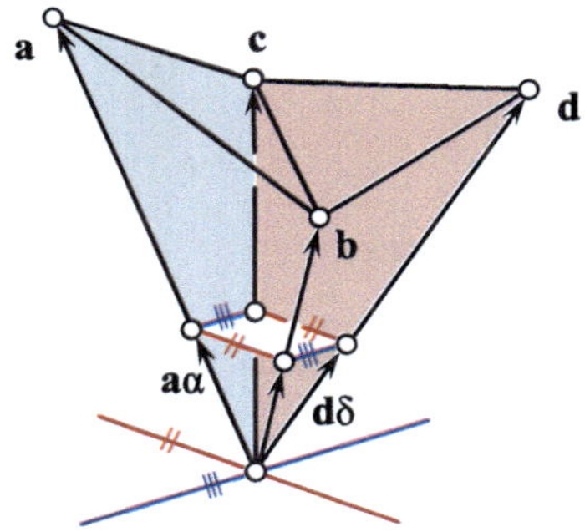

Fig. 6 Intersecting a quadrilateral cone in a parallelogram

Since $\mathbf{a}\alpha$, $\mathbf{c}\gamma$, $\mathbf{d}\beta$, $\mathbf{b}\delta$ is a parallelogram, it is an affine image of the unit square $[0, 1]^2$ meaning that the bi-linear map $\mathbf{x}(s, t)$ in (4) actually is a linear map. Consequently, if we reparametrize the pairs of triangular patches with these maps, we obtain quadrilateral patches in Sect. 7 that have G^r contacts with projective transition functions. However, in general, the scaling factors $\alpha, \beta, \gamma, \delta$ can not be chosen consistently for adjacent quadrilaterals and inconsistent scaling factors mean that common boundary curves of adjacent patches are equal, but with different parametrizations. Their Bézier points are the same though, yet they have different weights, see Remark 6.

9 Conclusions

So far, quadrilateral orbifold splines have been known only with low smoothness orders and certain restrictions on the patch layout. In this paper, we overcome these restrictions by a general construction that gives quadrilateral orbifold splines with an arbitrarily high smoothness order for any patch layout. This construction is based on rational triangular orbifold splines, which have a piecewise homogenous polynomial representation.

Piecewise polynomial triangular splines can be built using splitting techniques as, e.g., Powell-Sabin splits or by requiring supersmoothness at the vertices or by solving linear systems of C^k conditions, typically in addition to interpolation and fairness conditions. In general, we do not know the minimum degree needed for the latter solutions. The examples in [2, 3] show that the degree is considerably smaller than $4k + 1$ for which it is known that supersmooth solutions always exist.

For the quadrilateral orbifold splines in this paper, we need triangular splines with supersmoothness at the vertices but, on the other hand, they can be non-smooth or even discontinuous along the edges corresponding to the diagonals that split the quadrilaterals into triangles. Thus the triangular splines that we propose to use for building quadrilateral orbifold splines might have similar low degrees than the ones in [2, 3]. We plan to discuss such practical issues and to present the outcomes of our implementations in a forthcoming paper.

Acknowledgements C. Beccari is a member of the INdAM Research group GNCS, which has partially supported this work.

References

1. Alfeld, P., Neamtu, M., Schumaker, L.L.: Bernstein-Bézier polynomials on spheres and sphere-like surfaces. Comput. Aided Geom. Des. **13**, 333–349 (1996)
2. Beccari, C.V., Gonsor, D., Neamtu, M.: RAGS: rational geometric splines for surfaces of arbitrary topology. Comput. Aided Geom. Des. **31**, 97–110 (2014)
3. Beccari, C.V., Neamtu, M.: On constructing RAGS via homogeneous splines. Comput. Aided Geom. Des. **43**, 109–122 (2016)
4. Boehm, W., Prautzsch, H.: Geometric Concepts for Geometric Design. A K Peters/CRC Press, New York (1994)
5. de Boor, C.: B-form basics. In: Farin (ed.), Geometric Modeling: Algorithms and New Trends, pp. 131–148. Society for Industrial and Applied Mathematics, Philadelphia (1987)
6. Ferguson, H., Rockwood, A.: Multiperiodic functions for surface design. Comput. Aided Geom. Des. **10**, 315–328 (1993)
7. Gu, X., He, Y., Qin, H.: Manifold splines. Graph. Models **68**, 237–254 (2006)
8. Gu, X., He, Y., Jin, M., Luo, F., Qin, H., Yau, S.T.: Manifold splines with a single extraordinary point. Comput. Aided Des. **40**, 676–690 (2008)
9. Hilbert, D.: Grundlagen der Geometrie. Teubner (1903)
10. Hilbert, D., Cohn-Vossen, S.: Anschauliche Geometrie, 2nd ed. Springer (1996)
11. Kasten, H., Vogel, D.: Grundlagen der ebenen Geometrie. Springer Spektrum (2018)
12. Kowol, G.: Projektive Geometrie und Cayley-Klein Geometrien der Ebene. Birkhuser (2009)
13. Peters J., Fan, J.: The projective linear transition map for constructing smooth surfaces. In: Proceedings of the SMI 2010–international conference on shape modeling and applications, 2010, Art. no. 5521455, pp. 124–130 (2010)
14. Peters, J., Sarow, M.: Polynomial spline surfaces with rational linear transitions. Comput. Graph. **51**, 43–51 (2015)
15. Prautzsch, H.: Freeform splines. Comput. Aided Geom. Des. **14**, 201–206 (1997)
16. Prautzsch, H: Spline orbifold constructions. Talk at the SIAM Conference on Industrial and Applied Geometry, 10–12 July 2017, Pittsburgh
17. Prautzsch, H., Boehm, W., Paluszny, M.: Bézier and B-Spline Techniques. Springer, Berlin, Heidelberg (2002)
18. Prautzsch, H., Reif, U.: Degree estimates for C^k-piecewise polynomial subdivision surfaces. Adv. Comput. Math. **10**, 209–217 (1999)
19. Pottmann, H., Wallner, J.: Spline orbifolds. In: Le Méhauté, A., Rabut, C., Schumaker, L. (eds.) Curves and Surfaces with Applications in CAGD, pp. 445–464. Vanderbilt University Press, Nashville, TN (1997)
20. Ratcliffe, J.G.: Foundations of Hyperbolic Manifolds, 2nd ed. Grad. Texts in Mathematics, vol. 149. Springer, New York (2007)
21. Reif, U.: A degree estimate for subdivision surfaces of higher regularity. Proc. AMS **124**, 2167–2174 (1996)
22. Reif, U.: TURBS—topologically unrestricted rational B-splines. Constr. Approx. **14**, 57–77 (1998)
23. Sarov, M., Peters, J.: Refinable polycube G-splines. Comput. Graph. **58**, 92–101 (2016)
24. Wallner, J.: Geometric contributions to surface modeling. Ph.D. thesis. Vienna University of Technology (1996)

B-Symmetric Univariate Splines and Euler Numbers

Timothy Boafo-Adade, Daniel Larison, Irwin Paredes Escobar, and Tatyana Sorokina

Abstract We define a new class of periodic univariate splines of minimal defect over a uniform partition in $\mathbb{R}$, and study its interpolating, and approximating properties. Unexpectedly, the matrix associated with the corresponding interpolation problem is the key to a new closed form expression for Euler numbers.

1 Introduction

This study arose as an outset of modeling symmetric functions using piecewise polynomials, or splines. The univariate construction of spline models led to an unexpected discovery of a new closed form expression for Euler numbers. We begin by defining the standard 1-dimensional lattice generated by an interval.

Definition 1 Given an interval $[x_0, x_1]$, a lattice $\Delta_{[x_0,x_1]}$ is a union of intervals

$$\tau_i := [x_0 + i(x_1 - x_0), x_0 + (i+1)(x_1 - x_0)], \quad i \in \mathbb{Z}.$$

Next we introduce a subspace of univariate functions symmetric relative to the lattice. Essentially, it is a "highly-symmetric" subspace of continuous $(x_0, 2x_1 - x_0)$-periodic functions in one variable.

Definition 2 Given a lattice $\Delta_{[x_0,x_1]}$ as in Definition 1, let m_i be the midpoint of τ_i, $i \in \mathbb{Z}$. Then the space $\mathscr{F}(\Delta_{[x_0,x_1]})$ of *lattice-symmetric functions* f is defined as follows:

T. Boafo-Adade · D. Larison · I. Paredes Escobar · T. Sorokina (✉)
Department of Mathematics, Towson University, Towson, MD, USA
e-mail: tsorokina@towson.edu

T. Boafo-Adade
e-mail: tboafo1@students.towson.edu

D. Larison
e-mail: dlaris2@students.towson.edu

I. Paredes Escobar
e-mail: ipared1@students.towson.edu

C. Manni and H. Speleers (eds.), *Geometric Challenges in Isogeometric Analysis*, Springer INdAM Series 49, https://doi.org/10.1007/978-3-030-92313-6_2

1. $f(x + (2x_1 - 2x_0)) = f(x)$ for all $x \in \mathbb{R}$;
2. $f|_{\tau_i}$ is symmetric about m_i for all $i \in \mathbb{Z}$;
3. $f \in C^0(\mathbb{R})$.

The goal is to construct a family of smooth piecewise polynomials (or splines, see [1]) with nodes at the lattice points, that preserve the symmetric shape of the functions in $\mathscr{F}(\Delta_{[x_0,x_1]})$ described in (b) of Definition 2; and interpolate at the midpoints m_i for all $i \in \mathbb{Z}$. Thus, such splines are symmetric with respect to the geometry of the partition, the shape of the functions, and the data to be interpolated. If the shape requirement (b) is removed, the problem falls into the category of the so-called Marsden-type interpolation, see [3] and [4].

The paper is organized as follows. In Sect. 2, we define boundary-symmetric or B-symmetric polynomials, and prove a result from matrix theory that we use later in the paper. In Sect. 3, we review the so-called Euler numbers, and prove a new closed form formula for their computation. The family of B-symmetric splines is defined in Sect. 4. There we also show that they can be used to interpolate lattice-symmetric functions while preserving the shape symmetry. In Sect. 5, we show that B-symmetric splines approximate sufficiently smooth lattice-symmetric functions with order 2. Finally, we re-establish the results of our work in Sect. 6.

2 Preliminary Results

The most elegant way to define boundary-symmetric or B-symmetric polynomials is through barycentric coordinates. Note that this approach allows for easy generalization to the multivariate case. Since the theory of univariate splines, see [1], was developed prior to the use of barycentric coordinates in the field, our main reference [2] is concerned with splines in two and three variables. All relevant results in [2] hold true for univariate splines in much simpler forms. Let $[x_0, x_1]$ be a non-degenerate interval in $\mathbb{R}$. Then the barycentric coordinates b_0 and b_1 are defined as

$$b_0 = \frac{x_1 - x}{x_1 - x_0}, \quad b_1 = \frac{x - x_0}{x_1 - x_0}.$$

Then every $p \in \mathbb{P}_d$, the linear space of univariate polynomials of degree $\leq d$, can be written in its so-called Bernstein-Bézier (BB) form, see e.g. [2, Theorem 2.4].

Theorem 1 *Given an interval* $[x_0, x_1]$*, a polynomial* $p \in \mathbb{P}_d$ *can be uniquely written in its BB-form as*

$$p = \sum_{i+j=d} c_{ij} B_{ij}^d,$$

where

$$B_{ij}^d = \frac{d!}{i!j!} b_0^i b_1^j,$$

and b_0 and b_1 are the barycentric coordinates relative $[x_0, x_1]$.

Note that if $[x_0, x_1] = [0, 1]$, then $B^d_{ij} =: B^d_{d-i,i}$ are the standard Bernstein basis polynomials.

Definition 3 A polynomial $p \in \mathbb{P}_d$ is B-symmetric relative $[x_0, x_1]$ if $c_{ij} = c_{ji}$ for all $i + j = d$ in its BB-form of Theorem 1.

Let $\overline{\mathbb{P}}_d([x_0, x_1])$ be the space of B-symmetric polynomials of degree $\leq d$ relative $[x_0, x_1]$. Our next result shows that each such polynomial can be obtained by a change of variable from an even degree polynomial in Cartesian coordinates. First we show that there are no odd degree polynomials in $\overline{\mathbb{P}}_d([x_0, x_1])$.

Lemma 1 *For all* $k \in \mathbb{N} \cup \{0\}$, $\overline{\mathbb{P}}_{2k+1}([x_0, x_1]) = \overline{\mathbb{P}}_{2k}([x_0, x_1])$.

Proof Let $d = 2k + 1$, and $p \in \overline{\mathbb{P}}_{2k+1}([x_0, x_1])$. Then

$$
\begin{aligned}
p &= \sum_{i=0}^{k} c_{i,2k+1-i}\left(B^{2k+1}_{i,2k+1-i} + B^{2k+1}_{2k+1-i,i}\right)\\
&= \sum_{i=0}^{k} \binom{2k+1}{i} c_{i,2k+1-i} b_0^i b_1^i \left(b_1^{2k-2i+1} + b_0^{2k-2i+1}\right)\\
&= \sum_{i=0}^{k} \binom{2k+1}{i} c_{i,2k+1-i} b_0^i b_1^i (b_0 + b_1) \sum_{m=0}^{2k-2i} (-1)^m b_0^m b_1^{2k-2i-m}\\
&= \sum_{i=0}^{k} \binom{2k+1}{i} c_{i,2k+1-i} \sum_{m=0}^{2k-2i} (-1)^m b_0^{i+m} b_1^{2k-i-m} =: \sum_{i=0}^{k} \binom{2k+1}{i} c_{i,2k+1-i} q_i,
\end{aligned}
$$

where for each $i = 0, \ldots, k$, the corresponding q_i can be rewritten as

$$
q_i = \sum_{m=0}^{k-i-1} (-1)^m \left(b_0^{i+m} b_1^{2k-i-m} + b_0^{2k-i-m} b_1^{i+m}\right) + (-1)^{k-i} b_0^k b_1^k,
$$

and thus belongs to $\overline{\mathbb{P}}_{2k}([x_0, x_1])$. Since $\overline{\mathbb{P}}_{2k}([x_0, x_1])$ is a linear space, the proof is complete. □

Theorem 2 *Every B-symmetric polynomial relative* $[x_0, x_1]$ *is a shift of an even polynomial by* $m = \frac{x_0+x_1}{2}$.

Proof From Lemma (1), we know that for any $d = 2k$, a B-symmetric polynomial can be written as

$$
\sum_{i=0}^{k} \binom{2k}{i} c_{i,2k-i} b_0^i b_1^i \left(b_1^{2k-2i} + b_0^{2k-2i}\right). \tag{1}
$$

On the other hand, taking into account that

$$x - \frac{x_0 + x_1}{2} = \frac{x_1 - x_0}{2}(b_1 - b_0) =: l(b_1 - b_0),$$

every shift of an even polynomial of degree $2k$ by $m := \frac{x_0+x_1}{2}$ can be written as

$$\sum_{i=0}^{k} a_i (x - m)^{2i} = \sum_{i=0}^{k} a_i l^{2i} (b_1 - b_0)^{2i} = \sum_{i=0}^{k} a_i l^{2i} (b_1 - b_0)^{2i} (b_0 + b_1)^{2k-2i}. \quad (2)$$

Both (1) and (2) represent symmetric polynomials of degree $2k$ in two variables—b_0 and b_1—written in two different bases. In (1), the symmetric Bernstein basis is used, and in (2), we use the basis $\{(b_1 - b_0)^{2i}(b_1 + b_0)^{2k-2i})\}_{i=1}^{k}$, which is a linearly independent set since its restriction on $b_0 + b_1 = 1$ is a standard basis for univariate even polynomials of degree $2k$. □

Corollary 1 *For all* $k \in \mathbb{N}$, $\dim \overline{\mathbb{P}}_{2k} = k + 1$.

We also need a simple result from linear algebra to use later in the paper. The following example demonstrates the result for 3×3 matrices.

$$M = \begin{bmatrix} a_{11} & a_{12} & a_{13} \\ a_{21} & a_{22} & a_{23} \\ a_{31} & a_{32} & a_{33} \end{bmatrix}, \quad M^* = \begin{bmatrix} -a_{11} & a_{12} & -a_{13} \\ a_{21} & -a_{22} & a_{23} \\ -a_{31} & a_{32} & -a_{33} \end{bmatrix} \quad \Rightarrow \quad \det M^* = -\det M.$$

Lemma 2 *Let* $M := (m_{ij})$ *and* $M^* := (m^*_{ij})$ *be square matrices of size n, and let* $m^*_{ij} = (-1)^{i+j+1} m_{ij}$. *Then* $\det M^* = (-1)^n \det M$.

Proof Let $L := (l_{ij})$ be a square diagonal matrix of size n, where $l_{ii} = (-1)^i$; and let $R := -L$. Note that $\det(L) = (-1)^{\lceil \frac{n}{2} \rceil}$. Then $M^* = LMR$, and

$$\det(M^*) = \det(LMR) = (-1)^n [\det(L)]^2 \det(M) = (-1)^n \det(M).$$

□

3 Euler Numbers

The Euler numbers are the coefficients E_n in the following expansion, see e.g. [5],

$$\frac{1}{\cosh z} = \sum_{n=0}^{\infty} E_n \frac{z^n}{n!}, \quad |z| < \frac{\pi}{2}.$$

Note that E_{2n-1} vanish for all $n \in \mathbb{N}$, making the closed form expression for E_{2n} a classical problem in several areas of mathematics. The first few Euler numbers are:

$$E_0 = 1, \quad E_2 = -1, \quad E_4 = 5, \quad E_6 = -61, \quad E_8 = 1385, \quad E_{10} = -50521.$$

The Euler numbers also appear in the Taylor series of the secant function as well as in combinatorics, see [5] for more details. There are multiple ways to define the sequence of Euler numbers, however the one of particular interest for us is proven in [5, Theorem 1.1]. We restate it below.

Theorem 3 *For all $n \in \mathbb{N}$, the Euler number $E_{2n} = (-1)^n \det \mathscr{E}_{2n}$, where $\mathscr{E}_{2n} = (e_{ij})$ is a $(2n) \times (2n)$ matrix whose entries are given by*

$$e_{ij} = \binom{i}{j-1} \cos\left((i-j+1)\frac{\pi}{2}\right), \quad i = 1, \ldots, 2n, \quad j = 1, \ldots, 2n. \tag{3}$$

Our next theorem provides a new closed form expression for E_{2n}, $n \in \mathbb{N}$.

Theorem 4 *For all $n \in \mathbb{N}$, let $\widehat{\mathscr{E}}_n$ be an $n \times n$ matrix obtained from $\mathscr{E}_{2n}$ by removing all its odd-numbered rows and all its even-numbered columns. Let $\widetilde{\mathscr{E}}_n$ be an $n \times n$ matrix whose entries are equal to the absolute values of the entries in $\widehat{\mathscr{E}}_n$. Then*

$$\det \mathscr{E}_{2n} = (-1)^n \det \widehat{\mathscr{E}}_n = \det \widetilde{\mathscr{E}}_n.$$

Proof We begin by writing $\mathscr{E}_{2n}$ explicitly:

$$\begin{bmatrix}
0 & \binom{1}{1} & 0 & 0 & 0 & \cdots & 0 & 0 \\
-\binom{2}{0} & 0 & \binom{2}{2} & 0 & 0 & \cdots & 0 & 0 \\
0 & -\binom{3}{1} & 0 & \binom{3}{3} & 0 & \cdots & 0 & 0 \\
\binom{4}{0} & 0 & -\binom{4}{2} & 0 & \binom{4}{4} & \cdots & 0 & 0 \\
0 & \binom{5}{1} & 0 & -\binom{5}{3} & 0 & \cdots & 0 & 0 \\
\vdots & \vdots & \vdots & \vdots & \vdots & \ddots & \vdots & \vdots \\
0 & (-1)^{n-1}\binom{2n-1}{1} & 0 & (-1)^{n-2}\binom{2n-1}{3} & 0 & \cdots & 0 & \binom{2n-1}{2n-1} \\
(-1)^n\binom{2n}{0} & 0 & (-1)^{n-1}\binom{2n}{2} & 0 & (-1)^{n-2}\binom{2n}{4} & \cdots & -\binom{2n}{2n-2} & 0
\end{bmatrix}.$$

Note that the matrix has many zero entries. Indeed, if $i - j$ is even, then $\cos\left((i-j+1)\frac{\pi}{2}\right) = 0$. Thus, when i and j are both either even or odd, the corresponding e_{ij} vanishes. Also, for all $i < j - 1$, we have $\binom{i}{j-1} = 0$. Thus, all entries e_{ij} located above the first superdiagonal of $\mathscr{E}_{2n}$ vanish as well.

We focus on the superdiagonal entry $e_{2r-1,2r} = \binom{2r-1}{2r-1}$, $r = 1, \ldots, n$. It is located in an odd-numbered row $2r - 1$, and the entries to the right of it are all equal to zero. Additionally, the entries in the odd-numbered columns to the left of $e_{2r-1,2r}$ are also all equal to zero, as observed above.

We compute the determinant of $\mathscr{E}_{2n}$ by performing cofactor expansion across the first row. Then $\det \mathscr{E}_{2n} = -e_{12} \det \mathscr{E}_{2n-1}^{(1)}$, where $\mathscr{E}_{2n-1}^{(1)} = (e^{(1)})_{ij}$. Then

$$\mathscr{E}^{(1)}_{2n-1} = \begin{bmatrix} -\binom{2}{0} & \binom{2}{2} & 0 & 0 & \dots & 0 & 0 \\ 0 & 0 & \binom{3}{3} & 0 & \dots & 0 & 0 \\ \binom{4}{0} & -\binom{4}{2} & 0 & \binom{4}{4} & \dots & 0 & 0 \\ 0 & 0 & -\binom{5}{3} & 0 & \dots & 0 & 0 \\ \vdots & \vdots & \vdots & \vdots & \ddots & \vdots & \vdots \\ 0 & 0 & (-1)^{n-2}\binom{2n-1}{3} & 0 & \dots & 0 & \binom{2n-1}{2n-1} \\ (-1)^n\binom{2n}{0} & (-1)^{n-1}\binom{2n}{2} & 0 & (-1)^{n-2}\binom{2n}{4} & \dots & -\binom{2n}{2n-2} & 0 \end{bmatrix}.$$

Note that $\mathscr{E}^{(1)}_{2n-1}$ is obtained from $\mathscr{E}_{2n}$ by removing its first row and second column. Now we perform cofactor expansion across the second row of $\mathscr{E}^{(1)}_{2n-1}$. The second row of $\mathscr{E}^{(1)}_{2n-1}$ is the third row of $\mathscr{E}_{2n}$ without e_{32}. The only non-zero entry in this row is $e^{(1)}_{23} = e_{34}$. Then $\det \mathscr{E}_{2n} = (-1)^2 e_{34} \det \mathscr{E}^{(2)}_{2n-2}$, where $\mathscr{E}^{(2)}_{2n-2} = (e^{(2)})_{ij}$. Then

$$\mathscr{E}^{(2)}_{2n-2} = \begin{bmatrix} -\binom{2}{0} & \binom{2}{2} & 0 & \dots & 0 & 0 \\ \binom{4}{0} & -\binom{4}{2} & \binom{4}{4} & \dots & 0 & 0 \\ 0 & 0 & 0 & \dots & 0 & 0 \\ \vdots & \vdots & \vdots & \ddots & \vdots & \vdots \\ 0 & 0 & 0 & \dots & 0 & \binom{2n-1}{2n-1} \\ (-1)^n\binom{2n}{0} & (-1)^{n-1}\binom{2n}{2} & (-1)^{n-2}\binom{2n}{4} & \dots & -\binom{2n}{2n-2} & 0 \end{bmatrix}.$$

Note that $\mathscr{E}^{(2)}_{2n-2}$ is obtained from $\mathscr{E}_{2n}$ by removing its first and third row and its second and fourth column.

Continuing this pattern gives us $\mathscr{E}^{(l)}_{2n-l}$, $1 \leq l \leq n$, from $\mathscr{E}_{2n}$ by removing all odd-numbered rows up to row $2l$ and even-numbered columns up to column $2l+1$ of the original matrix. Then

$$\det \mathscr{E}_{2n} = (-1)^l e_{2l-1,2l} \det \mathscr{E}^{(l)}_{2n-l},$$

since all entries in the odd-numbered columns to the left of $e_{2l-1,2l}$ have been removed, and all the remaining entries are equal to zero. Note that $e_{2l-1,2l} = 1$, $1 \leq l \leq n$.

The last row and column we can remove using cofactor expansion are row $2n-1$ and column $2n$. Then $\det \mathscr{E}_{2n} = (-1)^n \det \widehat{\mathscr{E}_n}$, where

$$\widehat{\mathscr{E}_n} = \begin{bmatrix} -\binom{2}{0} & \binom{2}{2} & 0 & \dots & 0 \\ \binom{4}{0} & -\binom{4}{2} & \binom{4}{4} & \dots & 0 \\ \vdots & \vdots & \vdots & \ddots & \vdots \\ (-1)^n\binom{2n}{0} & (-1)^{n-1}\binom{2n}{2} & (-1)^{n-2}\binom{2n}{4} & \dots & -\binom{2n}{2n-2} \end{bmatrix}.$$

By Lemma 2, we obtain

$$\det \widehat{\mathscr{E}_n} = (-1)^n \det \widetilde{\mathscr{E}_n},$$

where

$$\widetilde{\mathscr{E}_n} = \begin{bmatrix} \binom{2}{0} & \binom{2}{2} & 0 & \dots & 0 \\ \binom{4}{0} & \binom{4}{2} & \binom{4}{4} & \dots & 0 \\ \vdots & \vdots & \vdots & \ddots & \vdots \\ \binom{2n}{0} & \binom{2n}{2} & \binom{2n}{4} & \dots & \binom{2n}{2n-2} \end{bmatrix}.$$

This completes the proof. □

4 B-Symmetric Splines of Minimal Defect

Lattices described in Definition 1 give rise to piecewise polynomials (or splines) over a uniform partition of $\mathbb{R}$. Our interest is in those splines that are associated with the concept of a univariate crystal lattice, i.e., the symmetry is not only in the uniformity of the partition but also in the "atomic structure": assume we have an atom located in the middle of each τ_i; then, in order to keep the uniformity of the partition, we must assume that there are only two types of alternating atoms, and thus one piece of data per each interval. This yields the following spline space.

Definition 4 Given a lattice $\Delta_{[x_0,x_1]}$, and an integer $k \geq 1$, the space of B-symmetric splines of minimal defect is defined as follows:

$$\mathbb{B}_{2k}(\Delta_{[x_0,x_1]}) := \{s \in C^{2k-1}(\mathbb{R}), \quad s|_{\tau_i} = s|_{\tau_{i+2}} \in \overline{\mathbb{P}}_{2k}, \quad \forall i \in \mathbb{Z}\}.$$

Note that $\mathbb{B}_{2k}(\Delta_{[x_0,x_1]})$ is a subspace of $\mathscr{F}(\Delta_{[x_0,x_1]})$, see Definition 2. Clearly, every spline s in such space is uniquely defined by two neighboring B-symmetric polynomials p and q defined over two intervals τ_l and τ_{l+1}, respectively. Next, we develop conditions on the coefficients of the neighboring B-symmetric polynomials p and q to ensure smoothness at the point of intersection of two intervals τ_l and τ_{l+1}. In view of Theorem 2, we can restrict ourselves to the intervals $[-2, 0]$ and $[0, 2]$, and write both q and p in Cartesian coordinates.

Lemma 3 *Let* $p = \sum_{i=0}^{k} a_i (x-1)^{2i}$, $x \in [0,2]$, *and let* $q = \sum_{i=0}^{k} b_i (x+1)^{2i}$, $x \in [-2, 0]$, *be the B-symmetric polynomials forming a spline* $s \in \mathbb{B}_{2k}(\Delta_{[0,2]})$. *Then* $b_i = -a_i$ *for all* $i = 1, \dots, k$.

Proof We begin by taking the $(2k-1)$-order derivatives of p and q:

$$p^{(2k-1)}(x) = a_k (2k-1)! \binom{2k}{2k-1} (x-1),$$
$$q^{(2k-1)}(x) = b_k (2k-1)! \binom{2k}{2k-1} (x+1).$$

Setting $p^{(2k-1)}(0) = q^{(2k-1)}(0)$ shows that $b_k = -a_k$.

Next, we take the $(2k-3)$-order derivatives of p and q:

$$p^{(2k-3)}(x) = (2k-3)! \Big[a_{k-1} \binom{2k-2}{2k-3} (x-1) + a_k \binom{2k}{2k-3} (x-1)^3 \Big],$$
$$q^{(2k-3)}(x) = (2k-3)! \Big[b_{k-1} \binom{2k-2}{2k-3} (x+1) + b_k \binom{2k}{2k-3} (x+1)^3 \Big].$$

Setting $p^{(2k-3)}(0) = q^{(2k-3)}(0)$ and taking into account that $b_k = -a_k$ implies $b_{k-1} = -a_{k-1}$. Continuing this process yields the following: when we set $p^{(2j-1)}(0) = q^{(2j-1)}(0)$ and substitute $a_l = -b_l$, for $l > j$, we obtain $b_j = -a_j$. This process terminates at the first order derivative, which completes the proof. □

Lemma 3 takes care of the smoothness conditions associated with odd-order derivatives. Our next result takes care of the ones associated with even order derivatives.

Theorem 5 *Let* $p = \sum_{i=0}^{k} a_i (x-1)^{2i}$, $x \in [0,2]$, *and let* $q = b_0 - \sum_{i=1}^{k} a_i (x+1)^{2i}$, $x \in [-2, 0]$, *be the B-symmetric polynomials forming a spline* $s \in \mathbb{B}_{2k}(\Delta_{[0,2]})$. *Furthermore, let* $a_0 = p(1)$ *and* $b_0 = q(-1)$ *be given. Then the remaining coefficients* a_i, $i = 1, \dots, k$, *are uniquely defined by the smoothness conditions.*

Proof Setting $p(0) = q(0)$ and using Lemma 3 we obtain:

$$\sum_{i=1}^{k} a_i = \frac{b_0 - a_0}{2}. \tag{4}$$

Using Lemma 3 and the differentiation formulas

$$p^{(2m)}(x) = (2m)! \sum_{i=m}^{k} a_i \binom{2i}{2m} (x-1)^{2(i-m)}, \quad m = 1, \ldots, k-1,$$

$$q^{(2m)}(x) = -(2m)! \sum_{i=m}^{k} a_i \binom{2i}{2m} (x+1)^{2(i-m)}, \quad m = 1, \ldots, k-1,$$

we set $p^{(2m)}(0) = q^{(2m)}(0), \quad m = 1, \cdots, k-1$, and obtain

$$\sum_{i=m}^{k} a_i \binom{2i}{2m} = 0, \quad m = 1, \cdots, k-1. \tag{5}$$

The system of k linear equations (4) and (5) with k unknowns gives rise to a $k \times k$ matrix $D_k = (d_{ij})$, whose entries are given by

$$d_{ij} = \binom{2j}{2(i-1)}, \quad 1 \leq i, j \leq k.$$

Note that D_k is the transpose of a non-singular matrix $\widetilde{\mathscr{E}}_k$ in Theorem 4. Thus, the associated system of k linear equations has a unique solution. □

Figure 1 illustrates our next result. In this figure, the B-symmetric splines of degrees 2 and 8 interpolate $f(-1) = 1$ and $f(1) = 2$. The proof of the corollary immediately follows from Theorem 5.

Corollary 2 *Let* $k \geq 1$ *be an integer. Then* $\dim \mathbb{B}_{2k}(\Delta_{[0,2]}) = 2$. *Moreover, for each* $f \in \mathscr{F}(\Delta_{[x_0,x_1]})$ *there exists a unique interpolating spline* $s_f \in \mathbb{B}_{2k}(\Delta_{[0,2]})$ *such that*

$$s_f(m_i) = f(m_i), \quad \forall i \in \mathbb{Z}.$$

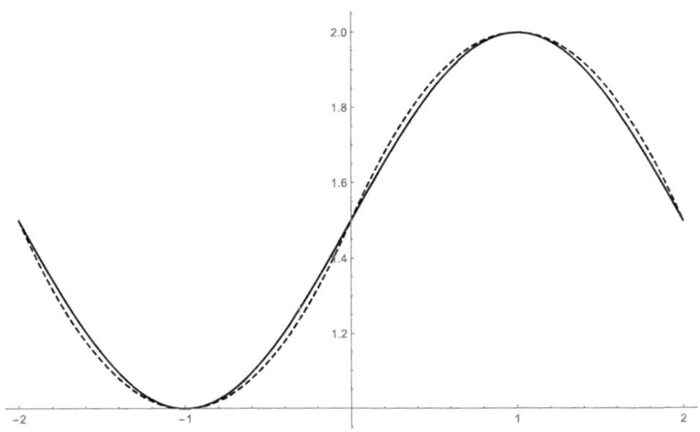

Fig. 1 B-symmetric interpolating splines of degrees 2 (dotted line) and 8 (solid line)

5 Approximation Order

Although it is the shape reproduction property (b) in Definition 2 of the B-symmetric splines that are of main interest and importance, our next result shows that such splines do enjoy some approximation power as well. Since refinability is an issue for any kind of periodic functions, the result below reflects an accuracy of approximation rather than the rate of convergence.

Theorem 6 *Let $k \geq 2$ be an integer, and with the standard Sobolev norm $W^2 := W^{2,\infty}$, let $s : \mathcal{F}(\Delta_{[x_0,x_1]}) \cap W^2([x_0, 2x_1 - x_0]) \to \mathbb{B}_{2k}(\Delta_{[x_0,x_1]})$ be the linear operator defined by Corollary 2. Then there exists a constant $K \in \mathbb{R}$ that does not depend on $|x_1 - x_0|$ such that*

$$\|f - s_f\| \leq K|x_1 - x_0|^2,$$

where $\|\cdot\|$ is the maximum norm.

Proof Without loss of generality, assume that $x \in [x_0, x_1]$. Let $a_0 = f(m_0)$, where $m_0 = \frac{x_0+x_1}{2}$. Since f is symmetric about m_0, the constant function $t(x) \equiv a_0$ is the first-order Taylor polynomial of f centered at m_0. Moreover, the unique B-symmetric spline $s_t \in \mathbb{B}_{2k}(\Delta_{[x_0,x_1]})$ interpolating the values of t at the midpoints m_i of each interval $\tau_i \in \Delta_{[x_0,x_1]}$ is, clearly, $t(x)$ itself. Thus,

$$\begin{aligned} |f(x) - s_f(x)| &= |f(x) - t(x) + t(x) - s_f(x)| = |f(x) - t(x) + s_t(x) - s_f(x)| \\ &\leq |f(x) - t(x)| + \|s\||f(x) - t(x)| \leq K|x_1 - x_0|^2, \end{aligned}$$

where $K = 1 + \|s\|$. We only need to show now that the operator norm $\|s\|$ does not depend on $|x_1 - x_0|$. To this end, we turn our attention to the BB-form of s_f. As it is shown in [2, Theorem 2.6], the BB-form is stable, i.e., to bound the norm of s_f it suffices to bound the BB-coefficients of s_f. In order to do that, we need to rewrite the interpolating and the smoothness conditions of Lemma 3 and Theorem 5 in barycentric coordinates. Since for the midpoint of any $[x_0, x_1]$ we have $b_0 = 1/2$ and $b_1 = 1/2$, the interpolating conditions clearly do not depend on the length of $[x_0, x_1]$. Moreover, the smoothness conditions in BB-form between two polynomials defined over two neighboring intervals depend only on the ratio of the respective lengths of the intervals, see e.g. [2, Theorem 2.13]. In our case, all intervals of $\Delta_{[x_0,x_1]}$ have the same length. Thus, $\|s\|$ depends on the determinant of the matrix formed by the interpolating and smoothness conditions that do not depend on the length of the interval $[x_0, x_1]$, and the proof is complete. □

6 Conclusions

We define a new family of univariate boundary-symmetric or B-symmetric polynomial splines of minimal defect. Such splines have even degree $2k$ and smoothness $2k-1$. Then we apply this family to solve an associated interpolation problem for a class of continuous periodic lattice-symmetric functions. The solution of the interpolation problem reveals a new closed-form formula for the computation of Euler numbers. We prove that B-symmetric splines approximate sufficiently smooth lattice-symmetric functions with order 2.

Acknowledgements T. Boafo-Adade is partially supported by Towson University Fisher Foundation Research grant. I. Paredes Escobar is partially supported by Towson University Undergraduate Summer Research grant. T. Sorokina is partially supported by the grant from the Simons Foundation #235411.

References

1. de Boor, C.: A Practical Guide to Splines. Springer, New York (1978)
2. Lai, M.-J., Schumaker, L.L.: Spline Functions on Triangulations. Cambridge University Press, Cambridge (2007)
3. Marsden, M.J.: Quadratic spline interpolation. Bull. Am. Math. Soc. **80**, 903–906 (1974)
4. Volkov, Y.S.: Interpolation by splines of even degree according to Subbotin and Marsden. Ukrainian Math. J. **66**, 994–1012 (2014)
5. Wei, C., Qi, F.: Several closed expressions for the Euler numbers. J. Inequal. Appl. **219**, 1–8 (2015)

A Collocation IGA-BEM for 3D Potential Problems on Unbounded Domains

Antonella Falini, Carlotta Giannelli, Tadej Kanduč, M. Lucia Sampoli, and Alessandra Sestini

Abstract In this paper the numerical solution of potential problems defined on 3D unbounded domains is addressed with Boundary Element Methods (BEMs), since in this way the problem is studied only on the boundary, and thus any finite approximation of the infinite domain can be avoided. The isogeometric analysis (IGA) setting is considered and in particular B-splines and NURBS functions are taken into account. In order to exploit all the possible benefits from using spline spaces, an important point is the development of specific cubature formulas for weakly and nearly singular integrals. Our proposal for this aim is based on spline quasi-interpolation and on the use of a spline product formula. Besides that, a robust singularity extraction procedure is introduced as a preliminary step and an efficient function-by-function assembly phase is adopted. A selection of numerical examples confirms that the numerical solutions reach the expected convergence orders.

A. Falini
Dipartimento di Informatica, Università Degli Studi di Bari Aldo Moro, Bari, Italy
e-mail: antonella.falini@uniba.it

C. Giannelli · A. Sestini (✉)
Dipartimento di Matematica e Informatica Ulisse Dini, Università degli Studi di Firenze, Florence, Italy
e-mail: alessandra.sestini@unifi.it

C. Giannelli
e-mail: carlotta.giannelli@unifi.it

T. Kanduč
Faculty of Mathematics and Physics, University of Ljubljana, Ljubljana, Slovenia
e-mail: tadej.kanduc@fmf.uni-lj.si

M. L. Sampoli
Dipartimento di Ingegneria dell'Informazione e Scienze Matematiche, Università degli Studi di Siena, Siena, Italy
e-mail: marialucia.sampoli@unisi.it

C. Manni and H. Speleers (eds.), *Geometric Challenges in Isogeometric Analysis*, Springer INdAM Series 49, https://doi.org/10.1007/978-3-030-92313-6_4

1 Introduction

The framework of Isogeometric Analysis (IGA), introduced in 2005 by Thomas J. R. Hughes and coauthors, has emerged as a very attractive approach to solve differential problems [13, 19]. IGA bridges the gap between Computational Mechanics and Computer Aided Design (CAD) by employing a unified representation of geometric objects and physical quantities. In particular, the spline basis functions commonly used for CAD geometry parameterization are also used to span the discretization space adopted for the numerical solution. This not only eliminates the need for slow and error-prone conversion processes between the geometry and simulation models, but also enables the possibility of a widespread use of design optimization tools in simulation-based engineering.

While IGA has achieved great success within the finite element paradigm by attracting most of the IGA researchers, a remaining critical challenge is how to efficiently and exactly parameterize a volumetric domain from its boundary [3, 9, 25]. Three dimensional problems need trivariate NURBS solids for analysis, but CAD systems use a boundary representation where only bivariate NURBS are involved. A possible solution to this issue consists in adopting boundary element methods (BEMs) [6, 26], whose isogeometric formulation is briefly referred to as IGA-BEM and has recently been succesfully applied to several classical problems, e.g., Laplace and Helmholtz equation [14, 29], elastostatics and elasticity [5, 24, 27, 31], potential flow [18] and wave-resistance problems [17]. BEMs reformulate the original differential problem through the fundamental solution into boundary integral equations. Thus, adopting a boundary representation, they can be easily coupled with standard CAD techniques and ensure dimension reduction of the computational domain where the numerical simulation is developed. Besides these key points, they are attractive because offering an easy treatment of problems on unbounded domains, see for example the potential or acoustic problems considered in [10, 23]. Within this kind of problems, screen or crack problems are particularly important for applications, being in this case the domain given by the whole Euclidean space external to an open limited arc in 2D or surface in 3D.

Relying on the general NURBS parametric description of the domain boundary and on (refinements of) the related spline space for the approximation of the unknowns, in this paper we confirm the effectiveness of the IGA-BEM in the 3D setting, previously experimented in 2D [2]. In particular, we focus on a collocation discretization strategy for 3D exterior potential problems. The combination of the superior smoothness of spline basis functions with the low computational cost and simplicity of collocation techniques seems to constitute an optimal basis for accurately modeling complex and computationally demanding problems, at least when smooth geometries are dealt with, as also outlined in [29]. However it is worth mentioning that considering a Galerkin based variant of the developed method could be of great interest in the future, either because the method could become more robust for treating problems with low regular solutions [28] or even from a more theoretical point of view, since most of the convergence results available in the literature refer to the Galerkin approach.

Since an effective implementation of BEMs surely requires suitable rules for the approximation of arising singular integrals, in this paper we focus on such aspect. In particular, the cubature rules for bivariate weakly singular integrals developed in [16] are here firstly applied to the 3D IGA-BEM context. These rules are based on spline Quasi-Interpolation (QI) and on the usage of a spline product formula [22]. Specifically formulated for weakly singular integrals containing a tensor-product B-spline factor in the integrand, their peculiarity consists in being applicable on its whole support. Since tensor product B-splines are here adopted to span the discretization space, this feature implies that the assembly phase of the system matrix can be done function-by function. This is a clear advantage of our approach, as already outlined in the 2D setting [1]. Furthermore, as a preliminary step the multiplicative singularity extraction procedure considered in [16] is here improved with the aim to apply the QI operator adopted by our rules to sufficiently smooth functions. Note that in the IGA-BEM setting, according to the distance between the source point and the integration domain, not all the integrals to be dealt with are singular, some being near singular and others regular. The final accuracy of our IGA-BEM implementation gained from treating near singular integrals with the same strategy adopted for singular ones. For regular integrals, a simplified formulation of the mentioned cubature rules has been developed, relying on the same QI operator, but possibly selecting for it a different spline degree and/or knot spacing.

The structure of the paper is as follows. Section 2 recalls the indirect integral formulation of a potential problem on a 3D unbounded domain, and then it introduces the related discretization based on an isogeometric collocation BEM. Section 3 presents cubature rules based on quasi-interpolation and on the spline product formula. Section 4 reports the results obtained with a Matlab implementation of the scheme for a screen problem and for two different problems on the unbounded domain external to a torus. Finally, some conclusions are given in Sect. 5.

2 Isogeometric Indirect BEMs on 3D Unbounded Domains

As well as for interior problems, even for general exterior problems different integral formulations are available in the literature, coming from different simple or double layer representation formulas. BEMs derived from an integral equation which directly involves the unknown Cauchy datum are called (simple or double layer) “direct” methods. As an alternative, (simple or double layer) “indirect” approaches can also be adopted. In such case the unknown appearing in the integral equation and in the associated representation formula is a different function, usually referred to as “density” function. In this paper we adopt a BEM indirect formulation, since it is the only one applicable also to screen/crack problems [11]. In particular we focus on such formulation for 3D Dirichlet exterior potential problems,

$$\begin{cases} \Delta u = 0\,, \ \mathbf{x} \in \mathbb{R}^3 \setminus \overline{\Omega}\,, \\ u = g\,, \ \mathbf{x} \in \Gamma\,, \\ \left.\begin{array}{l} |u(\mathbf{x})| \to 0 \\ \|\nabla u(\mathbf{x})\| = O(\|\mathbf{x}\|^{-2}) \end{array}\right\} \text{ for } \|\mathbf{x}\| \to +\infty\,, \end{cases} \tag{1}$$

where g is the given Dirichlet datum and $\Gamma := \partial\Omega$, with Ω denoting a bounded domain in $\mathbb{R}^3$. Note that for screen problems Ω has to be replaced with Γ, where Γ is a given open and limited surface. Observe also that, since we deal with unbounded domains, it is necessary to specify the required behavior of the harmonic solution at infinity to ensure uniqueness [26]. The boundary integral formulation of (1) is given by the following BIE,

$$V\psi(\mathbf{x}) := \int_\Gamma U(\mathbf{x}, \mathbf{y})\psi(\mathbf{y})\, \mathrm{d}\Gamma_{\mathbf{y}} = g(\mathbf{x})\,, \qquad \mathbf{x} \in \Gamma\,, \tag{2}$$

where the unknown density function ψ represents the jump across the boundary of the normal derivative of u (in the outward direction from $\mathbb{R}^3 \setminus \overline{\Omega}$ to Ω) and the kernel U is the fundamental solution of the Laplace equation in 3D,

$$U(\mathbf{x}\,,\, \mathbf{y}) := \frac{1}{4\pi}\, \frac{1}{\|\mathbf{x} - \mathbf{y}\|_2}\,. \tag{3}$$

The operator V defined in (2) is an elliptic isomorphism between its functional domain $H^{-\frac{1}{2}}(\Gamma)$ and its codomain $H^{\frac{1}{2}}(\Gamma)$, while the BIE also defined in (2) is a Fredholm integral equation of the first kind with weakly singular kernel.

Once the density function ψ has been (approximately) computed, the (approximated) expression of u at any point in $\mathbb{R}^3 \setminus \overline{\Omega}$ can be derived by relying on its simple layer representation formula,

$$u(\mathbf{x}) = \int_\Gamma U(\mathbf{x}\,,\mathbf{y})\, \psi(\mathbf{y})\, \mathrm{d}\Gamma_{\mathbf{y}}\,, \qquad \mathbf{x} \in \mathbb{R}^3 \setminus \overline{\Omega}\,. \tag{4}$$

Note that this formula automatically ensures the required decay behavior of u at infinity if ψ is continuous in Γ [26], and also that its numerical approximation can be challenging only when $\mathbf{x}$ is very close to Γ.

To compute a numerical solution of the BIE in (2) we adopt collocation, since this is probably the easiest discretization method suited for smooth problems (i.e., problems with a smooth Γ and sufficiently smooth Dirichlet datum). However, we observe that the related theory is not yet fully developed in particular for the 3D case; for example in Sects. 9.2 and 9.3 of [4] the available theoretical results can be applied just for the direct approach which is not usable for screen problems, as we already remarked. In general, after having numerically solved the considered BIE with a BEM approach, the representation formula associated with it has to be considered in order to compute values of the approximated solution inside the domain. However there are several applications, e.g. the simulation for screen or crack problems, which do not require such post-processing step, since their physically relevant quantity is the unknown Cauchy datum on the boundary itself, see for example [11].

We assume that the geometry is CAD-generated, that is $\overline{\Gamma}$ has a parametric representation, $\overline{\Gamma} = \text{Image}(\mathbf{F})$, where $\mathbf{F} : R \to \mathbb{R}^3$ is a one-to-one mapping[1] defined on the parametric rectangular domain $R := [a_1 , b_1] \times [a_2 , b_2]$ in the following standard NURBS form,

$$\mathbf{F}(\mathbf{t}) := \frac{\displaystyle\sum_{i=0}^{n_1}\sum_{j=0}^{n_2} w_{i,j}\, \mathbf{Q}_{i,j}\, B_{i,T_1}^{d_1}(t_1) B_{j,T_2}^{d_2}(t_2)}{\displaystyle\sum_{i=0}^{n_1}\sum_{j=0}^{n_2} w_{i,j}\, B_{i,T_1}^{d_1}(t_1) B_{j,T_2}^{d_2}(t_2)}, \quad \mathbf{t} \in R, \tag{5}$$

where $\mathbf{t} := (t_1, t_2)$. Each set of functions $B_{i,T_j}^{d_j}$, $i = 0, \ldots, n_j$, $j = 1, 2$ is made up of univariate B-spline functions defined in $[a_j , b_j]$, of degree d_j and with extended knot vector T_j. We assume that the mapping $\mathbf{F}$ is at least differentiable. The set $\{\mathbf{Q}_{i,j} \in \mathbb{R}^3, i = 0, \ldots, n_1, j = 0, \ldots, n_2\}$ defines the net of control points[2] and each $w_{i,j}$ is a positive weight used to obtain additional shape control (obviously, if the weights are all equal to a common constant value, $\mathbf{F}$ becomes just a vector tensor-product polynomial spline represented in B-spline form). This surface representation is often adopted in CAD environments to design free-form surfaces and it has the facility of involving just one patch, that is a unique continuous vector function $\mathbf{F}$ is used to define the whole considered surface. On this concern we note that clearly this representation can be not flexible enough for complex geometries characterized for example by linkage curves with just geometric continuity. Note also that, even if the NURBS form was originally introduced for guaranteeing exact representation of conic sections and quadric patches, it cannot be used for example for representing closed surfaces with zero genus without singularities in the parameterization. For example in reference [29] the one-patch NURBS parameterization of the sphere is singular at the poles and this has required the development of special cubature formulas to obtain accurate results. In order to deal with surface representations more flexible and more suitable for the analysis, multi-patch NURBS representation could be considered, as well as different kinds of spline spaces, e.g., multi-degree polar splines [30].

Focusing on the basic standard single patch NURBS representation and referring for brevity only to the case of clamped knot vectors T_j, $j = 1, 2$, let us define $\mathscr{T}_j$ as a possible refinement of T_j, $j = 1, 2$, obtained by inserting new breakpoints in $[a_j , b_j]$, or by increasing the multiplicity of a knot already included in T_j. Denoting with h_j the maximal distance between successive knots in $\mathscr{T}_j$, $j = 1, 2$ and with S_{h_j} the univariate spline space of dimension $m_j + 1$, $m_j \geq n_j$, spanned by the B-

---

[1] With the exception of the boundary of the parametric domain in case of a closed Γ or with cylinder-like shape.

[2] In order to avoid degenerate nets we assume the control net composed by distinct points.

splines of degree d_j associated with $\mathcal{T}_j$, we can define the tensor-product spline space $\mathcal{S}^h := S_{h_1} \otimes S_{h_2}$ of dimension $D := (m_1 + 1)(m_2 + 1)$, where $h := \max\{h_1, h_2\}$. Note that in the following we simplify the notation by denoting as B_j the bivariate B-spline $B^{d_1}_{i_1,\mathcal{T}_1} B^{d_2}_{i_2,\mathcal{T}_2}$ generating $\mathcal{S}^h$, where $j = i_2(m_1 + 1) + i_1$. Thus, the density function ψ is approximated in a finite dimensional space $\hat{\mathcal{S}}^h$ derived from $\mathcal{S}^h$ by lifting all the B-splines to the physical space

$$\hat{\mathcal{S}}^h := \text{span}\{B_j \circ \mathbf{F}^{-1}, j = 0, \ldots, D - 1\}.$$

Using collocation, the coefficient vector $\boldsymbol{\alpha}_h = (\alpha_0, \cdots, \alpha_{D-1})^T \in \mathbb{R}^D$ associated with the IGA-BEM discrete solution $\psi_h \in \hat{\mathcal{S}}^h$, $\psi_h(\mathbf{x}) = \sum_{j=0}^{D-1} \alpha_j (B_j \circ \mathbf{F}^{-1})(\mathbf{x})$, is determined by solving the following linear system with D equations,

$$A_h \boldsymbol{\alpha}_h = \mathbf{g}_h, \tag{6}$$

where

$$(A_h)_{i,j} := \int_\Gamma U(\mathbf{x}_i, \mathbf{y})\, (B_j \circ \mathbf{F}^{-1})(\mathbf{y})\, d\Gamma_\mathbf{y}, \tag{7}$$

and $(\mathbf{g}_h)_i := g(\mathbf{x}_i)$.

The points $\mathbf{x}_i$, $i = 0, \ldots, D - 1$, are D distinct collocation points belonging to Γ defined as the image through $\mathbf{F}$ of the Cartesian product of two sets of distinct abscissas, defined in $[a_1, b_1]$ and in $[a_2, b_2]$, respectively, as for example Greville abscissas (or suitable variants, see for example [31], to avoid collocation on the boundary of Γ). Thus, for each collocation point $\mathbf{x}_i \in \Gamma$, $i = 1, \ldots, D$, there exists $\mathbf{s}_i \in R$ such that $\mathbf{x}_i = \mathbf{F}(\mathbf{s}_i)$, with $\mathbf{s}_i \neq \mathbf{s}_j$, $i \neq j$.

Then, moving to the parametric domain, the entries of the matrix A_h can be written as follows,

$$(A_h)_{i,j} = \int_{R_j} U(\mathbf{F}(\mathbf{s}_i), \mathbf{F}(\mathbf{t}))\, B_j(\mathbf{t})\, J(\mathbf{t})\, dt_1 dt_2, \tag{8}$$

where $R_j := \text{supp}(B_j)$ and where J represents the infinitesimal surface area element,

$$J(\cdot) := \left\| \frac{\partial \mathbf{F}}{\partial t_1}(\cdot) \times \frac{\partial \mathbf{F}}{\partial t_2}(\cdot) \right\|_2.$$

Regarding the matrix A_h, we observe that it is a full matrix whose characteristics also depend on the distribution of the collocation points in Γ. On this concern we mention that the nonsingularity of the collocation matrix descending from a second kind Fredholm BIE has been theoretically studied in [4]. However our A_h comes from a Fredholm integral equation of the first kind and, up to our knowledge, sufficient conditions ensuring such property are not available in this case, unless for very special situations [12]. We can only add that in all our experiments we have always dealt with positive definite matrices, see also the comment about the condition number in the numerical experiments presented in Sect. 4.

In order to be able to compute the discrete solution efficiently and accurately, it is important to have an efficient and stable method for solving the linear system in (6) but even more to consider suitable cubature rules for weakly singular integrals to be used in the assembly phase. In this paper we focus on this second issue, extending to cubature the quadrature rules previously developed for analogous univariate weakly singular integrals. As already mentioned in the introduction, the use of such rules (which are specific for integrals containing an explicit B-spline factor in the integrand) allows us to avoid the element-by-element assembly strategy, since they are directly applied on the whole support of each B_j.

3 Cubature Rules

In this section we present the cubature rules introduced in [16] and here firstly applied in the 3D IGA-BEM context for the numerical approximation of all the integrals on the right hand side of (8). Such rules rely on spline quasi-interpolation and are an extension to the bivariate case of those firstly studied in the univariate setting [8] and successfully tested also working in adaptive spline spaces [15].

We first observe that the integrals defining the entries of A_h are weakly singular only if the source point $\mathbf{s}_i$ belongs to $\overline{R}_j$. If this is not the case, they are nearly singular or regular, depending on the distance between $\mathbf{s}_i$ and the integration domain R_j, clearly taking also into account if Γ is a closed surface. This initial classification is important because nearly singular integrals need as much care as weakly singular ones. In our implementation we adopt for them the same strategy used for weakly singular integrals, while a simplified version of the developed rules is used to approximate regular integrals, possibly using also higher QI spline degree in such a case.

Referring first of all to singular and nearly singular integrals, we observe that, as well as in the 2D case [1, 15], a preliminary singularity extraction procedure is necessary in order to increase the performance of our cubature rule. In [16] an analysis has been developed to show that two singularity extraction procedures, based on a Taylor expansion of $\mathbf{F}(\mathbf{t})$ about $\mathbf{s}$, are possible, one using an additive splitting of U and the other of multiplicative type. Considering that the first splitting requires the numerical computation of two integrals instead of one, here we have chosen to rely on the multiplicative technique. In this case the idea is to rewrite the entries of the matrix A_h given in (8) in the following equivalent form,

$$(A_h)_{i,j} := \int_{R_j} K_{\mathbf{F}}(\mathbf{s}_i, \mathbf{t})\ \rho_{\mathbf{s}_i,\mathbf{F}}(\mathbf{t}) B_j(\mathbf{t})\ J(\mathbf{t})\ \mathrm{d}t_1 \mathrm{d}t_2\,, \tag{9}$$

where

$$\rho_{\mathbf{s},\mathbf{F}}(\mathbf{t}) := \frac{U(\mathbf{F}(\mathbf{s}), \mathbf{F}(\mathbf{t}))}{K_{\mathbf{F}}(\mathbf{s}, \mathbf{t})}\,,$$

and $K_{\mathbf{F}}$ is a suitably locally defined weakly singular kernel such that the function $\rho_{\mathbf{s},\mathbf{F}}$ is at least continuous. In particular in [16] we considered the following definition of such reference kernel,

$$K_{\mathbf{F}} := K_{\mathbf{F}}(\mathbf{s},\mathbf{t}) := \frac{1}{4\pi}\frac{1}{\sqrt{P_{\mathbf{s},\mathbf{F}}(\mathbf{t})}}, \tag{10}$$

with $P_{\mathbf{s},\mathbf{F}}$ denoting the bivariate homogeneous quadratic polynomial vanishing with its first derivatives at $\mathbf{t} = \mathbf{s}$ and obtained by replacing $\mathbf{F}(\mathbf{t})$ with its first degree Taylor expansion about $\mathbf{s}$ in $\|\mathbf{F}(\mathbf{t}) - \mathbf{F}(\mathbf{s})\|_2^2$. Note anyway that for a general surface this definition just ensures continuity to $\rho_{\mathbf{s},\mathbf{F}}$ at $\mathbf{t} = \mathbf{s}$. However, if $\mathbf{F}$ is sufficiently smooth at the singular point $\mathbf{s}$, it is possible to use higher order expansions for $\mathbf{F}$, in order to ensure higher regularity to $\rho_{\mathbf{s},\mathbf{F}}$. In particular here we consider an expansion with up to 3 terms, which ensures up to C^2 smoothness to $\rho_{\mathbf{s},\mathbf{F}}$. Details about this improved singularity extraction procedure will be presented in a forthcoming paper,[3] since they form a technical part necessary for an exhaustive explanation but too long for being here reported. We just mention that in this case $K_{\mathbf{F}}$ is defined with the following expression which reduces to (10) for $n = 1$,

$$K_{\mathbf{F}}(\mathbf{s},\mathbf{t}) := \frac{1}{4\pi}\sum_{\ell=1}^{n}\sqrt{P_{\mathbf{s},\mathbf{F}}(\mathbf{t})^{-2\ell+1}}\,P_{3\ell-3}^{(\mathbf{s},\mathbf{F})}(\mathbf{t}), \tag{11}$$

where $P_{3\ell-3}^{(\mathbf{s},\mathbf{F})}$ are appropriate homogeneous polynomials of total degree $3\ell - 3$, $\ell = 1,\ldots,n$. Concerning this improved singularity extraction procedure, finally we add that, when $n > 1$ and the computational mesh is not fine enough, a correction term $\eta\|\mathbf{s} - \mathbf{t}\|_2^2$ for some suitably small $\eta > 0$ has to be added on the right hand side of (11) to avoid the introduction of new singularities in $\rho_{\mathbf{s},\mathbf{F}}$.

After this preliminary rewriting of the entries of the matrix A_h, each of them is approximated as follows,

$$(A_h)_{i,j} \approx \int_{R_j} K_{\mathbf{F}}(\mathbf{s}_i,\mathbf{t})\,\left(Q_{\mathbf{p},\Delta_j}\left(J\rho_{\mathbf{s}_i,\mathbf{F}}\right)\right)(\mathbf{t})\; B_j(\mathbf{t})\; \mathrm{d}t_1\mathrm{d}t_2\,, \tag{12}$$

where Δ_j is a uniform partition of R_j and $Q_{\mathbf{p},\Delta_j}$ denotes a tensor-product QI operator which approximates a bivariate function with a tensor-product spline of bi-degree $\mathbf{p} = (p_1, p_2)$ with respect to the partition Δ_j.

In order to compute the integral on the right hand side of (12), the spline $Q_{\mathbf{p},\Delta_j}\left(J\ \rho_{\mathbf{s}_i,\mathbf{F}}\right)$ is multiplied by the B-spline B_j, using for this aim the extension to the tensor-product setting of the Mørken formula [22] for spline product. As a result, the whole product $\left(J\ B_j\ \rho_{\mathbf{s}_i,\mathbf{F}}\right)$ is approximated with a local tensor-product spline function defined in R_j, having bi-degree $\mathbf{p} + \mathbf{d}$, and breakpoints in R_j. We

[3] T. Kanduč isoparametric singularity extraction technique for 3D potential problems in BEM, submitted, 2021.

observe that, in order to implement the so derived cubature rules, the evaluation of the following *modified moments* is firstly necessary,

$$\mu_{K_{\mathbf{F}},B}(\mathbf{s}) := \int_{\text{supp}(B)} K_{\mathbf{F}}(\mathbf{s},\mathbf{t})\ B(\mathbf{t})\ \mathrm{d}t_1 \mathrm{d}t_2\,, \tag{13}$$

with B denoting any tensor-product B-spline. This computation is done by relying on the recursion formula for B-splines and on the preliminary analytic computation of element integrals whose integrand is the product of a monomial times $K_{\mathbf{F}}$, procedure possible for a kernel $K_{\mathbf{F}}$ defined as in (10) or more generally as in (11).

When the integral defining the (i, j)-th entry of A_h is classified as a regular one, we do not need the preliminary computation of the modified moments and the strategy is simplified by directly setting

$$(A_h)_{i,j} \approx \int_{R_j} \left(Q_{\mathbf{q},\Lambda_j}\left(U(\mathbf{F}(\mathbf{s}_i),\cdot)J\right)\right)(\mathbf{t})\ B_j(\mathbf{t})\ \mathrm{d}t_1 \mathrm{d}t_2\,, \tag{14}$$

where $\mathbf{q} = (q_1\,,\ q_2)$ denotes a bi-degree for the local splines possibly different from $\mathbf{p}$ and Λ_j is a uniform partition of R_j, also possibly different from Δ_j.

Concerning the QI spline operator used by the proposed cubature rules, for our experiments we have always considered the uniform tensor-product formulation of the derivative free variant of the Hermite QI operator introduced in [20], see for example [7] for an introduction to its formulation in the bivariate setting and for an application in the context of adaptive bivariate spline approximation. This univariate spline QI operator has approximation order $p + 1$ and in the context of numerical integration it has special interest because its application to both regular and weakly singular quadrature is superconvergent when even spline degrees are considered on uniform knot distributions [8, 21]. Thus for smooth functions $Q_{\mathbf{p},\Delta_j}$ has approximation order $\min\{p_1, p_2\} + 1$ and it inherits the superconvergence feature when analogously applied for cubature. Observe however that, in our approach, choosing values of p_k, $k = 1, 2$, greater than 2 is not useful, because of the limited regularity of the function $\rho_{\mathbf{s}_i,\mathbf{F}}$. It can be profitable instead to select $q_k > 2$, $k = 1, 2$, if $\mathbf{F}$ is smooth in R_j. Considering the assumed analytic expression of $\mathbf{F}$, this is surely true when the interior of R_j does not contain any straight line $t_j = \tau_j$, with τ_j *geometric knot* in the j-th parametric direction, that is $\tau_j \in T_j$, $j = 1, 2$, where T_j, $j = 1, 2$ are the extended knot vectors used to define $\mathbf{F}$.

4 Numerical Simulations

In this section we test the uniform formulation of the proposed IGA boundary element model for the numerical solution of two Laplace problems on two different 3D unbounded domains. Furthermore a nonuniform implementation of the scheme is also tested for solving a more difficult Laplace problem exterior to the first considered domain. In all the examples the improved singularity extraction procedure is used within the developed singular cubature rules, always setting the parameter η (see Sect. 3) equal to 1. Concerning the discretization, in all the presented experiments approximate densities are constructed on a given initial mesh and also on successive finer ones obtained from a dyadic h-refinement procedure which inserts a simple knot at the midpoint between every pair of consecutive breakpoints in each parametric direction.

For the first two considered examples the density ψ is known, while the Dirichlet datum g is not available in closed form. Thus, the entries of the right-hand side in (6) are computed numerically,

$$(\mathbf{g}_h)_i = \int_R U(\mathbf{F}(\mathbf{s}_i), \mathbf{F}(\mathbf{t}))\,(\psi \circ \mathbf{F})(\mathbf{t})\,J(\mathbf{t})\,\mathrm{d}t_1\mathrm{d}t_2\,, \tag{15}$$

by applying the same approach adopted to compute the system matrix. For these examples the accuracy of the scheme is measured by the $L^2(\Gamma)$ norm of the absolute density error $(\psi_h - \psi)$, in particular considering its dependency on the mesh size h. The third exterior Laplace problem is based on Test 5.3 of [14] where the more general Helmholtz equation is solved. Since in this case the analytic expression of the exact potential u is known, the Dirichlet datum is given and the accuracy in the potential approximation $u_h \approx u$ can be measured, provided that the representation formula given in (4) is previously applied to ψ_h to define u_h in the considered unbounded domain $\mathbb{R}^3 \setminus \overline{\Omega}$.

4.1 Problem on an Unbounded Domain Exterior to a Torus

We consider a problem exterior to a toroidal surface Γ with a major radius $\rho_M = 3$ and a minor radius $\rho_m = 1$. The geometry is represented in NURBS form by defining as in (5) the vector function $\mathbf{F}$ on the parametric domain $R = [-3\,,\,3] \times [-1\,,\,1]$ with $\mathbf{d} = (2, 2)$, fixing the following knot vectors,

$$T_1 = (-3,\ -3,\ -3,\ -1.5,\ -1.5,\ -1.5,\ 0,\ 0,\ 0,\ 1.5,\ 1.5,\ 1.5,\ 3,\ 3,\ 3),$$
$$T_2 = (-1,\ -1,\ -1,\ -0.5,\ -0.5,\ -0.5,\ 0,\ 0,\ 0,\ 0.5,\ 0.5,\ 0.5,\ 1,\ 1,\ 1),$$

see solid lines in Fig. 1a. In this form the torus is composed by the 16 rational biquadratic sub-patches shown in Fig. 1c (separated by solid lines), which are locally represented in rational Bernstein form, since all the breakpoints have maximal multiplicity (equal to 3) in both the knot vectors T_j, $j = 1, 2$. The associated control net has the size 12×12; for clarity in Fig. 1b we show a sub-patch and the related 3×3 sub-net of control points with the corresponding weights. Dashed lines in Fig. 1a, c represent additional inserted knots for the initial computational mesh. Note that, despite the maximal multiplicity of the breakpoints in T_j, $j = 1, 2$, the vector function $\mathbf{F}$ is C^1 continuous because of the symmetry of the considered control net and of the related weights. We remark that, even if not strictly necessary, maximal multiplicity is used for the geometric knots to ensure to $\mathbf{F}$ maximal regularity when restricted to the support of a trial function B_j. As explained at the end of the previous section, this is useful for the accuracy of our cubature rules when regular integrals are dealt with, since the support of a trial function can never cross a line $t_j = \tau_j \in T_j$, $j = 1, 2$, i.e., solid lines in Fig. 1a.

The exact function $\psi \circ \mathbf{F}$ is set to

$$(\psi \circ \mathbf{F})(\mathbf{t}) = \sin(4/3\ t_1\, t_2)$$

and its shape in the parametric domain is depicted in Fig. 1d. Each rational sub-patch of the torus contains the same number of collocation points obtained from the Cartesian product of the improved Greville abscissas [31], defined on the corresponding sub-rectangles of R, and mapped to Γ by $\mathbf{F}$.

Two different experiments have been performed for this example. The first one is a standard test where initially $\mathscr{T}_2 = T_2$ and $\mathscr{T}_1$ is obtained by merging T_1 with $\{-2.5\,, -2\,, -1\,, -0.5\,, 0.5\,, 1\,, 2\,, 2.5\}$. Since in all the simulations the geometric knots keep maximal multiplicity, at any refinement level also discontinuous functions belong to the space $\mathscr{S}_h$. In the second experiment initially we define $\mathscr{T}_2$ still from T_2 but by decreasing the multiplicity of each inner knot by one. $\mathscr{T}_1$ is initialised as in the other case but again by reducing by one the multiplicity of all the geometric knots. Thus 3×3 degrees of freedom are eliminated with respect to the previous experiment and as a consequence the simulation space is always included in $C^0(R)$. Note that anyway we keep unchanged the set of collocation points so we end up with an overdetermined linear system which is solved in the least squares sense.

In both experiments the results shown are obtained by deriving the reference kernel $K_{\mathbf{F}}$ by using the first two terms of the Taylor expansion of $\mathbf{F}$. For the weakly or near singular cubature the QI bi-degree $\mathbf{p}$ is set to $\mathbf{p} = (2, 2)$, using in each parametric direction 13 uniform nodes on the support of any B-spline trial function. Due to higher global smoothness of integrands, for regular integrals we can profitably set $\mathbf{q} = (4, 4)$ and just use 7 uniform nodes on the support of any B-spline in each parametric direction.

Convergence plots of the behaviour of approximation error $\|\psi - \psi_h\|_{L^2(\Gamma)}$ using uniform h-refinement strategy together with the theoretical convergence order 3 is shown in Fig. 3a. Looking at the figure, we see that both the plots have the expected behaviour and also that the C^0 one is very near to the other. The system matrix A is well conditioned for all tests (in the last step its condition number for the first experiment is about $5.1 \cdot 10^3$).

4.2 A Screen Problem for a Curved Obstacle

As a second example we consider a screen problem (1) with Γ taken as the limited portion of a saddle open surface parametrized by $\mathbf{F} : R \rightarrow \overline{\Gamma}$, with $R = [-1\,, 1]^2$

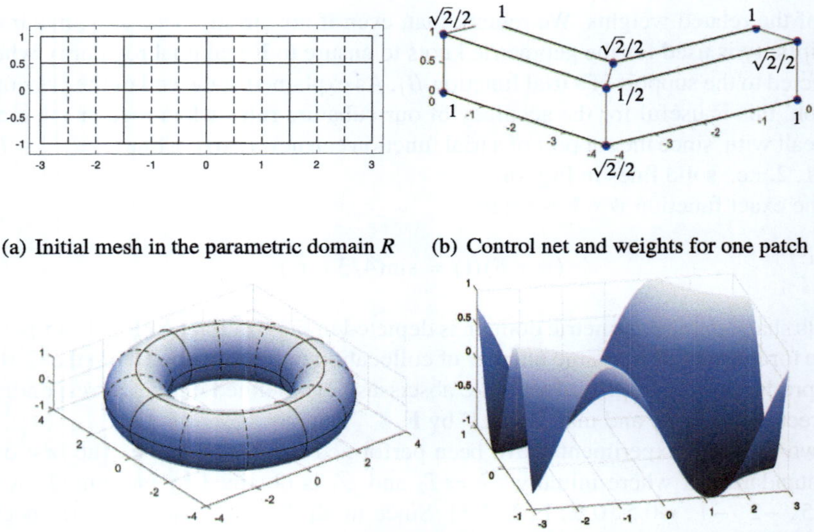

(a) Initial mesh in the parametric domain R (b) Control net and weights for one patch

(c) Surface $\Gamma = \partial\Omega = \mathbf{F}(R)$ and initial mesh (d) Graph of composition with $\mathbf{F}$ of the exact BIE solution ψ

Fig. 1 Example 4.1: problem exterior to a torus

and $\mathbf{F}(t_1, t_2) = (t_1\ ,\ t_2\ ,\ t_1^2 - t_2^2)$. The Dirichlet datum g is numerically computed such that the composition with $\mathbf{F}$ of the exact density function, solution of integral Eq. (2), is given by

$$(\psi \circ \mathbf{F})(\mathbf{t}) = \sqrt{1 + 4t_1^2 - 0.5t_2^2}.$$

In Fig. 2 we show the geometry and $\psi \circ \mathbf{F}$. Note that the considered $\mathbf{F}$ can be written as in (5) with $\mathbf{d} = (2, 2)$ and $T_j = (-1, -1, -1, 1, 1, 1)\ ,\ j = 1, 2\ ,$ since polynomials (restricted to R) are a subset of NURBS. Thus the problem is discretised by using quadratic B-splines and setting $\mathbf{p} = (2, 2)$ for the QI approximation of singular and nearly singular integrals, with 7 quadrature uniform nodes on the support of any B-spline trial function in each parametric direction. The reference kernel $K_{\mathbf{F}}$ is obtained in this case using the first three terms in the Taylor expansion of $\mathbf{F}$, while the improved Greville abscissas [31] are employed as collocation points also in this case.

In Fig. 3b we show the convergence plots for $\mathbf{q}$ varying from $(2, 2)$ to $(4, 4)$, using from 5 to 7 uniform quadrature nodes per B-spline support in each parametric direction for regular integrals. Better accuracy of the numerical solution corresponds to higher $\mathbf{q}$ but the order 3 is observed in all the cases. Finally, the condition number for the system matrix A has order 10^2 at the last level.

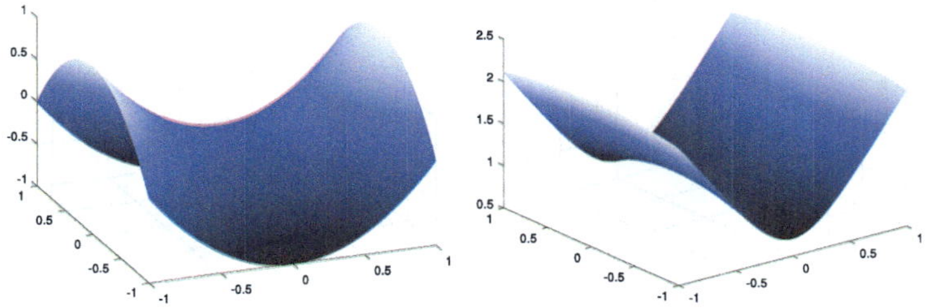

(a) Surface $\Gamma = \partial\Omega = \mathbf{F}(R)$ (b) Graph of composition with $\mathbf{F}$ of the exact BIE solution ψ

Fig. 2 Example 4.2: screen problem

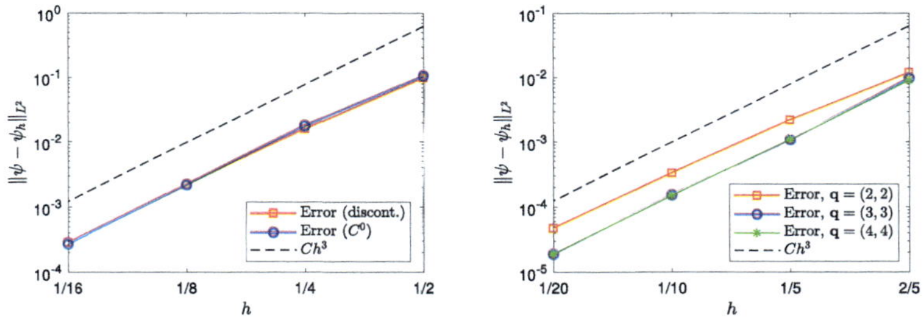

(a) Problem exterior to a torus $(\mathbf{d} = (2,2))$ (b) Screen problem $(\mathbf{d} = (2,2))$

Fig. 3 Convergence plots in $L^2(\Gamma)$ norm of the density error $\psi - \psi_h$ for Examples 4.1 and 4.2

4.3 A Nearly Singular Potential Problem Exterior to a Torus

We consider the same toroidal surface presented in Example 4.1, solving the problem (1) with the Dirichlet datum g chosen as,

$$g(\mathbf{x}) := \frac{1}{\|\mathbf{x} - \mathbf{v}\|_2} \quad \text{for } \mathbf{x} \in \Gamma\,, \tag{16}$$

which is also the known analytic expression of the exact potential u when $\mathbf{x} \in \mathbb{R}^3 \setminus \overline{\Omega}$. In particular in the reported experiments we fix $\mathbf{v} = (-3, 0, 0) \in \Omega$, see Fig. 4a for the graph of $g \circ \mathbf{F}$ with this setting. This type of Dirichlet datum gives rise to sharp gradients of the corresponding function $\psi \circ \mathbf{F}$ near two opposite boundary edges of the parametric domain, see Fig. 4b where we can only show the graph of $\psi_h \circ \mathbf{F}$, since the exact expression of ψ is unknown.

A suitable mesh, which is nonuniform in the horizontal parametric direction, has been manually constructed in order to get better accuracy. Clearly it would be advisable to rely on a fully automatic adaptive mesh generation strategy but this is out of the scope of the presented paper. The initial mesh used for the reported

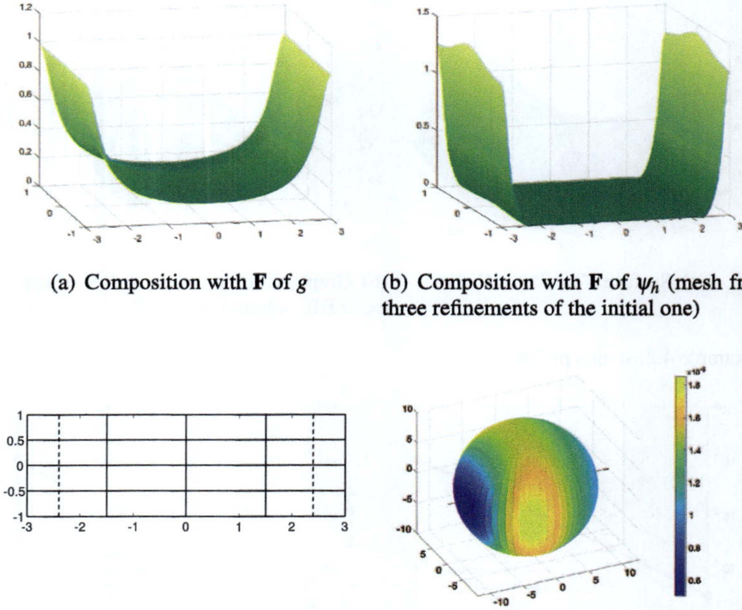

(a) Composition with $\mathbf{F}$ of g

(b) Composition with $\mathbf{F}$ of ψ_h (mesh from three refinements of the initial one)

(c) Initial mesh in the parameter domain

(d) Distribution on S^2 of the potential error $|u-u_h|$ (mesh from three refinements of the initial one)

Fig. 4 Example 4.3: another problem exterior to the torus considered in Example 4.1

experiments is shown in Fig. 4c; as in Example 4.1, solid lines correspond to the knots of the vectors T_1 and T_2 which are defined as before, while dashed lines correspond to additional inserted knots. As done in the second subtest described in Sect. 4.1, the multiplicity of each inner geometry knot is decreased by one in order to deal with continuous spline spaces. The collocation points are obtained from the improved Greville abscissas and their number matches the global number of degrees of freedom characterising the discretization space.

Using the representation formula (4), for this test we also evaluate the approximated potential at some point $\mathbf{x}$ of the unbounded domain,

$$u_h(\mathbf{x}) := V\psi_h(\mathbf{x}) \quad \text{for } \mathbf{x} \in \mathbb{R}^3 \setminus \overline{\Omega}. \tag{17}$$

In particular, in order to evaluate a posteriori the accuracy of our scheme, we compute the potential error $u - u_h$ at 2^{m+2} points uniformly sampled on a spherical surface S^2 with center located at the origin and with radius equal to 10 and so exterior to the considered toroidal surface. Note that here m denotes the dyadical refinement step, where $m = 1$ for the initial mesh and in particular $m = 4$ was adopted to get the error distribution in S^2 shown in Fig. 4d. Looking at the figure, it can be observed that the

Table 1 Example 4.3: degrees of freedom and norms of the potential error $u - u_h$ obtained at different refinement level m setting $\mathbf{d} = \mathbf{p} = \mathbf{q} = (2, 2)$

m	Ndof_1	Ndof_2	Ndof	$\|u - u_h\|_{L^2(S^2)}$	$\|u - u_h\|_\infty$
1	11	9	99	$1.24 \cdot 10^{-2}$	$5.88 \cdot 10^{-4}$
2	17	13	221	$3.58 \cdot 10^{-3}$	$2.21 \cdot 10^{-4}$
3	29	21	609	$1.51 \cdot 10^{-4}$	$1.00 \cdot 10^{-5}$
4	53	37	1961	$4.68 \cdot 10^{-7}$	$1.82 \cdot 10^{-8}$

potential error has a nice constant order of magnitude on the sphere equal to 10^{-8} and that its distribution is symmetric with respect to the x axis where the point $\boldsymbol{\nu}$ is located (it is not shown in the figure since it is inside the sphere).

The accuracy achieved at different refinement levels m is reported in Table 1 together with the corresponding global (Ndof) and directional (Ndof_1, Ndof_2) number of degrees of freedom. It is measured in both the maximum and the $L^2(S^2)$ norms, both approximated using the available discrete information on the sphere. The test has been performed by using quadratic discretization splines and selecting $\mathbf{p} = \mathbf{q} = (2, 2)$. Concerning the cubature rules, 7 and 13 uniform nodes on the support of every basis function are used in each parametric direction, respectively for the evaluation of singular/nearly singular and regular integrals. It is worth noticing that, since the evaluation of u_h is done at points located far enough from the surface Γ, the integral in (17) is non singular. Hence, the same cubature rule constructed for the evaluation of the regular integrals has been employed for the representation formula. Finally for completeness we mention that also for this example the coefficient matrix in (6) is well conditioned in all the developed numerical experiments, having a condition number about equal to $3.3 \cdot 10^3$ at the last considered level.

5 Conclusion

In this paper a collocation IGA-BEM scheme relying on the indirect integral formulation of Laplace problems is proposed and tested for 3D exterior problems, obtaining the expected convergence order in the density approximation. Cubature rules based on spline quasi-interpolation and on a spline product formula are applied for the numerical approximation of both weakly and nearly singular bivariate integrals appearing in the 3D IGA-BEM scheme. They are combined with a robust multiplicative singularity extraction procedure to increase the accuracy of the rules. A variant of such rules is proposed for the regular integrals to be also dealt with in the assembly phase. This can be efficiently implemented in a function-by-function form thanks to the peculiarity of the considered cubature.

Acknowledgements A. Falini, C. Giannelli, M. L. Sampoli, and A. Sestini are members of the INdAM Research group GNCS. The INdAM support through GNCS and Finanziamenti Premiali SUNRISE is gratefully acknowledged. A. Falini was also supported by the INdAM-GNCS project "Finanziamenti Giovani Ricercatori 2019-2020".

References

1. Aimi, A., Calabrò, F., Diligenti, M., Sampoli, M., Sangalli, G., Sestini, A.: Efficient assembly based on B-spline tailored quadrature rules for the IgA-SGBEM. Comput. Methods Appl. Mech. Eng. **331**, 327–342 (2018)
2. Aimi, A., Diligenti, M., Sampoli, M.L., Sestini, A.: Isogeometric analysis and symmetric Galerkin BEM: A 2D numerical study. Appl. Math. Comput. **272**, 173–186 (2016)
3. Al Akhras, H., Elguedj, T., Gravouil, A., Rochette, M.: Isogeometric analysis-suitable trivariate NURBS models from standard B-Rep models. Comput. Methods Appl. Mech. Eng. **307**, 256–274 (2016)
4. Atkinson, K.E.: The Numerical Solution of Integral Equations of the Second Kind. Cambridge University Press (2009)
5. Beer, G., Mallardo, V., Ruocco, E., Marussig, B., Zechner, J., Dünser, C., Fries, T.P.: Isogeometric boundary element analysis with elasto-plastic inclusions. Part 2: 3-D problems. Comput. Methods Appl. Mech. Eng. **315**, 418–433 (2017)
6. Beer, G., Marussig, B., Duenser, C.: The Isogeometric Boundary Element Method. Springer (2020)
7. Bracco, C., Giannelli, C., Mazzia, F., Sestini, A.: Bivariate hierarchical Hermite spline quasi-interpolation. BIT Numer. Math. **56**, 1165–1188 (2016)
8. Calabrò, F., Falini, A., Sampoli, M.L., Sestini, A.: Efficient quadrature rules based on spline quasi-interpolation for application to IgA-BEMs. J. Comput. Appl. Math. **338**, 153–167 (2018)
9. Cohen, E., Martin, T., Kirby, R., Lyche, T., Riesenfeld, R.F.: Analysis-aware modeling: understanding quality considerations in modeling for isogeometric analysis. Comput. Methods Appl. Mech. Eng. **199**, 334–356 (2010)
10. Coox, L., Atak, O., Vandepitte, D., Desmet, W.: An isogeometric indirect boundary element method for solving acoustic problems in open-boundary domains. Comput. Methods Appl. Mech. Eng. **316**, 186–208 (2017)
11. Costabel, M.: Principles of Boundary Element Methods. Techn. Hochsch, Fachbereich Mathematik (1986)
12. Costabel, M., Penzel, F., Schneider, R.: Error analysis of a boundary element collocation method for a screen problem in $\mathbb{R}^3$. Math. Comput. **58**, 575–586 (1992)
13. Cottrell, J.A., Hughes, T.J.R., Bazilevs, Y.: Isogeometric Analysis: Toward Integration of CAD and FEA. Wiley (2009)
14. Dölz, J., Harbrecht, H., Kurz, S., Schöps, S., Wolf, F.: A fast isogeometric BEM for the three dimensional Laplace and Helmholtz problems. Comput. Methods Appl. Mech. Eng. **330**, 83–101 (2018)
15. Falini, A., Giannelli, C., Kanduč, T., Sampoli, M.L., Sestini, A.: An adaptive IgA-BEM with hierarchical B-splines based on quasi-interpolation quadrature schemes. Int. J. Numer. Methods Eng. **117**, 1038–1058 (2019)
16. Falini, A., Kanduč, T., Sampoli, M.L., Sestini, A.: Cubature rules based on bivariate spline quasi-interpolation for weakly singular integrals. In: Fasshauer, G.E., Neamtu, M., Schumaker, L.L. (eds.), Approximation Theory XVI, Springer in Mathematics & Statistics, vol. 336, pp. 73–86. Springer (2020)
17. Ginnis, A.I., Kostas, K.V., Politis, C.G., Kaklis, P.D., Belibassakis, K.A., Gerostathis, T.P., Scott, M.A., Hughes, T.J.R.: Isogeometric boundary-element analysis for the wave-resistance problem using T-splines. Comput. Methods Appl. Mech. Eng. **279**, 425–439 (2014)

18. Gong, Y.P., Dong, C.Y., Qin, X.C.: An isogeometric boundary element method for three dimensional potential problems. J. Comput. Appl. Math. **313**, 454–468 (2017)
19. Hughes, T.J.R., Cottrell, J.A., Bazilevs, Y.: Isogeometric analysis: CAD, finite elements, NURBS, exact geometry and mesh refinement. Comput. Methods Appl. Mech. Eng. **194**, 4135–4195 (2005)
20. Mazzia, F., Sestini, A.: The BS class of Hermite spline quasi-interpolants on nonuniform knot distributions. BIT Numerical Mathematics **49**, 611–628 (2009)
21. Mazzia, F., Sestini, A.: Quadrature formulas descending from BS Hermite spline quasi-interpolation. J. Comput. Appl. Math. **236**, 4105–4118 (2012)
22. Mørken, K.: Some identities for products and degree raising of splines. Constr. Approx. **7**, 195–208 (1991)
23. Nguyen, B.H., Tran, H.D., Anitescu, C., Zhuang, X., Rabczuk, T.: An isogeometric symmetric Galerkin boundary element method for two-dimensional crack problems. Comput. Methods Appl. Mech. Eng. **306**, 252–275 (2016)
24. Nguyen, B.H., Zhuang, X., Wriggers, P., Rabczuk, T., Mear, M.E., Tran, H.D.: Isogeometric symmetric Galerkin boundary element method for three-dimensional elasticity problems. Comput. Methods Appl. Mech. Eng. **323**, 132–150 (2017)
25. Pan, M., Chen, F., Tong, W.: Volumetric spline parameterization for isogeometric analysis. Comput. Methods Appl. Mech. Eng. **359**, 112–769 (2020)
26. Sauter, S.A., Schwab, C.: Boundary Element Methods, Springer Series in Computational Mathematics, vol. 39. Springer (2011)
27. Simpson, R.N., Bordas, S.P.A., Trevelyan, J., Rabczuk, T.: A two-dimensional isogeometric boundary element method for elastostatic analysis. Comput. Methods Appl. Mech. Eng. **209–212**, 87–100 (2012)
28. Sutradhar, A., Paulino, G., Gray, L.J.: Symmetric Galerkin Boundary Element Method. Springer Science & Business Media (2008)
29. Taus, M., Rodin, G.J., Hughes, T.J.R.: Isogeometric analysis of boundary integral equations: high-order collocation methods for the singular and hyper-singular equations. Math. Models Methods Appl. Sci. **26**, 1447–1480 (2016)
30. Toshniwal, D., Speleers, H., Hiemstra, R.R., Hughes, T.J.R.: Multi-degree smooth polar splines: a framework for geometric modeling and isogeometric analysis. Comput. Methods Appl. Mech. Eng. **316**, 1005–1061 (2017)
31. Wang, Y.J., Benson, D.J.: Multi-patch nonsingular isogeometric boundary element analysis in 3D. Comput. Methods Appl. Mech. Eng. **293**, 71–91 (2015)

DPB-Splines: The Decoupled Basis of Patchwork Splines

Nora Hemelmayr and Bert Jüttler

Abstract We consider partitions of the d-dimensional unit cube into patches with an associated tensor-product spline space for each of them. The spline spaces possess the same multi-degree $\mathbf{p} = (p_1, \ldots, p_d)$ and the same maximum smoothness $C^{\mathbf{p}-1}$, but the choice of the knots is very flexible. Under certain assumptions, we show how to construct Decoupled Patchwork B-splines (DPB-splines) that span the corresponding patchwork spline space $\mathbb{P}$. More precisely, we generate a basis for the space $\mathbb{P}$ formed by all $C^{\mathbf{p}-1}$ smooth functions that admit patch-wise representations in the associated spline spaces. Based on the framework of decoupled tensor-product B-splines [31, 32], we obtain a basis that is algebraically complete, forms a convex partition of unity, and preserves the coefficients of the local B-spline representations. Furthermore, we present an adaptive refinement algorithm for surface approximation generating partitions that satisfy the required assumptions and hence can be equipped with a DPB-spline basis.

1 Introduction

Tensor-product B-splines, which are the standard for describing free-form geometries in Computer-Aided Design [14] possess a fundamental limitation: Their tensor-product structure does not support local refinement. Therefore, adaptive spline constructions have been developed in order to overcome this restriction and provide more flexibility in design and analysis. Forsey and Bartels were the first to introduce hierarchical B-spline refinement [15]. Their hierarchical construction was extended by Kraft, who established a selection mechanism that defines a basis for the hierarchical spline space [27].

T-splines [35] have been developed as an adaptive construction on meshes with T-joints. In general these splines are not linearly independent and therefore, analysis-suitable (AS) T-splines and AS++ T-splines have been introduced and discussed in [2, 30, 33, 41]. Moreover, polynomial splines over hierarchical T-meshes [25, 29] and

N. Hemelmayr · B. Jüttler (✉)
Institute of Applied Geometry, Johannes Kepler University, Linz, Austria
e-mail: bert.juettler@jku.at

C. Manni and H. Speleers (eds.), *Geometric Challenges in Isogeometric Analysis*, Springer INdAM Series 49, https://doi.org/10.1007/978-3-030-92313-6_3

locally refined B-splines [9] have been established. Recently, the latter construction has been generalized to LR T-splines [8], where local refinement is performed on an initial T-mesh instead of a tensor-product mesh.

In order to restore the partition of unity property for hierarchical splines, a truncation mechanism has been introduced in [18]. The resulting truncated hierarchical B-splines (THB-splines) possess good mathematical properties, see [19, 21, 32, 36, 38] for a detailed analysis of stability, completeness and approximation power. Applications of (T)HB-splines include surface approximation [5, 22, 26] and isogeometric analysis [1, 3, 4, 6, 28, 34]. Aspects of the implementation have been discussed in [16, 17]. Moreover, the hierarchical principle has been applied to other constructions, such as Powell-Sabin splines [37], triangular splines [40], box splines [20, 23, 42], B-splines on triangulations [24], T-splines [7, 13], and subdivision functions [39, 43].

While THB-splines possess good mathematical properties, they are not as flexible as other constructions like T-splines or LR B-splines. Hierarchical B-splines rely on a sequence of *nested* spline spaces and therefore, the choice of possible refinement strategies is limited. Recently, we generalized hierarchical splines to Patchwork B-splines (PB-splines), which are defined on a sequence of partially nested spline spaces and enable the use of independent refinement strategies in different parts of the domain [10, 12]. An adapted version of Kraft's selection mechanism generates a basis for the patchwork spline space under certain assumptions. Similar to the hierarchical splines, a truncation mechanism has been established in order to obtain the basis of truncated PB-splines (TPB-splines) [10] that forms a non-negative partition of unity. Applications of the PB-splines in surface approximation and to lofting B-spline curves [11] have shown the potential of this new construction.

Although the (T)PB-splines provide increased flexibility, there are still certain limitations on the available refinement strategies. For instance, a region that separates two coarser spline spaces is required to have a certain degree-dependent width, see Fig. 1, bottom row. Additionally, the spaces that are present in a degree-dependent neighborhood of any cell need to be (upwards or downwards) compatible with the chosen space for this cell if TPB-splines are to be generated. Consequently, a region that separates two finer (but non-nested) refinement regions is also required to possess a certain degree-dependent width, see Fig. 1, top row. For instance, also the PB-spline hierarchies considered in [11] for solving the lofting problem are not suitable for TPB-splines.

Finally we note that (T)PB-splines are not algebraically complete in general. This is similar to THB-splines, see [31], where the use of decoupled B-splines has led to significant improvements in this regard. We build on this idea to define the new framework of *Decoupled Patchwork B-splines* (DPB-splines). More precisely, we define local basis functions, called patch B-splines, that are obtained by restricting and decoupling tensor-product B-splines.

The remainder of this paper is organized as follows: Sect. 2 introduces the general framework consisting of patches, the local spline spaces and a patchwork spline space. Moreover, we observe that the patchwork spline space is equal to the full spline space, i.e., a basis for this space is algebraically complete. A truncation and

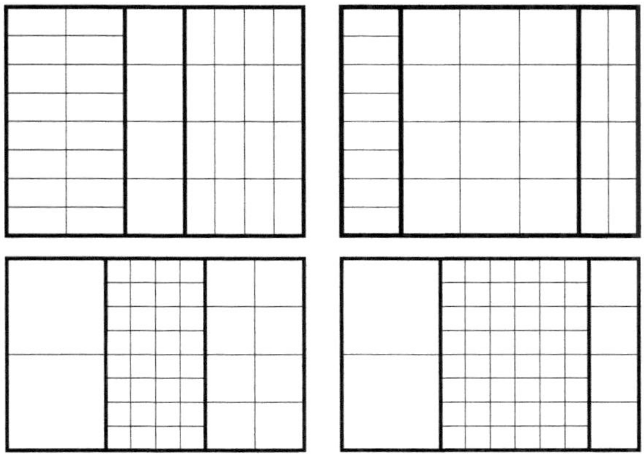

Fig. 1 Invalid (left) and valid (right) meshes for bicubic PB-splines (bottom) and TPB-splines (top)

selection mechanism are then used to define the DPB-splines in Sect. 3. We identify assumptions that are needed in order to guarantee a certain order of smoothness of the basis functions and for characterizing the obtained spline space. These assumptions affect only a patch and its neighbors. The mathematical properties of the DPB-splines are analyzed in Sect. 4. In particular, the basis functions possess continuous values and derivatives up to a certain order, they span the patchwork spline space and form a convex partition of unity. Furthermore, we identify a simple sufficient condition for a certain class of patches that guarantees that the assumptions are satisfied. Section 5 provides several examples that compare the flexibility of DPB-splines with TPB- and THB-splines. We introduce an improved refinement algorithm for least-squares fitting in Sect. 6, which generates hierarchies that can be equipped with a Decoupled Patchwork and a (truncated) Patchwork B-spline basis. Two examples are used to illustrate the advantages of this new strategy. Finally, we conclude the paper.

2 The Patchwork Hierarchy

For constructing DPB-splines we generate a hierarchy consisting of patches and corresponding spaces, which are not necessarily nested. Furthermore, the collection of patches and spline spaces allows the definition of a patchwork spline space.

2.1 Patches

We consider a finite sequence of *patches* $\{\pi^\ell\}_{\ell=1,\ldots,N}$, which are closed subsets of the d–dimensional unit cube $[0, 1]^d$. The upper index ℓ will be called the *level*. The patches possess a mutually disjoint interior, $\mathrm{int}(\pi^\ell) \cap \mathrm{int}(\pi^k) = \emptyset$, for $k \neq \ell$, and their union defines the *domain*,

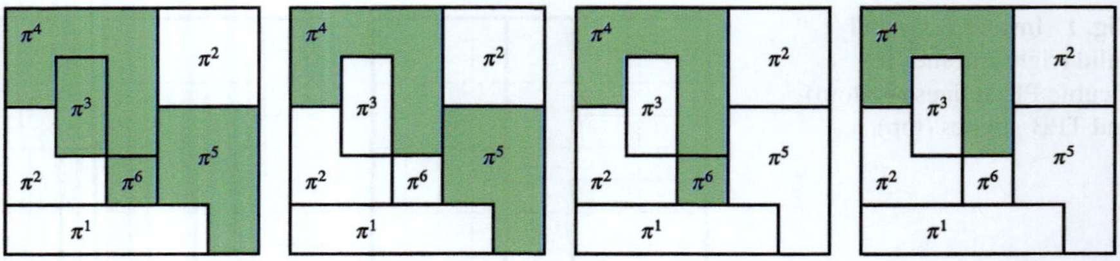

(a) $\Delta_{\geq 3}$ (b) $\Delta_{\mathscr{D},\geq 3}$ (c) $\Delta_{\mathscr{N}^3}$ (d) $\Delta_{\mathscr{N}^3_{\mathscr{D}}}$

Fig. 2 Example 1—Various subdomains for $\mathscr{D} = \{2, 4, 5\}$

$$\Omega = \bigcup_{\ell=1}^{N} \pi^{\ell}.$$

Not only the entire domain but certain subsets thereof will be used in the remainder of the paper. An index set $\mathscr{D} \subseteq \{1, \ldots, N\}$ generates the *subdomain*,

$$\Delta_{\mathscr{D}} = \bigcup_{\ell \in \mathscr{D}} \pi^{\ell} \subseteq \Omega. \tag{1}$$

In particular, we define $\Delta_{\mathscr{D},\geq \ell} = \Delta_{\mathscr{D}} \cap \Delta_{\geq \ell}$, with the index set $\geq \ell = \{\ell, \ldots, N\}$, as the subdomain formed by all patches in $\Delta_{\mathscr{D}}$ with level greater or equal to ℓ. Furthermore, we introduce the neighboring index set

$$\mathscr{N}^{\ell} = \{k \;:\; k > \ell \text{ and } \pi^k \cap \pi^{\ell} \neq \emptyset\},$$

that comprises the levels $k > \ell$ of higher level neighboring patches of π^{ℓ}. The index set of higher level neighboring patches in $\Delta_{\mathscr{D}}$ will be denoted as

$$\mathscr{N}^{\ell}_{\mathscr{D}} = \mathscr{N}^{\ell} \cap \mathscr{D}.$$

The two index sets $\mathscr{N}^{\ell}$ and $\mathscr{N}^{\ell}_{\mathscr{D}}$ define the subdomains $\Delta_{\mathscr{N}^{\ell}}$ and $\Delta_{\mathscr{N}^{\ell}_{\mathscr{D}}}$, respectively.

Example 1 We consider the subdivision of the unit square into six patches, $\pi^1, \ldots, \pi^6$. Note that the patches might possess several connected components, as it is the case for patch π^2 in this example. Figure 2 depicts several subdomains of Ω. The subdomain $\Delta_{\geq 3}$ is formed by the patches $\pi^3, \ldots, \pi^6$ as illustrated in Fig. 2a. Consider $\mathscr{D} = \{2, 4, 5\}$, thus $\Delta_{\mathscr{D},\geq 3} = \pi^4 \cup \pi^5$, see Fig. 2b. The higher level neighbors of π^3 define the index set $\mathscr{N}^3 = \{4, 6\}$, see Fig. 2c. The intersection of $\Delta_{\mathscr{N}^3}$ and $\Delta_{\mathscr{D}}$ gives the subdomain $\Delta_{\mathscr{N}^3_{\mathscr{D}}}$, which consists of π^4, see Fig. 2d. ◊

Except for trivial situations, each patch shares parts of its boundary $\partial\pi^{\ell}$ with patches of a different level. The part of the boundary that intersects with patches of a *higher* level will be called the *constraining boundary*,

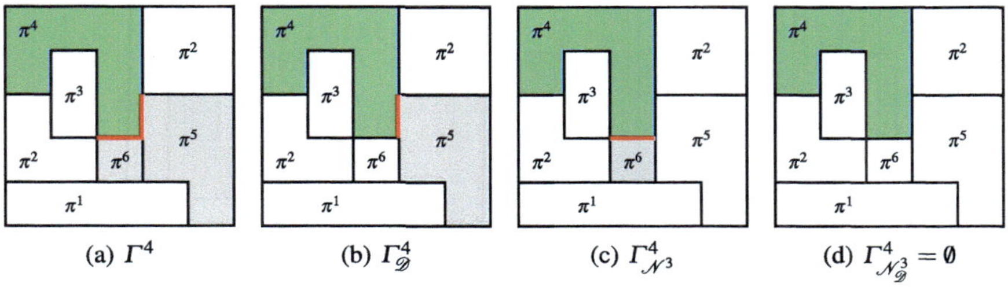

Fig. 3 Example 2—Various constraining boundaries (shown in red)

$$\Gamma^{\ell} = \Delta_{\geq \ell+1} \cap \pi^{\ell}.$$

In addition we define the constraining boundary

$$\Gamma^{\ell}_{\mathscr{D}} = \Gamma^{\ell} \cap \Delta_{\mathscr{D}} = \Delta_{\mathscr{D},\geq \ell+1} \cap \pi^{\ell},$$

with respect to the subdomain $\Delta_{\mathscr{D}}$.

Example 2 We re-visit the previous example. Figure 3 shows the constraining boundaries Γ^4, $\Gamma^4_{\mathscr{D}}$, $\Gamma^4_{\mathscr{N}^3}$ and $\Gamma^4_{\mathscr{N}^3_{\mathscr{D}}} = \emptyset$. ◊

2.2 *Spline Spaces*

For each level ℓ, we choose a globally[1] defined tensor-product spline basis $\hat{B}^{\ell}$ that consists of individual basis functions $\hat{\beta}^{\ell} \in \hat{B}^{\ell}$, spanning a certain spline space $\hat{\mathbb{V}}^{\ell}$. Note that we consider sets $\hat{B}^{\ell}$ of basis functions and use the symbol $\hat{\beta}^{\ell}$ to represent their elements. The upper index indicates the level but does not identify individual basis functions.

More precisely, the basis functions are tensor-product B-splines of degree $\mathbf{p} = (p_1, \ldots, p_d)$ defined on d open knot vectors with $(p_i + 1)$-fold boundary knots 0 and 1 for $i = 1, \ldots, d$. All inner knots possess multiplicity 1. Hence, we obtain B-splines of maximal smoothness, $C^{\mathbf{s}}(\Omega)$, with $\mathbf{s} = \mathbf{p} - \mathbf{1}$. In other words, the basis functions possess continuous values and partial derivatives up to order $p_i - 1$ in each variable x_i separately. The supports[2] of the functions $\hat{\beta}^{\ell}$ are axis-aligned open boxes in $[0, 1]^d$.

In addition to the B-spline bases, which are independent of the patches π^{ℓ}, we define *local* spline bases B^{ℓ} on π^{ℓ} that consist of *patch B-splines* β^{ℓ} and span the

[1] i.e., with domain $[0, 1]^d$.

[2] Note that we use the set-theoretic definition of the support, i.e., for a function $f : \Omega \mapsto \mathbb{R}^d$ the support of f is defined as $\operatorname{supp} f = \{\mathbf{x} \in \Omega \ : \ f(\mathbf{x}) \neq 0\}$.

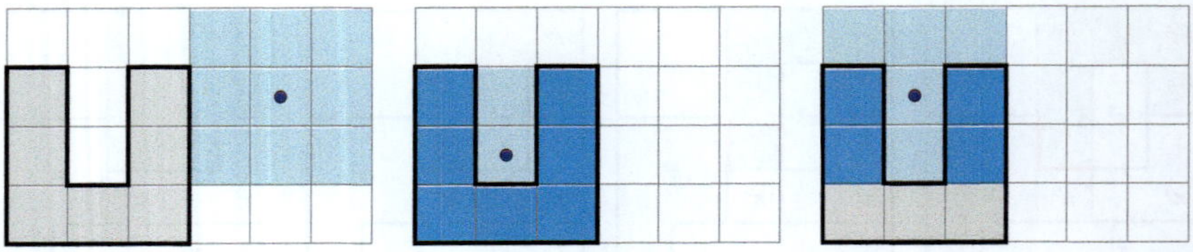

Fig. 4 Example 3—Restriction of three tensor-product B-splines to a U-shaped patch

spaces $\mathbb{V}^\ell$. The basis functions are constructed by *decoupling* the restricted function $\hat{\beta}^\ell|_{\pi^\ell}$, see [31]. More precisely, for each connected component ϕ of $\mathrm{supp}\hat{\beta}^\ell \cap \pi^\ell$ we obtain a basis function

$$\beta^\ell = \begin{cases} \hat{\beta}^\ell & \text{on } \phi \\ 0 & \text{on } \pi^\ell \setminus \phi. \end{cases} \tag{2}$$

The knot hyperplanes of the space $\mathbb{V}^\ell$ divide the patch π^ℓ into cells $z \in Z^\ell$, where Z^ℓ denotes the set of all cells of level ℓ.

For later reference we note that

$$\mathrm{supp}\beta^k \cap \Gamma^\ell = \mathrm{supp}\beta^k \cap \Gamma^\ell_{\mathscr{D}}, \tag{3}$$

for $\beta^k \in B^k$ and any index set $\mathscr{D}$ with $k \in \mathscr{D}$ and $\ell < k$.

Example 3 We consider three different biquadratic tensor-product B-splines $\hat{\beta}^\ell$ and discuss the resulting patch B-splines with respect to a U-shaped patch π^ℓ, represented as gray area in Fig. 4. The knot vectors of level ℓ define the grid that is illustrated in the three pictures. The supports of the B-splines are depicted as blue squares, and the blue circles indicate their Greville points. Depending on the location of the support with respect to the patch we obtain zero, one or two patch B-splines. ◊

The functions in B^ℓ possess good mathematical properties on π^ℓ: They are linearly independent, non-negative, form a partition of unity and have continuous values and partial derivatives up to order $p_i - 1$ in each variable x_i.

2.3 *The Patchwork Spline Space*

We conclude this section by introducing the *patchwork spline space*. On a general subdomain $\Delta_{\mathscr{D}}$ we define

$$\mathbb{P}_{\mathscr{D}} = \{f \in C^{\mathbf{s}}(\Delta_{\mathscr{D}}) \ : \ f|_{\pi^k} \in \mathbb{V}^k, \ \text{for } k \in \mathscr{D}\}.$$

Note that the tensor-product polynomials of degree $\mathbf{p}$ restricted to the subdomain $\Delta_{\mathscr{D}}$ are contained in the patchwork spline space $\mathbb{P}_{\mathscr{D}}$. With $\mathbb{P}_{\mathscr{D},\geq\ell}$ we denote the patchwork spline space on the subdomain $\Delta_{\mathscr{D},\geq\ell}$, while $\mathbb{P} = \mathbb{P}_{\geq 1}$ is the patchwork spline space on the entire domain.

In addition to the patchwork spline space $\mathbb{P}_{\mathscr{D}}$ we consider the full spline space

$$\mathbb{F}_{\mathscr{D}} = \{f \in C^{\mathbf{s}}(\Delta_{\mathscr{D}}) \ : f|_z \in \Pi^{\mathbf{p}} \ \ \forall z \in Z^{\ell},\ \ell \in \mathscr{D}\},$$

where $\Pi^{\mathbf{p}}$ is the space of tensor-product polynomials of degree $\mathbf{p}$. We obtain the following result:

Corollary 1 *The patchwork spline space* $\mathbb{P}_{\mathscr{D}}$ *is equivalent to the full spline space* $\mathbb{F}_{\mathscr{D}}$.

Proof Theorem 2.12. of [32] implies that $\mathbb{P}_{\mathscr{D}} = \mathbb{F}_{\mathscr{D}}$. □

Thus, a basis that spans the patchwork spline spaces is *algebraically complete*.

3 Decoupled Patchwork B-Splines

We define a basis for the patchwork spline space by suitably adapting Kraft's selection mechanism [27] to the patch structure.

3.1 Preliminaries

Our construction is based on the natural assumption that the spline spaces associated with neighboring patches are nested.

Assumption NNC—Nested Neighbor Condition

Any two globally defined spline spaces satisfy $\hat{\mathbb{V}}^{\ell} \subseteq \hat{\mathbb{V}}^{k}$ if $k \in \mathscr{N}^{\ell}$.

We introduce the *extension operator*, $\mathscr{E} : B^{\ell} \mapsto \hat{B}^{\ell}$, that transforms a patch B-spline $\beta^{\ell} \in B^{\ell}$ into the globally defined tensor-product B-spline $\mathscr{E}(\beta^{\ell}) = \hat{\beta}^{\ell} \in \hat{B}^{\ell}$ such that

$$\beta^{\ell} = \hat{\beta}^{\ell}|_{\mathrm{supp}\beta^{\ell}}. \tag{4}$$

The extension of a patch B-spline β^{ℓ} is always denoted by $\hat{\beta}^{\ell}$ in the remainder of this paper. Note that different patch B-splines β^{ℓ} and β'^{ℓ} with disjoint supports possess the same extension $\hat{\beta}^{\ell} = \hat{\beta}'^{\ell}$ if they are derived from the same tensor-product B-spline. It follows that $\hat{\beta}^{\ell}|_{\pi^{\ell}} \in \mathbb{V}^{\ell}$ since

$$\hat{\beta}^\ell|_{\pi^\ell} = \sum_{\substack{\beta'^\ell \in B^\ell \\ \hat{\beta}'^\ell = \hat{\beta}^\ell}} \beta'^\ell.$$

The dual functionals that generate the coefficients of the local representation

$$f(\mathbf{x}) = \sum_{\beta^\ell \in B^\ell} \lambda_{\beta^\ell}(f)\beta^\ell(\mathbf{x}), \quad \text{for } \mathbf{x} \in \pi^\ell,$$

of a function $f \in \mathbb{V}^\ell$ will be denoted by $\lambda_{\beta^\ell}(.)$. If the spaces spanned by a pair of spline bases $\hat{B}^\ell$ and $\hat{B}^k$ are nested, i.e., $\hat{\mathbb{V}}^\ell \subseteq \hat{\mathbb{V}}^k$, for $\ell < k$, then it holds that $\hat{\beta}^\ell|_{\pi^k} \in \mathbb{V}^k$ and

$$\lambda_{\beta^k}(\hat{\beta}^\ell) = \lambda_{\hat{\beta}^k}(\hat{\beta}^\ell), \tag{5}$$

where the values $\lambda_{\hat{\beta}^k}(\hat{\beta}^\ell)$ of the dual functionals are the coefficients obtained by B-spline refinement.

3.2 Recursive Selection

On the subdomain $\Delta_{\mathscr{D}}$ the *Decoupled Patchwork B-splines* $P_{\mathscr{D}}$ are computed recursively, starting at the highest level N and using the initialization $P_{\mathscr{D}}^{N+1} = \emptyset$. Since the index set $\mathscr{D}$ does not necessarily contain all levels, we distinguish between two cases: When proceeding to the next lower level ℓ, the DPB-splines remain unchanged if $\ell \notin \mathscr{D}$. Otherwise, we *select* functions of levels higher than ℓ that vanish on the constraining boundary,

$$S_{\mathscr{D}}^{\ell+1} = \{\tilde{\beta}_{\mathscr{D}}^k \in P_{\mathscr{D}}^{\ell+1} \; : \; \tilde{\beta}_{\mathscr{D}}^k = 0 \quad \text{on} \quad \Gamma_{\mathscr{D}}^\ell\},$$

and apply *truncation*—indicated by the tilde operator—to the extensions of the patch B-splines of level ℓ,

$$T_{\mathscr{D}}^\ell = \{\tilde{\beta}_{\mathscr{D}}^\ell \; : \; \beta^\ell \in B^\ell\}.$$

The DPB-splines of level ℓ are then defined as the union of *selected* and *truncated* patch B-splines,

$$P_{\mathscr{D}}^\ell = \begin{cases} S_{\mathscr{D}}^{\ell+1} \cup T_{\mathscr{D}}^\ell & \text{if } \ell \in \mathscr{D} \\ P_{\mathscr{D}}^{\ell+1} & \text{otherwise.} \end{cases}$$

Since truncation is applied to the extension of the patch B-splines, it extends them to the full domain. More precisely, the *truncated patch B-splines* $\tilde{\beta}_{\mathscr{D}}^\ell$, for $\ell \in \mathscr{D}$, are equal to the patch B-splines on π^ℓ and extend smoothly (as we shall see later) into $\Delta_{\mathscr{D},\geq\ell} \setminus \pi^\ell$,

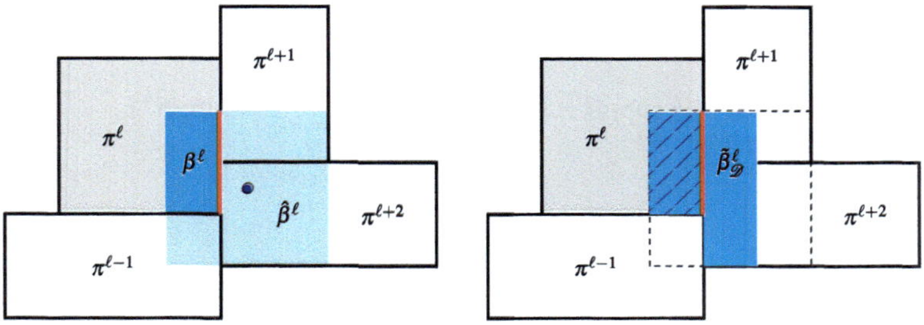

Fig. 5 Left: The supports of a patch B-spline β^{ℓ} (blue) and its extension $\hat{\beta}^{\ell}$ (light blue). Right: The support (blue) of the corresponding truncated patch B-spline $\tilde{\beta}^{\ell}_{\mathscr{D}}$. In the hatched area, it is equal to the patch B-spline and it is smoothly extended in the non-hatched part

$$\tilde{\beta}^{\ell}_{\mathscr{D}} = \begin{cases} \beta^{\ell} & \text{on } \pi^{\ell} \\ \sum\limits_{\substack{\tilde{\beta}^{k}_{\mathscr{D}} \in P^{\ell+1}_{\mathscr{D}} \\ \operatorname{supp}\beta^{\ell} \cap \operatorname{supp}\beta^{k} \neq \emptyset}} \lambda_{\beta^{k}}(\hat{\beta}^{\ell})\, \tilde{\beta}^{k}_{\mathscr{D}} & \text{on } \Delta_{\mathscr{D},\geq \ell} \setminus \pi^{\ell} \\ 0 & \text{elsewhere.} \end{cases} \tag{6}$$

The support condition $\operatorname{supp}\beta^{\ell} \cap \operatorname{supp}\beta^{k} \neq \emptyset$ implies that the truncated functions are linear combinations of non-selected basis functions of higher levels on $\Delta_{\mathscr{D},\geq \ell} \setminus \pi^{\ell}$, i.e., of basis functions that do not vanish on the constraining boundary $\Gamma^{\ell}_{\mathscr{D}}$, see Fig. 5. Note that the sum in (6) considers only basis functions of patches π^{k}, with $k > \ell$, that are neighbors of π^{ℓ}, i.e., $k \in \mathscr{N}^{\ell} \cap \mathscr{D}$. According to NNC it follows that $\hat{\beta}^{\ell}|_{\pi^{k}} \in \mathbb{V}^{k}$ and therefore, the coefficients $\lambda_{\beta^{k}}(\hat{\beta}^{\ell})$ exist and are non-negative due to the properties of B-spline refinement, see Eq. (5).

After completing the recursion we obtain $P_{\mathscr{D}} = P^{1}_{\mathscr{D}}$ and $P = P_{\geq 1}$.

Lemma 1 *The truncated patch B-splines* $\tilde{\beta}^{k}_{\mathscr{D}} \in T^{k}_{\mathscr{D}}$ *satisfy*

$$\operatorname{supp}\tilde{\beta}^{k}_{\mathscr{D}} \subseteq \operatorname{supp}\hat{\beta}^{k}. \tag{7}$$

Proof We prove this statement by induction starting from level $n = \max \mathscr{D}$, which is decreased until we arrive at $n = 1$. According to the piecewise definition of $\tilde{\beta}^{n}_{\mathscr{D}}$ it follows immediately that

$$\operatorname{supp}\tilde{\beta}^{n}_{\mathscr{D}} = \operatorname{supp}\beta^{n} \subseteq \operatorname{supp}\hat{\beta}^{n}.$$

Now we assume that (7) is satisfied for $k \geq \bar{k} \in \mathscr{D}$ and it remains to be shown that the statement holds as well for $\ell = \max\{m \in \mathscr{D} \ : \ m < \bar{k}\}$. We consider a basis function $\tilde{\beta}^{\ell}_{\mathscr{D}} \in T^{\ell}_{\mathscr{D}}$. Note that

$$\{\tilde{\beta}^{k}_{\mathscr{D}} \in P^{\ell+1}_{\mathscr{D}} \ : \ \operatorname{supp}\beta^{k} \cap \operatorname{supp}\beta^{\ell} \neq \emptyset\}$$
$$\subseteq \{\tilde{\beta}^{k}_{\mathscr{D}} \in T^{k}_{\mathscr{D}} \ : \ \ell < k \in \mathscr{D} \wedge \operatorname{supp}\beta^{k} \cap \operatorname{supp}\beta^{\ell} \neq \emptyset\}.$$

Thus, it follows from the definition of $\tilde{\beta}^{\ell}_{\mathscr{D}}$ and the induction hypothesis that

$$\operatorname{supp}\tilde{\beta}^{\ell}_{\mathscr{D}} \subseteq \operatorname{supp}\beta^{\ell} \cup \Big(\bigcup_{k\in\mathscr{D},\, k>\ell} \bigcup_{\substack{\beta^k \in B^k \\ \operatorname{supp}\beta^k \cap \operatorname{supp}\beta^{\ell} \neq \emptyset \\ \lambda_{\beta^k}(\hat{\beta}^{\ell}) \neq 0}} \operatorname{supp}\hat{\beta}^k \Big), \tag{8}$$

where the coefficients $\lambda_{\beta^k}(\hat{\beta}^{\ell})$ exist according to NNC. For non-zero coefficients $\lambda_{\hat{\beta}^k}(\hat{\beta}^{\ell}) = \lambda_{\beta^k}(\hat{\beta}^{\ell}) \neq 0$ it holds that the supports of $\hat{\beta}^k$ and $\hat{\beta}^{\ell}$ are nested, i.e., $\operatorname{supp}\hat{\beta}^k \subseteq \operatorname{supp}\hat{\beta}^{\ell}$. Therefore, the supports of the B-splines $\hat{\beta}^k$ in (8) are contained in the support of $\hat{\beta}^{\ell}$. Since also $\operatorname{supp}\beta^{\ell} \subseteq \operatorname{supp}\hat{\beta}^{\ell}$ we conclude that $\operatorname{supp}\tilde{\beta}^{\ell}_{\mathscr{D}} \subseteq \operatorname{supp}\hat{\beta}^{\ell}$. □

4 Properties of DPB-Splines

Besides NNC, we impose two further assumptions in order to ensure good mathematical properties of the DPB-splines.

4.1 Assumptions and Main Result

First, we introduce a condition that concerns the intersection of a patch B-spline with the constraining boundaries of lower level patches. To be more precise, if a patch B-spline $\beta^n \in B^n$ does not vanish on a pair of constraining boundaries Γ^{ℓ} and Γ^k, where $\ell < n$ and $k < n$, then a non-empty intersection of the patch B-spline and the two constraining boundaries is assumed to exist, see also Fig. 6.

Assumption IPC—Intermediate Patch Condition

If $\beta^n \in B^n$ satisfies

$$\operatorname{supp}\beta^n \cap \Gamma^{\ell} \neq \emptyset,\ \ell < n \ \text{ and } \ \operatorname{supp}\beta^n \cap \Gamma^k \neq \emptyset,\ k < n,$$

then

$$\operatorname{supp}\beta^n \cap \Gamma^{\ell} \cap \Gamma^k \neq \emptyset.$$

The second assumption is needed to guarantee that the truncated basis functions are C^s-smooth. We impose a condition on the intersection of a patch B-spline with a lower level constraining boundary, see Fig. 7.

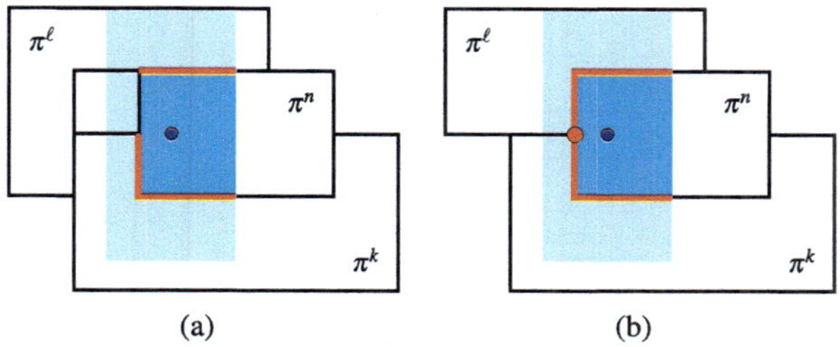

Fig. 6 Intersections (red) of a patch B-spline's support (dark blue) with Γ^ℓ and Γ^k. The support of the associated extension $\hat{\beta}^n$ is also shown (light blue). IPC is violated in (**a**) and satisfied in (**b**)

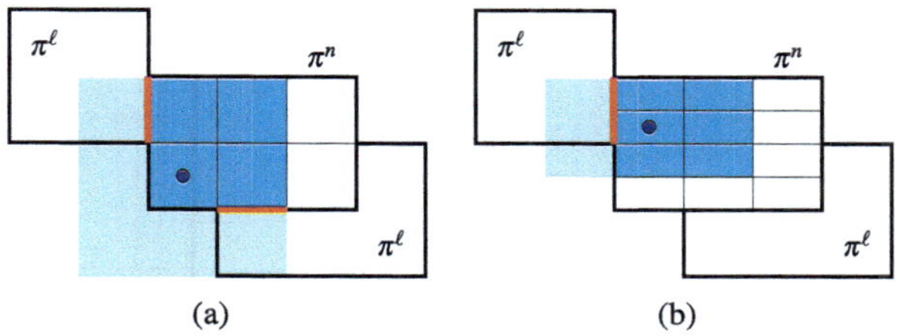

Fig. 7 Intersections (red) of a patch B-spline's support (dark blue) with a constraining boundary Γ^ℓ, $\ell < n$. SIC is violated in (**a**) and satisfied in (**b**)

Assumption SIC—Support Intersection Condition

The support of a patch B-spline $\beta^n \in B^n$ intersected with a constraining boundary Γ^ℓ,

$$\text{supp}\beta^n \cap \Gamma^\ell,$$

where $\ell < n$, is connected.

The following theorem shows that the DPB-splines form a basis and characterizes the space that is spanned by them.

Theorem 1 *Assume NNC, IPC and SIC are satisfied. We consider DPB-splines $P_{\mathscr{D}}$ on any subdomain $\Delta_{\mathscr{D}} \subseteq \Omega$.*

(i) The functions $\tilde{\beta}^\ell_{\mathscr{D}} \in P_{\mathscr{D}}$ possess continuous values and partial derivatives up to order $p - 1$ in each variable separately.
(ii) The DPB-splines form a basis of the patchwork spline space $\mathbb{P}_{\mathscr{D}}$.
(iii) The DPB-splines satisfy the coefficient preservation property, i.e., for any function $f \in \mathbb{P}_{\mathscr{D}}$ it holds that

$$f = \sum_{\tilde{\beta}^k_{\mathscr{D}} \in P_{\mathscr{D}}} \lambda_{\beta^k}(f)\, \tilde{\beta}^k_{\mathscr{D}}, \quad on\ \Delta_{\mathscr{D}}. \tag{9}$$

(iv) The DPB-spline functions are non-negative and form a partition of unity.

The proof of this theorem together with several technical lemmas is presented in the next section. Readers not interested in the details of the proof may skip this section and continue directly with Sect. 5.

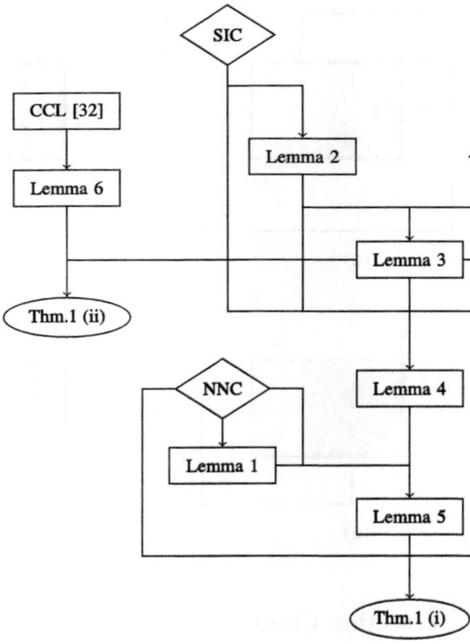

Fig. 8 Connections between Assumptions (diamonds), Lemmas (rectangles), and DPB-spline properties (ellipses)

4.2 *Proof of Theorem 1*

In the following we assume that NNC, IPC and SIC are always satisfied. Equation (3) implies that IPC and SIC are then also fulfilled for constraining boundaries $\Gamma_{\mathscr{D}}^{\ell}$ with respect to index sets $\mathscr{D}$, see (1).

4.2.1 Some Preliminary Observations

The proof of Theorem 1 relies on several technical results, see Fig. 8. The first observation will help us to analyze the truncation defined in (6).

Lemma 2 *Two patch B-splines $\beta^k \in B^k$ and $\beta^\ell \in B^\ell$ with intersecting supports,*

$$\emptyset \neq \operatorname{supp}\beta^k \cap \operatorname{supp}\beta^\ell \subseteq \Gamma_{\mathscr{D}}^{\ell}, \text{ with } \ell < k \in \mathscr{D},$$

satisfy

$$\operatorname{supp}\beta^k \cap \Gamma_{\mathscr{D}}^{\ell} \subseteq \operatorname{supp}\beta^\ell \cap \Gamma_{\mathscr{D}}^{\ell}, \tag{10}$$

if $\lambda_{\beta^k}(\hat{\beta}^\ell) \neq 0$.

Proof The existence of non-zero coefficients $\lambda_{\beta^k}(\hat{\beta}^\ell) \neq 0$ implies $\mathrm{supp}\beta^k \subseteq \mathrm{supp}\hat{\beta}^\ell$, hence

$$\mathrm{supp}\beta^k \cap \Gamma_{\mathscr{D}}^\ell \subseteq \mathrm{supp}\hat{\beta}^\ell \cap \Gamma_{\mathscr{D}}^\ell. \tag{11}$$

We distinguish between several cases: First, we assume that $\mathrm{supp}\hat{\beta}^\ell \cap \pi^\ell$ is connected. This leads immediately to (10) since $\beta^\ell = \hat{\beta}^\ell|_{\pi^\ell}$ and $\Gamma_{\mathscr{D}}^\ell \subseteq \pi^\ell$.

Second, we consider the case that $\mathrm{supp}\hat{\beta}^\ell \cap \pi^\ell$ possesses two connected components. Thus, besides β^ℓ, there exists another patch B-spline β'^ℓ such that $\mathrm{supp}\beta^\ell \cap \mathrm{supp}\beta'^\ell = \emptyset$ and

$$\mathrm{supp}\hat{\beta}^\ell \cap \pi^\ell = \mathrm{supp}\beta^\ell \cup \mathrm{supp}\beta'^\ell. \tag{12}$$

The intersection of $\mathrm{supp}\beta^k \cap \mathrm{supp}\beta^\ell$ belongs to the constraining boundary $\Gamma_{\mathscr{D}}^\ell$, thus

$$\emptyset \neq \mathrm{supp}\beta^k \cap \mathrm{supp}\beta^\ell = \mathrm{supp}\beta^k \cap \mathrm{supp}\beta^\ell \cap \Gamma_{\mathscr{D}}^\ell. \tag{13}$$

According to SIC and Eq. (3), $\mathrm{supp}\beta^k \cap \Gamma_{\mathscr{D}}^\ell$ is connected. Since the two parts $\mathrm{supp}\beta^\ell \cap \Gamma_{\mathscr{D}}^\ell$ and $\mathrm{supp}\beta'^\ell \cap \Gamma_{\mathscr{D}}^\ell$ of the constraining boundary are disjoint it follows that

$$\mathrm{supp}\beta^k \cap \mathrm{supp}\beta'^\ell \cap \Gamma_{\mathscr{D}}^\ell = \emptyset. \tag{14}$$

We observe that (11) is equivalent to

$$(\mathrm{supp}\beta^k \cap \Gamma_{\mathscr{D}}^\ell) \cap (\mathrm{supp}\hat{\beta}^\ell \cap \Gamma_{\mathscr{D}}^\ell) = \mathrm{supp}\beta^k \cap \Gamma_{\mathscr{D}}^\ell. \tag{15}$$

We use (12) $\Gamma_{\mathscr{D}}^\ell \subseteq \pi^\ell$ to rewrite the left-hand side of this equation as

$$\mathrm{supp}\beta^k \cap \mathrm{supp}\hat{\beta}^\ell \cap \Gamma_{\mathscr{D}}^\ell = \mathrm{supp}\beta^k \cap (\mathrm{supp}\beta^\ell \cup \mathrm{supp}\beta'^\ell) \cap \Gamma_{\mathscr{D}}^\ell.$$

We simplify the result with the help of (13) and (14),

$$\mathrm{supp}\beta^k \cap (\mathrm{supp}\beta^\ell \cup \mathrm{supp}\beta'^\ell) \cap \Gamma_{\mathscr{D}}^\ell = \mathrm{supp}\beta^k \cap \mathrm{supp}\beta^\ell \cap \Gamma_{\mathscr{D}}^\ell.$$

Therefore, (15) is equivalent to

$$(\mathrm{supp}\beta^k \cap \Gamma_{\mathscr{D}}^\ell) \cap (\mathrm{supp}\beta^\ell \cap \Gamma_{\mathscr{D}}^\ell) = \mathrm{supp}\beta^k \cap \Gamma_{\mathscr{D}}^\ell.$$

This confirms that

$$\mathrm{supp}\beta^k \cap \Gamma_{\mathscr{D}}^\ell \subseteq \mathrm{supp}\beta^\ell \cap \Gamma_{\mathscr{D}}^\ell.$$

Finally we note that the proof in case two can be extended to more than two connected components. □

The next Lemma characterizes the intersection of a truncated patch B-spline $\tilde{\beta}_{\mathscr{D}}^k$, for $k \in \mathscr{D}$, with a lower level constraining boundary $\Gamma_{\mathscr{D}}^\ell$, see Fig. 9.

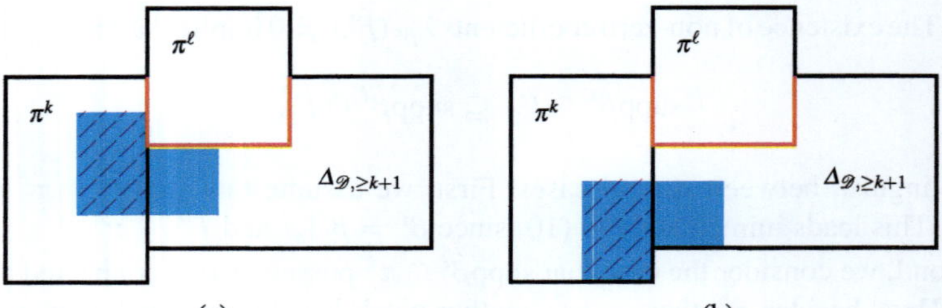

Fig. 9 According to Lemma 3, the supports $\operatorname{supp}\beta^k$ (hatched blue) and $\operatorname{supp}\tilde{\beta}^k_{\mathscr{D}}$ (hatched blue and blue) either intersect both with $\Gamma^\ell_{\mathscr{D}}$ (red) or not, see (**a**) and (**b**), respectively

Lemma 3 *The truncated patch B-splines* $\tilde{\beta}^k_{\mathscr{D}} \in T^k$ *satisfy*

$$\operatorname{supp}\tilde{\beta}^k_{\mathscr{D}} \cap \Gamma^\ell_{\mathscr{D}} \neq \emptyset \Leftrightarrow \operatorname{supp}\beta^k \cap \Gamma^\ell_{\mathscr{D}} \neq \emptyset$$

if $k > \ell$ *and* $k \in \mathscr{D}$.

Proof According to the definition of the truncated functions, $\operatorname{supp}\beta^k \cap \Gamma^\ell_{\mathscr{D}} \neq \emptyset$ implies that $\operatorname{supp}\tilde{\beta}^k_{\mathscr{D}} \cap \Gamma^\ell_{\mathscr{D}} \neq \emptyset$. In order to prove the other implication, we assume that there exists a $\tilde{\beta}^k_{\mathscr{D}} \in T^k$ such that

$$\operatorname{supp}\tilde{\beta}^k_{\mathscr{D}} \cap \Gamma^\ell_{\mathscr{D}} \neq \emptyset, \quad \text{for } \ell < k,$$

but

$$\operatorname{supp}\beta^k \cap \Gamma^\ell_{\mathscr{D}} = \emptyset. \tag{16}$$

It follows that

$$\operatorname{supp}\tilde{\beta}^k_{\mathscr{D}} \cap (\Delta_{\mathscr{D},\geq k+1} \setminus \pi^k) \cap \Gamma^\ell_{\mathscr{D}} \neq \emptyset,$$

since $\tilde{\beta}^k_{\mathscr{D}}$ takes non-zero values only on the associated patch and on the subdomain $\Delta_{\mathscr{D},\geq k+1}$. Among all such functions we pick one where k is maximal. The representation

$$\tilde{\beta}^k_{\mathscr{D}} = \sum_{\substack{\tilde{\beta}^m_{\mathscr{D}} \in P^{k+1}_{\mathscr{D}} \\ \operatorname{supp}\beta^m \cap \operatorname{supp}\beta^k \neq \emptyset}} \lambda_{\beta^m}(\hat{\beta}^k)\, \tilde{\beta}^m_{\mathscr{D}} \quad \text{on } \Delta_{\mathscr{D},\geq k+1},$$

see (6), implies that there exists a function $\tilde{\beta}^m_{\mathscr{D}} \in P^{k+1}_{\mathscr{D}}$ with $\operatorname{supp}\beta^m \cap \operatorname{supp}\beta^k \neq \emptyset$ such that

$$\operatorname{supp}\tilde{\beta}^m_{\mathscr{D}} \cap \Gamma^\ell_{\mathscr{D}} \neq \emptyset.$$

It follows immediately that

$$\operatorname{supp}\beta^m \cap \Gamma^\ell_{\mathscr{D}} \neq \emptyset,$$

since k is the largest level with the property (16). Furthermore, $\operatorname{supp}\beta^m \cap \operatorname{supp}\beta^k \neq \emptyset$ implies that $\operatorname{supp}\beta^m \cap \Gamma^k_{\mathscr{D}} \neq \emptyset$, and hence

$$\emptyset \neq \operatorname{supp}\beta^m \cap \Gamma^k_{\mathscr{D}} \cap \Gamma^\ell_{\mathscr{D}} \subseteq \operatorname{supp}\beta^k \cap \Gamma^k_{\mathscr{D}} \cap \Gamma^\ell_{\mathscr{D}} = \emptyset,$$

where the first equation holds according to IPC, the second statement follows from Lemma 2, and the last equality is obtained by (16). This is a contradiction. □

Lemma 3 enables us to reformulate the selection mechanism. More precisely, instead of considering the truncated patch B-spline $\tilde{\beta}^k_{\mathscr{D}}$ if suffices to analyze whether the associated patch B-spline β^k vanishes on the constraining boundary,

$$S^{\ell+1}_{\mathscr{D}} = \{\tilde{\beta}^k_{\mathscr{D}} \in P^{\ell+1}_{\mathscr{D}} \ : \ \beta^k = 0 \ \text{ on } \ \Gamma^\ell_{\mathscr{D}}\}.$$

Next, we observe that the truncated patch B-splines $\tilde{\beta}^k_{\mathscr{D}} \in T^k$ inherit the support intersection property:

Lemma 4 *The intersections of the supports of truncated patch B-splines* $\tilde{\beta}^k_{\mathscr{D}} \in T^k$ *with constraining boundaries*

$$\operatorname{supp}\tilde{\beta}^k_{\mathscr{D}} \cap \Gamma^\ell_{\mathscr{D}} \ \textit{where} \ \ell < k \tag{17}$$

are connected.

Proof Assume there exists a truncated patch B-spline $\tilde{\beta}^k_{\mathscr{D}}$ whose support intersects the lower level constraining boundary $\Gamma^\ell_{\mathscr{D}}$ in more than one connected component. Among all those functions we choose one where $k \in \mathscr{D}$ is maximal. The intersection splits into two disjoint sets $\delta_1 \neq \emptyset$ and $\delta_2 \neq \emptyset$ with $\delta_1 \cap \delta_2 = \emptyset$ such that

$$\operatorname{supp}\tilde{\beta}^k_{\mathscr{D}} \cap \Gamma^\ell_{\mathscr{D}} = \delta_1 \cup \delta_2. \tag{18}$$

SIC requires that $\operatorname{supp}\beta^k \cap \Gamma^\ell_{\mathscr{D}}$ is connected. Since $\operatorname{supp}\beta^k \subseteq \operatorname{supp}\tilde{\beta}^k_{\mathscr{D}}$, this implies that $\operatorname{supp}\beta^k \cap \Gamma^\ell_{\mathscr{D}}$ is a subset of one δ_i whereas its intersection with the other one is empty. Without loss of generality we assume

$$\operatorname{supp}\beta^k \cap \Gamma^\ell_{\mathscr{D}} \subseteq \delta_1 \quad \text{and} \quad \operatorname{supp}\beta^k \cap \delta_2 = \emptyset. \tag{19}$$

According to the latter observation, the truncated patch B-spline $\tilde{\beta}^k_{\mathscr{D}}$ intersects δ_2 in patches of levels higher than k, i.e.,

$$\operatorname{supp}\tilde{\beta}^k_{\mathscr{D}} \cap (\Delta_{\mathscr{D},\geq k+1} \setminus \pi^k) \cap \delta_2 \neq \emptyset.$$

Due to the definition of the truncation (6), there exists a function $\tilde{\beta}^m_{\mathscr{D}} \in P^{k+1}_{\mathscr{D}}$ with $\operatorname{supp}\beta^m \cap \operatorname{supp}\beta^k \neq \emptyset$ that satisfies

$$\operatorname{supp}\tilde{\beta}^m_{\mathscr{D}} \cap \delta_2 \neq \emptyset. \tag{20}$$

Moreover, we have

$$\operatorname{supp}\tilde{\beta}^m_{\mathscr{D}} \cap \Gamma^\ell_{\mathscr{D}} \subseteq \operatorname{supp}\tilde{\beta}^k_{\mathscr{D}} \cap \Gamma^\ell_{\mathscr{D}} = \delta_1 \cup \delta_2.$$

Recall that we assumed k to be maximal. Thus $\operatorname{supp}\tilde{\beta}^m_{\mathscr{D}} \cap \Gamma^\ell_{\mathscr{D}}$ is connected and we obtain

$$\operatorname{supp}\tilde{\beta}^m_{\mathscr{D}} \cap \delta_1 = \emptyset. \tag{21}$$

From $\operatorname{supp}\beta^m \cap \operatorname{supp}\beta^k \neq \emptyset$ it follows that $\operatorname{supp}\beta^m \cap \Gamma^k_{\mathscr{D}} \neq \emptyset$. Furthermore, (20) leads to

$$\emptyset \neq \operatorname{supp}\tilde{\beta}^m_{\mathscr{D}} \cap \delta_2 \subseteq \operatorname{supp}\tilde{\beta}^m_{\mathscr{D}} \cap \Gamma^\ell_{\mathscr{D}},$$

since (18) implies that $\delta_2 \subseteq \Gamma^\ell_{\mathscr{D}}$. According to Lemma 3 $\operatorname{supp}\tilde{\beta}^m_{\mathscr{D}} \cap \Gamma^\ell_{\mathscr{D}} \neq \emptyset$ is equivalent to $\operatorname{supp}\beta^m \cap \Gamma^\ell_{\mathscr{D}} \neq \emptyset$. Applying IPC leads to

$$\operatorname{supp}\beta^m \cap \Gamma^k_{\mathscr{D}} \cap \Gamma^\ell_{\mathscr{D}} \neq \emptyset.$$

Furthermore, Lemma 2 states that $\operatorname{supp}\beta^m \cap \Gamma^k_{\mathscr{D}} \subseteq \operatorname{supp}\beta^k \cap \Gamma^k_{\mathscr{D}}$. Combining these observations with simple subset relations and (19) results in the following equations,

$$\emptyset \neq \operatorname{supp}\beta^m \cap \Gamma^k_{\mathscr{D}} \cap \Gamma^\ell_{\mathscr{D}} \subseteq \operatorname{supp}\beta^k \cap \Gamma^k_{\mathscr{D}} \cap \Gamma^\ell_{\mathscr{D}} \subseteq \operatorname{supp}\beta^k \cap \Gamma^\ell_{\mathscr{D}} \subseteq \delta_1.$$

However, this implies that

$$\emptyset \neq \operatorname{supp}\beta^m \cap \delta_1 \subseteq \operatorname{supp}\tilde{\beta}^m_{\mathscr{D}} \cap \delta_1,$$

which contradicts (21), and thereby completes the proof. □

Finally we present an extension of Lemma 2.

Lemma 5 *The supports of truncated patch B-splines $\tilde{\beta}^k_{\mathscr{D}} \in P^{\ell+1}_{\mathscr{D}}$ and patch B-splines $\beta^\ell \in B^\ell$ of lower level $\ell < k$ satisfy*

$$\operatorname{supp}\tilde{\beta}^k_{\mathscr{D}} \cap \Gamma^\ell_{\mathscr{D}} \subseteq \operatorname{supp}\beta^\ell \cap \Gamma^\ell_{\mathscr{D}},$$

if $\operatorname{supp}\beta^k \cap \operatorname{supp}\beta^\ell \neq \emptyset$ *and* $\lambda_{\beta^k}(\hat{\beta}^\ell) \neq 0$.

Proof The proof works similar to the proof of Lemma 2. We do not provide all details but show how to adapt it to the current situation.

Recall that $\lambda_{\beta^k}(\hat{\beta}^\ell) \neq 0$ implies $\operatorname{supp}\hat{\beta}^k \subseteq \operatorname{supp}\hat{\beta}^\ell$. From Lemma 1 it follows that $\operatorname{supp}\tilde{\beta}^k_{\mathscr{D}} \subseteq \operatorname{supp}\hat{\beta}^\ell$ and thus,

$$\operatorname{supp}\tilde{\beta}^k_{\mathscr{D}} \cap \Gamma^\ell_{\mathscr{D}} \subseteq \operatorname{supp}\hat{\beta}^\ell \cap \Gamma^\ell_{\mathscr{D}}.$$

As in Lemma 2, the proof follows immediately for the case that $\operatorname{supp}\hat{\beta}^\ell \cap \pi^\ell$ possesses one connected component. For the other case we note that

$$\emptyset \neq \mathrm{supp}\beta^k \cap \mathrm{supp}\beta^\ell \cap \Gamma^\ell_{\mathscr{D}} \subseteq \mathrm{supp}\tilde{\beta}^k_{\mathscr{D}} \cap \mathrm{supp}\beta^\ell \cap \Gamma^\ell_{\mathscr{D}},$$

according to the assumption that $\mathrm{supp}\beta^k \cap \mathrm{supp}\beta^\ell \neq \emptyset$. Using Lemma 4 allows us to proceed as in the proof of Lemma 2 using $\tilde{\beta}^k_{\mathscr{D}}$ instead of β^k. □

We introduce the notion of *homogeneous boundary conditions*: A $C^{\mathbf{s}}$–smooth function f satisfies homogeneous boundary conditions in a point $\mathbf{x}$ if the values of the function and of all its partial derivatives up to order $s_k = p_k - 1$ in all variables separately[3] are equal to zero,

$$\vartheta f(\mathbf{x}) = 0,$$

with

$$[\vartheta f(\mathbf{x})]_{\mathbf{i}} = \partial_{\mathbf{i}} f(\mathbf{x}), \quad \mathbf{i} = (i_1, \ldots, i_d), \quad i_k \leq p_k - 1.$$

and partial derivative operators

$$(\partial_{\mathbf{i}} f)(\mathbf{x}) = \frac{\partial^{i_1}}{\partial x_1^{i_1}} \cdots \frac{\partial^{i_d}}{\partial x_d^{i_d}} f(x_1, \ldots, x_d).$$

Since π^ℓ can be divided into a set of cells $z \in Z^\ell$, the boundary of a patch π^ℓ naturally splits into facets ξ of different dimensions. E.g., the facets can be edges or vertices of cells of level ℓ for dimension $d = 2$, see also the Appendix of [10].

Lemma 6 *If a function $f \in \mathbb{V}^\ell$ satisfies homogeneous boundary conditions on a boundary facet ξ of π^ℓ, then $\lambda_{\beta^\ell}(f) = 0$ if $\xi \subseteq \mathrm{supp}\beta^\ell$.*

Proof We note that for a cell $z \in Z^\ell$ and every $\beta^\ell \in B^\ell$ that does not vanish on that cell, i.e., $z \subseteq \mathrm{supp}\beta^\ell$, it holds that

$$0 \neq \beta^\ell|_z = \hat{\beta}^\ell|_z,$$

according to (4). Suitably adapting the proof of Lemma 2 in [10], which is based on the Contact Characterization Lemma (CCL) in [32], confirms this fact. □

4.2.2 Proof of the Theorem

Finally, we can prove the results stated in Theorem 1. We prove the statements (i)–(iv) for all subdomains $\Delta_{\mathscr{D}} \subseteq \Delta_{\geq \ell}$, i.e., $\mathscr{D} \subseteq \{\ell, \ldots, N\}$, by induction over ℓ, proceeding from the maximum level N down to 1.

For $\ell = N$ we note that $\Delta_{\geq N} = \pi^N$. Therefore, a subdomain of $\Delta_{\geq N}$ is either the empty set (trivial case) or $\mathscr{D} = \{N\}$. Then $P_{\mathscr{D}} = T^N_{\mathscr{D}}$ consists of functions

[3] For instance, $f = f_x = f_y = f_{xx} = f_{xy} = f_{yy} = f_{xxy} = f_{xyy} = f_{xxyy} = 0$ for $\mathbf{x} = (x, y)$ and $\mathbf{s} = (2, 2)$.

$$\tilde{\beta}^N_{\mathscr{D}} = \begin{cases} \beta^N & \text{on } \pi^N \\ 0 & \text{elsewhere} \end{cases}.$$

Therefore, we can conclude that (i)–(iv) are satisfied for any subdomain $\Delta_{\mathscr{D}} \subseteq \Delta_{\geq N}$. Now we assume that (i)–(iv) are satisfied for any subdomain $\Delta_{\mathscr{D}} \subseteq \Delta_{\geq \ell+1}$. We show that the properties extend to any subdomain $\Delta_{\mathscr{D}} \subseteq \Delta_{\geq \ell}$. We consider only the case where $\ell \in \mathscr{D}$, hence $\mathscr{D} \subseteq \{\ell, \ldots, N\}$. Otherwise, the result follows immediately from the induction hypothesis.

First Statement (i)**:** We have that

$$P_{\mathscr{D}} = P^1_{\mathscr{D}} = P^\ell_{\mathscr{D}} = T^\ell_{\mathscr{D}} \cup S^{\ell+1}_{\mathscr{D}}.$$

For functions in $S^{\ell+1}_{\mathscr{D}}$ the smoothness on $\Delta_{\mathscr{D}} = \Delta_{\mathscr{D},\geq \ell}$ follows from the induction hypothesis combined with the fact that the functions are zero on π^ℓ and satisfy homogeneous boundary conditions on $\Gamma^\ell_{\mathscr{D}}$ according to the selection mechanism. Next we consider a function $\tilde{\beta}^\ell_{\mathscr{D}} \in T^\ell_{\mathscr{D}}$. We note that $\tilde{\beta}^\ell_{\mathscr{D}}$ obviously possesses the required smoothness on π^ℓ and also on $\Delta_{\mathscr{D},\geq \ell+1} \setminus \pi^\ell$ (considered separately). Indeed, its restriction to the latter domain is a linear combination of functions in $P^{\ell+1}_{\mathscr{D}}$, which satisfy the smoothness conditions according to the induction hypothesis.

Now it remains to be shown that $\tilde{\beta}^\ell_{\mathscr{D}}$ possesses the required order of smoothness on the constraining boundary $\Gamma^\ell_{\mathscr{D}}$. According to NNC it holds that $\hat{\beta}^\ell \in \mathbb{P}_{\mathscr{N}^\ell_{\mathscr{D}}}$. From the induction hypothesis (ii) it follows that there exists a representation,

$$\hat{\beta}^\ell = \sum_{\tilde{\beta}^k_{\mathscr{N}^\ell_{\mathscr{D}}} \in P_{\mathscr{N}^\ell_{\mathscr{D}}}} \lambda_{\beta^k}(\hat{\beta}^\ell) \tilde{\beta}^k_{\mathscr{N}^\ell_{\mathscr{D}}} \quad \text{on } \Delta_{\mathscr{N}^\ell_{\mathscr{D}}}.$$

We use it to define the auxiliary function

$$\hat{\beta}^\ell_{\mathscr{N}^\ell_{\mathscr{D}}} = \sum_{\substack{\tilde{\beta}^k_{\mathscr{N}^\ell_{\mathscr{D}}} \in P_{\mathscr{N}^\ell_{\mathscr{D}}} \\ \operatorname{supp}\beta^k \cap \operatorname{supp}\beta^\ell \neq \emptyset}} \lambda_{\beta^k}(\hat{\beta}^\ell) \tilde{\beta}^k_{\mathscr{N}^\ell_{\mathscr{D}}}, \tag{22}$$

by considering only functions near the boundary to $\operatorname{supp}\beta^\ell$ and observe that

$$\vartheta \hat{\beta}^\ell_{\mathscr{N}^\ell_{\mathscr{D}}} = \vartheta \hat{\beta}^\ell = \vartheta \beta^\ell \quad \text{on } \Gamma^\ell_{\mathscr{D}} \cap \operatorname{supp}\beta^\ell.$$

Furthermore, it holds that

$$\vartheta \hat{\beta}^\ell_{\mathscr{N}^\ell_{\mathscr{D}}} = \vartheta \beta^\ell = 0 \quad \text{on } \Gamma^\ell_{\mathscr{D}} \setminus \operatorname{supp}\beta^\ell,$$

since $\operatorname{supp}\tilde{\beta}^k_{\mathscr{N}^\ell_{\mathscr{D}}} \cap \Gamma^\ell_{\mathscr{N}^\ell_{\mathscr{D}}} \subseteq \operatorname{supp}\beta^\ell \cap \Gamma^\ell_{\mathscr{N}^\ell_{\mathscr{D}}}$ according to Lemma 5 and $\Gamma^\ell_{\mathscr{N}^\ell_{\mathscr{D}}} = \Gamma^\ell_{\mathscr{D}}$. Therefore, we obtain that $\vartheta \hat{\beta}^\ell_{\mathscr{N}^\ell_{\mathscr{D}}} = \vartheta \beta^\ell$ on $\Gamma^\ell_{\mathscr{D}}$.

Now we show that the auxiliary function shares values and derivatives with the truncated function along the constraining boundary, $\vartheta \hat{\beta}^{\ell}_{\mathcal{N}^{\ell}_{\mathcal{D}}} = \vartheta \tilde{\beta}^{\ell}_{\mathcal{D}}$ on $\Gamma^{\ell}_{\mathcal{D}}$. More precisely, we prove that

$$\vartheta \Big(\sum_{\substack{\tilde{\beta}^k_{\mathcal{N}^{\ell}_{\mathcal{D}}} \in P_{\mathcal{N}^{\ell}_{\mathcal{D}}} \\ \mathrm{supp}\beta^k \cap \mathrm{supp}\beta^{\ell} \neq \emptyset}} \lambda_{\beta^k}(\hat{\beta}^{\ell}) \tilde{\beta}^k_{\mathcal{N}^{\ell}_{\mathcal{D}}} \Big) = \vartheta \Big(\sum_{\substack{\tilde{\beta}^k_{\mathcal{D}} \in P^{\ell+1}_{\mathcal{D}} \\ \mathrm{supp}\beta^k \cap \mathrm{supp}\beta^{\ell} \neq \emptyset}} \lambda_{\beta^k}(\hat{\beta}^{\ell}) \tilde{\beta}^k_{\mathcal{D}} \Big), \quad \text{on } \Gamma^{\ell}_{\mathcal{D}}. \tag{23}$$

We note that

$$\mathrm{supp}\beta^k \cap \mathrm{supp}\beta^{\ell} \neq \emptyset \;\Rightarrow\; \beta^k|_{\Gamma^{\ell}_{\mathcal{D}}} \neq 0, \quad \text{if } k > \ell.$$

The proof of (23) is based on two technical Lemmas. The first one states that the selection mechanism always selects the same functions for $\mathcal{D}$ and $\mathcal{N}^{\ell}_{\mathcal{D}}$.

Lemma 7 *For all $\beta^k \in B^k$ with $\beta^k \neq 0$ on $\Gamma^{\ell}_{\mathcal{D}}$ it holds that*

$$\tilde{\beta}^k_{\mathcal{N}^{\ell}_{\mathcal{D}}} \in P^m_{\mathcal{N}^{\ell}_{\mathcal{D}}} \Leftrightarrow \tilde{\beta}^k_{\mathcal{D}} \in P^m_{\mathcal{D}}, \tag{24}$$

if $k \geq m > \ell$.

Proof On the one hand, we consider a function $\tilde{\beta}^k_{\mathcal{D}} \in P^m_{\mathcal{D}}$ with $\mathrm{supp}\beta^k \cap \Gamma^{\ell}_{\mathcal{D}} \neq \emptyset$. Hence, $k \in \mathcal{N}^{\ell}_{\mathcal{D}}$. We note that

$$P^m_{\mathcal{D}} = \bigcup_{k \in \mathcal{D}, k \geq m} \{\tilde{\beta}^k_{\mathcal{D}} \in T^k_{\mathcal{D}} \;:\; \mathrm{supp}\beta^k \cap \Gamma^r_{\mathcal{D}} = \emptyset, \quad \text{for all } r \in \mathcal{D},\; m \leq r < k\} \tag{25}$$

and

$$P^m_{\mathcal{N}^{\ell}_{\mathcal{D}}} = \bigcup_{k \in \mathcal{N}^{\ell}_{\mathcal{D}}, k \geq m} \{\tilde{\beta}^k_{\mathcal{N}^{\ell}_{\mathcal{D}}} \in T^k_{\mathcal{N}^{\ell}_{\mathcal{D}}} \;:\; \mathrm{supp}\beta^k \cap \Gamma^r_{\mathcal{N}^{\ell}_{\mathcal{D}}} = \emptyset \text{ for all } r \in \mathcal{N}^{\ell}_{\mathcal{D}},\; m \leq r < k\},$$

according to the selection mechanism and Lemma 3. Since $\Gamma^r_{\mathcal{N}^{\ell}_{\mathcal{D}}} \subseteq \Gamma^r_{\mathcal{D}}$ for $r \in \mathcal{N}^{\ell}_{\mathcal{D}}$, the first condition $\mathrm{supp}\beta^k \cap \Gamma^r_{\mathcal{D}} = \emptyset$ implies $\mathrm{supp}\beta^k \cap \Gamma^r_{\mathcal{N}^{\ell}_{\mathcal{D}}} = \emptyset$ if $m \leq r < k$. Therefore, we conclude that $\tilde{\beta}^k_{\mathcal{N}^{\ell}_{\mathcal{D}}} \in P^m_{\mathcal{N}^{\ell}_{\mathcal{D}}}$.

On the other hand, assume there exists a function $\tilde{\beta}^k_{\mathcal{N}^{\ell}_{\mathcal{D}}} \in P^m_{\mathcal{N}^{\ell}_{\mathcal{D}}}$ with $\tilde{\beta}^k_{\mathcal{D}} \notin P^m_{\mathcal{D}}$ and $\beta^k \neq 0$ on $\Gamma^{\ell}_{\mathcal{D}}$. We invoke the characterization (25) to conclude that there exists a level $r \in \mathcal{D} \setminus \mathcal{N}^{\ell}_{\mathcal{D}}$ and $m \leq r < k$ such that $\mathrm{supp}\beta^k \cap \Gamma^r_{\mathcal{D}} \neq \emptyset$. The considered function satisfies $\mathrm{supp}\beta^k \cap \Gamma^{\ell}_{\mathcal{D}} \neq \emptyset$ and hence it follows that $\mathrm{supp}\beta^k \cap \Gamma^{\ell}_{\mathcal{D}} \cap \Gamma^r_{\mathcal{D}} \neq \emptyset$ by IPC. However, since π^r is not a neighbor of π^{ℓ} it holds that $\Gamma^r_{\mathcal{D}} \cap \Gamma^{\ell}_{\mathcal{D}} \subseteq \pi^r \cap \pi^{\ell} = \emptyset$ which implies that $\mathrm{supp}\beta^k \cap \Gamma^{\ell}_{\mathcal{D}} \cap \Gamma^r_{\mathcal{D}} = \emptyset$. This is a contradiction. □

The second lemma confirms that the functions considered in the previous lemma possess the same values and derivatives along the constraining boundary.

Lemma 8 *For all functions in* (24) *with* $\tilde{\beta}^k_{\mathscr{D}} \in P^m_{\mathscr{D}}$ *it holds that*

$$\vartheta \tilde{\beta}^k_{\mathscr{D}} = \vartheta \tilde{\beta}^k_{\mathscr{N}^\ell_{\mathscr{D}}} \quad \text{on } \Gamma^\ell_{\mathscr{D}} \cap \Delta_{\mathscr{D},\geq m}, \tag{26}$$

if $k \geq m > \ell$.

Proof We use induction over the level m from n down to $\ell + 1$ with $n = \max\{k \in \mathscr{N}^\ell_{\mathscr{D}}\}$. It suffices to consider these functions since levels higher than n are not active on $\Gamma^\ell_{\mathscr{D}}$.

For level n it holds that

$$\tilde{\beta}^n_{\mathscr{D}} = \beta^n = \tilde{\beta}^n_{\mathscr{N}^\ell_{\mathscr{D}}}, \quad \text{on } \pi^n,$$

for all $\tilde{\beta}^n_{\mathscr{D}} \in P^n_{\mathscr{D}}$. Thus, $\vartheta \tilde{\beta}^n_{\mathscr{D}} = \vartheta \tilde{\beta}^n_{\mathscr{N}^\ell_{\mathscr{D}}}$ on $\Gamma^\ell_{\mathscr{D}} \cap \Delta_{\mathscr{D},\geq n}$. In order to complete the proof by induction, we assume that (26) is satisfied for $m + 1$ and show that this extends to m.

First, we consider functions $\tilde{\beta}^k_{\mathscr{D}} \in P^m_{\mathscr{D}}$ with $k > m$. The induction hypothesis implies that

$$\vartheta \tilde{\beta}^k_{\mathscr{D}} = \vartheta \tilde{\beta}^k_{\mathscr{N}^\ell_{\mathscr{D}}}, \quad \text{on } \Gamma^\ell_{\mathscr{D}} \cap \Delta_{\mathscr{D},\geq m+1}.$$

Furthermore, the selection ensures that these functions vanish on π^m, hence

$$\vartheta \tilde{\beta}^k_{\mathscr{D}} = \vartheta \tilde{\beta}^k_{\mathscr{N}^\ell_{\mathscr{D}}} = 0, \quad \text{on } \Gamma^\ell_{\mathscr{D}} \cap \pi^m.$$

The equality on $\Gamma^\ell_{\mathscr{D}} \cap \Delta_{\mathscr{D},\geq m}$ now follows from

$$\Gamma^\ell_{\mathscr{D}} \cap \Delta_{\mathscr{D},\geq m} = (\Gamma^\ell_{\mathscr{D}} \cap \pi^m) \cup (\Gamma^\ell_{\mathscr{D}} \cap \Delta_{\mathscr{D},\geq m+1}). \tag{27}$$

Second, we consider the remaining functions $\tilde{\beta}^m_{\mathscr{D}} \in T^m_{\mathscr{D}}$. Obviously, it holds that

$$\vartheta \tilde{\beta}^m_{\mathscr{D}} = \vartheta \tilde{\beta}^m_{\mathscr{N}^\ell_{\mathscr{D}}} = \vartheta \beta^m, \quad \text{on } \Gamma^\ell_{\mathscr{D}} \cap \pi^m.$$

Furthermore,

$$\tilde{\beta}^m_{\mathscr{D}} = \sum_{\substack{\tilde{\beta}^r_{\mathscr{D}} \in P^{m+1}_{\mathscr{D}} \\ \operatorname{supp}\beta^r \cap \operatorname{supp}\beta^m \neq \emptyset}} \lambda_{\beta^r}(\hat{\beta}^m) \tilde{\beta}^r_{\mathscr{D}} \quad \text{on } \Delta_{\mathscr{D},\geq m+1}.$$

The previous lemma (see (24)) and the induction hypothesis imply

$$\vartheta\Big(\sum_{\substack{\tilde{\beta}^r_{\mathscr{D}} \in P^{m+1}_{\mathscr{D}} \\ \operatorname{supp}\beta^r \cap \operatorname{supp}\beta^m \neq \emptyset}} \lambda_{\beta^r}(\hat{\beta}^m) \tilde{\beta}^r_{\mathscr{D}} \Big) = \vartheta\Big(\sum_{\substack{\tilde{\beta}^r_{\mathscr{N}^\ell_{\mathscr{D}}} \in P^{m+1}_{\mathscr{N}^\ell_{\mathscr{D}}} \\ \operatorname{supp}\beta^r \cap \operatorname{supp}\beta^m \neq \emptyset}} \lambda_{\beta^r}(\hat{\beta}^m) \tilde{\beta}^r_{\mathscr{N}^\ell_{\mathscr{D}}} \Big) = \vartheta \tilde{\beta}^m_{\mathscr{N}^\ell_{\mathscr{D}}}$$

on $\Gamma^\ell_{\mathcal{D}} \cap \Delta_{\mathcal{D},\geq m+1}$. Invoking the decomposition (27) confirms that (26) is satisfied for all $\tilde{\beta}^m_{\mathcal{D}} \in T^m_{\mathcal{D}}$. □

Now we are able to prove the original statement (23). Applying both Lemmas with $m = \ell + 1$ confirms that $\vartheta \hat{\beta}^\ell_{\mathcal{N}^\ell_{\mathcal{D}}} = \vartheta \beta^\ell = \vartheta \tilde{\beta}^\ell_{\mathcal{D}}$ on $\Gamma^\ell_{\mathcal{D}} = \Gamma^\ell_{\mathcal{D}} \cap \Delta_{\mathcal{D},\geq \ell}$ since $P_{\mathcal{N}^\ell_{\mathcal{D}}} = P^1_{\mathcal{N}^\ell_{\mathcal{D}}} = P^{\ell+1}_{\mathcal{N}^\ell_{\mathcal{D}}}$. Hence, we conclude that the functions $\tilde{\beta}^\ell_{\mathcal{D}} \in T^\ell_{\mathcal{D}}$ possess the required order of smoothness on $\Delta_{\mathcal{D},\geq \ell}$.

Second Statement (*ii*): We consider a function $f \in \mathbb{P}_{\mathcal{D}}$. Its restriction to π^ℓ has a representation

$$f = \sum_{\beta^\ell \in B^\ell} \lambda_{\beta^\ell}(f)\, \beta^\ell = \sum_{\beta^\ell \in B^\ell} \lambda_{\beta^\ell}(f)\, \tilde{\beta}^\ell_{\mathcal{D}} \quad \text{on } \pi^\ell, \tag{28}$$

since $\beta^\ell = \tilde{\beta}^\ell_{\mathcal{D}}$ on π^ℓ. We now define

$$\tilde{f} = f - \sum_{\beta^\ell \in B^\ell} \lambda_{\beta^\ell}(f)\, \tilde{\beta}^\ell_{\mathcal{D}} \quad \text{on } \Delta_{\mathcal{D}}. \tag{29}$$

Clearly, $\tilde{f}$ is equal to zero on π^ℓ and satisfies homogeneous boundary conditions on $\Gamma^\ell_{\mathcal{D}}$. We observe that $\tilde{f} \in \mathbb{P}^{\ell+1}_{\mathcal{D}}$. Indeed, the definition of the patchwork spline space $\mathbb{P}_{\mathcal{D}}$ gives $f \in \mathbb{P}^{\ell+1}_{\mathcal{D}}$, and the functions $\tilde{\beta}^\ell_{\mathcal{D}} \in T^\ell_{\mathcal{D}}$ are linear combinations of functions in $\mathbb{P}^{\ell+1}_{\mathcal{D}}$. According to the induction hypothesis there exists a representation

$$\tilde{f} = \sum_{\tilde{\beta}^k_{\mathcal{D}} \in P^{\ell+1}_{\mathcal{D}}} \lambda_{\beta^k}(\tilde{f})\, \tilde{\beta}^k_{\mathcal{D}} \quad \text{on } \Delta_{\mathcal{D},\geq \ell+1}. \tag{30}$$

We consider a function $\tilde{\beta}^k_{\mathcal{D}} \in P^{\ell+1}_{\mathcal{D}} \setminus S^{\ell+1}_{\mathcal{D}}$ in the above sum. The definition of $S^{\ell+1}_{\mathcal{D}}$ and Lemma 3 imply that

$$\text{supp}\beta^k \cap \Gamma^\ell_{\mathcal{D}} \neq \emptyset.$$

There exists a facet ξ of $\pi^k \cap \pi^\ell \subseteq \Gamma^\ell_{\mathcal{D}}$ with $\xi \subseteq \text{supp}\beta^k$ where $\tilde{f}$ satisfies the homogeneous boundary conditions. Therefore, we can apply Lemma 6 and obtain that $\lambda_{\beta^k}(\tilde{f}) = 0$.

Consequently, only functions in $S^{\ell+1}_{\mathcal{D}}$ contribute to the representation in (30) and thus, we can extend this representation to $\Delta_{\mathcal{D}}$,

$$\tilde{f} = \sum_{\tilde{\beta}^k_{\mathcal{D}} \in S^{\ell+1}_{\mathcal{D}}} \lambda_{\beta^k}(\tilde{f})\, \tilde{\beta}^k_{\mathcal{D}} \quad \text{on } \Delta_{\mathcal{D}}. \tag{31}$$

Therefore, we can combine equations (29) and (31) to represent f with respect to $P_{\mathcal{D}}$ on $\Delta_{\mathcal{D}}$,

$$f = \sum_{\tilde{\beta}^k \in P_{\mathcal{D}}} \mu_{\tilde{\beta}^k_{\mathcal{D}}}(f)\, \tilde{\beta}^k_{\mathcal{D}} \quad \text{on } \Delta_{\mathcal{D}}, \tag{32}$$

with coefficients

$$\mu_{\tilde{\beta}^k_{\mathscr{D}}}(f) = \begin{cases} \lambda_{\beta^\ell}(f) & \text{if } k = \ell \\ \lambda_{\beta^k}(\tilde{f}) & \text{otherwise.} \end{cases}$$

Finally, we note that the functions in $P_{\mathscr{D}}$ are linearly independent. Indeed, $T^\ell_{\mathscr{D}}$ is linearly independent on π^ℓ, since $\tilde{\beta}^\ell_{\mathscr{D}} = \beta^\ell$ on π^ℓ and the functions β^ℓ are linearly independent on π^ℓ. Moreover, the remaining functions in $S^{\ell+1}_{\mathscr{D}} \subseteq P^{\ell+1}_{\mathscr{D}}$ vanish on π^ℓ while being linearly independent on $\Delta_{\mathscr{D},\geq \ell+1}$ according to the induction hypothesis.

Third Statement (*iii*)**:** We rewrite Eq. (32) by using the definition of the DPB-splines $P_{\mathscr{D}}$ and obtain

$$f = \sum_{\tilde{\beta}^\ell_{\mathscr{D}} \in T^\ell_{\mathscr{D}}} \mu_{\tilde{\beta}^\ell_{\mathscr{D}}}(f)\, \tilde{\beta}^\ell_{\mathscr{D}} + \sum_{\tilde{\beta}^k_{\mathscr{D}} \in S^{\ell+1}_{\mathscr{D}}} \mu_{\tilde{\beta}^k_{\mathscr{D}}}(f)\, \tilde{\beta}^k_{\mathscr{D}} \quad \text{on } \Delta_{\mathscr{D}}. \tag{33}$$

This confirms (9) with respect to the patch B-splines on π^ℓ since $\mu_{\tilde{\beta}^\ell_{\mathscr{D}}}(f) = \lambda_{\beta^\ell}(f)$.

Next we consider the restriction of f to $\Delta_{\mathscr{D},\geq \ell+1}$. On the one hand, it satisfies

$$f = \sum_{\tilde{\beta}^k_{\mathscr{D}} \in P^{\ell+1}_{\mathscr{D}}} \lambda_{\beta^k}(f)\, \tilde{\beta}^k_{\mathscr{D}} \quad \text{on } \Delta_{\mathscr{D},\geq \ell+1}, \tag{34}$$

according to the induction hypothesis (*ii*). On the other hand, we may rewrite (33) as

$$f = \underbrace{\sum_{\tilde{\beta}^\ell_{\mathscr{D}} \in T^\ell_{\mathscr{D}}} \lambda_{\beta^\ell}(f) \sum_{\substack{\tilde{\beta}^k_{\mathscr{D}} \in P^{\ell+1}_{\mathscr{D}} \\ \operatorname{supp}\beta^\ell \cap \operatorname{supp}\beta^k \neq \emptyset}} \lambda_{\beta^k}(\hat{\beta}^\ell)\, \tilde{\beta}^k_{\mathscr{D}}}_{(\star)} + \sum_{\tilde{\beta}^k_{\mathscr{D}} \in S^{\ell+1}_{\mathscr{D}}} \mu_{\tilde{\beta}^k_{\mathscr{D}}}(f)\, \tilde{\beta}^k_{\mathscr{D}}$$

on $\Delta_{\mathscr{D},\geq \ell+1}$, according to the definition of the truncation operator. Note that none of the functions $\tilde{\beta}^k_{\mathscr{D}}$ that appear in the double sum ($\star$) belong to the set $S^{\ell+1}_{\mathscr{D}}$ of selected functions, since $\operatorname{supp}\beta^\ell \cap \operatorname{supp}\beta^k \neq \emptyset$ implies that $\beta^k \neq 0$ on $\Gamma^\ell_{\mathscr{D}}$. Therefore, comparing the coefficients with (34) confirms that $\mu_{\tilde{\beta}^k_{\mathscr{D}}}(f) = \lambda_{\beta^k}(f)$ for $\tilde{\beta}^k_{\mathscr{D}} \in S^{\ell+1}_{\mathscr{D}}$.

Fourth Statement (*iv*)**:** First, we show that the basis functions in $P^\ell_{\mathscr{D}} = T^\ell_{\mathscr{D}} \cup S^{\ell+1}_{\mathscr{D}}$ are non-negative on $\Delta_{\mathscr{D}}$.

Since $S^{\ell+1}_{\mathscr{D}} \subseteq P^{\ell+1}_{\mathscr{D}}$ the induction hypothesis implies that functions in $S^{\ell+1}_{\mathscr{D}}$ are non-negative on $\Delta_{\mathscr{D},\geq \ell+1}$. This extends to $\Delta_{\mathscr{D}}$ as these functions vanish on π^ℓ according to the selection mechanism. The functions in $T^\ell_{\mathscr{D}}$ are non-negative on π^ℓ since there $\tilde{\beta}^\ell_{\mathscr{D}} = \beta^\ell$ and the functions β^ℓ are non-negative. Moreover, their representation with respect to $P^{\ell+1}_{\mathscr{D}} \setminus S^{\ell+1}_{\mathscr{D}}$ on $\Delta_{\mathscr{D},\geq \ell+1}$ is a linear combination of non-negative functions with non-negative coefficients.

Second, we note that the partition of unity is obtained by applying (9) to the function $f = 1$.

This completes the proof of Theorem 1.

5 Comparison with THB- and TPB-Splines

This section is used for comparing the flexibility of DPB-, TPB- and THB-splines. First, we observe that the new construction of DPB-splines complements the other two bases. To be more precise, on some hierarchies we can construct only one of the bases while on others we can define two or even all three of them.

Example 4 We consider several instances of hierarchical meshes for spline spaces of degree $\mathbf{p} = (2, 2)$ and discuss if the hierarchies possess a DPB-spline basis, see Fig. 10. Note that IPC and SIC are automatically satisfied for a patch π^ℓ if it shares its boundary with higher level patches only, e.g., as it is the case for the first patch π^1. Furthermore, if a patch of level ℓ shares its boundary with only one patch of lower level $k < \ell$ then IPC and SIC are satisfied for π^ℓ. Moreover, we analyze if we can construct a TPB-spline basis or a THB-spline basis on these meshes:

- A hierarchy is feasible for THB-splines if the spline spaces of all levels are nested [18].
- For TPB-splines the FSC ("full shadow compatibility") and CBA ("constrained boundary alignment") assumption, see [10], have to be fulfilled in order to obtain a basis. Note that CBA is satisfied if all patches are aligned with the knot lines of the corresponding spline space.

We observe that the new construction of DPB-splines complements the other two bases.

Although there exist hierarchies that are valid for TPB- and not for DPB-splines we conjecture that this set is rather small compared to the set of hierarchies that possess a DPB- and no TPB-spline basis. The following example supports this assumption.

Example 5 We consider the regular subdivision of the unit square into 4×4 patches and randomly assign values from 0 to 3 to each patch. The spline spaces of the patches remain unchanged, are refined dyadically in x_1-direction, in x_2-direction and in both directions for the values 0, 1, 2 and 3, respectively. The patches are sorted according to these values, and patches with the same value are ordered randomly. We compute $120,000$ samples of such random hierarchies and analyze how many of them are feasible for TPB- and DPB-splines. The results are reported in Table 1. Most of the samples ($114,605$) do not satisfy NNC and are unsuitable for all constructions. From the remaining $5,395$ hierarchies, 588 (approx. 11%) can be equipped with a DPB-spline basis, but only 4 (approx. 0.07%) admit a TPB-spline basis. 3 hierarchies are valid for TPB- and DPB-splines. The same experiment was performed for bicubic

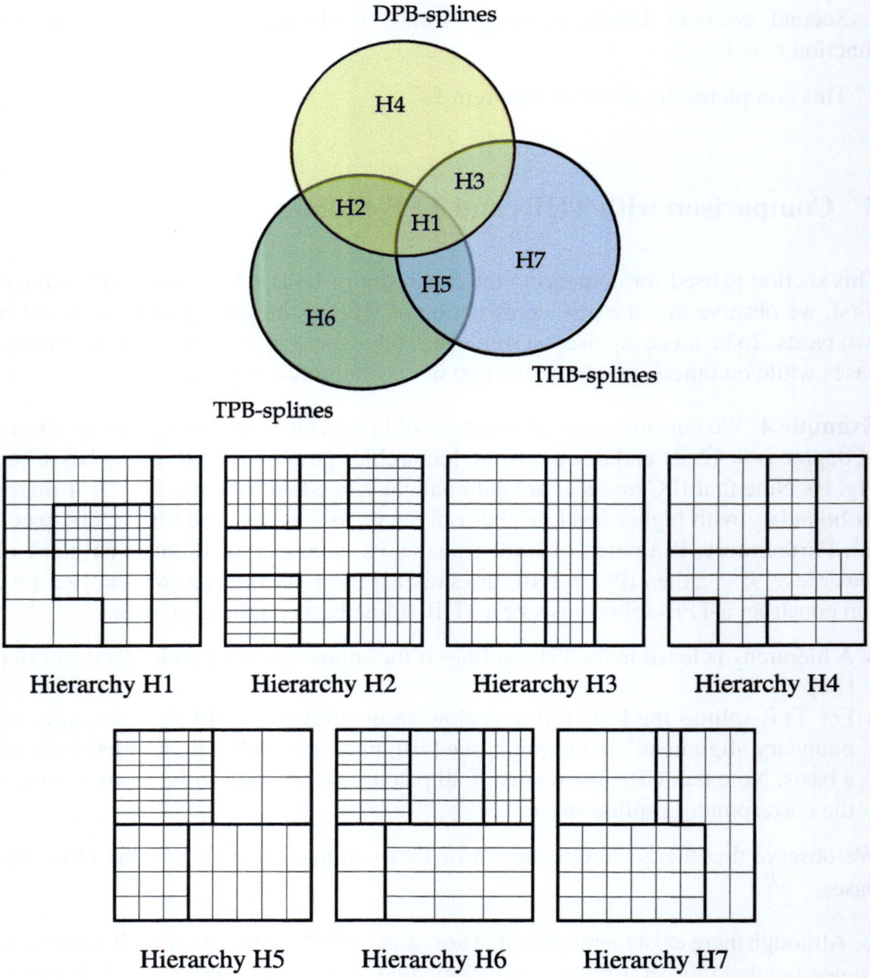

Fig. 10 Venn diagram and hierarchical meshes for DPB-, TPB- and THB-splines

splines, on 3×3 patches with triadic refinement, see Table 1 last row. Finally, we note that very few (2 and 231) of the 120, 000 samples did not use anisotropic refinement, i.e., not the refinement values 1 and 2.

Table 1 Example 5—Experimental quantification of patch structures suitable for TPB- and DPB-splines

	Nested	DPB	TPB	Both
$\mathbf{p} = (2, 2), 4 \times 4$ patches	$5, 395$	588	4	3
$\mathbf{p} = (3, 3), 3 \times 3$ patches	$25, 169$	$11, 636$	$2, 011$	$1, 757$

6 Macro Element Refinement for Adaptive Surface Reconstruction

We present an adaptive refinement algorithm for surface approximation that generates a two-dimensional patchwork hierarchy where NNC, IPC and SIC are satisfied. Thus, a DPB-spline fitting surface can be constructed on the patch structure. In contrast to our previous algorithm in [10], this new procedure avoids additional refinement and expensive sorting by constructing a valid hierarchy directly.

6.1 *Sufficient Condition*

A hierarchy of patches π^ℓ and spaces $\mathbb{V}^\ell$, for $\ell = 1, \ldots, N$, will be called *feasible* if the corresponding patchwork spline space can be equipped with a DPB-spline basis. Therefore, in order to analyze whether a given hierarchy is feasible, we have to check if NNC, IPC and SIC are satisfied for all patches and corresponding spline spaces. Verifying the latter two assumptions might be expensive since it requires us to go through all patch B-splines. However, for certain types of patches we can formulate conditions for IPC and SIC in the bivariate case, i.e., $d = 2$, that analyze only the boundary of a patch.

We define a *lower level boundary component* (LLBC) of a patch π^ℓ as a connected component of

$$\pi^k \cap \pi^\ell \neq \emptyset, \quad k < \ell.$$

Clearly, *IPC and SIC are satisfied for all patches with at most one LLBC.*

In addition to this simple observation, we present a necessary and sufficient condition for a certain class of patches. A *fat* patch π^ℓ is the union of mutually disjoint boxes composed of $p_1 \times p_2$ cells in Z^ℓ, where adjacent boxes share either a horizontal or a vertical boundary segment of p_1/p_2 knot spans, see Fig. 11. We consider single LLBCs of π^ℓ and connected components of Γ^ℓ. These parts of the boundary are denoted as *critical boundary components* if they are enclosed by LLBC(s) at both ends.

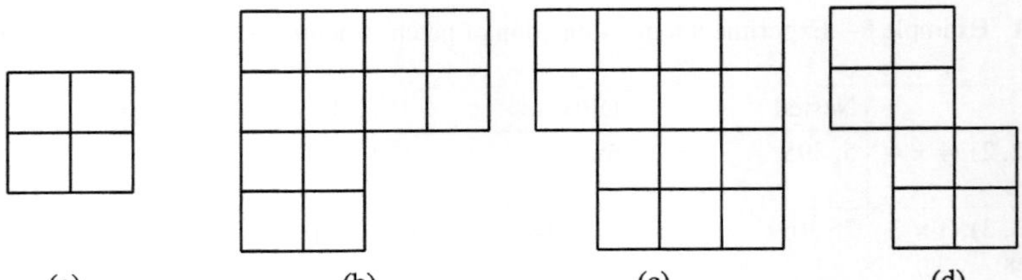

(a) (b) (c) (d)

Fig. 11 Two fat patches (**a**) and (**b**) for $\mathbf{p} = (2, 2)$. The patch shown in (**c**) is not a fat patch, since not all disjoint boxes are composed of 2×2 cells and (**d**) is not a fat patch, since the shared segment between the north-western and the south-eastern boxes consisting of 2×2 cells is too short

Corollary 2 *IPC and SIC are satisfied for a fat patch π^ℓ if and only if the width or the height of each critical component δ equals at least p_1 or p_2 knot spans with respect to $\mathbb{V}^\ell$.*

Proof First, we show that the condition on the critical components is sufficient. The intersection $\text{supp}\beta^\ell \cap \partial\pi^\ell$, for $\beta^\ell \in B^\ell$ is either empty, a horizontal or vertical line segment that is $p_1 + 1$ or $p_2 + 1$ knot spans long with respect to $\mathbb{V}^\ell$ or a polygon composed of a horizontal and vertical line segment with a maximum length of p_1 and p_2 knot spans, respectively. Therefore, if a patch B-spline β^ℓ intersects two LLBCs of π^ℓ, γ_1 and γ_2, then it holds that $\text{supp}\beta^\ell \cap \gamma_1 \cap \gamma_2 \neq \emptyset$. This implies IPC and SIC for π^ℓ.

Second, we prove that the condition is also necessary for IPC and SIC. We assume that there exists a critical boundary component δ that is joined to the LLBCs $\gamma_k \subseteq \pi^k$ and $\gamma_m \subseteq \pi^m$ at each end and does not meet the required assumption. Consequently, there exists a patch B-spline $\beta^\ell \in B^\ell$ such that

$$\text{supp}\beta^\ell \cap \gamma_k \neq \emptyset \text{ and } \text{supp}\beta^\ell \cap \gamma_m \neq \emptyset.$$

If $k = m$ then SIC is not fulfilled. For $k \neq m$ IPC can be only satisfied if there is another LLBC γ' of π^k or π^m such that $\text{supp}\beta^\ell \cap \gamma_m \cap \gamma' \neq \emptyset$ or $\text{supp}\beta^\ell \cap \gamma_k \cap \gamma' \neq \emptyset$, respectively. However, then SIC is violated either for level k or m. □

6.2 Macro Elements

We consider bivariate tensor-product spline spaces of fixed degree $\mathbf{p} = (p_1, p_2)$ defined by uniform knot vectors obtained by performing r_i-times p_i-adic refinement in i–th direction, $i = 1, 2$. To be more precise, the spline space

$$\mathbb{S}^{\mathbf{r}} = \mathbb{S}^{(r_1, r_2)}$$

possesses the knot vector

$$\left(0,\ldots,0,\frac{1}{(p_i)^{r_i}},\ldots,\frac{(p_i)^{r_i}-1}{(p_i)^{r_i}},1,\ldots,1\right),$$

in the i–th coordinate direction. These spline spaces are collected in a set, which we denote by $\mathfrak{S}$. A local spline space $\mathbb{V}^\ell$ is then defined as the restriction of a tensor-product spline space $\hat{\mathbb{V}}^\ell \in \mathfrak{S}$ to the corresponding patch. Consequently, there is a function

$$\mathbf{r} : \{1,\ldots,N\} \to \mathbb{N}\times\mathbb{N}$$

such that $\hat{\mathbb{V}}^\ell = \mathbb{S}^{\mathbf{r}(\ell)}$ and

$$\mathbb{V}^\ell = \mathbb{S}^{\mathbf{r}(\ell)}|_{\pi^\ell}.$$

The knot lines of a spline space $\mathbb{S}^{\mathbf{r}}$ divide the domain into a set of cells $\hat{Z}^{\mathbf{r}}$. Hence, we obtain the local cell set

$$Z^\ell = \hat{Z}^{\mathbf{r}(\ell)}|_{\pi^\ell}.$$

We introduce a set of *macro elements* $\mathscr{M}^{\mathbf{r}}$, which contains the disjoint axis-aligned boxes consisting of $p_1 \times p_2$ cells in $\hat{Z}^{\mathbf{r}}$. To be more precise, the set $\mathscr{M}^{\mathbf{r}}$ contains the $p_1 \times p_2$ subgrids of $\hat{Z}^{\mathbf{r}}$ with lower left corners

$$\left(\frac{mp_1}{(p_1)^{r_1}},\frac{np_2}{(p_2)^{r_2}}\right),\quad \text{for } m,n\in\mathbb{Z}.$$

6.3 Iterative Refinement and its Application to Surface Approximation

We consider an iterative refinement procedure consisting of several *refinement steps*. The input is a hierarchy where the patches are equal to the macro elements of an initial tensor-product spline space $\mathbb{S}^{\mathbf{r}}$. In each refinement step we add knots to the spline spaces of certain marked patches, thereby generating patches with finer spline spaces. A *refinement indicator* marks the cells that should be refined and a *direction indicator* specifies the refinement direction, i.e., whether we perform knot insertion in the first (or x_1-) coordinate direction, second (or x_2-) coordinate direction, or both. The concrete choice of these indicators depends on the application. For the case of surface approximation, it will be discussed below.

We refine such that the tensor-product spline spaces of neighboring patches are nested, which guarantees that NNC is fulfilled. As a consequence the refinement is not according to the direction indicator in some cases. Furthermore, all cells of a marked patch will be refined to the same spline space. Hence, a marked patch is replaced by at least p_1, p_2 and $p_1 \times p_2$ patches, which are again macro elements, for refinement in x_1-, x_2- and both directions, respectively. More precisely we apply the following procedure.

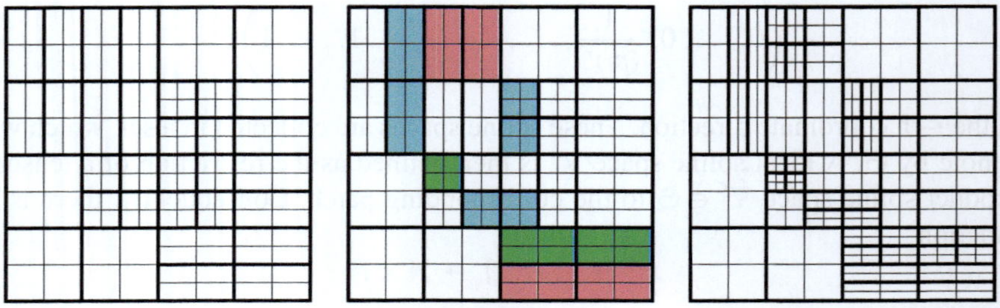

Fig. 12 Example 6—Illustration of one refinement step

- *Determining a Refinement Order*: We sort the marked macro elements $\mu \in \mathscr{M}^{\mathbf{r}(\ell)}$, for $\ell = 1, \ldots, N$, lexicographically with respect to

$$(r_1(\ell) + r_2(\ell), r_2(\ell), \delta(\mu)).$$

  The elements are then refined one by one in this order.
- *Refinement of the Elements*: The corresponding spline space of an element $\mu \in \mathscr{M}^{\mathbf{r}(\ell)}$ is refined to a new spline space $\mathbb{V}^{\mathbf{r}(k)}$ that satisfies

  (i) $r_i(k) \geq r_i(\ell)$, for $i = 1, 2$,
  (ii) $\mathbb{S}^{\mathbf{r}(k)}$ is nested with the current spline spaces of neighboring patches, and
  (iii) $|r_1(k) - r_2(k)| \leq 3$.

  Note that the property *(iii)* controls the aspect ratio of the elements. If refinement according to the direction indicator does not result in a space that fulfills these properties then we choose the smallest $\mathbf{r}(k)$ with respect to the lexicographical ordering $(r_1 + r_2, r_2)$ that satisfies them.
- *Final Ordering*: We sort all macro elements lexicographically with respect to $(r_1 + r_2, r_2)$ and apply the selection mechanism.

Example 6 We consider the hierarchy in Fig. 12 on the left for $\mathbf{p} = (2, 2)$. The first five patches (nos. 1–5, south-west and north-east) are macro elements of the spline space $\mathbb{S}^{(3,3)}$, followed by eight macro elements (nos. 6–13, north-west) from $\mathscr{M}^{(4,3)}$ and eight elements (nos. 14–21, south-east) from $\mathscr{M}^{(3,4)}$. The remaining 12 patches (nos. 22–33, center) are elements of $\mathscr{M}^{(4,4)}$. Within each group of macro elements the chosen ordering does not affect the outcome of the algorithm in this case. The refinement and direction indicators mark certain patches for refinement as illustrated in the middle picture. Blue, red and green cells are marked for refinement in x_1-, x_2- and both directions, respectively. The image on the right shows the resulting mesh after the corresponding knot (local) insertion. Note that the macro element with lower left corner $(\frac{1}{4}, \frac{3}{4})$ is marked for refinement in x_2-direction, which would result in non-nested spaces of neighboring patches. Therefore, we refine this patch in both directions instead. ◊

A hierarchy that is constructed by this refinement procedure is feasible for DPB-splines. NNC is satisfied by the definition of the refinement algorithm. Moreover, all patches are boxes of size $p_1 \times p_2$ and according to the refinement mechanism, a patch shares an entire edge of its boundary either with a single patch of lower level or several patches of higher levels. Therefore, Corollary 2 guarantees that IPC and SIC are satisfied for all levels.

Furthermore, the so-constructed hierarchies also admit (T)PB-spline bases. Since all patches are aligned with the knot lines of the corresponding spline spaces CBA is satisfied. Moreover, the tensor-product basis functions $\hat{\beta}^\ell$ that do not vanish on the corresponding patch π^ℓ intersect only with neighboring patches of lower level or higher level patches where the corresponding spline space is a superspace of $\hat{\mathbb{V}}^\ell$. This implies that FSC is satisfied.

The remainder of this section presents some details concerning the indicators used to perform *surface approximation*. More precisely, we use the refinement procedure in an iterative algorithm for approximating data with a DPB-spline surface by a standard regularized least-squares fitting, cf. [26]. The algorithm stops if the error does not exceed a user-defined threshold ε in a certain percentage of data points (usually between 95% and 99%) or the numbers of degrees of freedom or iteration steps exceed a predefined maximum. The *refinement algorithm* is guided by the following indicators.

- *Refinement Indicator:* We mark all patches that contain points with an error exceeding a certain threshold.
- *Direction Indicator:* The desired refinement direction for the marked elements is determined by a local-fitting method, see [10]. The direction indicator $\delta(\mu)$ takes the values 1, 2 or 3 for refinement in x_1-, x_2- and (x_1, x_2)-direction, respectively.

Example 7 We approximate the data set of Example 6 in [10] by biquadratic splines, i.e., $\mathbf{p} = (2, 2)$, and compare the hierarchies obtained from the macro element refinement (MER) and the original, simple patch refinement (SPR) algorithm presented in [10].

For both algorithms we choose the initial spline space $\mathbb{S}^{(3,3)}$, thus we obtain 4×4 initial macro elements/patches for MER. For SPR, the initial number of patches is user-defined and we decide to start from a single patch. The regularized least-squares fitting is solved with smoothness parameter $\lambda = 1e - 7$ and the iteration stops if the error in $\geq 98\%$ of the points is lower or equal to the threshold $\varepsilon = 1e - 4$. Figure 13 depicts the resulting surface (note that the surfaces are visually identical for both algorithms and therefore we display only one of them) and meshes of both algorithms. Table 2 presents the number of degrees of freedom, some error statistics and computation times for both methods in rows 1 and 4, respectively. We observe that MER achieves a similar good result with fewer degrees of freedom than SPR. Furthermore, the sorting and possible additional refinement that is necessary for generating a feasible hierarchy in SPR is not required by MER, which is reflected in the lower computational times. The more patches are present in a hierarchy the more pronounced the difference becomes.

Table 2 Example 7

	No. of dofs	% of dofs (%)	Max. error	Avg. error	Total time
SPR—1 patch	13, 579	100	6.9e−4	2.24e−5	8 min 34 s
SPR—4 patches	21, 607	159	1e−3	2.28e−5	1h 8 min
MER—(2, 2)	9, 732	71.7	6.4e−4	2.44e−5	5 min 46 s
MER—(3, 3)	9, 732	71.7	6.4e−4	2.44e−5	5 min 40 s
MER—(4, 4)	9, 732	71.7	6.4e−4	2.44e−5	4 min 52 s

Moreover, we observed that SPR is very sensitive to the choice of the initial setting. If we choose the same initial spline space but start from four patches we obtain significantly worse results, see Table 2, second row. Therefore, in order to obtain good results in terms of number of degrees of freedom and computation time it is crucial to find a suitable initial setting for SPR. In contrast to this, the initial setting for MER is defined by the initial spline space only. We obtained nearly equivalent results for sufficiently coarse initial layouts, see row 3 and 5 of Table 2. ◊

Example 8 We reconstruct a real-world aircraft engine blade from ≈ 300, 000 data points, which were obtained by optical scanning. Many industrial applications require C^2-smooth surfaces and thus, the data set is approximated by bicubic splines. We set $\lambda = 5d - 7$, $\varepsilon = 2d - 5$ and stop the iteration if $\geq 95\%$ of the data points are below the threshold. Again we compare the results of the SPR and MER algorithm, see Figs. 14 and 15 for images of the geometry and the meshes and Table 3 for statistics.

The SPR algorithm uses dyadic knot refinement and starts from the initial hierarchy consisting of 8 patches with initial spline spaces defined by the periodic knot vectors

$$\left(-\frac{3}{32}, -\frac{2}{32}, -\frac{1}{32}, 0, \frac{1}{32}, \ldots, \frac{31}{32}, 1, \frac{33}{32}, \frac{34}{32}, \frac{35}{32}\right)$$

in both directions. The resulting mesh from the iterative fitting process is depicted in Fig. 15 (left) and additional information is given in the first row of Table 3. Changing the initial setting for SPR again leads to a drastically different result, see row 2 of Table 3.

For MER we start from the initial spline space $\mathbb{S}^{(3,3)}$ with periodic knot vectors and perform triadic knot refinement. The resulting mesh is depicted in Fig. 15 on the right. As before, the algorithm yields a result with fewer degrees of freedom and less computation time but comparable errors, see fourth row of Table 3. Furthermore, MER leads to a surface with a better visual quality compared to the result obtained by SPR, see Fig. 14. The SPR based surface (center) possesses oscillations that are not present or less pronounced in the MER based surface (right). Choosing the initial spline spaces $\mathbb{S}^{(2,2)}$ and $\mathbb{S}^{(4,4)}$ again leads to similar results, see rows 3 and 5 of Table 3. Although the numbers of degrees of freedom vary, all three initial settings need similar computation time and lead to similar approximation errors. ◊

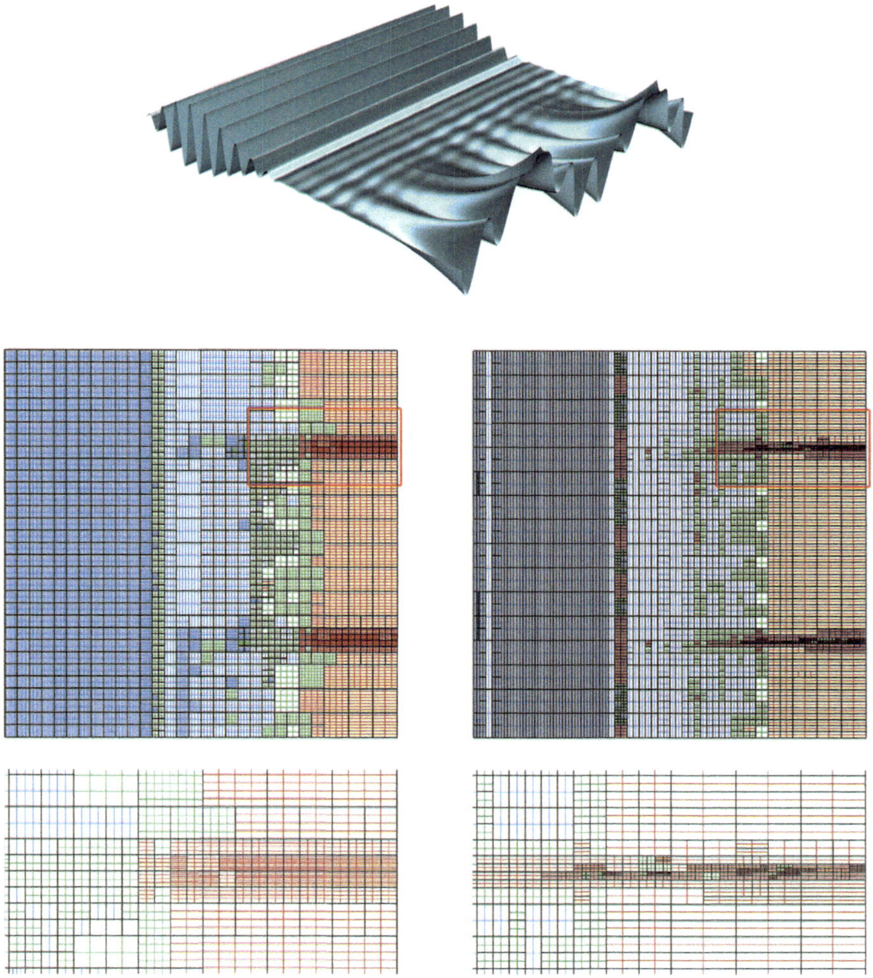

Fig. 13 Example 7—Approximated surface (top) and corresponding meshes for SPR (left) and MER (right). Patch boundaries are black. The knot lines of a spline space $\mathbb{S}^{(r_1,r_2)}$ are colored according to the dominant refinement direction, from blue ($r_1 > r_2$) via green ($r_1 = r_2$) to red ($r_1 < r_2$)

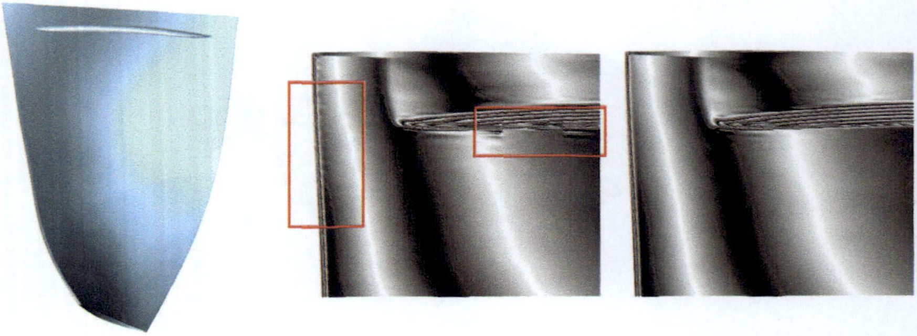

Fig. 14 Example 8—Approximated surface (left) and reflection line analysis for SPR (middle) and MER (right)

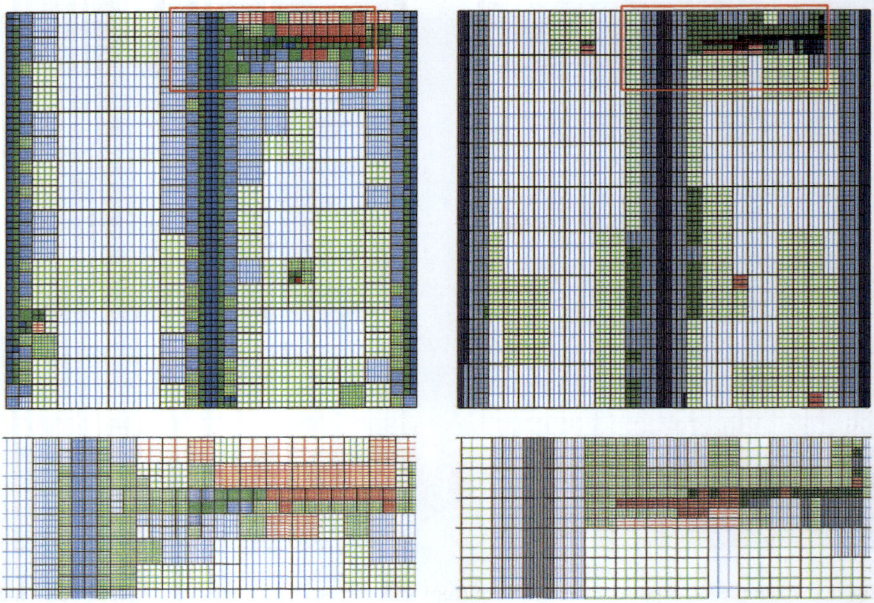

Fig. 15 Example 8—Meshes for SPR (left) and MER (right)

7 Conclusion

The definition of a TPB-spline basis requires a strong nestedness assumption, which limits the choice of possible refinement strategies. The new construction of DPB-splines, which was presented in this paper, substantially increases the flexibility for anisotropic refinement. The use of locally defined, decoupled basis functions enables the definition of a basis under three assumptions (NNC, IPC and SIC) that affect only

Table 3 Example 8

	No. of dofs	% of dofs (%)	Max. error	Avg. error	Total time
SPR—16 patches	19, 823	100	$5.0e{-}5$	$3.32e{-}6$	17 min 38 s
SPR—64 patches	31, 184	157.3	$1.1e{-}3$	$3.51e{-}6$	54 min 40 s
MER—(2, 2)	9, 774	49.3	$5.6e{-}5$	$3.84e{-}6$	11 min 47 s
MER—(3, 3)	13, 284	67	$5.6e{-}5$	$3.33e{-}6$	10 min 20 s
MER—(4, 4)	24, 498	123.6	$5.6e{-}5$	$3.01e{-}6$	14 min

the direct neighborhood of a patch. In particular, only the spaces associated with neighboring patches have to be nested. Furthermore, for a certain class of patches we identified a particularly simple sufficient condition for IPC and SIC.

The DPB-splines possess the required order of smoothness, they span the patchwork spline space, form a non-negative partition of unity and preserve the coefficients of the local B-spline representations. Moreover, the DPB-splines are algebraically complete, i.e., the patchwork spline space is equal to the full spline space. This property is not necessarily satisfied for the (T)PB-splines. Although there exist some hierarchies that are valid for TPB-splines but cannot be equipped with a DPB-spline basis, this set is rather small compared to the hierarchies that are feasible for DPB-splines and do not allow the definition of a TPB-spline basis.

Finally, the definition of DPB-splines inspired us to introduce a new refinement algorithm for least-squares fitting. The resulting hierarchies admit a PB-, TPB- and DPB-spline basis. Furthermore, this macro element-based refinement offers the significant advantage that a feasible hierarchy is automatically generated without the additional sorting and refinement that is required for the previous SPR algorithm presented in [10]. This enabled us to reduce the computation time and the number of degrees of freedom compared to SPR, while preserving the high quality of the results.

Acknowledgements The financial support of the FWF (NFN S117 "Geometry + Simulation" and Doctoral Program W1214 "Computational Mathematics") and of the ERC (GA no. 694515) is gratefully acknowledged. Special thanks go to the geometry department at MTU Aero Engines for providing the data of the turbine blade example.

References

1. Atri, H.R., Shojaee, S.: Truncated hierarchical B-splines in isogeometric analysis of thin shell structures. Steel Compos. Struct. **26**, 171–182 (2018)
2. Beirão da Veiga, L., Buffa, A., Cho, D., Sangalli, G., Analysis-suitable T-splines are dual-compatible: Comput. Methods Appl. Mech. Eng. **249–252**, 42–51 (2012)
3. Bornemann, P.B., Cirak, F.: A subdivision-based implementation of the hierarchical B-spline finite element method. Comput. Methods Appl. Mech. Eng. **253**, 584–598 (2013)

4. Bracco, C., Buffa, A., Giannelli, C., Vázquez, R.: Adaptive isogeometric methods with hierarchical splines: an overview. Disc. Contin. Dyn. Syst.—Ser. A **39**, 241–262 (2019)
5. Bracco, C., Giannelli, C., Großmann, D., Sestini, A.: Adaptive fitting with THB-splines: error analysis and industrial applications. Comput. Aided Geom. Des. **62**, 239–252 (2018)
6. Buffa, A., Giannelli, C.: Adaptive isogeometric methods with hierarchical splines: error estimator and convergence. Math. Models Methods Appl. Sci. **26**, 1–25 (2016)
7. Chen, L., de Borst, R.: Adaptive refinement of hierarchical T-splines. Comput. Methods Appl. Mech. Eng. **337**, 220–245 (2018)
8. Chen, L., de Borst, R.: Locally refined T-splines. Int. J. Numer. Methods Eng. **114**, 637–659 (2018)
9. Dokken, T., Lyche, T., Pettersen, K.F.: Polynomial splines over locally refined box-partitions. Comput. Aided Geom. Des. **30**, 331–356 (2013)
10. Engleitner, N., Jüttler, B.: Patchwork B-spline refinement. Comput.-Aided Des. **90**, 168–179 (2017)
11. Engleitner, N., Jüttler, B.: Lofting with patchwork B-splines. In: Giannelli, C., Speleers, H. (eds.), Advanced methods for geometric modeling and numerical simulation. Springer INdAM Series, vol. 35, pp. 77–98. Springer (2019)
12. Engleitner, N., Jüttler, B., Zore, U.: Partially nested hierarchical refinement of bivariate tensor-product splines with highest order smoothness. In: Floater, M. et al. (eds.), Mathematical methods for curves and surfaces. Lecture Notes in Computer Science, vol. 10521, pp. 126–144. Springer (2017)
13. Evans, E.J., Scott, M.A., Li, X., Thomas, D.C.: Hierarchical T-splines: analysis-suitability, Bézier extraction, and application as an adaptive basis for isogeometric analysis. Comput. Methods Appl. Mech. Eng. **284**, 1–20 (2015)
14. Farin, G., Hoschek, J., Kim, M.S. (eds.): Handbook of Computer Aided Geometric Design. Elsevier (2002)
15. Forsey, D.R., Bartels, R.H.: Hierarchical B-spline refinement. ACM SIGGRAPH Comput. Graph. **22**, 205–212 (1988)
16. Garau, E.M., Vázquez, R.: Algorithms for the implementation of adaptive isogeometric methods using hierarchical B-splines. Appl. Numer. Math. **123**, 58–87 (2018)
17. Giannelli, C., Jüttler, B., Kleiss, S., Mantzaflaris, A., Simeon, B., Špeh, J.: THB-splines: an effective mathematical technology for adaptive refinement in geometric design and isogeometric analysis. Comput. Methods Appl. Mech. Eng. **299**, 337–365 (2016)
18. Giannelli, C., Jüttler, B., Speleers, H.: THB-splines: the truncated basis for hierarchical splines. Comput. Aided Geom. Des. **29**, 485–498 (2012)
19. Giannelli, C., Jüttler, B., Speleers, H.: Strongly stable bases for adaptively refined multilevel spline spaces. Adv. Comput. Math. **40**, 459–490 (2014)
20. Giannelli, C., Kanduč, T., Pelosi, F., Speleers, H.: An immersed-isogeometric model: application to linear elasticity and implementation with THBox-splines. J. Comput. Appl. Math. **349**, 410–423 (2019)
21. Giust, A., Jüttler, B., Mantzaflaris, A.: Local (T)HB-spline projectors via restricted hierarchical spline fitting. Comput. Aided Geom. Des. **80**, 101–865 (2020)
22. Greiner, G., Hormann, K.: Interpolating and approximating scattered 3D-data with hierarchical tensor product B-splines. In: Le Méhauté, A., Rabut, C., Schumaker, L.L. (eds.) Surface Fitting and Multiresolution Methods, pp. 163–172. Vanderbilt University Press, Nashville, TN (1997)
23. Kanduč, T., Giannelli, C., Pelosi, F., Speleers, H.: Adaptive isogeometric analysis with hierarchical box splines. Comput. Methods Appl. Mech. Eng. **316**, 817–838 (2017)
24. Kang, H., Chen, F., Deng, J.: Hierarchical B-splines on regular triangular partitions. Graph. Models **76**, 289–300 (2014)
25. Kang, H., Xu, J., Chen, F., Deng, J.: A new basis for PHT-splines. Graph. Models **82**, 149–159 (2015)
26. Kiss, G., Giannelli, C., Zore, U., Jüttler, B., Großmann, D., Barner, J.: Adaptive CAD model (re-)construction with THB-splines. Graph. Models **76**, 273–288 (2014)

27. Kraft, R.: Adaptive und linear unabhängige Multilevel B-splines und ihre Anwendungen. Ph.D. thesis, University Stuttgart (1998)
28. Kuru, G., Verhoosel, C.V., van der Zeeb, K.G., van Brummelen, E.H.: Goal-adaptive isogeometric analysis with hierarchical splines. Comput. Methods Appl. Mech. Eng. **270**, 270–292 (2014)
29. Li, X., Deng, J., Chen, F.: Polynomial splines over general T-meshes. Vis. Comput. **26**, 277–286 (2010)
30. Li, X., Zheng, J., Sederberg, T.W., Hughes, T.J.R., Scott, M.A.: On linear independence of T-spline blending functions. Comput. Aided Geom. Des. **29**, 63–76 (2012)
31. Mokriš, D., Jüttler, B.: TDHB-splines: the truncated decoupled basis of hierarchical tensor-product splines. Comput. Aided Geom. Des. **31**, 531–544 (2014)
32. Mokriš, D., Jüttler, B., Giannelli, C.: On the completeness of hierarchical tensor-product B-splines. J. Comput. Appl. Math. **271**, 53–70 (2014)
33. Morgenstern, P., Peterseim, D.: Analysis-suitable adaptive T-mesh refinement with linear complexity. Comput. Aided Geom. Des. **34**, 50–66 (2015)
34. Schillinger, D., Dedé, L., Scott, M.A., Evans, J.A., Borden, M.J., Rank, E., Hughes, T.J.R.: An isogeometric design-through-analysis methodology based on adaptive hierarchical refinement of NURBS, immersed boundary methods, and T-spline CAD surfaces. Comput. Methods Appl. Mech. Eng. **249–252**, 116–150 (2012)
35. Sederberg, T.W., Zheng, J., Bakenov, A., Nasri, A.: T-splines and T-NURCCS. ACM Trans. Graph. **22**, 477–484 (2003)
36. Speleers, H.: Hierarchical spline spaces: Quasi-interpolants and local approximation estimates. Adv. Comput. Math. **43**, 235–255 (2017)
37. Speleers, H., Dierckx, P., Vandewalle, S.: Quasi-hierarchical Powell-Sabin B-splines. Comput. Aided Geom. Des. **26**, 174–191 (2009)
38. Speleers, H., Manni, C.: Effortless quasi-interpolation in hierarchical spaces. Numerische Mathematik **132**, 155–184 (2016)
39. Wei, X., Zhang, Y., Hughes, T.J.R., Scott, M.A.: Extended truncated hierarchical Catmull-Clark subdivision. Comput. Methods Appl. Mech. Eng. **299**, 316–336 (2016)
40. Yvart, A., Hahmann, S., Bonneau, G.P.: Hierarchical triangular splines. ACM Trans. Graph. **24**, 1374–1391 (2005)
41. Zhang, J., Li, X.: Local refinement for analysis-suitable++ T-splines. Comput. Methods Appl. Mech. Eng. **342**, 32–45 (2018)
42. Zore, U., Jüttler, B.: Adaptively refined multilevel spline spaces from generating systems. Comput. Aided Geom. Des. **31**, 545–566 (2014)
43. Zore, U., Jüttler, B., Kosinka, J.: On the linear independence of truncated hierarchical generating systems. J. Comput. Appl. Math. **306**, 200–216 (2016)

Simplex-Splines on the Clough-Tocher Split with Arbitrary Smoothness

Tom Lyche, Jean-Louis Merrien, and Tomas Sauer

Abstract The space of piecewise polynomials of smoothness r and degree $3r$ is considered on the Clough-Tocher split of a triangle. For any $r \geq 1$ we give a basis of simplex splines for this space, then a Marsden-like identity, which is proved explicitly for $r \leq 3$ and symbolically for $4 \leq r \leq 6$. In addition, generalizing results for $r = 1$, we prove for $r = 2, 3$ a geometry independent bound for the condition number in the infinity norm of this basis, and conditions to connect two triangles with smoothness r.

1 Introduction

Splines over triangulations have applications in several branches of the sciences ranging from finite element analysis, surfaces in computer aided design and other engineering problems, see for example [7, 8, 17]. For many of these applications, piecewise linear $\mathscr{C}^0$ surfaces do not suffice. In some cases, we need smoother elements for modeling, or higher polynomial degrees to increase the approximation order.

In this paper, we are interested in the space of polynomial splines over a triangulation Δ of a polygonal domain Ω of $\mathbb{R}^2$,

$$\mathbb{S}_d^r(\Delta) := \left\{ f \in \mathscr{C}^r(\Omega) : f|_{\mathscr{T}} \in \mathbb{P}_d, \ \text{for all } \mathscr{T} \in \Delta \right\},$$

T. Lyche (✉)
University of Oslo, Oslo, Norway
e-mail: tom@math.uio.no

J.-L. Merrien
University of Rennes, INSA Rennes, CNRS, Rennes, France
e-mail: jmerrien@insa-rennes.fr

T. Sauer
University of Passau, Passau, Germany
e-mail: tomas.sauer@uni-passau.de

C. Manni and H. Speleers (eds.), *Geometric Challenges in Isogeometric Analysis*, Springer INdAM Series 49, https://doi.org/10.1007/978-3-030-92313-6_5

where $d > r > 0$ are given integers, and $\mathbb{P}_d$ is the space of bivariate polynomials of total degree $\leq d$. The dimension of this finite dimensional vector space is difficult to determine in general [17].

We can obtain smoothness with lower degree if we are willing to split each triangle into a number of subtriangles. The most well known examples are the Clough-Tocher split [5], and the Powell-Sabin 6 and 12 splits [22]. For these splits each triangle is divided into 3, 6 and 12 subtriangles, respectively. For material on these splits and B-spline like bases for splines on triangulations see [2, 3, 6, 9, 10, 12–19, 24–36].

Here we consider the Clough-Tocher triangulation Δ_{CT}, see [4, 5], where each triangle in the original triangulation Δ is split into 3 subtriangles by connecting the vertices of each triangle to its barycenter, see Fig. 1. In [14], Hermite interpolation problems were considered for super-spline subspaces of $\mathbb{S}^r_{3r+1}(\Delta_{CT})$ and $\mathbb{S}^r_{3r}(\Delta_{CT})$ for r even and odd, respectively. It was also stated that these degrees are minimal for global $\mathscr{C}^r$. See also [24]. In [15] stable local bases were constructed for even smaller super-spline subspaces of $\mathbb{S}^r_{3r+1}(\Delta_{CT})$ and $\mathbb{S}^r_{3r}(\Delta_{CT})$.

In this paper, we consider for any $r \in \mathbb{N}$, the spaces $\mathbb{S}^r_{3r}(\triangle)$ on the refinement $\triangle$ of a single (macro)triangle for which we construct a B-spline like basis made out of simplex splines. They constitute a basis for the space since we show that their number agrees with the dimension of the space and that they are linearly independent. For the latter the differentiation formula for simplex splines is used. This extends and generalizes the case $r = 1$ that was considered in [18]. For more on the Clough-Tocher split see [1, 4, 11, 13–15, 17, 20, 27, 30]. Looking in more detail at the cases $r = 2, 3$, corresponding to degrees $d = 6, 9$, we moreover give explicit formulas for connecting two neighboring triangles in a $\mathscr{C}^r$ fashion across an edge using Bernstein-Bézier techniques, and give an upper bound for the L_∞ condition number of the basis. This upper bound is independent of the shape of the triangle $\mathscr{T}$. We also give a Marsden-like identity for the reproduction of polynomials which is proved for $r \leq 3$ and shown symbolically for $r \leq 6$. We conjecture that it holds for any r.

It was shown in [14, 15] that global $\mathscr{C}^2$ continuity cannot be achieved for $d = 6$ for a general triangulation refined by Clough-Tocher splits into Δ_{CT}. However, for r odd it follows, again from [14, 15], that global $\mathscr{C}^r$ continuity holds for $d = 3r$. This means in particular that the formulas for $\mathscr{C}^3$ continuity across an edge in Sect. 4.2 can be used to compute with elements in $\mathbb{S}^3_9(\Delta_{CT})$, using the simplex spline basis on each triangle in $\mathbb{S}^3_9(\Delta)$ in the usual Bernstein-Bézier fashion.

Some results in this paper are based on symbolic computation. The first author can provide code in Mathematica for specific results upon request.

The paper is organized as follows. Since it depends heavily on properties of Bernstein polynomials and simplex splines, we recall some well known facts about these functions in the next section. In Sect. 3 we define a collection of simplex splines and show that they constitute a basis for $\mathbb{S}^r_{3r}(\triangle)$, $r \geq 1$. In Sect. 4, we consider the cases of global $\mathscr{C}^2$ and $\mathscr{C}^3$ regularity in more detail. In Sect. 5, we give a generalization of the barycentric Marsden-like identity for $3 < r$.

2 Preliminaries

In this section we recall some properties of Bernstein polynomials on a triangle and bivariate simplex splines. Here we use the notation $d \in \mathbb{N}_0 := \mathbb{N} \cup \{0\}$, and let $\langle \mathscr{S} \rangle$ denote the **convex hull** of the set $\mathscr{S} \subset \mathbb{R}^2$.

2.1 Bernstein Polynomials

For a given nondegenerate triangle $\mathscr{T} := \langle\{\boldsymbol{p}_1, \boldsymbol{p}_2, \boldsymbol{p}_3\}\rangle \in \mathbb{R}^2$, and $i, j, k \in \mathbb{N}_0$, the Bernstein polynomial $B^d_{ijk} : \mathbb{R}^2 \to \mathbb{R}$ of degree $d := i + j + k \in \mathbb{N}_0$, is defined by

$$B^d_{ijk}(x, y) = B^d_{ijk}(\beta_1, \beta_2, \beta_3) := \frac{d!}{i!j!k!}\beta_1^i \beta_2^j \beta_3^k,$$

where $\beta = (\beta_1, \beta_2, \beta_3)$ are the **barycentric coordinates** of $\boldsymbol{x} = (x, y) \in \mathbb{R}^2$ with respect to $\mathscr{T}$, i.e.,

$$\boldsymbol{x} = \beta_1 \boldsymbol{p}_1 + \beta_2 \boldsymbol{p}_2 + \beta_3 \boldsymbol{p}_3, \quad \beta_1 + \beta_2 + \beta_3 = 1.$$

The barycentric form of **Marsden's identity** for Bernstein polynomials is simply the multinomial expansion

$$\begin{gathered}(u_1\beta_1 + u_2\beta_2 + u_3\beta_3)^d = \sum_{i+j+k=d} u_1^i u_2^j u_3^k B^d_{ijk}(\beta_1, \beta_2, \beta_3), \\ (u_1, u_2, u_3) \in \mathbb{R}^3, \quad \beta_1 + \beta_2 + \beta_3 = 1,\end{gathered} \tag{1}$$

where in the expression $\sum_{i+j+k=d}$ it is understood that $i, j, k \in \mathbb{N}_0$, which is consistent with the convention that $B^d_{ijk} = 0$ if one of the indices becomes negative.

Taking partial derivatives of order $\nu, \lambda, \kappa \in \mathbb{N}_0$ with respect to u_1, u_2, u_3, respectively, in (1) and setting $\boldsymbol{u} = (1, 1, 1)$ we obtain

$$\begin{gathered}\beta_1^\nu \beta_2^\lambda \beta_3^\kappa = \sum_{i+j+k=d} \gamma_{ijk}(\beta_1^\nu \beta_2^\lambda \beta_3^\kappa) B^d_{ijk}(\beta_1, \beta_2, \beta_3), \\ \nu + \lambda + \kappa \le d, \quad \gamma_{ijk}(\beta_1^\nu \beta_2^\lambda \beta_3^\kappa) \in \mathbb{R}.\end{gathered} \tag{2}$$

We find

$$\begin{aligned}1 &= \sum_{i+j+k=d} B^d_{ijk}(\beta_1, \beta_2, \beta_3) \\ (\beta_1, \beta_2, \beta_3) &= \sum_{i+j+k=d} \boldsymbol{b}^*_{ijk} B^d_{ijk}(\beta_1, \beta_2, \beta_3), \quad \boldsymbol{b}^*_{ijk} = \left(\frac{i}{d}, \frac{j}{d}, \frac{k}{d}\right).\end{aligned} \tag{3}$$

The vector $\boldsymbol{b}_{ijk}^*$ is called the **barycentric form** of the **domain point** of B_{ijk}^d. From (2) and (3) it follows that the elements in the set

$$\mathscr{B}^d := \{B_{ijk}^d : i, j, k \geq 0, \ i + j + k = d\}$$

form a **partition of unity basis** for $\mathbb{P}_d$. Indeed, the number of elements $\#\mathscr{B}^d$ of $\mathscr{B}^d$ equals $\binom{d+2}{2}$, the dimension of $\mathbb{P}_d$. We refer to [17] for further properties of B_{ijk}^d.

2.2 *Bivariate Simplex Splines*

For our purpose it is convenient to work with **area normalized bivariate simplex splines** [19] of degree $d \geq 0$ with **knots**

$$K := \{\boldsymbol{k}_1, \ldots, \boldsymbol{k}_{d+3}\}, \quad \boldsymbol{k}_j \in \mathbb{R}^2, \quad j = 1, \ldots, d+3.$$

We can consider K either as a multiset or as a matrix $K \in \mathbb{R}^{2\times(d+3)}$.

The simplex spline $Q[K] : \mathbb{R}^2 \to \mathbb{R}$, is now defined as $Q[K](\boldsymbol{x}) = 0$ for all $\boldsymbol{x} \in \mathbb{R}^2$ if area$(\langle K \rangle) = 0$, and otherwise

$$Q[K] := \frac{\text{area}\,(\mathscr{T})}{\binom{d+2}{2}} M[K],$$

where $\mathscr{T}$ is a fixed reference triangle in the original triangulation Δ. Here area$(\mathscr{S})$ is the area in $\mathbb{R}^2$ of a set $\mathscr{S} \in \mathbb{R}^2$. The function $M[K]$ is a **bivariate unit integral normalized simplex spline**, defined as a linear functional $M[K] : C(\mathbb{R}^2) \to \mathbb{R}$ given by

$$M[K](\varphi) := (d+2)! \int_{\mathscr{S}_{d+2}} \varphi\Big(\sum_{j=1}^{d+3} \boldsymbol{k}_j t_j\Big) \mathrm{d}t_1 \cdots \mathrm{d}t_{d+2}, \quad \varphi \in C(\mathbb{R}^2), \tag{4}$$

with $\mathscr{S}_n := \{(t_1, \ldots, t_{n+1}) \in \mathbb{R}^{n+1} : t_i \geq 0, \sum_{i=1}^{n+1} t_i = 1\}$, the standard simplex in $\mathbb{R}^n$, $n \in \mathbb{N}$. If area$(\langle K \rangle) \neq 0$ then $M[K]$ can be identified with a function $M[K] : \mathbb{R}^2 \to \mathbb{R}$, and we write (4) in the form

$$\int_{\mathbb{R}^s} M[K](\boldsymbol{x})\varphi(\boldsymbol{x})\mathrm{d}\boldsymbol{x} = (d+2)! \int_{\mathscr{S}_{d+2}} \varphi\Big(\sum_{j=1}^{d+3} \boldsymbol{k}_j t_j\Big) \mathrm{d}t_1 \cdots \mathrm{d}t_{d+2}, \quad \varphi \in C(\mathbb{R}^2).$$

We mention the following well-known properties of $M[K]$ [21, 23] and $Q[K]$.

1. $Q[K]$ and $M[K]$ are **piecewise polynomials** of degree $d = \#K - 3$ with support $\langle K \rangle$.

2. **Local Smoothness**: Across a **knot line**, which is a line in the complete graph associated with K, we have that $M[K], Q[K] \in \mathscr{C}^{d+1-m}$, where m is the number of knots on that knot line, including multiplicities.
3. **Differentiation Formula**: For $\boldsymbol{u} = (u_1, u_2) \in \mathbb{R}^2$ and any choice of $a_1, \ldots, a_{d+3}$ such that $\sum_j a_j \boldsymbol{k}_j = \boldsymbol{u}$, $\sum_j a_j = 0$, one has

$$D_{\boldsymbol{u}} M[K] = (d+2) \sum_{j=1}^{d+3} a_j \, M[K \setminus \boldsymbol{k}_j], \quad D_{\boldsymbol{u}} Q[K] = d \sum_{j=1}^{d+3} a_j \, Q[K \setminus \boldsymbol{k}_j],$$

   where $D_{\boldsymbol{u}} := u_1 D_1 + u_2 D_2$ and D_1, D_2 denotes partial derivatives. (**A-recurrence**)
4. **Recurrence Relation**: For any $\boldsymbol{x} \in \mathbb{R}^2$ and any $b_1, \ldots, b_{d+3}$ such that $\sum_j b_j \boldsymbol{k}_j = \boldsymbol{x}$, $\sum_j b_j = 1$, one has

$$M[K](\boldsymbol{x}) = \frac{d+2}{d} \sum_{j=1}^{d+3} b_j M[K \setminus \boldsymbol{k}_j](\boldsymbol{x}), \quad Q[K](\boldsymbol{x}) = \sum_{j=1}^{d+3} b_j Q[K \setminus \boldsymbol{k}_j](\boldsymbol{x}).$$

   (**B-recurrence**)
5. **Knot Insertion Formula**: For any $\boldsymbol{y} \in \mathbb{R}^2$ and any $c_1, \ldots, c_{d+3}$ such that $\sum_j c_j \boldsymbol{k}_j = \boldsymbol{y}$, $\sum_j c_j = 1$, one has

$$M[K] = \sum_{j=1}^{d+3} c_j M[K \cup \boldsymbol{y} \setminus \boldsymbol{k}_j], \quad Q[K] = \sum_{j=1}^{d+3} c_j Q[K \cup \boldsymbol{y} \setminus \boldsymbol{k}_j]. \tag{5}$$

   (**C-recurrence**)
6. **Degree Zero**: For $K = \{\boldsymbol{k}_1, \boldsymbol{k}_2, \boldsymbol{k}_3\}$

$$M[K](\boldsymbol{x}) := \frac{1}{\text{area}(\langle K \rangle)}, \quad Q[K](\boldsymbol{x}) := \frac{\text{area}(\mathscr{T})}{\text{area}(\langle K \rangle)}, \quad \boldsymbol{x} \in \langle K \rangle^o,$$
$$M[K](\boldsymbol{x}) := Q[K](\boldsymbol{x}) = 0, \quad \boldsymbol{x} \notin \langle K \rangle,$$

   where $\mathscr{S}^o$ is the interior of the set $\mathscr{S}$. The values of $M[K]$ and $Q[K]$ on the boundary of $\langle K \rangle$ has to be dealt with separately, see below.

We refer to [21, 23] for further properties of $M[K]$.

3 The Clough-Tocher Split and a Basis for $\mathbb{S}_{3r}^r$

Given a nondegenerate triangle $\mathscr{T}$ in $\mathbb{R}^2$, we connect the vertices $\boldsymbol{p}_1, \boldsymbol{p}_2, \boldsymbol{p}_3$ to the barycenter $\boldsymbol{p}_T := (\boldsymbol{p}_1 + \boldsymbol{p}_2 + \boldsymbol{p}_3)/3$. With this construction, known as the **Clough-Tocher split** $\triangle$, we obtain three subtriangles $\mathscr{T}_1 := \langle \boldsymbol{p}_T, \boldsymbol{p}_2, \boldsymbol{p}_3 \rangle$, $\mathscr{T}_2 := \langle \boldsymbol{p}_T, \boldsymbol{p}_3, \boldsymbol{p}_1 \rangle$ and $\mathscr{T}_3 := \langle \boldsymbol{p}_T, \boldsymbol{p}_1, \boldsymbol{p}_2 \rangle$, see Fig. 1.

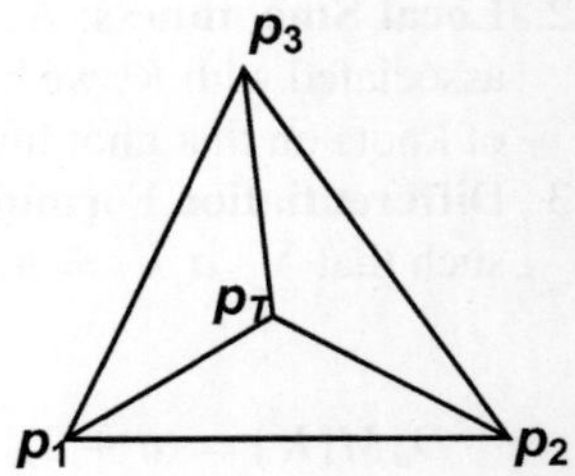

Fig. 1 The Clough-Tocher split, $\mathcal{T} = \langle \boldsymbol{p}_1, \boldsymbol{p}_2, \boldsymbol{p}_3 \rangle$, $\mathcal{T}_1 := \langle \boldsymbol{p}_T, \boldsymbol{p}_2, \boldsymbol{p}_3 \rangle$, $\mathcal{T}_2 := \langle \boldsymbol{p}_T, \boldsymbol{p}_3, \boldsymbol{p}_1 \rangle$ and $\mathcal{T}_3 := \langle \boldsymbol{p}_T, \boldsymbol{p}_1, \boldsymbol{p}_2 \rangle$

We consider the spline space $\mathbb{S}_d^r(\triangle)$ with respect to these three subtriangles on $\mathcal{T}$. To obtain a unique function value at each point in $\mathcal{T}$ we associate the half open edges

$$\langle \boldsymbol{p}_i, \boldsymbol{p}_T \rangle := \{(1-t)\boldsymbol{p}_i + t\boldsymbol{p}_T : 0 \le t < 1\}, \quad i = 1, 2, 3,$$

to the three subtriangles of $\mathcal{T}$ as follows

$$\langle \boldsymbol{p}_1, \boldsymbol{p}_T \rangle \in \mathcal{T}_2, \quad \langle \boldsymbol{p}_2, \boldsymbol{p}_T \rangle \in \mathcal{T}_3, \quad \langle \boldsymbol{p}_3, \boldsymbol{p}_T \rangle \in \mathcal{T}_1,$$

and we somewhat arbitrarily associate the point $\boldsymbol{p}_T$ to $\mathcal{T}_2$.

The dimension of the space $\mathbb{S}_d^r(\triangle)$ is given in the following proposition. It follows from Theorem 9.3 in [17] with $n = m_v = 3$.

Proposition 1 *We have*

$$\dim \mathbb{S}_d^r(\triangle) = \binom{r+2}{2} + 3\binom{d-r+1}{2} + \sum_{j=1}^{d-r} (r+1-2j)_+.$$

We focus on a special collection of simplex splines on the Clough-Tocher split of a single triangle in the following way: for integers i, j, k, ℓ, we consider the simplex spline $Q\left[\boldsymbol{p}_1^{\{i\}}, \boldsymbol{p}_2^{\{j\}}, \boldsymbol{p}_3^{\{k\}}, \boldsymbol{p}_{\mathcal{T}}^{\{\ell\}}\right]$, where $\boldsymbol{p}_j^{\{m\}}$, means that the vertex $\boldsymbol{p}_j$ has multiplicity m, i.e., is repeated m times. For brevity, we will also use two alternative notations, a more compact, and a more illustrative one, namely,

$$Q\left[\boldsymbol{p}_1^{\{i\}}, \boldsymbol{p}_2^{\{j\}}, \boldsymbol{p}_3^{\{k\}}, \boldsymbol{p}_{\mathcal{T}}^{\{\ell\}}\right] =: \triangle_{ijk\ell} =:$$ 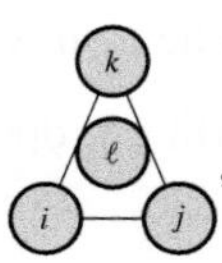

$$, \quad i, j, k, \ell \in \mathbb{N}_0. \qquad (6)$$

We note that $\triangle_{ijk\ell}$ is a simplex spline of degree $d = i + j + k + \ell - 3$, and by the local smoothness property has smoothness $d + 1 - i - \ell$ across the knotline $\langle \boldsymbol{p}_1, \boldsymbol{p}_T \rangle$ for $\ell > 0$. In this notation, the Bernstein polynomials $B_{ijk}(u, v, w) := \frac{(i+j+k)!}{i!j!k!} u^i v^j w^k$ of degree $d = i + j + k$ have the form

$$B_{ijk} = \triangle_{i+1,j+1,k+1,0} =$$

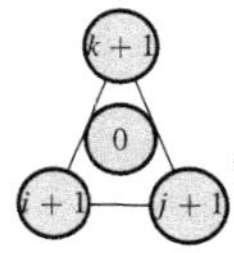

$$, \quad i+j+k=d.$$

We will use the more graphic form on the right hand side of (6) whenever possible to make the basic ideas more accessible, but it is convenient to use the more compact $\triangle$ notations in computations. For $i, j, k \in \mathbb{N}$, the *knot insertion formula* (5) for the insertion of a knot at the barycenter can be written

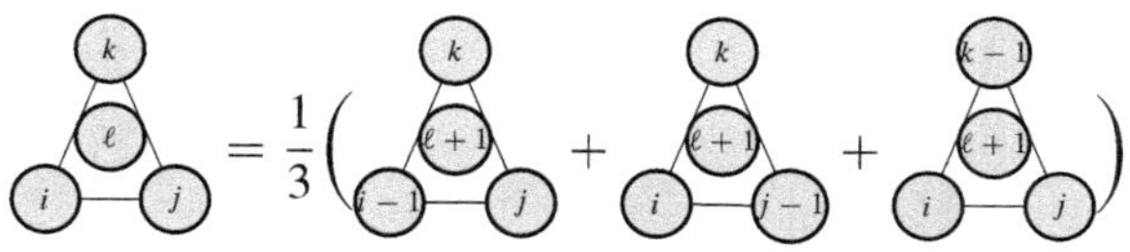

(7)

Definition 1 The number $\mu := d+1-r$ is called the maximum *multiplicity* of an interior knot line of the simplex spline $\triangle_{ijk\ell}$.

Some obvious properties for simplex splines to be in $\mathbb{S}_d^r(\triangle)$ are listed in the following lemma.

Lemma 1 *If* $\triangle_{ijk\ell} \in \mathbb{S}_d^r(\triangle)$ *then*

1. $i+j+k+\ell = d+3 = \mu+r+2$,
2. *if, in addition,* $\ell \neq 0$ *then regularity* $\mathscr{C}^r$ *implies* $\max\{i, j, k\} + \ell \leq \mu$.

Next, let us turn to splines of degree $3r$, and focus on specific elements of $\mathbb{S}_{3r}^r(\triangle)$ that will turn out to be useful. The three types are categorized by whether $\max\{i, j, k\} + \ell$ equals μ, exceeds it or is strictly less than μ, where the second case occurs only for $\ell = 0$.

Definition 2 Let Σ_{3r}^r denote the set of elements $\triangle_{ijk\ell}$ of $\mathbb{S}_{3r}^r(\triangle)$ that consists of the elements of the following three types:

Type (1): $\max\{i, j, k\} + \ell = \mu$ and $\min\{i, j, k\} \geq 1$,
Type (2): $\ell = 0$ and $\max\{i, j, k\} > \mu$,
Type (3): $\max\{i, j, k\} + \ell < \mu$ and $\min\{i, j, k\} = 1$.

Remark 1 The types are symmetric with respect to i, j, k, i.e., all subclasses are closed under permutation of the indices.

For $r = 1, 2, 3, 4$, Definition 2 results in the following set of splines.

Example 1 For $r = 1, d = 3, \mu = 3, \dim \mathbb{S}_3^1(\triangle) = 12$ we have

1. Type (1): 9 elements

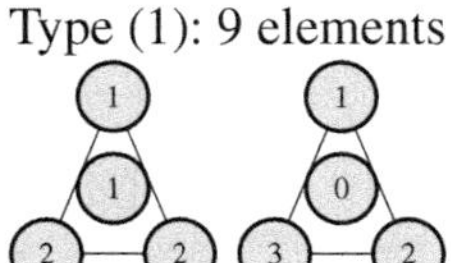

and symmetries,

2. Type (2): 3 elements

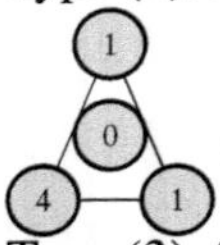

and symmetries,

3. Type (3): 0 elements.

Example 2 For $r = 2$, $d = 6$, $\mu = 5$, $\dim \mathbb{S}_6^2(\triangle) = 37$ the different types look as follows:

1. Type (1): 25 elements

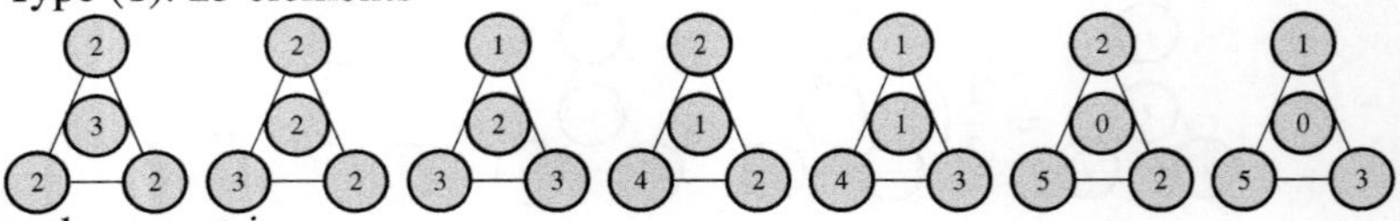

and symmetries,

2. Type (2): 9 elements

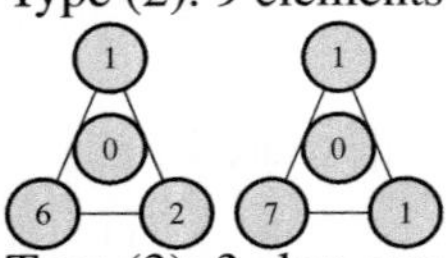

and symmetries,

3. Type (3): 3 elements

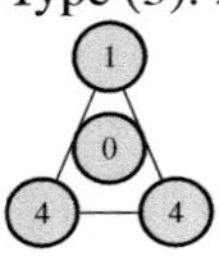

and symmetries.

Example 3 In the case $r = 3$, $d = 9$, $\mu = 7$, $\dim \mathbb{S}_9^3(\triangle) = 75$ we get

1. Type (1): 48 elements

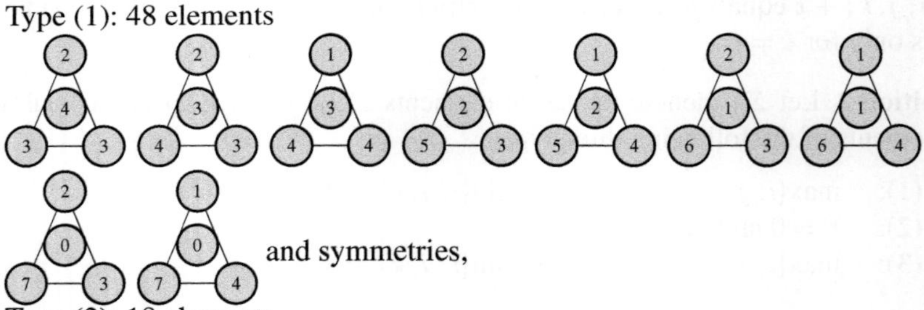

and symmetries,

2. Type (2): 18 elements

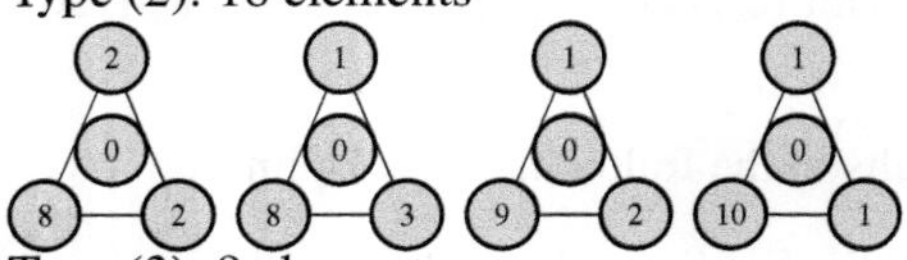

and symmetries,

3. Type (3): 9 elements

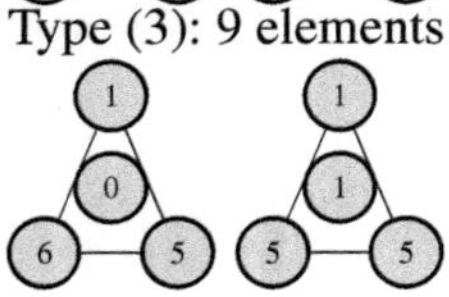

and symmetries.

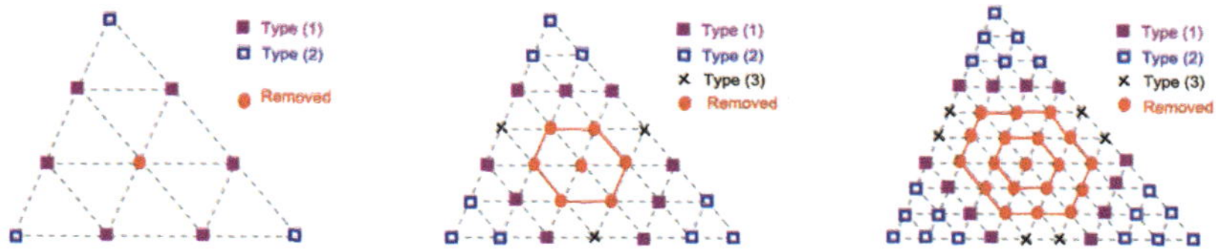

Fig. 2 Domain points of the Bernstein polynomials used in the CTS basis, left: $r = 1, d = 3$, middle: $r = 2, d = 6$, right: $r = 3, d = 9$. "Removed" means that the respective Bernstein polynomials are of none of the Types (1), (2) or (3)

Example 4 For $r = 4,\ d = 12,\ \mu = 9,\ \dim \mathbb{S}^4_{12}(\triangle) = 127$, the elements look as follows:

1. Type (1): 79 elements

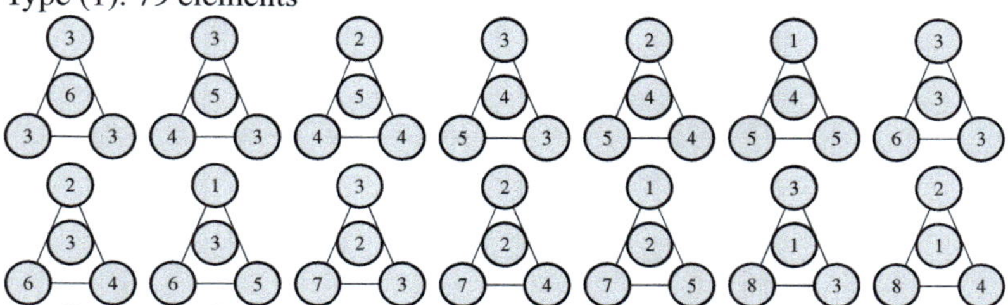

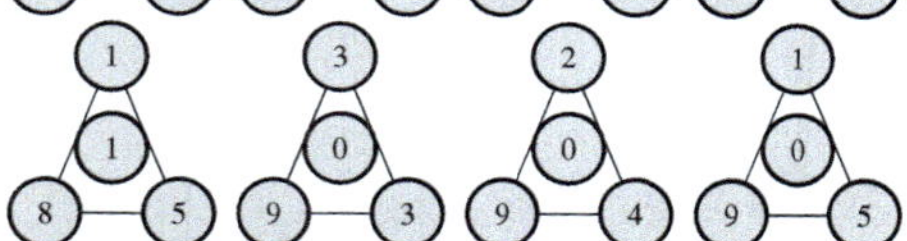

and symmetries,

2. Type (2): 30 elements

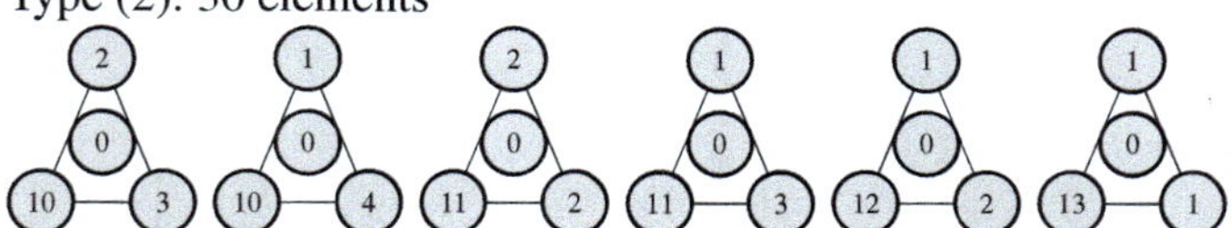

and symmetries,

3. Type (3): 18 elements

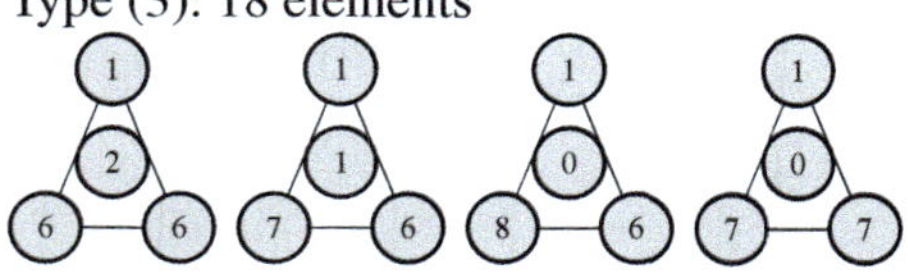

and symmetries.

Remark 2 Some Bernstein polynomials are **not** of any of the three types and thus are not in the basis. A Bernstein polynomial in the basis can be of Type (1), (2) or (3), see Fig. 2 and the previous examples.

One central result of this paper, stated in Theorem 1, is that Σ^r_{3r} is in fact a *basis* for $\mathbb{S}^r_{3r}(\triangle)$. The proof of this fact will consist of showing that Σ^r_{3r} is a subset the space $\mathbb{S}^r_{3r}(\triangle)$ containing $\dim(\mathbb{S}^r_{3r}(\triangle))$ linearly independent elements.

To count the number of elements of Σ^r_{3r}, we start with some bounds of ℓ with respect to the different types of functions in Σ^r_{3r}.

Lemma 2 *For Type (1) we have $\ell \le d - 3r/2$ while for Type (3) $\ell \le d - 2r - 2$ holds.*

Proof. For a Type (1) element, assume that $i + \ell = \mu$, then $j + \ell \le \mu$ and $k + \ell \le \mu$, hence $i + j + k + 3\ell \le 3\mu$. Since $i + j + k + \ell = d + 3$ and $\mu = d - r + 1$, we thus deduce that $d + 3 + 2\ell \le 3d + 3 - 3r$ or $\ell \le d - 3r/2$.

For Type (3), we assume that $k = 1$, hence $i + \ell \le \mu - 1$ and $j + \ell \le \mu - 1$, then $i + j + k + 2\ell \le 2\mu - 1$. Since $i + j + k + \ell = d + 3$ and $\mu = d - r + 1$, it follows that $d + 3 + \ell \le 2d - 2r + 2 - 1$ or $\ell \le d - 2r - 2$. □

Next we prove that the sum of the numbers of elements of the three types is exactly the dimension of $\dim \mathbb{S}^r_{3r}(\triangle)$.

Proposition 2 *For $r \ge 1$ and $d = 3r$, we have the following table according to the parity of r.*

r	$2s,\ s > 0$	$2s+1,\ s \ge 0$
d	$6s$	$6s+3$
$\dim \mathbb{S}^r_d(\triangle)$	$27s^2 + 9s + 1$	$3(9s^2 + 12s + 4)$
$\#Type(1)$	$3s(5s+3) + 1$	$3(s+1)(5s+3)$
$\#Type(2)$	$3s(2s+1)$	$3(s+1)(2s+1)$
$\#Type(3)$	$3s(2s-1)$	$3s(2s+1)$

(8)

Hence,

$$\#Type(1) + \#Type(2) + \#Type(3) = \dim \mathbb{S}^r_{3r}(\triangle), \quad r \ge 1. \tag{9}$$

Proof. We begin with r of *even* parity, i.e., $r = 2s$, $d = 3r = 6s$, $\mu = 4s + 1$, $i + j + k + \ell = 6s + 3$ and count the basis elements of different types.

1. Type (1): According to Lemma 2 we have $\ell \le d - 3r/2 = 3s$ and a generic element of Type (1) is of the form

$$\triangle[i = s + p + 1, j, k = 2s + 2 - j; 3s - p], \quad 1 \le j \le i,\ 1 \le k \le i,$$

   from which it follows that $s - p + 1 \le j \le 2s + 1$ and $p = 0, \dots, 3s$. We count the number of elements with respect to p:

   a. For $p = 0$: 1 element.
   The only possible choice is $\triangle[s + 1, s + 1, s + 1; 3s]$.
   b. For $1 \le p \le s$: $6p$ elements.
   The elements $\triangle[i, j, k; 3s - p]$ are of the form

$$\triangle[s + p + 1, s - p + 1 + q, s + p + 1 - q; 3s - p], \quad q = 0, \dots, 2p - 1, \tag{10}$$

   i.e., $i = s + p + 1$, $j = s - p + 1 + q$, $s + p + 1 - q$, and with the permutations $[i, j, k]$, $[j, k, i]$ and $[k, j, i]$, which gives 3 elements for any q in (10). We notice that in two cases we obtain the same three elements up to the permutation, namely, for $q = 0$ the values $i = k = s + p + 1$, $j = s - p + 1$, and for $q = 2p$ the values $i = j = s + p + 1, k = s - p + 1$. In the other cases, $0 < q < 2p$, we have $j = s - p + 1 + q < i = s + p + 1$ and $k = s + p + 1 - q < i = s + p + 1$, so that the elements are different.

c. For $s + 1 \le p \le 3s$: $6s + 3$ elements consisting of

$$\triangle[s + p + 1, q + 1, 2s - q + 1; 3s - p], \quad q = 0, \ldots, 2s.$$

and again the permutations $[i, j, k], [j, k, i], [k, j, i]$.

Hence the total number of elements of Type (1) is

$$1 + \sum_{p=1}^{s} 6p + \sum_{p=s+1}^{3s} (6s + 3) = 1 + 3s(s + 1) + 2s(6s + 3) = 1 + 3s(5s + 3)$$

as listed in (8).

2. Type (2) requires $\ell = 0$ by definition and the generic elements of the form

$$\triangle[i = 4s + 2 + p, j, k = s + 1 - j - p; 0], \quad 1 \le j \le i, \ i \le k \le i,$$

with the additional constraint $i \ge \mu + 1 = 4s + 2$ that leads to $1 \le j \le 2s - p$, $p = 0, \ldots, 2s - 1$. This gives $3(2s - p)$ elements

$$\triangle[4s + 2 + p, q + 1, 2s - p - q; 0], \quad q = 0, \ldots, 2s - p - 1,$$

and the respective permutations so that the total number of elements of Type (2) is

$$\sum_{p=1}^{2s-1} 3(2s - p) = \sum_{q=1}^{2s} q = 3s(2s + 1)$$

in this case.

3. Type (3): up to symmetry we can assume that $k = 1$ and, by Lemma 2 that $\ell \le d - 2r - 2 = 2s - 2$. The generic element is of the form

$$\begin{gathered}\triangle[i, j = 4s + 4 + p - i, k = 1; \ell = 2s - 2 - p],\\ i + \ell \le \mu - 1 = 4s, \ j + \ell \le 4s, \ p = 0, \ldots, 2s - 2,\end{gathered}$$

since $i + j + k + \ell = d + 3 = 6s + 3$. Hence, for $0 \le p \le 2s - 2$ we get $3(p + 1)$ elements

$$\triangle[2s + 2 + p - q, 2s + 2 + q, 1; 2s - 2 - p], \quad q = 0, \ldots, p,$$

and the permutations $[i, j, k], [j, k, i], [k, j, i]$, leading to a total of

$$\sum_{p=0}^{2s-2} 3(p+1) = 3s(2s-1)$$

elements of Type (3).

In the case of odd parity, i.e., $r = 2s+1$, $d = 3r = 6s+3$, $\mu = 4s+3$, $i+j+k+\ell = 6s+6$, we proceed in the same way and distinguish by types.

1. For Type (1) we have the bound $\ell \leq 3s+1$ and the generic element

$$\triangle[i = s+p+2, j, k = 2s+3-j; 3s+1-p],$$
$$1 \leq j \leq i,\ 1 \leq k \leq i,\ p = 0, \ldots, 3s+1.$$

Again, Type (1) request the distinction of several cases according to p.

a. For $0 \leq p \leq s$: $6p+3$ elements
The generic elements are

$$\triangle[s+p+2, s+1-p+q, s+2+p-q; 3s+1-p], \quad q = 0, \ldots, 2p,$$

and the permutations $[i, j, k], [j, k, i], [k, j, i]$. Again, we notice that we obtain the same three elements up to the permutations for $q = 0$, namely $i = k = s+p+2$, $j = s-p+1$, and for $q = 2p+1$, namely $i = j = s+p+2$, $k = s-p+1$, respectively. For $0 < q < 2p$, on the other hand, we have $j = s+1-p+q < i = s+p+2$ and $k = s+2+p-q < i = s+p+2$ so that all the elements are different again, just like in the case of even r.

b. For $s+1 \leq p \leq 3s+1$: $6(s+1)$ elements
based on the generic element

$$\triangle[s+p+2, q+1, 2s-q+2; 3s+1-p], \quad q = 0, \ldots, 2s+1,$$

and its permutations.

Therefore, the total number of elements of Type (1) is

$$\sum_{p=0}^{s}(6p+3) + \sum_{p=s+1}^{3s+1} 6(s+1) = 3(s+1)(5s+3).$$

2. Type (2) again requests $\ell = 0$ and leads to $3(2s+1-p)$ elements based on the generic element

$$\triangle[4s+4+p, q+1, 2s+1-p-q; 0], \quad q = 0, \ldots, 2s-p,$$

and its permutations, so that the total number of elements of Type (2) is

$$\sum_{p=0}^{2s} 3(2s+1-p) = 3(s+1)(2s+1).$$

3. For Type (3) we again assume that $k = 1$, note $\ell \le d - 2r - 2 = 2s - 1$ and obtain $3(p+1)$ elements from the generic element

$$\triangle[2s+3+p-q, 2s+3+q, 1; 2s-1-p], \quad q = 0, \dots, p$$

and its permutations totalling up to

$$\sum_{p=0}^{2s-1} 3(p+1) = 3s(2s+1)$$

elements of Type (3).

Having completed the table in (8) the claim (9) follows from summing up the columns of the table. □

Theorem 1 *Σ_{3r}^r is a basis of $\mathbb{S}_{3r}^r$.*

We prove Theorem 1 by verifying in Proposition 3 that the elements of Σ_{3r}^r are linearly independent. Since we already know from Proposition 2 that $\#\Sigma_{3r}^r = \dim \mathbb{S}_{3r}^r$, this indeed shows that they are a basis of the spline space. Consequently, Σ_{3r}^r spans the space of all simplex splines contained in $\mathbb{S}_{3r}^r$.

To prove linear independence, we need the following technical tool concerning particular derivatives of simplex splines.

Lemma 3 *For $i > 0$ we have that*

$$D_{p_1 - x}\triangle[i, j, k; \ell](x) = d\left(\triangle[i-1, j, k; \ell](x) - \triangle[i, j, k; \ell](x)\right), \tag{11}$$

while

$$\begin{aligned} D_{p_1 - x}\triangle[0, j, k; \ell](x) &= d\big(3\triangle[0, j, k; \ell-1](x) \\ &- \triangle[0, j-1, k; \ell](x) - \triangle[0, j, k-1; \ell](x) - \triangle[0, j, k; \ell](x)\big). \end{aligned} \tag{12}$$

We remark that in the special case $i + j + k + \ell = 3$, the degree is $d = 0$, and then the derivatives in (11) and (12) are equal to zero.

Proof. Write $x = \sum \alpha_j p_j$, with $\sum \alpha_j = 1$. Then the derivative formula and the recurrence yield that

$$\begin{aligned}
D_{p_1-x}\triangle[i,j,k;\ell](x) &= d\big((1-\alpha_1)\triangle[i-1,j,k;\ell](x)\\
&\quad -\alpha_2\triangle[i,j-1,k;\ell](x)-\alpha_3\triangle[i,j,k-1;\ell](x)\big)\\
&= d\,\triangle[i-1,j,k;\ell](x)-d\,\big(\alpha_1\triangle[i-1,j,k;\ell](x)\\
&\quad +\alpha_2\triangle[i,j-1,k;\ell](x)+\alpha_3\triangle[i,j,k-1;\ell](x)\big)\\
&= d\big(\triangle[i-1,j,k;\ell](x)-\triangle[i,j,k;\ell](x)\big),
\end{aligned}$$

which is (11). For the second identity we note that $p_T = \frac{1}{3}(p_1+p_2+p_3)$ implies $p_1 = 3p_T - p_2 - p_3$ and hence, writing $x = \alpha_T p_T + \alpha_2 p_2 + \alpha_3 p_3$ with $\sum \alpha_j = 1$, we obtain that

$$\begin{aligned}
D_{p_1-x}\triangle[0,j,k;\ell](x) &= d\big((3-\alpha_T)\triangle[0,j,k;\ell-1](x)\\
&\quad -(1+\alpha_2)\triangle[0,j-1,k;\ell](x)-(1+\alpha_3)\triangle[0,j,k-1;\ell](x)\big)
\end{aligned}$$

which can be recombined as above to yield (12). □

Since for $i, j, k, l \in \mathbb{N}_0$, and $x \in \mathscr{T}$ with barycentric coordinates $\beta_1, \beta_2, \beta_3$ with respect to p_1, p_2, p_3, we have

$$\begin{aligned}
\triangle[0,j+1,k+1;\ell+1](x) &= \frac{(j+k+\ell)!}{j!k!\ell!}(\beta_2-\beta_1)^j(\beta_3-\beta_1)^k(3\beta_1)^\ell, \quad x\in\mathscr{T}_1,\\
\triangle[i+1,0,k+1;\ell+1](x) &= \frac{(i+k+\ell)!}{i!k!\ell!}(\beta_1-\beta_2)^i(\beta_3-\beta_2)^k(3\beta_2)^\ell, \quad x\in\mathscr{T}_2,\\
\triangle[i+1,j+1,0;\ell+1](x) &= \frac{(i+j+\ell)!}{i!j!\ell!}(\beta_1-\beta_3)^i(\beta_2-\beta_3)^j(3\beta_3)^\ell, \quad x\in\mathscr{T}_3,\\
\triangle[i+1,j+1,k+1;0](x) &= \frac{(i+j+k)!}{i!j!k!}\beta_1^i\beta_2^j\beta_3^k = B_{ijk}^d(x), \quad x\in\mathscr{T},
\end{aligned} \tag{13}$$

we can also obtain (12) by applying the directional derivative

$$D_{p_1-x} = (1-\beta_1)\frac{\partial}{\partial\beta_1} - \beta_2\frac{\partial}{\partial\beta_2} - \beta_3\frac{\partial}{\partial\beta_3}$$

to the first line of (13).

Proposition 3 *The elements of Σ_{3r}^r are linearly independent.*

Proof. Denote by

$$\mathscr{I}^r := I(\Sigma_{3r}^r) = \big\{(i,j,k,\ell) : \triangle[i,j,k;\ell] \in \Sigma_{3r}^r\big\} \tag{14}$$

the set of all knot multiplicities of splines in Σ_{3r}^r. Now assume that there exist coefficients $a_{ijk\ell}$ such that

$$s := \sum_{(i,j,k,\ell)\in\mathscr{I}^r} a_{ijk\ell}\,\triangle[i,j,k;\ell] = 0.$$

On the boundary $\langle \boldsymbol{p}_2, \boldsymbol{p}_3 \rangle$ we have $\triangle[i, j, k; \ell] \neq 0$ if and only if $i = 1$ and $\ell = 0$ and the splines $\triangle[1, j, k; 0]$ reduce to univariate Bernstein polynomials that can with $\ell = 0$, be classified as follows:

1. Type (1): $\triangle[1, \mu, r+1; 0]$,
2. Type (2): $\triangle[1, \mu+m, r+1-m; 0]$, $1 \leq m \leq r$,
3. Type (3): $\triangle[1, \mu-m, r+1+m; 0]$, $1 \leq m \leq \frac{\mu-r-1}{2}$,

together with their symmetric elements where j and k are interchanged. Recall that if these symmetries coincide they are considered as only one element in Σ^r_{3r}, hence they are only counted once. Therefore, the Bernstein polynomials from the above list are linearly independent within the same type as the above construction prohibits repetitions. Between types, they must be linearly independent since no index combination can appear in two different types: the maximal multiplicity is $= \mu$ in Type (1), $> \mu$ in Type (2) and $< \mu$ in Type (3). Therefore, restricting s to the boundary $\langle \boldsymbol{p}_2, \boldsymbol{p}_3 \rangle$, it follows that $a_{ijk\ell} = 0$ for $(i, j, k, \ell) = (1, j, k, 0)$. Considering the other boundaries of $\mathscr{T}$, we can thus conclude that $a_{ijk\ell} = 0$ whenever $\min\{i, j, k\} = 1$ and $\ell = 0$.

Starting from this observation, we prove by induction on $m = 1, 2, \ldots$ that

$$a_{ijk\ell} = 0, \quad \min\{i, j, k\} + \ell = m, \tag{15}$$

where the case $m = 1$ has been treated in the first part of this proof. We will treat the case $m = 2$ explicitly as the general procedure will become clear by then. We assume that i is the minimal value, consider the identity

$$0 = D_{\boldsymbol{p}_1 - \boldsymbol{x}} s(x), \quad \boldsymbol{x} \in \langle \boldsymbol{p}_2, \boldsymbol{p}_3 \rangle,$$

and find by Lemma 3 that there are only two types of splines which are nonzero on the boundary. The first is

$$D_{\boldsymbol{p}_1 - \boldsymbol{x}} \triangle[2, j, k; 0](\boldsymbol{x}) = \triangle[1, j, k; 0](\boldsymbol{x}) - \triangle[2, j, k; 0](\boldsymbol{x}),$$

which coincides with $\triangle[1, j, k; 0](\boldsymbol{x})$ on $\langle \boldsymbol{p}_2, \boldsymbol{p}_3 \rangle$. The second is

$$D_{\boldsymbol{p}_1 - \boldsymbol{x}} \triangle[1, j, k; 1](\boldsymbol{x}) = \triangle[0, j, k; 1](\boldsymbol{x}) - \triangle[1, j, k; 1](\boldsymbol{x}),$$

which by (13) coincides with $\triangle[0, j, k; 1] = \triangle[1, j, k; 0]$ on $\langle \boldsymbol{p}_2, \boldsymbol{p}_3 \rangle$. Again, these splines are linearly independent within the types by construction and across the types by the different values of the maximal multiplicity, and (15) for $m = 2$ follows by considering all three faces of $\mathscr{T}$ by a symmetry argument.

The general induction step proceeds in exactly this way by assuming that $i = \min\{i, j, k\}$ and $i + \ell = m + 1$ and applying (11) i times we find from (13) with $\boldsymbol{x} = (\beta_1, \beta_2, \beta_3)$ and $\ell = m + 1 - i$

$$D^i_{p_1-x}\triangle[i,j,k;\ell](x) = C\triangle[0,j,k;\ell](x) = \frac{(j+k+\ell)!}{j!k!\ell!}(\beta_2-\beta_1)^j(\beta_3-\beta_1)^k(3\beta_1)^\ell$$

for some constant C. Differentiating ℓ times with respect to β_1 and setting $\beta_1 = 0$ we find

$$\frac{\partial^\ell}{\partial\beta_1^\ell}D^i_{p_1-x}\triangle[i,j,k;\ell](x) = C\beta_2^j\beta_3^k,$$

where again C is some nonzero constant, and where the admissible values for j, k lead to linearly independent polynomials on the boundary. This completes the proof of Proposition 3. □

Example 5 We illustrate the elimination procedure of Proposition 3 for the case $r = 3$.

The elimination process starts by considering the Bernstein polynomials

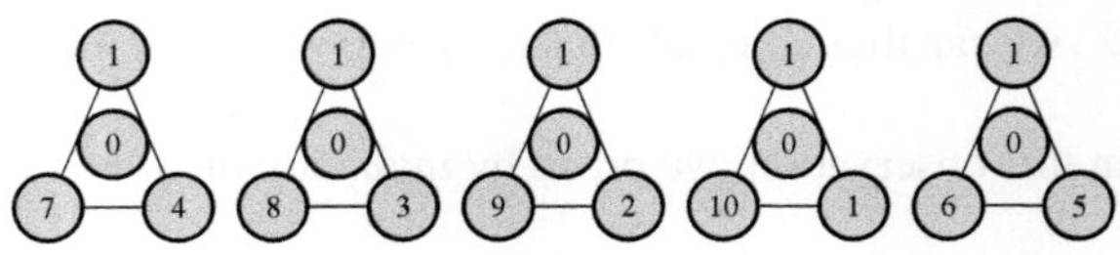

on the "lower" edge of the triangle where alle other simplex splines from our list vanish. Note that the three types are then distinguished by whether the maximum equals $\mu = 7$ (the element on the left), exceeds this values (the three elements in the middle) or is strictly smaller (the element on the right). Therefore, the coefficients of these splines and all their symmetries have to vanish which deals with the first "ring" of coefficients on the boundary.

Applying the differential operator D_{p_1-x} once, gives us the Bernstein polynomials

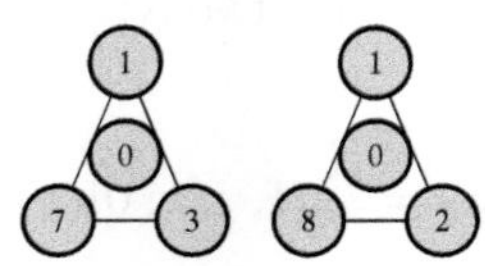

of Type (1) and (2). In addition, we get the splines

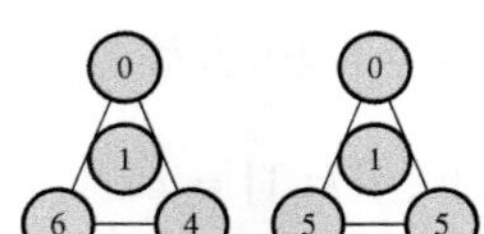

which also reduce to nonvanishing univariate Bernstein polynomials on the boundary. This finishes off all elements with $\min\{i, j, k\} + \ell = 2$.

The application of $D^2_{p_1-x}$ gives us

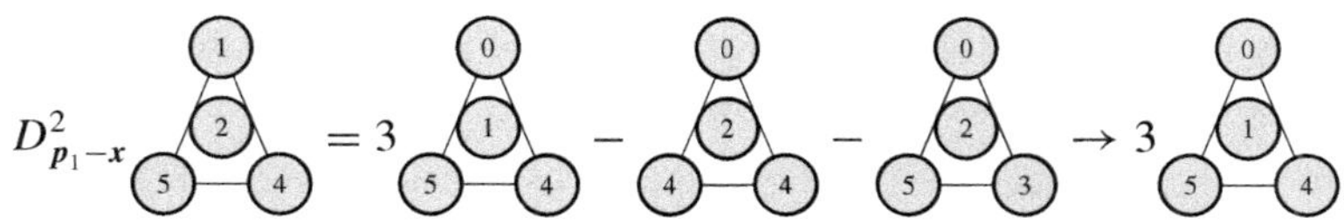

since the two Bernstein polynomias which are subtracted vanish on the boundary.

The other nonvanishing element is 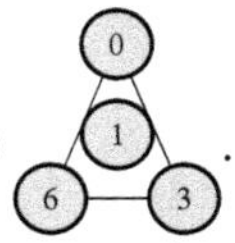 .

Now it should be clear how the process is completed: $D^3_{p_1-x}$ extracts the nonzero polynomials

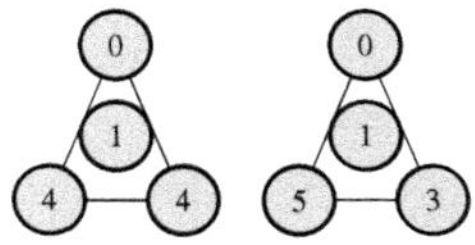

together with their symmetries and applying $D^4_{p_1-x}$ and $D^5_{p_1-x}$ we get

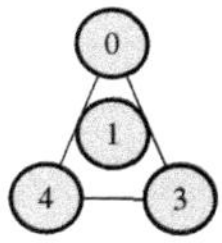 and 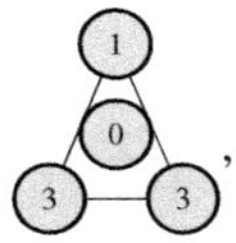 ,

respectively. Observe that in each step of the process always Bernstein polynomials of the same fixed degree are considered on the boundary.

4 Marsden-Like Identity, Domain Points, Stability and $\mathscr{C}^r$-Connection

A Marsden-like identity allows us to derive *explicit* formulas for the representation of polynomials of degree up to $3r$ in the simplex spline basis of $\mathbb{S}^r_{3r}(\triangle)$. The basis Σ^r_{3r} does not provide a partition of unity (for $r = 2s$). Hence, instead of having redundancy or a null component in the sums, we remove one further element according to the following definition.

Definition 3 Depending on the parity of r, we define the following sets and spaces:

1. If $r = 2s + 1$ with $s \geq 0$, then

$$\bar{\Sigma}^r_{3r} := \Sigma^r_{3r}, \quad \bar{\mathbb{S}}^r_{3r}(\triangle) = \mathbb{S}^r_{3r}(\triangle).$$

2. If $r = 2s$ then

$$\bar{\Sigma}^r_{3r} := \Sigma^r_{3r} \setminus \{\triangle[s+1, s+1, s+1; 3s]\}, \quad \bar{\mathbb{S}}^r_{3r}(\triangle) := \operatorname{span} \bar{\Sigma}^r_{3r}.$$

In analogy with (14), the set of indices (i, j, k, ℓ) of the basis $\bar{\Sigma}_{3r}^r$ will be written as $\bar{\mathscr{I}}^r$. Moreover, for the Bernstein polynomials in $\bar{\Sigma}_{3r}^r$ (cf. Fig. 2), we define the index sets

$$\begin{aligned}\mathscr{I}^r_{\text{removed}} &:= \{(i,j,k) : i+j+k = 3r,\ \max\{i,j,k\} \le 2r-1,\ \min\{i,j,k\} \ge 1\},\\ \mathscr{I}^r_{\text{Bernstein}} &:= \{(i,j,k) : i,j,k \in \mathbb{N}_0,\ i+j+k = 3r\} \setminus \mathscr{I}^r_{\text{removed}}.\end{aligned} \tag{16}$$

It immediately follows that

$$\#\mathscr{I}^r_{\text{Bernstein}} = \frac{3}{2}r(r+5), \quad \#\mathscr{I}^r_{\text{removed}} = 3r(r-1)+1,$$
$$\#\mathscr{I}^r_{\text{Bernstein}} + \#\mathscr{I}^r_{\text{removed}} = \binom{3r+2}{2} = \dim \mathbb{P}_{3r}.$$

We now consider the cases $r = 2, 3$ in more detail. We prove, using knot insertion, a Marsden-like identity for $r = 2$, and state it in a form valid for any $r \ge 1$. It is verified symbolically for $r \le 6$.

4.1 The $\mathscr{C}^2$ Elements, $\bar{\Sigma}_6^2$

According to Definition 3, we use only 36 elements of $\bar{\Sigma}_6^2$ with indices from the set $\bar{\mathscr{I}}^2$, which corresponds to removing $\triangle_{2223}$ from Σ_6^2. In Definition 4 we list these elements, normalized to ensure partition of unity. The set $\bar{\Sigma}_6^2$ consists of Bernstein polynomials as depicted in Fig. 2, and 15 other simplex splines with at least one knot at the barycenter of the triangle.

Definition 4 (Functions $\bar{\Sigma}_6^2$) The functions $\bar{\Sigma}_6^2$ consist of the 21 Bernstein polynomials

$$S^6_{i+1,j+1,k+1,0} := \triangle_{i+1,j+1,k+1;0} = B^6_{ijk}, \quad (i,j,k) \in \mathscr{I}^2_{\text{Bernstein}},$$

where $\mathscr{I}^2_{\text{Bernstein}}$ is given by (16), and the additional simplex splines

$$\begin{aligned}
S^6_{22} &:= S^6_{4311} := \frac{1}{3}\triangle_{4311}, & S^6_{23} &:= S^6_{3411} := \frac{1}{3}\triangle_{3411}, & S^6_{24} &:= S^6_{1431} := \frac{1}{3}\triangle_{1431},\\
S^6_{25} &:= S^6_{1341} := \frac{1}{3}\triangle_{1341}, & S^6_{26} &:= S^6_{3141} := \frac{1}{3}\triangle_{3141}, & S^6_{27} &:= S^6_{4131} := \frac{1}{3}\triangle_{4131},\\
S^6_{28} &:= S^6_{4221} := \frac{2}{3}\triangle_{4221}, & S^6_{30} &:= S^6_{2421} := \frac{2}{3}\triangle_{2421}, & S^6_{32} &:= S^6_{2241} := \frac{2}{3}\triangle_{2241},\\
S^6_{29} &:= S^6_{3312} := \frac{1}{3}\triangle_{3312}, & S^6_{31} &:= S^6_{1332} := \frac{1}{3}\triangle_{1332}, & S^6_{33} &:= S^6_{3132} := \frac{1}{3}\triangle_{3132},\\
S^6_{34} &:= S^6_{3222} := \frac{2}{3}\triangle_{3222}, & S^6_{35} &:= S^6_{2322} := \frac{2}{3}\triangle_{2322}, & S^6_{36} &:= S^6_{2232} := \frac{2}{3}\triangle_{2232}.
\end{aligned} \tag{17}$$

For the numbering of the functions $S_1^6, \ldots, S_{21}^6$ see Figs. 3 and 4 below. In the following we give explicit formulas for the simplex splines in (17). These will be used below for obtaining $\mathscr{C}^2$-continuity across an edge between two adjacent triangles in a global triangulation.

On $\mathscr{T}_1$ we have

$$\begin{pmatrix}
22 \ \{4,3,1,1\} & \beta_1^6 - 6\beta_2\beta_1^5 + 15\beta_2^2\beta_1^4 \\
23 \ \{3,4,1,1\} & -\beta_1^6 + 6\beta_2\beta_1^5 - 15\beta_2^2\beta_1^4 + 20\beta_2^3\beta_1^3 \\
24 \ \{1,4,3,1\} & S_{24}^6(1) \\
25 \ \{1,3,4,1\} & S_{25}^6(1) \\
26 \ \{3,1,4,1\} & -\beta_1^6 + 6\beta_3\beta_1^5 - 15\beta_3^2\beta_1^4 + 20\beta_3^3\beta_1^3 \\
27 \ \{4,1,3,1\} & \beta_1^6 - 6\beta_3\beta_1^5 + 15\beta_3^2\beta_1^4 \\
28 \ \{4,2,2,1\} & 4\beta_1^6 - 12\beta_2\beta_1^5 - 12\beta_3\beta_1^5 + 60\beta_2\beta_3\beta_1^4 \\
29 \ \{3,3,1,2\} & 9\beta_1^6 - 36\beta_2\beta_1^5 + 45\beta_2^2\beta_1^4 \\
30 \ \{2,4,2,1\} & S_{30}^6(1) \\
31 \ \{1,3,3,2\} & S_{31}^6(1) \\
32 \ \{2,2,4,1\} & S_{32}^6(1) \\
33 \ \{3,1,3,2\} & 9\beta_1^6 - 36\beta_3\beta_1^5 + 45\beta_3^2\beta_1^4 \\
34 \ \{3,2,2,2\} & 36\beta_1^6 - 72\beta_2\beta_1^5 - 72\beta_3\beta_1^5 + 180\beta_2\beta_3\beta_1^4 \\
35 \ \{2,3,2,2\} & -72\beta_1^6 + 216\beta_2\beta_1^5 + 108\beta_3\beta_1^5 - 180\beta_2^2\beta_1^4 - 360\beta_2\beta_3\beta_1^4 + 360\beta_2^2\beta_3\beta_1^3 \\
36 \ \{2,2,3,2\} & -72\beta_1^6 + 108\beta_2\beta_1^5 + 216\beta_3\beta_1^5 - 180\beta_3^2\beta_1^4 - 360\beta_2\beta_3\beta_1^4 + 360\beta_2\beta_3^2\beta_1^3
\end{pmatrix}$$

where the first entry denotes the position of the spline with respect to aforementioned ordering, the second the multiplicity of the knots and the third one the explicit expression on $\mathscr{T}_1$, using

$$\begin{aligned}
S_{24}^6(1) &= -10\beta_1^6 + 36\beta_2\beta_1^5 + 24\beta_3\beta_1^5 - 45\beta_2^2\beta_1^4 - 15\beta_3^2\beta_1^4 - 90\beta_2\beta_3\beta_1^4 + 20\beta_2^3\beta_1^3 \\
&\quad + 60\beta_2\beta_3^2\beta_1^3 + 120\beta_2^2\beta_3\beta_1^3 - 90\beta_2^2\beta_3^2\beta_1^2 - 60\beta_2^3\beta_3\beta_1^2 + 60\beta_2^3\beta_3^2\beta_1, \\
S_{25}^6(1) &= -10\beta_1^6 + 24\beta_2\beta_1^5 + 36\beta_3\beta_1^5 - 15\beta_2^2\beta_1^4 - 45\beta_3^2\beta_1^4 - 90\beta_2\beta_3\beta_1^4 + 20\beta_3^3\beta_1^3 \\
&\quad + 120\beta_2\beta_3^2\beta_1^3 + 60\beta_2^2\beta_3\beta_1^3 - 60\beta_2\beta_3^3\beta_1^2 - 90\beta_2^2\beta_3^2\beta_1^2 + 60\beta_2^2\beta_3^3\beta_1, \\
S_{30}^6(1) &= 8\beta_1^6 - 36\beta_2\beta_1^5 - 12\beta_3\beta_1^5 + 60\beta_2^2\beta_1^4 + 60\beta_2\beta_3\beta_1^4 - 40\beta_2^3\beta_1^3 - 120\beta_2^2\beta_3\beta_1^3 + 120\beta_2^3\beta_3\beta_1^2, \\
S_{31}^6(1) &= 90\beta_1^6 - 216\beta_2\beta_1^5 - 216\beta_3\beta_1^5 + 135\beta_2^2\beta_1^4 + 135\beta_3^2\beta_1^4 + 540\beta_2\beta_3\beta_1^4 \\
&\quad - 360\beta_2\beta_3^2\beta_1^3 - 360\beta_2^2\beta_3\beta_1^3 + 270\beta_2^2\beta_3^2\beta_1^2, \\
S_{32}^6(1) &= 8\beta_1^6 - 12\beta_2\beta_1^5 - 36\beta_3\beta_1^5 + 60\beta_3^2\beta_1^4 + 60\beta_2\beta_3\beta_1^4 - 40\beta_3^3\beta_1^3 - 120\beta_2\beta_3^2\beta_1^3 + 120\beta_2\beta_3^3\beta_1^2.
\end{aligned}$$

Here, $S_k^6(1)$ is used to indicate the restriction of S_k^6 to the triangle $\mathscr{T}_1$. On $\mathscr{T}_2$ we have, in the same fashion,

$$\begin{pmatrix}
22\ \{4,3,1,1\} & -\beta_2^6 + 6\beta_1\beta_2^5 - 15\beta_1^2\beta_2^4 + 20\beta_1^3\beta_2^3 \\
23\ \{3,4,1,1\} & \beta_2^6 - 6\beta_1\beta_2^5 + 15\beta_1^2\beta_2^4 \\
24\ \{1,4,3,1\} & \beta_2^6 - 6\beta_3\beta_2^5 + 15\beta_3^2\beta_2^4 \\
25\ \{1,3,4,1\} & -\beta_2^6 + 6\beta_3\beta_2^5 - 15\beta_3^2\beta_2^4 + 20\beta_3^3\beta_2^3 \\
26\ \{3,1,4,1\} & S_{26}^6(2) \\
27\ \{4,1,3,1\} & S_{27}^6(2) \\
28\ \{4,2,2,1\} & S_{28}^6(2) \\
29\ \{3,3,1,2\} & 9\beta_2^6 - 36\beta_1\beta_2^5 + 45\beta_1^2\beta_2^4 \\
30\ \{2,4,2,1\} & 4\beta_2^6 - 12\beta_1\beta_2^5 - 12\beta_3\beta_2^5 + 60\beta_1\beta_3\beta_2^4 \\
31\ \{1,3,3,2\} & 9\beta_2^6 - 36\beta_3\beta_2^5 + 45\beta_3^2\beta_2^4 \\
32\ \{2,2,4,1\} & S_{32}^6(2) \\
33\ \{3,1,3,2\} & S_{33}^6(2) \\
34\ \{3,2,2,2\} & -72\beta_2^6 + 216\beta_1\beta_2^5 + 108\beta_3\beta_2^5 - 180\beta_1^2\beta_2^4 - 360\beta_1\beta_3\beta_2^4 + 360\beta_1^2\beta_3\beta_2^3 \\
35\ \{2,3,2,2\} & 36\beta_2^6 - 72\beta_1\beta_2^5 - 72\beta_3\beta_2^5 + 180\beta_1\beta_3\beta_2^4 \\
36\ \{2,2,3,2\} & -72\beta_2^6 + 108\beta_1\beta_2^5 + 216\beta_3\beta_2^5 - 180\beta_3^2\beta_2^4 - 360\beta_1\beta_3\beta_2^4 + 360\beta_1\beta_3^2\beta_2^3
\end{pmatrix}$$

where

$$\begin{aligned}
S_{26}^6(2) &= -10\beta_2^6 + 24\beta_1\beta_2^5 + 36\beta_3\beta_2^5 - 15\beta_1^2\beta_2^4 - 45\beta_3^2\beta_2^4 - 90\beta_1\beta_3\beta_2^4 + 20\beta_3^3\beta_2^3 \\
&\quad + 120\beta_1\beta_3^2\beta_2^3 + 60\beta_1^2\beta_3\beta_2^3 - 60\beta_1\beta_3^3\beta_2^2 - 90\beta_1^2\beta_3^2\beta_2^2 + 60\beta_1^2\beta_3^3\beta_2, \\
S_{27}^6(2) &= -10\beta_2^6 + 36\beta_1\beta_2^5 + 24\beta_3\beta_2^5 - 45\beta_1^2\beta_2^4 - 15\beta_3^2\beta_2^4 - 90\beta_1\beta_3\beta_2^4 + 20\beta_1^3\beta_2^3 \\
&\quad + 60\beta_1\beta_3^2\beta_2^3 + 120\beta_1^2\beta_3\beta_2^3 - 90\beta_1^2\beta_3^2\beta_2^2 - 60\beta_1^3\beta_3\beta_2^2 + 60\beta_1^3\beta_3^2\beta_2, \\
S_{28}^6(2) &= 8\beta_2^6 - 36\beta_1\beta_2^5 - 12\beta_3\beta_2^5 + 60\beta_1^2\beta_2^4 + 60\beta_1\beta_3\beta_2^4 - 40\beta_1^3\beta_2^3 - 120\beta_1^2\beta_3\beta_2^3 + 120\beta_1^3\beta_3\beta_2^2, \\
S_{32}^6(2) &= 8\beta_2^6 - 12\beta_1\beta_2^5 - 36\beta_3\beta_2^5 + 60\beta_3^2\beta_2^4 + 60\beta_1\beta_3\beta_2^4 - 40\beta_3^3\beta_2^3 - 120\beta_1\beta_3^2\beta_2^3 + 120\beta_1\beta_3^3\beta_2^2, \\
S_{33}^6(2) &= 90\beta_2^6 - 216\beta_1\beta_2^5 - 216\beta_3\beta_2^5 + 135\beta_1^2\beta_2^4 + 135\beta_3^2\beta_2^4 + 540\beta_1\beta_3\beta_2^4 \\
&\quad - 360\beta_1\beta_3^2\beta_2^3 - 360\beta_1^2\beta_3\beta_2^3 + 270\beta_1^2\beta_3^2\beta_2^2,
\end{aligned}$$

and finally on $\mathscr{T}_3$

$$\begin{pmatrix}
22\ \{4,3,1,1\} & S_{22}^6(3) \\
23\ \{3,4,1,1\} & S_{23}^6(3) \\
24\ \{1,4,3,1\} & -\beta_3^6 + 6\beta_2\beta_3^5 - 15\beta_2^2\beta_3^4 + 20\beta_2^3\beta_3^3 \\
25\ \{1,3,4,1\} & \beta_3^6 - 6\beta_2\beta_3^5 + 15\beta_2^2\beta_3^4 \\
26\ \{3,1,4,1\} & \beta_3^6 - 6\beta_1\beta_3^5 + 15\beta_1^2\beta_3^4 \\
27\ \{4,1,3,1\} & -\beta_3^6 + 6\beta_1\beta_3^5 - 15\beta_1^2\beta_3^4 + 20\beta_1^3\beta_3^3 \\
28\ \{4,2,2,1\} & S_{28}^6(3) \\
29\ \{3,3,1,2\} & S_{29}^6(3) \\
30\ \{2,4,2,1\} & S_{30}^6(3) \\
31\ \{1,3,3,2\} & 9\beta_3^6 - 36\beta_2\beta_3^5 + 45\beta_2^2\beta_3^4 \\
32\ \{2,2,4,1\} & 4\beta_3^6 - 12\beta_1\beta_3^5 - 12\beta_2\beta_3^5 + 60\beta_1\beta_2\beta_3^4 \\
33\ \{3,1,3,2\} & 9\beta_3^6 - 36\beta_1\beta_3^5 + 45\beta_1^2\beta_3^4 \\
34\ \{3,2,2,2\} & -72\beta_3^6 + 216\beta_1\beta_3^5 + 108\beta_2\beta_3^5 - 180\beta_1^2\beta_3^4 - 360\beta_1\beta_2\beta_3^4 + 360\beta_1^2\beta_2\beta_3^3 \\
35\ \{2,3,2,2\} & -72\beta_3^6 + 108\beta_1\beta_3^5 + 216\beta_2\beta_3^5 - 180\beta_2^2\beta_3^4 - 360\beta_1\beta_2\beta_3^4 + 360\beta_1\beta_2^2\beta_3^3 \\
36\ \{2,2,3,2\} & 36\beta_3^6 - 72\beta_1\beta_3^5 - 72\beta_2\beta_3^5 + 180\beta_1\beta_2\beta_3^4
\end{pmatrix} \tag{18}$$

with

$$\begin{aligned}
S_{22}^6(3) &= S_{22}^6|_{\mathcal{T}_3} = -10\beta_3^6 + 36\beta_1\beta_3^5 + 24\beta_2\beta_3^5 - 45\beta_1^2\beta_3^4 - 15\beta_2^2\beta_3^4 - 90\beta_1\beta_2\beta_3^4 + 20\beta_1^3\beta_3^3 \\
&\quad + 60\beta_1\beta_2^2\beta_3^3 + 120\beta_1^2\beta_2\beta_3^3 - 90\beta_1^2\beta_2^2\beta_3^2 - 60\beta_1^3\beta_2\beta_3^2 + 60\beta_1^3\beta_2^2\beta_3, \\
S_{23}^6(3) &= S_{23}^6|_{\mathcal{T}_3} = -10\beta_3^6 + 24\beta_1\beta_3^5 + 36\beta_2\beta_3^5 - 15\beta_1^2\beta_3^4 - 45\beta_2^2\beta_3^4 - 90\beta_1\beta_2\beta_3^4 + 20\beta_2^3\beta_3^3 \\
&\quad + 120\beta_1\beta_2^2\beta_3^3 + 60\beta_1^2\beta_2\beta_3^3 - 60\beta_1\beta_2^3\beta_3^2 - 90\beta_1^2\beta_2^2\beta_3^2 + 60\beta_1^2\beta_2^3\beta_3 = 60\beta_1^2\beta_2^3\beta_3 + O(\beta_3^2), \\
S_{28}^6(3) &= 8\beta_3^6 - 36\beta_1\beta_3^5 - 12\beta_2\beta_3^5 + 60\beta_1^2\beta_3^4 + 60\beta_1\beta_2\beta_3^4 - 40\beta_1^3\beta_3^3 - 120\beta_1^2\beta_2\beta_3^3 + 120\beta_1^3\beta_2\beta_3^2, \\
S_{29}^6(3) &= S_{29}^6|_{\mathcal{T}_3} = 90\beta_3^6 - 216\beta_1\beta_3^5 - 216\beta_2\beta_3^5 + 135\beta_1^2\beta_3^4 + 135\beta_2^2\beta_3^4 + 540\beta_1\beta_2\beta_3^4 \\
&\quad - 360\beta_1\beta_2^2\beta_3^3 - 360\beta_1^2\beta_2\beta_3^3 + 270\beta_1^2\beta_2^2\beta_3^2, \\
S_{30}^6(3) &= 8\beta_3^6 - 12\beta_1\beta_3^5 - 36\beta_2\beta_3^5 + 60\beta_2^2\beta_3^4 + 60\beta_1\beta_2\beta_3^4 - 40\beta_2^3\beta_3^3 - 120\beta_1\beta_2^2\beta_3^3 + 120\beta_1\beta_2^3\beta_3^2.
\end{aligned} \tag{19}$$

We next formulate and prove the announced Marsden identity.

Theorem 2 (Barycentric Marsden-Like Identity for $d = 6$) *For* $\boldsymbol{u} := [u_1, u_2, u_3]^T \in \mathbb{R}^3, \boldsymbol{\beta} := [\beta_1, \beta_2, \beta_3]^T \in \mathbb{R}^3$, *with* $\beta_i \geq 0,\ i = 1, 2, 3$ *and* $\beta_1 + \beta_2 + \beta_3 = 1$ *we have*

$$(\boldsymbol{u}^T\boldsymbol{\beta})^6 = \sum_{(i,j,k\ell)\in\bar{\mathcal{I}}^2} \rho_{ijk\ell}(\boldsymbol{u}) S_{ijk\ell}^6(\boldsymbol{\beta}),$$

where, with $\bar{u}_{m,n} := (u_m + u_n)/2$ *for* $m, n = 1, 2, 3$, *and* $\bar{u}_{123} := (u_1 + u_2 + u_3)/3$,

$$\rho_{i+1,j+1,k+1,0}(\boldsymbol{u}) = u_1^i u_2^j u_3^k, \quad (i, j, k) \in \mathcal{I}_{Bernstein}^2,$$

and

$$\begin{aligned}
&\rho_{4311}(\boldsymbol{u}) = u_1^3u_2^2u_3, && \rho_{4131}(\boldsymbol{u}) = u_1^3u_2u_3^2, && \rho_{3411}(\boldsymbol{u}) = u_1^2u_2^3u_3, \\
&\rho_{3141}(\boldsymbol{u}) = u_1^2u_2u_3^3, && \rho_{1431}(\boldsymbol{u}) = u_1u_2^3u_3^2, && \rho_{1341}(\boldsymbol{u}) = u_1u_2^2u_3^3, \\
&\rho_{4221}(\boldsymbol{u}) = u_1^3u_2u_3\bar{u}_{2,3}, && \rho_{2421}(\boldsymbol{u}) = u_1u_2^3u_3\bar{u}_{1,3}, && \rho_{2241}(\boldsymbol{u}) = u_1u_2u_3^3\bar{u}_{1,2} \\
&\rho_{3312}(\boldsymbol{u}) = u_1^2u_2^2u_3\bar{u}_{123}, && \rho_{3132}(\boldsymbol{u}) = u_1^2u_2u_3^2\bar{u}_{123}, && \rho_{1332}(\boldsymbol{u}) = u_1u_2^2u_3^2\bar{u}_{123}, \\
&\rho_{3222}(\boldsymbol{u}) = u_1^2u_2u_3\bar{u}_{2,3}\bar{u}_{123}, && \rho_{2322}(\boldsymbol{u}) = u_1u_2^2u_3\bar{u}_{1,3}\bar{u}_{123}, && \rho_{2232}(\boldsymbol{u}) = u_1u_2u_3^2\bar{u}_{1,2}\bar{u}_{123}.
\end{aligned} \tag{20}$$

Proof. The barycentric Marsden-like identity follows from the barycentric form (1) of the Marsden identity for Bernstein polynomials by expressing the removed Bernstein polynomials in terms of the elements in $\bar{\Sigma}_6^2$. Here are some details. For $(i, j, k) \in \mathcal{I}_{\text{removed}}^2$, where

$$\mathcal{I}_{\text{removed}}^2 := \{(1, 2, 3), (1, 3, 2), (2, 1, 3), (2, 3, 1), (3, 1, 2), (3, 2, 1), (2, 2, 2)\},$$

we insert knots at the barycenter using (7), (17), and find

$$
\begin{aligned}
B^6_{1,2,3} &= \triangle_{2340} = \big(\triangle_{1341} + \triangle_{2241} + \triangle_{2331}\big)/3 \\
&= \big(\triangle_{1341} + \triangle_{2241}\big)/3 + \big(\triangle_{1332} + \triangle_{2232} + \triangle_{2322}\big)/9, \\
B^6_{1,3,2} &= \big(\triangle_{1431} + \triangle_{2421}\big)/3 + \big(\triangle_{1332} + \triangle_{2232} + \triangle_{2322}\big)/9, \\
B^6_{2,1,3} &= \big(\triangle_{2241} + \triangle_{3141}\big)/3 + \big(\triangle_{3132} + \triangle_{2232} + \triangle_{3222}\big)/9, \\
B^6_{2,3,1} &= \big(\triangle_{2421} + \triangle_{3411}\big)/3 + \big(\triangle_{3312} + \triangle_{2322} + \triangle_{3222}\big)/9, \\
B^6_{3,1,2} &= \big(\triangle_{4131} + \triangle_{4221}\big)/3 + \big(\triangle_{3132} + \triangle_{2232} + \triangle_{3222}\big)/9, \\
B^6_{3,2,1} &= \big(\triangle_{4221} + \triangle_{4311}\big)/3 + \big(\triangle_{3312} + \triangle_{2322} + \triangle_{3222}\big)/9, \\
B^6_{2,2,2} &= \big(\triangle_{1332} + \triangle_{3132} + \triangle_{3312} + 2\triangle_{2232} + 2\triangle_{2322} + 2\triangle_{3222}\big)/9.
\end{aligned}
\tag{21}
$$

By (1)

$$
(\boldsymbol{u}^T\boldsymbol{\beta})^6 = \sum_{(i,j,k)\in\mathscr{I}^2_{\text{Bernstein}}} u_1^i u_2^j u_3^k B^6_{ijk}(\boldsymbol{\beta}) + \sum_{(i,j,k)\in\mathscr{I}^2_{\text{removed}}} u_1^i u_2^j u_3^k B^6_{ijk}(\boldsymbol{\beta}).
$$

For $(i,j,k) \in \mathscr{I}^2_{\text{Bernstein}}$ we have $B^6_{ijk}(\boldsymbol{\beta}) = S^6_{i+1,j+1,k+1,0}(\boldsymbol{\beta})$ and hence $\rho_{[i+1,j+1,k+1,0]} = u_1^i u_2^j u_3^k$. In the second sum we insert the expressions in (21) for B^6_{ijk}, and collect terms for each $\triangle_{ijk\ell}$ to obtain (20). We show this for three typical cases.

$$
\begin{aligned}
\sum_{(i,j,k)\in\mathscr{I}^2_{\text{removed}}} u_1^i u_2^j u_3^k B^6_{ijk}(\boldsymbol{\beta}) &= u_1u_2^2u_3^3\triangle_{1431}(\boldsymbol{\beta})/3 + \big(u_1u_2^2u_3^3 + u_1u_2^3u_3^2 + u_1^2u_2^2u_3^2\big)\triangle_{1332}(\boldsymbol{\beta})/9 \\
&+ \big(u_1u_2^2u_3^3 + u_1u_2^3u_3^2 + u_1^2u_2u_3^3 + u_1^3u_2u_3^2 + 2u_1^2u_2^2u_3^2\big)\triangle_{2232}(\boldsymbol{\beta})/9 + \cdots \\
&= u_1u_2^2u_3^3\triangle_{1431}(\boldsymbol{\beta})/3 + u_1u_2^2u_3^2\bar{u}_{123}\triangle_{1332}(\boldsymbol{\beta})/3 + 2u_1u_2u_3^2\bar{u}_{1,2}\bar{u}_{123}\triangle_{2232}(\boldsymbol{\beta})/3 + \cdots, \\
&= u_1u_2^2u_3^3 S^6_{1431}(\boldsymbol{\beta}) + u_1u_2^2u_3^2\bar{u}_{123}S^6_{1332}(\boldsymbol{\beta}) + u_1u_2u_3^2\bar{u}_{1,2}\bar{u}_{123}S^6_{2232}(\boldsymbol{\beta}) + \cdots,
\end{aligned}
$$

and (20) follows. □

Remark 3 Since $\rho_{ijk\ell}(1,1,1) = 1$ for any $ijk\ell$ it follows that

$$
\sum_{(i,j,k,\ell)\in\bar{\mathscr{I}}^2} S^6_{ijk\ell}(\boldsymbol{\beta}) = 1.
$$

Corollary 1 (Domain Points for $d = 6$) *The domain points $\boldsymbol{p}^*_{ijk\ell}$ in barycentric form, defined as the coefficients in the expansion*

$$
\boldsymbol{\beta} = \sum_{(i,j,k,\ell)\in\bar{\mathscr{I}}^2} \boldsymbol{p}^*_{ijk\ell} S^6_{ijk\ell}(\boldsymbol{\beta}).
$$

are given by

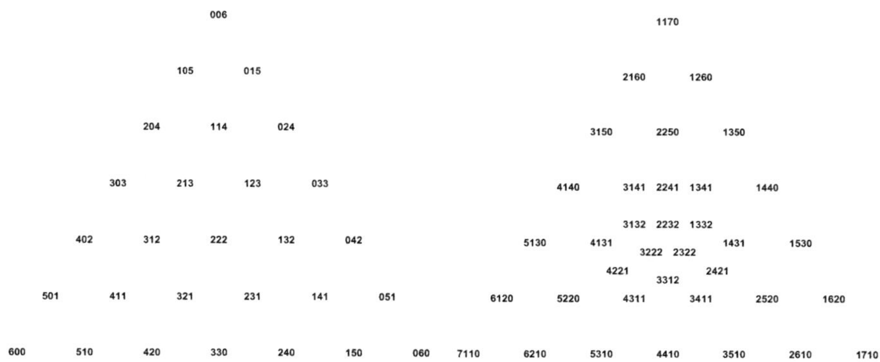

Fig. 3 Bernstein domain points on the left, Simplex splines domain points on the right

$$
\begin{array}{lll}
\boldsymbol{p}^*_{i+1,j+1,k+1,0} = (i,j,k)/6, & (i,j,k) \in \mathscr{I}^2_{Bernstein}, & \\
\boldsymbol{p}^*_{4311} = (3,2,1)/6, & \boldsymbol{p}^*_{4131} = (3,1,2)/6, & \boldsymbol{p}^*_{3411} = (2,3,1)/6, \\
\boldsymbol{p}^*_{3141} = (2,1,3)/6, & \boldsymbol{p}^*_{1431} = (1,3,2)/6, & \boldsymbol{p}^*_{1341} = (1,2,3)/6, \\
\boldsymbol{p}^*_{3312} = (7,7,4)/18, & \boldsymbol{p}^*_{3132} = (7,4,7)/18, & \boldsymbol{p}^*_{1332} = (4,7,7)/18, \\
\boldsymbol{p}^*_{4221} = (2,1,1)/4, & \boldsymbol{p}^*_{2421} = (1,2,1)/4, & \boldsymbol{p}^*_{2241} = (1,1,2)/4, \\
\boldsymbol{p}^*_{3222} = (14,11,11)/36, & \boldsymbol{p}^*_{2322} = (11,14,11)/36, & \boldsymbol{p}^*_{2232} = (11,11,14)/36.
\end{array} \tag{22}
$$

Proof. By the Marsden-like identity we have

$$
\beta_m = \sum_{(i,j,k,\ell)\in\bar{\mathscr{I}}^2} \frac{\partial}{\partial u_m} \rho_{ijk\ell}(1,1,1) S^6_{ijk\ell}(\boldsymbol{\beta}), \quad m = 1,2,3,
$$

and (22) follows after a straightforward calculation. □

The indices i, j, k and i, j, k, ℓ for the domain points are shown in Fig. 3.
Next, we define, the ∞-norm condition number of the $\bar{\Sigma}^r_{3r}$ bases for $\bar{\mathbb{S}}^r_{3r}(\triangle)$ by

$$
\kappa_{d,\infty}(\mathscr{T}) := \max_{c\neq 0} \frac{\|\boldsymbol{b}^T \boldsymbol{c}\|_{L_\infty(\mathscr{T})}}{\|\boldsymbol{c}\|_\infty} \max_{c\neq 0} \frac{\|\boldsymbol{c}\|_\infty}{\|\boldsymbol{b}^T \boldsymbol{c}\|_{L_\infty(\mathscr{T})}},
$$

where $d = 3r$ and $\boldsymbol{b}^T \boldsymbol{c} := \sum_{(i,j,k,\ell)\in\bar{\mathscr{I}}^r} c_{ijk\ell} S^d_{ijk\ell} \in \bar{\mathbb{S}}^r_d(\triangle)$. This number has a bound which is *independent* of the shape of the basis triangle $\mathscr{T}$.

Proposition 4 (Stability) *For any triangle $\mathscr{T}$ we have $\kappa_{6,\infty}(\mathscr{T}) < 1350$.*

Proof. Since the $S^6_{ijk\ell}$, $(i,j,k,\ell) \in \bar{\mathscr{I}}^2$ form a nonnegative partition of unity it follows that for any $\boldsymbol{x} \in \triangle$,

$$
|\boldsymbol{b}(\boldsymbol{x})^T \boldsymbol{c}| \le \sum_{(i,j,k,\ell)\in\bar{\mathscr{I}}^2} |c_{ijk\ell}|\,|S^6_{ijk\ell}(\boldsymbol{x})| \le \|\boldsymbol{c}\|_\infty \sum_{(i,j,k,\ell)\in\bar{\mathscr{I}}^2} S^6_{ijk\ell}(\boldsymbol{x}) = \|\boldsymbol{c}\|_\infty,
$$

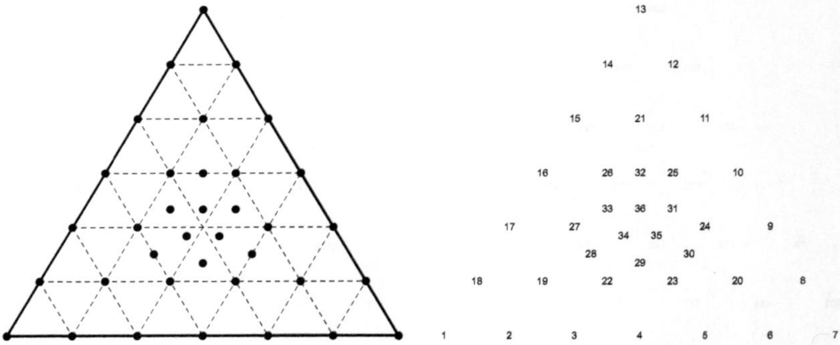

Fig. 4 The position, left, and ordering of the domain points and corresponding basis functions in Fig. 3, right

with an equality if all the $c_{ijk\ell} = 1$, so that $\max_{c \neq 0} \|b^T c\|_{L_\infty(\mathscr{T})}/\|c\|_\infty = 1$.

To bound the second part of $\kappa_{6,\infty}(\mathscr{T})$, we consider a spline $s \in \mathbb{S}_6^2$ interpolating given data at the 36 domain points. Using the ordering of domain points $\boldsymbol{p}_m^*$ and basis functions S_n^6, $m, n = 1, \ldots, 36$, shown in Fig. 4 we obtain a linear system

$$\boldsymbol{A}\boldsymbol{c} = \begin{bmatrix} S_1^6(\boldsymbol{p}_1^*) & \cdots & S_{36}^6(\boldsymbol{p}_1^*) \\ \vdots & \vdots & \\ S_1^6(\boldsymbol{p}_{36}^*) & \cdots & S_{36}^6(\boldsymbol{p}_{36}^*) \end{bmatrix} \begin{bmatrix} c_1 \\ \vdots \\ c_{36} \end{bmatrix} = \begin{bmatrix} s(\boldsymbol{p}_1^*) \\ \vdots \\ s(\boldsymbol{p}_{36}^*) \end{bmatrix} = \boldsymbol{s}^*$$

for the coefficients $\boldsymbol{c} := [c_1, \ldots, c_{36}]^T$ of $s := \sum_{n=1}^{36} c_n S_n^6$. Here $S_1^6, \ldots, S_{21}^6$ are Bernstein polynomials, and $S_n^6(\boldsymbol{p}_m^*) = 0$ for $m = 1, \ldots, 18$ and $n = 19, \ldots, 36$, by continuity properties of simplex splines. Thus $\boldsymbol{A}$ has the lower triangular block form $\boldsymbol{A} = \begin{bmatrix} \boldsymbol{A}_1 & \boldsymbol{0} \\ \boldsymbol{A}_2 & \boldsymbol{A}_3 \end{bmatrix}$, where $\boldsymbol{A}_m, \in \mathbb{R}^{18\times 18}$, for $m = 1, 2, 3$. Using symbolic computation it follows that $\boldsymbol{A}_1$ and $\boldsymbol{A}_2$ are nonsingular. Thus $\boldsymbol{A}$ is nonsingular with inverse $\boldsymbol{A}^{-1} = \begin{bmatrix} \boldsymbol{B}_1 & \boldsymbol{0} \\ \boldsymbol{B}_2 & \boldsymbol{B}_3 \end{bmatrix}$, and $\boldsymbol{B}_1 = \boldsymbol{A}_1^{-1}$, $\boldsymbol{B}_3 = \boldsymbol{A}_3^{-1}$, $\boldsymbol{B}_2 = -\boldsymbol{B}_3\boldsymbol{A}_2\boldsymbol{B}_1$. We compute

$$\|\boldsymbol{A}^{-1}\|_\infty = \frac{12\,209\,201\,545\,461}{9\,044\,604\,800} < 1350 - \frac{1}{9},$$

and the proposition follows. □

Finally, we consider next the problem of obtaining $\mathscr{C}^2$-continuity across an edge between two adjacent triangles in a global triangulation, making use of the local properties of our simplex spline basis.

We begin with a technical observation.

Lemma 4 *On $\mathscr{T}_3$, i.e., for $0 \leq \beta_3 \leq \beta_1 \leq 1$ and $0 \leq \beta_3 \leq \beta_2 \leq 1$,*

$$
\begin{aligned}
&S^6_{i+1,j+1,k+1,0} = B^6_{ijk}, \ \{(i,j,k) : i+j+k=6,\ i \geq 4 \ or \ j \geq 4\},\\
&S^6_{4311}|_{\mathscr{T}_3} = B^6_{321} - B^6_{312} - B^6_{222} + O(\beta_3^3),\\
&S^6_{3411}|_{\mathscr{T}_3} = B^6_{231} - B^6_{222} - B^6_{132} + O(\beta_3^3),\\
&S^6_{4221}|_{\mathscr{T}_3} = 2B^6_{312} + O(\beta_3^3),\ S^6_{3312}|_{\mathscr{T}_3} = 3B^6_{222} + O(\beta_3^3),\ S^6_{2421}|_{\mathscr{T}_3} = 2B^6_{132} + O(\beta_3^3),\\
&S^6_{3222}|_{\mathscr{T}_3} = 6B^6_{213} + O(\beta_3^4),\ S^6_{2322}|_{\mathscr{T}_3} = 6B^6_{123} + O(\beta_3^4).
\end{aligned}
\tag{23}
$$

while $S^6_{ijk\ell}|_{\mathscr{T}_3} = O(\beta_3^3)$ *for the remaining splines.*

Proof. The set $S^6_{i+1,j+1,k+1,0}$ with $\{(i,j,k) : i+j+k=6,\ i \geq 4 \text{ or } j \geq 4\}$ are $2 \times 6 = 12$ classical Bernstein polynomials on the triangle $\mathscr{T}$, located on the lower left and right corners in Fig. 3. Using the explicit forms (18) and (19), (23) follows by inspection. □

For

$$s = \sum_{(i,j,k,\ell)\in\bar{\mathscr{I}}^2} c_{ijk\ell} S^6_{ijk\ell}$$

let

$$
\begin{aligned}
\boldsymbol{c}_0 &= [c_{7110}, c_{6210}, c_{5310}, c_{4410}, c_{3510}, c_{2610}, c_{1710}]^T \in \mathbb{R}^7,\\
\boldsymbol{c}_1 &= [c_{6120}, c_{5220}, c_{4311}, c_{3411}, c_{2520}, c_{1620}]^T \in \mathbb{R}^6,\\
\boldsymbol{c}_2 &= [c_{5130}, c_{4221}, c_{3312}, c_{2421}, c_{1530}]^T \in \mathbb{R}^5,
\end{aligned}
\tag{24}
$$

be the coefficients involved in obtaining $\mathscr{C}^2$ continuity across an edge between two triangles, see Fig. 3.

Proposition 5 *Let*

$$s = \sum_{(i,j,k,\ell)\in\bar{\mathscr{I}}^2} c_{ijk\ell} S^6_{ijk\ell} \quad or \quad \tilde{s} = \sum_{(i,j,k,\ell)\in\bar{\mathscr{I}}^2} \tilde{c}_{ijk\ell} \tilde{S}^6_{ijk\ell},$$

respectively, be defined on the triangle $\mathscr{T} := \langle \boldsymbol{p}_1, \boldsymbol{p}_2, \boldsymbol{p}_3 \rangle$ *(resp.* $\tilde{\mathscr{T}} := \langle \boldsymbol{p}_1, \boldsymbol{p}_2, \tilde{\boldsymbol{p}}_3 \rangle$*). We suppose that* $\tilde{\boldsymbol{p}}_3 = \lambda_1 \boldsymbol{p}_1 + \lambda_2 \boldsymbol{p}_2 + \lambda_3 \boldsymbol{p}_3$ *with* $\lambda_1 + \lambda_2 + \lambda_3 = 1$*. The function* $s_+ = \begin{cases} s & on\ \mathscr{T} \\ \tilde{s} & on\ \tilde{\mathscr{T}} \end{cases}$ *is* $\mathscr{C}^r$ *with* $r \leq 2$ *if and only if*

$$\begin{bmatrix} \tilde{\boldsymbol{c}}_0 \\ \vdots \\ \tilde{\boldsymbol{c}}_r \end{bmatrix} = \begin{bmatrix} \boldsymbol{C}_{00} & \dots & \boldsymbol{C}_{0r} \\ \vdots & & \vdots \\ \boldsymbol{C}_{r0} & \dots & \boldsymbol{C}_{rr} \end{bmatrix} \begin{bmatrix} \boldsymbol{c}_0 \\ \vdots \\ \boldsymbol{c}_r \end{bmatrix},$$

where the $\tilde{\boldsymbol{c}}_0, \tilde{\boldsymbol{c}}_1, \tilde{\boldsymbol{c}}_2$ *are the coefficients of* $\tilde{s}$ *corresponding to* (24)*, and the matrices* $\boldsymbol{C}_{mn}$ *are defined by*

$$\boldsymbol{C}_{00} = \boldsymbol{I} \in \mathbb{R}^{7\times 7}, \quad \boldsymbol{C}_{01} = \boldsymbol{0} \in \mathbb{R}^{7\times 6}, \quad \boldsymbol{C}_{02} = \boldsymbol{0} \in \mathbb{R}^{7\times 5}, \tag{25}$$

$$C_{10} = \begin{bmatrix} \lambda_1 & \lambda_2 & 0 & \dots & 0 \\ 0 & \lambda_1 & \lambda_2 & \ddots & \vdots \\ \vdots & \ddots & \ddots & \ddots & 0 \\ 0 & \dots & 0 & \lambda_1 & \lambda_2 \end{bmatrix} \in \mathbb{R}^{6\times 7}, \quad C_{11} = \lambda_3 I \in \mathbb{R}^{6\times 6}, \quad C_{12} = \mathbf{0} \in \mathbb{R}^{6\times 5}, \tag{26}$$

and

$$\begin{aligned} C_{20} &= \begin{bmatrix} \lambda_1^2 & 2\lambda_1\lambda_2 & \lambda_2^2 & 0 & 0 & 0 & 0 \\ 0 & \lambda_1^2/2 & \lambda_1(1/2+\lambda_2) & \lambda_2(1+\lambda_2)/2 & 0 & 0 & 0 \\ 0 & 0 & \lambda_1(1+\lambda_1)/3 & (\lambda_1+\lambda_2+2\lambda_1\lambda_2)/3 & \lambda_2(1+\lambda_2)/3 & 0 & 0 \\ 0 & 0 & 0 & \lambda_1(1+\lambda_1)/2 & \lambda_2(1/2+\lambda_1) & \lambda_2^2/2 & 0 \\ 0 & 0 & 0 & 0 & \lambda_1^2 & 2\lambda_1\lambda_2 & \lambda_2^2 \end{bmatrix} \in \mathbb{R}^{5\times 7}, \\ C_{21} &= \begin{bmatrix} 2\lambda_1\lambda_3 & 2\lambda_2\lambda_3 & 0 & 0 & 0 & 0 \\ 0 & \lambda_1\lambda_3 & \lambda_3(1/2-\lambda_3/2+\lambda_2) & 0 & 0 & 0 \\ 0 & 0 & \lambda_3(1-\lambda_3+2\lambda_1)/3 & \lambda_3(1-\lambda_3+2\lambda_2)/3 & 0 & 0 \\ 0 & 0 & 0 & \lambda_3(1/2-\lambda_3/2+\lambda_1) & \lambda_2\lambda_3 & 0 \\ 0 & 0 & 0 & 0 & 2\lambda_1\lambda_3 & 2\lambda_2\lambda_3 \end{bmatrix} \in \mathbb{R}^{5\times 6}, \\ C_{22} &= \lambda_3^2 I \in \mathbb{R}^{5\times 5}. \end{aligned} \tag{27}$$

Proof. We begin by the $\mathscr{C}^r$-continuity for the Bézier surfaces using the Bernstein basis. Let $\sigma = \sum_{\nu+\mu+\kappa=6} \gamma_{\nu\mu\kappa} B^6_{\nu\mu\kappa} \in \mathbb{P}_6$ be defined on the triangle $\mathscr{T} := \langle \boldsymbol{p}_1, \boldsymbol{p}_2, \boldsymbol{p}_3 \rangle$ (respectively $\tilde{\sigma} = \sum_{\nu+\mu+\kappa=6} \tilde{\gamma}_{\nu\mu\kappa} \tilde{B}^6_{\nu\mu\kappa} \in \mathbb{P}_6$ on $\tilde{\mathscr{T}} := \langle \boldsymbol{p}_1, \boldsymbol{p}_2, \tilde{\boldsymbol{p}}_3 \rangle$) where $\tilde{B}^6_{\nu\mu\kappa}$ are the Bernstein polynomials with barycentric variables $\tilde{\beta}_1, \tilde{\beta}_2, \tilde{\beta}_3$.

We recall, see [17, Theorem 2.28], that the function $\sigma_+ = \begin{cases} \sigma & \text{on } \mathscr{T} \\ \tilde{\sigma} & \text{on } \tilde{\mathscr{T}} \end{cases}$ is $\mathscr{C}^r$ if and only if

$$[\tilde{\boldsymbol{\gamma}}_m]_{m=0,\dots,r} = [\boldsymbol{\Gamma}_{mn}]_{m,n=0,\dots,r} [\boldsymbol{\gamma}_n]_{n=0,\dots,r},$$

where

$$\begin{aligned} \boldsymbol{\gamma}_0 &= [\gamma_{600}, \gamma_{510}, \gamma_{420}, \gamma_{330}, \gamma_{240}, \gamma_{150}, \gamma_{160}]^T, \\ \boldsymbol{\gamma}_1 &= [\gamma_{501}, \gamma_{411}, \gamma_{321}, \gamma_{231}, \gamma_{141}, \gamma_{051}]^T, \\ \boldsymbol{\gamma}_2 &= [\gamma_{402}, \gamma_{312}, \gamma_{222}, \gamma_{132}, \gamma_{042}]^T, \end{aligned}$$

similarly for the $\tilde{\boldsymbol{\gamma}}_m$ and the matrices are defined by

$$\boldsymbol{\Gamma}_{00} = I \in \mathbb{R}^{7\times 7}, \quad \boldsymbol{\Gamma}_{01} = \mathbf{0} \in \mathbb{R}^{7\times 6}, \quad \boldsymbol{\Gamma}_{02} = \mathbf{0} \in \mathbb{R}^{7\times 5}, \tag{28}$$

$$\boldsymbol{\Gamma}_{10} = \begin{bmatrix} \lambda_1 & \lambda_2 & 0 & \dots & 0 \\ 0 & \lambda_1 & \lambda_2 & \ddots & \vdots \\ \vdots & \ddots & \ddots & \ddots & 0 \\ 0 & \dots & 0 & \lambda_1 & \lambda_2 \end{bmatrix} \in \mathbb{R}^{6\times 7}, \quad \boldsymbol{\Gamma}_{11} = \lambda_3 I \in \mathbb{R}^{6\times 6}, \quad \boldsymbol{\Gamma}_{12} = \mathbf{0} \in \mathbb{R}^{6\times 5}, \tag{29}$$

and

$$\boldsymbol{\Gamma}_{20} = \begin{bmatrix} \lambda_1^2 & 2\lambda_1\lambda_2 & \lambda_2^2 & 0 & \dots & 0 \\ 0 & \lambda_1^2 & 2\lambda_1\lambda_2 & \lambda_2^2 & \ddots & \vdots \\ \vdots & \ddots & \ddots & \ddots & \ddots & 0 \\ 0 & \dots & 0 & \lambda_1^2 & 2\lambda_1\lambda_2 & \lambda_2^2 \end{bmatrix} \in \mathbb{R}^{5\times 7},$$

$$\boldsymbol{\Gamma}_{21} = \begin{bmatrix} 2\lambda_1\lambda_3 & 2\lambda_2\lambda_3 & 0 & \dots & 0 \\ 0 & 2\lambda_1\lambda_3 & 2\lambda_2\lambda_3 & \ddots & \vdots \\ \vdots & \ddots & \ddots & \ddots & 0 \\ 0 & \dots & 0 & 2\lambda_1\lambda_3 & 2\lambda_2\lambda_3 \end{bmatrix} \in \mathbb{R}^{5\times 6}, \quad \boldsymbol{\Gamma}_{22} = \lambda_3^2 \boldsymbol{I} \in \mathbb{R}^{5\times 5}. \tag{30}$$

See also Fig. 3.

Consider now the function $s_+ = \begin{cases} s & \text{on } \mathscr{T} \\ \tilde{s} & \text{on } \tilde{\mathscr{T}} \end{cases}$ on $\mathscr{T} \cup \tilde{\mathscr{T}}$. To study the $\mathscr{C}^r$-continuity through the edge $\langle \boldsymbol{p}_1 \boldsymbol{p}_2 \rangle$, it is sufficient to consider s on $\mathscr{T}_3$ and $\tilde{s}$ on $\tilde{\mathscr{T}}_3$. On $\mathscr{T}_3$ (resp. $\tilde{\mathscr{T}}_3$), any $S^6_{ijk\ell}$ of the basis (resp. $\tilde{S}^6_{ijk\ell}$) is a polynomial of degree at most 6, $S^6_{ijk\ell} = \sum_{\nu+\mu+\kappa=6} w^{ijk\ell}_{\nu\mu\kappa} B^6_{\nu\mu\kappa}$ (resp. $\tilde{S}^6_{ijk\ell} = \sum_{\nu+\mu+\kappa=6} w^{ijk\ell}_{\nu\mu\kappa} \tilde{B}^6_{\nu\mu\kappa}$). So that $s|_{\mathscr{T}_3}$ and $\tilde{s}|_{\tilde{\mathscr{T}}_3}$ can also be written in the Bernstein bases

$$s|_{\mathscr{T}_3} = \sum_{(i,j,k,\ell)\in\bar{\mathscr{I}}^2} c_{ijk\ell} S^6_{ijk\ell}|_{\mathscr{T}_3} = \sum_{\nu+\mu+\kappa=6} \gamma_{\nu\mu\kappa} B^6_{\nu\mu\kappa} \text{ with } \gamma_{\nu\mu\kappa} = \sum_{(i,j,k,\ell)\in\bar{\mathscr{I}}^2} c_{ijk\ell} w^{ijk\ell}_{\nu\mu\kappa},$$

$$\tilde{s}|_{\tilde{\mathscr{T}}_3} = \sum_{(i,j,k,\ell)\in\bar{\mathscr{I}}^2} \tilde{c}_{ijk\ell} \tilde{S}^6_{ijk\ell}|_{\mathscr{T}_3} = \sum_{\nu+\mu+\kappa=6} \tilde{\gamma}_{\nu\mu\kappa} \tilde{B}^6_{\nu\mu\kappa} \text{ with } \tilde{\gamma}_{\nu\mu\kappa} = \sum_{(i,j,k,\ell)\in\bar{\mathscr{I}}^2} \tilde{c}_{ijk\ell} \tilde{w}^{ijk\ell}_{\nu\mu\kappa}.$$

From (23), we deduce the components $\tilde{w}^{ijk\ell}_{\nu\mu\kappa}$ for $\kappa = 0, 1, 2$ and we put forward the corresponding components

$$\begin{aligned} s|_{\mathscr{T}_3} &= c_{7110} B^6_{600} + c_{6210} B^6_{510} + c_{5310} B^6_{420} + c_{4410} B^6_{330} + c_{3510} B^6_{240} + c_{2610} B^6_{150} \\ &+ c_{1710} B^6_{060} + c_{6120} B^6_{501} + c_{5220} B^6_{411} + c_{4311} B^6_{321} + c_{3411} B^6_{231} + c_{2520} B^6_{141} \\ &+ c_{1620} B^6_{051} + c_{5130} B^6_{402} + (2c_{4221} - c_{4311}) B^6_{312} + (3c_{3312} - c_{4311} - c_{3411}) B^6_{222} \\ &+ (2c_{2421} - c_{3411}) B^6_{132} + c_{1530} B^6_{042} + O(\beta_3^3) \end{aligned}$$

and a similar expression for $\tilde{s}|_{\tilde{\mathscr{T}}_3}$.

The conditions for the regularity $\mathscr{C}^0$ of s_+, (25), is a consequence of (28) and similarly for $\mathscr{C}^1$ with also (26) coming from (29). To obtain $\mathscr{C}^2$, we add the conditions (30). They can be rewritten

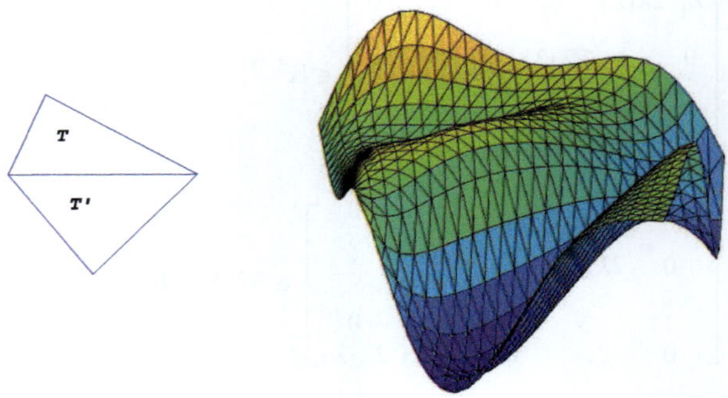

Fig. 5 A $\mathscr{C}^2$ surface on two triangles

$$\begin{aligned}
\tilde{c}_{5130} &= c_{7110}\lambda_1^2 + 2c_{6210}\lambda_1\lambda_2 + c_{5310}\lambda_2^2 + 2c_{5220}\lambda_2\lambda_3 + c_{5130}\lambda_3^2 + 2c_{6120}\lambda_1\lambda_3,\\
2\tilde{c}_{4221} - \tilde{c}_{4311} &= c_{6210}\lambda_1^2 + 2c_{5310}\lambda_1\lambda_2 + c_{4410}\lambda_2^2 + 2c_{4311}\lambda_2\lambda_3\\
&\quad + (2c_{4221} - c_{4311})\lambda_3^2 + 2c_{5220}\lambda_1\lambda_3,\\
3\tilde{c}_{3312} - \tilde{c}_{4311} - \tilde{c}_{3411} &= c_{5310}\lambda_1^2 + 2c_{4410}\lambda_1\lambda_2 + c_{3510}\lambda_2^2 + 2c_{3411}\lambda_2\lambda_3\\
&\quad + (3c_{3312} - c_{4311} - c_{3411})\lambda_3^2 + 2c_{4311}\lambda_1\lambda_3\\
2\tilde{c}_{2421} - \tilde{c}_{3411} &= c_{4410}\lambda_1^2 + 2c_{3510}\lambda_1\lambda_2 + c_{2610}\lambda_2^2 + 2c_{2520}\lambda_2\lambda_3\\
&\quad + (2c_{2421} - c_{3411})\lambda_3^2 + 2c_{3411}\lambda_1\lambda_3,\\
\tilde{c}_{1530} &= c_{3510}\lambda_1^2 + 2c_{2610}\lambda_1\lambda_2 + c_{1710}\lambda_2^2 + c_{1620}\lambda_2\lambda_3 + c_{1530}\lambda_3^2 + 2c_{2520}\lambda_1\lambda_3.
\end{aligned}$$

We already have $\tilde{c}_{4311}$ and $\tilde{c}_{3411}$ in (26) so that we deduce that the last components have to satisfy (27). □

Several examples have been considered for scattered data on the CT-split, see for example [11, 20]. Here, we consider a surface on two triangles, see Fig. 5. With the 18 conditions from Proposition 5, we obtain a $\mathscr{C}^2$ surface on the two triangles.

4.2 The $\mathscr{C}^3$ Elements, $\bar{\Sigma}_9^3 = \Sigma_9^3$

In this section we repeat the process of the preceding one for $r = 3$. The methods are the same, but the expression become more lengthy. For that reason, we essentially list the results.

The partition of unity basis $\bar{\Sigma}_9^3$ is constructed from the 75 elements $\triangle_{ijk\ell}$ defined in Example 3 as follows,

$$S^9_{ijk\ell} = c_{ijk\ell}\triangle_{ijk\ell}, \quad c_{ijk\ell} = \begin{cases} 1 \text{ if } \ell = 0,\\ 1/3 \text{ if } (\ell > 0 \text{ and } \min(i,j,k) = 1) \text{ or } \ell = 4,\\ 2/3 \text{ if } 0 < \ell < 4 \text{ and } \min(i,j,k) > 1,\end{cases}$$

for $\{i, j, k, \ell\} \in \bar{\mathscr{I}}^3$.

Theorem 3 (Barycentric Marsden-Like Identity for $d = 9$) *For $\boldsymbol{u} := [u_1, u_2, u_3]^T \in \mathbb{R}^3, \boldsymbol{\beta} := [\beta_1, \beta_2, \beta_3]^T \in \mathbb{R}^3$, with $\beta_i \geq 0,\ i = 1, 2, 3$ and $\beta_1 + \beta_2 + \beta_3 = 1$ we have*

$$(\boldsymbol{u}^T \boldsymbol{\beta})^9 = \sum_{(i,j,k,\ell) \in \bar{\mathscr{I}}^3} \rho_{ijk\ell}(\boldsymbol{u}) S^9_{ijk\ell}(\boldsymbol{\beta}), \tag{31}$$

where

$$\rho_{i+1,j+1,k+1,\ell}(\boldsymbol{u}) = \begin{cases} u_1^i u_2^j u_3^k, & (i, j, k) \in \mathscr{I}^3_{Bernstein},\ \ell = 0, \\ u_1^2 u_2^2 u_3^2 \bar{u}_{123}^3, & \ell = 4, \\ u_1^{\max(i,1)} u_2^{\max(j,1)} u_3^{\max(k,1)} \mu_{ijk\ell} \bar{u}_{123}^{\ell-1}, & \textit{otherwise}. \end{cases}$$

Here $\bar{u}_{123} := (u_1 + u_2 + u_3)/3$, and

$$\mu_{ijk\ell} := \begin{cases} 1, & \textit{if } 1, 2 \textit{ is not among (i,j,k)}, \\ (u_r + u_s)/2, & \textit{if } 1, 2 \textit{ is at position } r, s \textit{ in } (i, j, k). \end{cases}$$

Proof. As for $r = 2$ the barycentric Marsden-like identity follows from the barycentric form (1) of the Marsden identity for Bernstein polynomials by expressing the 19 removed Bernstein polynomials B^9_{ijk} in terms of the elements in $\bar{\Sigma}^3_9$. For $(i, j, k) \in \mathscr{I}^3_{\text{removed}}$

$$\begin{aligned}
B^9_{1,3,5} &= (\triangle_{1461} + \triangle_{2361})/3 + (\triangle_{1452} + \triangle_{2352})/9 + (\triangle_{1443} + \triangle_{2343} + \triangle_{2433})/27, \\
B^9_{1,5,3} &= (\triangle_{1641} + \triangle_{2631})/3 + (\triangle_{1542} + \triangle_{2532})/9 + (\triangle_{1443} + \triangle_{2433} + \triangle_{2343})/27, \\
B^9_{5,1,3} &= (\triangle_{6141} + \triangle_{6231})/3 + (\triangle_{5142} + \triangle_{5232})/9 + (\triangle_{4143} + \triangle_{4233} + \triangle_{3243})/27, \\
B^9_{5,3,1} &= (\triangle_{6411} + \triangle_{6321})/3 + (\triangle_{5412} + \triangle_{5322})/9 + (\triangle_{4413} + \triangle_{4323} + \triangle_{3423})/27, \\
B^9_{3,5,1} &= (\triangle_{4611} + \triangle_{3621})/3 + (\triangle_{4512} + \triangle_{3522})/9 + (\triangle_{4413} + \triangle_{4323} + \triangle_{3423})/27, \\
B^9_{3,1,5} &= (\triangle_{4161} + \triangle_{3261})/3 + (\triangle_{4152} + \triangle_{3252})/9 + (\triangle_{4143} + \triangle_{4233} + \triangle_{3243})/27, \\
B^9_{1,4,4} &= \triangle_{1551}/3 + (\triangle_{1452} + \triangle_{1542} + \triangle_{2352} + \triangle_{2532})/9 + 2(\triangle_{2343} + \triangle_{2433} + 2\triangle_{1443})/27, \\
B^9_{4,1,4} &= \triangle_{5151}/3 + (\triangle_{4152} + \triangle_{5142} + \triangle_{3252} + \triangle_{5232})/9 + 2(\triangle_{3243} + \triangle_{4233} + 2\triangle_{4143})/27, \\
B^9_{4,4,1} &= \triangle_{5511}/3 + (\triangle_{4512} + \triangle_{5412} + \triangle_{3522} + \triangle_{5322})/9 + 2(\triangle_{3423} + \triangle_{4323} + 2\triangle_{4413})/27,
\end{aligned} \tag{32}$$

$$
\begin{aligned}
B^9_{2,2,5} &= (\triangle_{2361} + \triangle_{3261})/3 + (\triangle_{2352} + \triangle_{3252})/9 \\
&\quad + (\triangle_{2343} + \triangle_{3243})/27 + (\triangle_{2334} + \triangle_{3234} + \triangle_{3324})/81, \\
B^9_{2,5,2} &= (\triangle_{2631} + \triangle_{3621})/3 + (\triangle_{2532} + \triangle_{3522})/9 \\
&\quad + (\triangle_{2433} + \triangle_{3423})/27 + (\triangle_{2334} + \triangle_{3324} + \triangle_{3234})/81, \\
B^9_{5,2,2} &= (\triangle_{6231} + \triangle_{6321})/3 + (\triangle_{5232} + \triangle_{5322})/9 \\
&\quad + (\triangle_{4233} + \triangle_{4323})/27 + (\triangle_{3234} + \triangle_{3324} + \triangle_{2334})/81,
\end{aligned}
\tag{33}
$$

$$
\begin{aligned}
B^9_{2,3,4} &= (\triangle_{1452} + 2\triangle_{2352} + \triangle_{3252} + \triangle_{2433})/9 \\
&\quad + (2\triangle_{1443} + 4\triangle_{2343} + 2\triangle_{3243} + \triangle_{3423} + \triangle_{2334} + \triangle_{3234} + \triangle_{3324})/27, \\
B^9_{2,4,3} &= (\triangle_{1542} + 2\triangle_{2532} + \triangle_{3522} + \triangle_{2343})/9 \\
&\quad + 2\triangle_{1443} + 4\triangle_{2433} + 2\triangle_{3423} + \triangle_{3243} + \triangle_{2334} + \triangle_{3234} + \triangle_{3324})/27, \\
B^9_{4,2,3} &= (\triangle_{5142} + 2\triangle_{5232} + \triangle_{5322} + \triangle_{3243})/9 \\
&\quad + (2\triangle_{4143} + 4\triangle_{4233} + 2\triangle_{4323} + \triangle_{2343} + \triangle_{2334} + \triangle_{3234} + \triangle_{3324})/27, \\
B^9_{4,3,2} &= (\triangle_{5412} + 2\triangle_{5322} + \triangle_{5232} + \triangle_{3423})/9 \\
&\quad + (2\triangle_{4413} + 4\triangle_{4323} + 2\triangle_{4233} + \triangle_{2433} + \triangle_{2334} + \triangle_{3234} + \triangle_{3324})/27, \\
B^9_{3,4,2} &= (\triangle_{4512} + 2\triangle_{3522} + \triangle_{2532} + \triangle_{4323})/9 \\
&\quad + (2\triangle_{4413} + 4\triangle_{3423} + 2\triangle_{2433} + \triangle_{4233} + \triangle_{2334} + \triangle_{3234} + \triangle_{3324})/27, \\
B^9_{3,2,4} &= (\triangle_{4152} + 2\triangle_{3252} + \triangle_{2352} + \triangle_{4233})/9 \\
&\quad + (2\triangle_{4143} + 4\triangle_{3243} + 2\triangle_{2343} + \triangle_{4323} + \triangle_{2334} + \triangle_{3234} + \triangle_{3324})/27, \\
B^9_{3,3,3} &= (\triangle_{2343} + \triangle_{2433} + \triangle_{3243} + \triangle_{3423} + \triangle_{4233} + \triangle_{4323})/9 \\
&\quad + (\triangle_{1443} + \triangle_{4143} + \triangle_{4413} + 2\triangle_{2334} + 2\triangle_{3234} + 2\triangle_{3324})/27,
\end{aligned}
\tag{34}
$$

By (1)

$$
\big(u_1\beta_1 + u_2\beta_2 + u_3\beta_3\big)^9 = \sum_{(i,j,k)\in\mathscr{I}^3_{\text{Bernstein}}} u_1^i u_2^j u_3^k B^9_{ijk}(\boldsymbol{\beta}) + \sum_{(i,j,k)\in\mathscr{I}^3_{\text{removed}}} u_1^i u_2^j u_3^k B^9_{ijk}(\boldsymbol{\beta}).
$$

For $(i,j,k) \in \mathscr{I}^3_{\text{Bernstein}}$ we have $B^9_{i,j,k}(\boldsymbol{\beta}) = S^9_{i+1,j+1,k+1,0}(\boldsymbol{\beta})$ and hence $\rho_{i+1,j+1,k+1,0}(\boldsymbol{u}) = u_1^i u_2^j u_3^k$. In the second sum we insert the expressions in (32)–(34) for B^9_{ijk}, and collect terms for each $\triangle_{ijk\ell}$ to obtain (31). We show this for 7 typical cases. Let $\bar{u}_{r,s} := (u_r + u_s)/2$ for $r, s = 1, 2, 3$. Then

$$
\begin{aligned}
\sum_{(i,j,k)\in\mathscr{I}^3_{\text{removed}}} & u_1^i u_2^j u_3^k B^9_{ijk}(\boldsymbol{\beta}) = u_1^1 u_2^4 u_3^4 \triangle_{1551}(\boldsymbol{\beta})/3 + \left(u_1^3 u_2^1 u_3^5 + u_1^2 u_2^2 u_3^5\right)\triangle_{3261}(\boldsymbol{\beta})/3 \\
& + \left(u_1^5 u_2^3 u_3^1 + u_1^4 u_2^4 u_3^1 + u_1^4 u_2^3 u_3^2\right)\triangle_{5412}(\boldsymbol{\beta})/9 \\
& + \left(u_1^3 u_2^5 u_3^1 + u_1^4 u_2^4 u_3^1 + u_1^2 u_2^5 u_3^2 + u_1^2 u_2^4 u_3^3 + 2u_1^3 u_2^4 u_3^2\right)\triangle_{3552}(\boldsymbol{\beta})/9 \\
& + \left(u_1^5 u_2^3 u_3^1 + u_1^3 u_2^5 u_3^1 + 2u_1^4 u_2^4 u_3^1 + 2u_1^4 u_2^3 u_3^2 + u_1^3 u_2^4 u_3^2 + u_1^3 u_2^3 u_3^3\right)\triangle_{4413}(\boldsymbol{\beta})/27 \\
& + \left(u_1^5 u_2^1 u_3^3 + u_1^3 u_2^1 u_3^5 + 2u_1^4 u_2^1 u_3^4 + u_1^5 u_2^2 u_3^2 + 4u_1^4 u_2^2 u_3^3\right. \\
& \left. + 2u_1^4 u_2^3 u_3^2 + u_1^3 u_2^4 u_3^2 + 3u_1^3 u_2^2 u_3^4 + 3u_1^3 u_2^3 u_3^3\right)\triangle_{4233}(\boldsymbol{\beta})/27 \\
& + \left(u_1^2 u_2^2 u_3^5 + u_1^2 u_2^5 u_3^2 + u_1^5 u_2^2 u_3^2 + 3u_1^2 u_2^3 u_3^4 + 3u_1^2 u_2^4 u_3^3 + 3u_1^4 u_2^2 u_3^3\right. \\
& \left. + 3u_1^4 u_2^3 u_3^2 + 3u_1^3 u_2^4 u_3^2 + 3u_1^3 u_2^2 u_3^4 + 6u_1^3 u_2^3 u_3^3\right)\triangle_{3324}(\boldsymbol{\beta})/81 + \cdots \\
= {} & u_1^1 u_2^4 u_3^4 \triangle_{1551}(\boldsymbol{\beta})/3 + +2u_1^2 u_2^1 u_3^5 \bar{u}_{1,2} \triangle_{3261}(\boldsymbol{\beta})/3 + u_1^4 u_2^3 u_3^1 \bar{u}_{123} \triangle_{5412}(\boldsymbol{\beta})/3 \\
& + 2u_1^2 u_2^4 u_3^1 \bar{u}_{1,3} \bar{u}_{123} \triangle_{3522}(\boldsymbol{\beta})/3 + u_1^3 u_2^3 u_3^1 \bar{u}_{123}^2 \triangle_{4413}(\boldsymbol{\beta})/3 \\
& + 2u_1^3 u_2^1 u_3^2 \bar{u}_{2,3} \bar{u}_{123}^2 \triangle_{4233}(\boldsymbol{\beta})/3 + u_1^2 u_2^2 u_3^2 \bar{u}_{123}^3 \triangle_{3324}(\boldsymbol{\beta})/3 + \cdots \\
= {} & u_1^1 u_2^4 u_3^4 S^9_{1551}(\boldsymbol{\beta}) + u_1^2 u_2^1 u_3^5 \bar{u}_{1,2} S^9_{3261}(\boldsymbol{\beta}) + u_1^4 u_2^3 u_3^1 \bar{u}_{123} S^9_{5412}(\boldsymbol{\beta}) \\
& + u_1^2 u_2^4 u_3^1 \bar{u}_{1,3} \bar{u}_{123} S^9_{3522}(\boldsymbol{\beta}) + u_1^3 u_2^3 u_3^1 \bar{u}_{123}^2 S^9_{4413}(\boldsymbol{\beta}) \\
& + u_1^3 u_2^1 u_3^2 \bar{u}_{2,3} \bar{u}_{123}^2 S^9_{4233}(\boldsymbol{\beta}) + u_1^2 u_2^2 u_3^2 \bar{u}_{123}^3 S^9_{3324}(\boldsymbol{\beta}) + \cdots
\end{aligned}
$$

and (31) follows. □

The barycentric form of the domain points are computed as explained in Corollary 1. For a plot see Fig. 6.

For the stability, the computation is similar to the proof of Proposition 4, except that since S^9_{3324}, S^9_{3234}, S^9_{2334} have the same domain points $(1/3, 1/3, 1/3)$. We replace this triple point by the three points $(3, 3, 1)/7$, $(3, 1, 3)/7$, $(1, 3, 3)/7$ and find (Fig. 7)

$$
\kappa_{9,\infty}(\mathscr{T}) := \max_{\boldsymbol{c}\neq\boldsymbol{0}} \frac{\|\boldsymbol{b}^T\boldsymbol{c}\|_{L_\infty(\mathscr{T})}}{\|\boldsymbol{c}\|_\infty} \max_{\boldsymbol{c}\neq\boldsymbol{0}} \frac{\|\boldsymbol{c}\|_\infty}{\|\boldsymbol{b}^T\boldsymbol{c}\|_{L_\infty(\mathscr{T})}} \simeq 159\,844.34\ldots
$$

$\mathscr{C}^3$-Continuity Through an Edge:
Let $s = \sum_{(i,j,k,\ell)\in\bar{\mathscr{I}}^3} c_{ijk\ell} S^9_{ijk\ell}$ (respectively $\tilde{s} = \sum_{(i,j,k,\ell)\in\bar{\mathscr{I}}^3} \tilde{c}_{ijk\ell} \tilde{S}^9_{ijk\ell}$) be defined on the triangle $\mathscr{T} := \langle \boldsymbol{p}_1, \boldsymbol{p}_2, \boldsymbol{p}_3\rangle$ (resp. $\tilde{\mathscr{T}} := \langle \boldsymbol{p}_1, \boldsymbol{p}_2, \tilde{\boldsymbol{p}}_3\rangle$). We suppose that $\tilde{\boldsymbol{p}}_3 = \lambda_1\boldsymbol{p}_1 + \lambda_2\boldsymbol{p}_2 + \lambda_3\boldsymbol{p}_3$ with $\lambda_1 + \lambda_2 + \lambda_3 = 1$.

With an extension of the notations of Proposition 5, we define

$$
\begin{aligned}
\boldsymbol{c}_0 &= [c_{10-i,i+1,1,0}]^T_{i=0,\ldots,9}, \\
\boldsymbol{c}_1 &= [c_{9120}, c_{8220}, c_{7320}, c_{6411}, c_{5511}, c_{4611}, c_{3720}, c_{2820}, c_{1920}]^T, \\
\boldsymbol{c}_2 &= [c_{8130}, c_{7230}, c_{6321}, c_{7412}, c_{4512}, c_{3621}, c_{2730}, c_{1830}]^T, \\
\boldsymbol{c}_3 &= [c_{7140}, c_{6231}, c_{5322}, c_{4413}, c_{43522}, c_{2631}, c_{1740}]^T.
\end{aligned}
$$

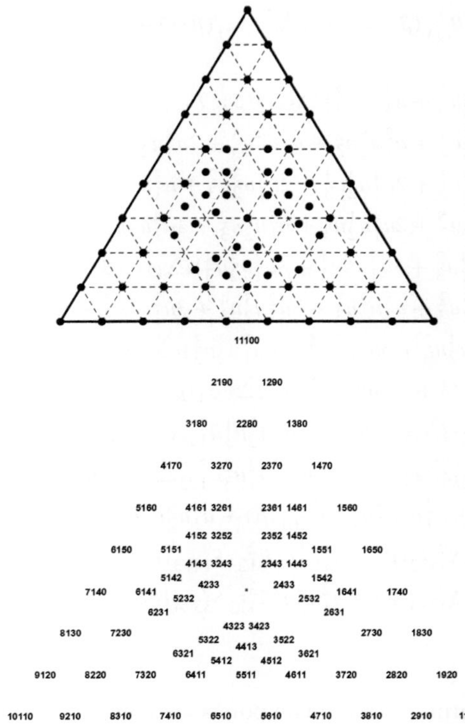

Fig. 6 Simplex splines domain points on the right, with their positions on the left. S^9_{3324}, S^9_{3234}, S^9_{2334} have the same domain point $(1/3, 1/3, 1/3)$ as indicated by a *

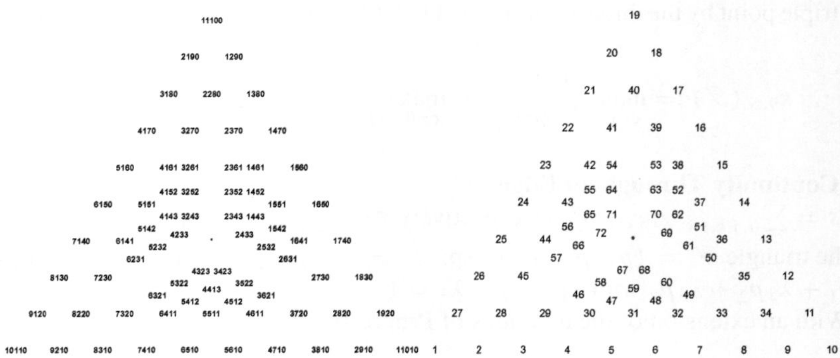

Fig. 7 Simplex splines domain points on the left, with their sorting on the left. S^9_{3324}, S^9_{3234}, S^9_{2334} indicated by a * have the numbers 73, 74, 75

For $r \leq 3$, we connect with smoothness $\mathscr{C}^r$ two adjacent triangles in the following proposition.

Proposition 6 *Let* $s=\sum_{(i,j,k,\ell)\in\mathscr{I}^2} c_{ijk\ell} S^9_{ijk\ell}$ *(respectively* $\tilde{s}=\sum_{(i,j,k,\ell)\in\mathscr{I}^2} \tilde{c}_{ijk\ell} \tilde{S}^9_{ijk\ell}$*) be defined on the triangle* $\mathscr{T} := \langle \boldsymbol{p}_1, \boldsymbol{p}_2, \boldsymbol{p}_3\rangle$ *(resp.* $\tilde{\mathscr{T}} := \langle \boldsymbol{p}_1, \boldsymbol{p}_2, \tilde{\boldsymbol{p}}_3\rangle$*). We suppose that* $\tilde{\boldsymbol{p}}_3 = \lambda_1 \boldsymbol{p}_1 + \lambda_2 \boldsymbol{p}_2 + \lambda_3 \boldsymbol{p}_3$ *with* $\lambda_1+\lambda_2+\lambda_3 = 1$.

The function $s_+ = \begin{cases} s & \text{on } \mathscr{T} \\ \tilde{s} & \text{on } \tilde{\mathscr{T}} \end{cases}$ *is* $\mathscr{C}^r$ *with* $r \leq 3$ *if and only if*

$$[\tilde{\boldsymbol{c}}_m]_{m=0,\dots,r} = [\boldsymbol{C}_{mn}]_{m,n=0,\dots,r}[\boldsymbol{c}_n]_{n=0,\dots,n}, \quad \boldsymbol{C}_{mn} = \in \mathbb{R}^{(10-m)\times(10-n)},$$

where the nonzero submatrices or components are written below in (35) (36), (37) *and* (38).

The proof of the proposition is a reproduction of the one of Proposition 5, firstly by connecting two polynomials written in the two corresponding Bernstein basis, then computing the Bernstein polynomials in the simplex-spline basis.

$$\boldsymbol{C}_{00} = \boldsymbol{I} \in \mathbb{R}^{10\times 10}, \tag{35}$$

$$\boldsymbol{C}_{10} = \begin{bmatrix} \lambda_1 & \lambda_2 & 0 & \dots & 0 \\ 0 & \lambda_1 & \lambda_2 & \ddots & \vdots \\ \vdots & \ddots & \ddots & \ddots & 0 \\ 0 & \dots & 0 & \lambda_1 & \lambda_2 \end{bmatrix} \in \mathbb{R}^{9\times 10}, \quad \boldsymbol{C}_{11} = \lambda_3 \boldsymbol{I} \in \mathbb{R}^{9\times 9}, \tag{36}$$

$$\begin{array}{lcccc}
\boldsymbol{C}_{20}(1,1:3) & = & \lambda_1^2 & 2\lambda_1\lambda_2 & {\lambda_2}^2 \\
\boldsymbol{C}_{20}(2,2:4) & = & \lambda_1^2 & 2\lambda_1\lambda_2 & {\lambda_2}^2 \\
\boldsymbol{C}_{20}(3,3:5) & = & \frac{\lambda_1^2}{2} & \lambda_1\lambda_2 + \frac{\lambda_1}{2} & \frac{\lambda_2^2}{2} + \frac{\lambda_2}{2} \\
\boldsymbol{C}_{20}(4,4:6) & = & \frac{\lambda_1^2}{3} + \frac{\lambda_1}{3} & \frac{2\lambda_1\lambda_2}{3} + \frac{\lambda_2}{3} + \frac{\lambda_1}{3} & \frac{\lambda_2^2}{3} + \frac{\lambda_2}{3} \\
\boldsymbol{C}_{20}(5,5:7) & = & \frac{\lambda_1^2}{3} + \frac{\lambda_1}{3} & \frac{2\lambda_1\lambda_2}{3} + \frac{\lambda_2}{3} + \frac{\lambda_1}{3} & \frac{\lambda_2^2}{3} + \frac{\lambda_2}{3} \\
\boldsymbol{C}_{20}(6,6:8) & = & \frac{\lambda_1^2}{2} + \frac{\lambda_1}{2} & \lambda_1\lambda_2 + \frac{\lambda_2}{2} & \frac{\lambda_2^2}{2} \\
\boldsymbol{C}_{20}(7,7:9) & = & \lambda_1^2 & 2\lambda_1\lambda_2 & \lambda_2^2 \\
\boldsymbol{C}_{20}(8,8:10) & = & \lambda_1^2 & 2\lambda_1\lambda_2 & \lambda_2^2
\end{array}$$

$$\begin{array}{lccc}
\boldsymbol{C}_{21}(1,1:2) = & 2\lambda_1\lambda_3 & 2\lambda_2\lambda_3 \\
\boldsymbol{C}_{21}(2,2:3) = & 2\lambda_1\lambda_3 & 2\lambda_2\lambda_3 \\
\boldsymbol{C}_{21}(3,3:4) = & \lambda_1\lambda_3 & \frac{2\lambda_2\lambda_3-\lambda_3^2}{2} + \frac{\lambda_3}{2} \\
\boldsymbol{C}_{21}(4,4:5) = & \frac{2\lambda_1\lambda_3-\lambda_3^2}{3} + \frac{\lambda_3}{3} & \frac{2\lambda_2\lambda_3-\lambda_3^2}{3} + \frac{\lambda_3}{3} \\
\boldsymbol{C}_{21}(5,5:6) = & \frac{2\lambda_1\lambda_3-\lambda_3^2}{3} + \frac{\lambda_3}{3} & \frac{2\lambda_2\lambda_3-\lambda_3^2}{3} + \frac{\lambda_3}{3} \\
\boldsymbol{C}_{21}(6,6:7) = & \frac{2\lambda_1\lambda_3-\lambda_3^2}{2} + \frac{\lambda_3}{2} & \lambda_2\lambda_3 \\
\boldsymbol{C}_{21}(7,7:8) = & 2\lambda_1\lambda_3 & 2\lambda_2\lambda_3 \\
\boldsymbol{C}_{21}(8,8:9) = & 2\lambda_1\lambda_3 & 2\lambda_2\lambda_3
\end{array}, \quad \boldsymbol{C}_{22} = \lambda_3^2 \boldsymbol{I} \in \mathbb{R}^{8\times 8}, \tag{37}$$

$$
\begin{array}{lcccc}
\boldsymbol{C}_{30}(1,1:4) = & \lambda_1^3 & 3\lambda_1^2\lambda_2 & 3\lambda_1\lambda_2^2 & \lambda_2^3 \\
\boldsymbol{C}_{30}(2,2:5) = & \frac{\lambda_1^3}{2} & \frac{3\lambda_1^2\lambda_2}{2} + \frac{\lambda_1^2}{2} & \frac{3\lambda_1\lambda_2^2}{2} + \lambda_1\lambda_2 & \frac{\lambda_2^3}{2} + \frac{\lambda_2^2}{2} \\
\boldsymbol{C}_{30}(3,3:6) = & \frac{\lambda_1^3}{6} + \frac{\lambda_1^2}{6} & \frac{\lambda_1^2\lambda_2}{2} + \frac{\lambda_1\lambda_2}{3} + \frac{\lambda_1^2}{3} + \frac{\lambda_1}{6} & \frac{\lambda_1\lambda_2^2}{2} + \frac{\lambda_2^2}{6} + \frac{2\lambda_1\lambda_2}{3} + \frac{\lambda_2}{6} + \frac{\lambda_1}{6} & \frac{\lambda_2^3}{6} + \frac{\lambda_2^2}{3} + \frac{\lambda_2}{6} \\
\boldsymbol{C}_{30}(4,4:7) = & \frac{\lambda_1^3}{9} + \frac{2\lambda_1^2}{9} + \frac{\lambda_1}{9} & \frac{\lambda_1^2\lambda_2}{3} + \frac{4\lambda_1\lambda_2}{9} + \frac{\lambda_2}{9} + \frac{2\lambda_1^2}{9} + \frac{2\lambda_1}{9} & \frac{\lambda_1\lambda_2^2}{3} + \frac{2\lambda_2^2}{9} + \frac{4\lambda_1\lambda_2}{9} + \frac{2\lambda_2}{9} + \frac{\lambda_1}{9} & \frac{\lambda_2^3}{9} + \frac{2\lambda_2^2}{9} + \frac{\lambda_2}{9} \\
\boldsymbol{C}_{30}(5,5:8) = & \frac{\lambda_1^3}{6} + \frac{\lambda_1^2}{3} + \frac{\lambda_1}{6} & \frac{\lambda_1^2\lambda_2}{2} + \frac{2\lambda_1\lambda_2}{3} + \frac{\lambda_2}{6} + \frac{\lambda_1^2}{6} + \frac{\lambda_1}{6} & \frac{\lambda_1\lambda_2^2}{2} + \frac{\lambda_2^2}{3} + \frac{\lambda_1\lambda_2}{3} + \frac{\lambda_2}{6} & \frac{\lambda_2^3}{6} + \frac{\lambda_2^2}{6} \\
\boldsymbol{C}_{30}(6,6:9) = & \frac{\lambda_1^3}{2} + \frac{\lambda_1^2}{2} & \frac{3\lambda_1^2\lambda_2}{2} + \lambda_1\lambda_2 & \frac{3\lambda_1\lambda_2^2}{2} + \frac{\lambda_2^2}{2} & \frac{\lambda_2^3}{2} \\
\boldsymbol{C}_{30}(7,7:10) = & \lambda_1^3 & 3\lambda_1^2\lambda_2 & 3\lambda_1\lambda_2^2 & \lambda_2^3
\end{array}
\tag{38}
$$

$$
\begin{array}{lcc}
\boldsymbol{C}_{31}(1,1:2) = & 3\lambda_1^2\lambda_3 & 6\lambda_1\lambda_2\lambda_3 \\
\boldsymbol{C}_{31}(1,3) = & 3\lambda_2^2\lambda_3 & \\
\boldsymbol{C}_{31}(2,2:3) = & \frac{3\lambda_1^2\lambda_3}{2} & 3\lambda_1\lambda_2\lambda_3 + \lambda_1\lambda_3 \\
\boldsymbol{C}_{31}(2,4) = & \frac{\lambda_3^3 - 3\lambda_2\lambda_3^2 + 3\lambda_2^2\lambda_3}{2} + \frac{2\lambda_2\lambda_3 - \lambda_3^2}{2} & \\
\boldsymbol{C}_{31}(3,3:4) = & \frac{\lambda_1^2\lambda_3}{2} + \frac{\lambda_1\lambda_3}{3} & \frac{2\lambda_3^3 - 3\lambda_2\lambda_3^2 - 3\lambda_1\lambda_3^2 + 6\lambda_1\lambda_2\lambda_3}{6} + \frac{2\lambda_2\lambda_3 - \lambda_3^2}{6} + \frac{2\lambda_1\lambda_3 - \lambda_3^2}{3} + \frac{\lambda_3}{6} \\
\boldsymbol{C}_{31}(3,5) = & \frac{\lambda_3^3 - 3\lambda_2\lambda_3^2 + 3\lambda_2^2\lambda_3}{6} + \frac{2\lambda_2\lambda_3 - \lambda_3^2}{3} + \frac{\lambda_3}{6} & \\
\boldsymbol{C}_{31}(4,4:5) = & \frac{\lambda_3^3 - 3\lambda_1\lambda_3^2 + 3\lambda_1^2\lambda_3}{9} + \frac{2(2\lambda_1\lambda_3 - \lambda_3^2)}{9} + \frac{\lambda_3}{9} & \frac{2\lambda_3^3 - 3\lambda_2\lambda_3^2 - 3\lambda_1\lambda_3^2 + 6\lambda_1\lambda_2\lambda_3}{9} + \frac{2(2\lambda_2\lambda_3 - \lambda_3^2)}{9} + \frac{2(2\lambda_1\lambda_3 - \lambda_3^2)}{9} + \frac{2\lambda_3}{9} \\
\boldsymbol{C}_{31}(4,6) = & \frac{\lambda_3^3 - 3\lambda_2\lambda_3^2 + 3\lambda_2^2\lambda_3}{9} + \frac{2(2\lambda_2\lambda_3 - \lambda_3^2)}{9} + \frac{\lambda_3}{9} & \\
\boldsymbol{C}_{31}(5,5:6) = & \frac{\lambda_3^3 - 3\lambda_1\lambda_3^2 + 3\lambda_1^2\lambda_3}{6} + \frac{2\lambda_1\lambda_3 - \lambda_3^2}{3} + \frac{\lambda_3}{6} & \frac{2\lambda_3^3 - 3\lambda_2\lambda_3^2 - 3\lambda_1\lambda_3^2 + 6\lambda_1\lambda_2\lambda_3}{6} + \frac{2\lambda_2\lambda_3 - \lambda_3^2}{3} + \frac{2\lambda_1\lambda_3 - \lambda_3^2}{6} + \frac{\lambda_3}{6} \\
\boldsymbol{C}_{31}(5,7) = & \frac{\lambda_2^2\lambda_3}{2} + \frac{\lambda_2\lambda_3}{3} & \\
\boldsymbol{C}_{31}(6,6:7) = & \frac{\lambda_3^3 - 3\lambda_1\lambda_3^2 + 3\lambda_1^2\lambda_3}{2} + \frac{2\lambda_1\lambda_3 - \lambda_3^2}{2} & \frac{3\lambda_2^2\lambda_3}{2} \\
\boldsymbol{C}_{31}(6,8) = & 3\lambda_1\lambda_2\lambda_3 + \lambda_2\lambda_3 & \\
\boldsymbol{C}_{31}(7,7:8) = & 3\lambda_1^2\lambda_3 & 6\lambda_1\lambda_2\lambda_3 \\
\boldsymbol{C}_{31}(7,9) = & 3\lambda_2^2\lambda_3 &
\end{array}
$$

$$
\begin{array}{lcc}
\boldsymbol{C}_{32}(1,1:2) = & 3\lambda_1\lambda_3^2 & 3\lambda_2\lambda_3^2 \\
\boldsymbol{C}_{32}(2,2:3) = & \frac{3\lambda_1\lambda_3^2}{2} & \frac{6\lambda_2\lambda_3^2 - 2\lambda_3^3}{2} + \lambda_3^2 \\
\boldsymbol{C}_{32}(3,3:4) = & \frac{6\lambda_1\lambda_3^2 - 2\lambda_3^3}{6} + \frac{\lambda_3^2}{3} & \frac{9\lambda_2\lambda_3^2 - 6\lambda_3^3}{6} + \lambda_3^2 \\
\boldsymbol{C}_{32}(4,4:5) = & \frac{9\lambda_1\lambda_3^2 - 6\lambda_3^3}{9} + \frac{2\lambda_3^2}{3} & \frac{9\lambda_2\lambda_3^2 - 6\lambda_3^3}{9} + \frac{2\lambda_3^2}{3} \\
\boldsymbol{C}_{32}(5,5:6) = & \frac{9\lambda_1\lambda_3^2 - 6\lambda_3^3}{6} + \lambda_3^2 & \frac{6\lambda_2\lambda_3^2 - 2\lambda_3^3}{6} + \frac{\lambda_3^2}{3} \\
\boldsymbol{C}_{32}(6,6:7) = & \frac{6\lambda_1\lambda_3^2 - 2\lambda_3^3}{2} + \lambda_3^2 & \frac{3\lambda_2\lambda_3^2}{2} \\
\boldsymbol{C}_{32}(7,7:8) = & 3\lambda_1\lambda_3^2 & 3\lambda_2\lambda_3^2
\end{array}
, \quad \boldsymbol{C}_{33} = \lambda_3^3 \boldsymbol{I} \in \mathbb{R}^{7\times 7}.
$$

5 Conclusion

For any $r \geq 1$, we have built a B-spline like basis made out of simplex splines for the space $\mathbb{S}^r_{3r}(\triangle)$ of splines on the Clough-Tocher split $\triangle$ on a single triangle. For even values of r, we removed one of the elements in order to obtain the partition of unity and a Marsden-like identity proved for $r \leq 3$ and shown symbolically for $r \leq 6$. Looking in more detail at the cases $r = 2, 3$, corresponding to degrees $d = 6, 9$, we gave explicit formulas for connecting two neighboring triangles in a $\mathscr{C}^r$ fashion across an edge using Bernstein-Bézier techniques, and gave an upper bound for the L_∞ condition number of the basis.

For $r = 4, 5, 6$ the domain points can be computed as for $r \leq 3$ using the Marsden like-identity shown below. This gives interpolation points to study the stability. The coefficients to obtain the $\mathscr{C}^r$**-connection** between two triangles $\mathscr{T} := \langle \boldsymbol{p}_1, \boldsymbol{p}_2, \boldsymbol{p}_3 \rangle$

and $\tilde{\mathscr{T}} := \langle \boldsymbol{p}_1, \boldsymbol{p}_2, \tilde{\boldsymbol{p}}_3 \rangle$ can be found by a computation in the Bernstein basis of the polynomials in $\langle \boldsymbol{p}_1, \boldsymbol{p}_2, \boldsymbol{p}_T \rangle$ and $\langle \boldsymbol{p}_1, \boldsymbol{p}_2, \tilde{\boldsymbol{p}}_T \rangle$.

We end by restating the Marsden-like identity in a general form. It is proved for $r = 1, 2, 3$ and symbolically for $r = 4, 5, 6$. It is a conjecture for $r > 6$.

Theorem 4 (Barycentric Marsden-Like Identity for Degree $3r$, $r \leq 6$) *For $r \in \mathbb{N}$, $d = 3r$, $u_i, \beta_i \in \mathbb{R}$, with $\beta_i \geq 0$, $i = 1, 2, 3$, and $\beta_1 + \beta_2 + \beta_3 = 1$ we have*

$$(u_1\beta_1 + u_2\beta_2 + u_3\beta_3)^d = \sum_{(i,j,k,\ell)\in\bar{\mathscr{I}}^r} \rho_{ijk\ell}(u_1, u_2, u_3) S^d_{ijk\ell}(\beta_1, \beta_2, \beta_3),$$

where the index set $\bar{\mathscr{I}}^r$ is given in Definition 3 and

$$S^{3r}_{ijk\ell} := c_{ijk\ell} \triangle_{ijk\ell}, \text{ where } c_{ijk\ell} := \begin{cases} 1, & \text{if } \ell = 0, \\ 2/3 & \text{if } \max(\epsilon_1, \epsilon_2, \epsilon_3) = 1/2, \\ 1/3 & \text{otherwise,} \end{cases}$$
$$\rho_{ijk\ell}(u_1, u_2, u_3) := u_1^{i-1} u_2^{j-1} u_3^{k-1} \bar{u}_{123}^{\lambda-1} (\epsilon_1 u_1 + \epsilon_2 u_2 + \epsilon_3 u_3 + \delta_{0\ell})^{1-\delta_{0\ell}}. \tag{39}$$

Here, $\lambda := \max(\ell, 1)$, $\bar{u}_{123} := (u_1 + u_2 + u_3)/3$, $\nu := \max(i, j, k)$, *and*

$$\epsilon_1 := \begin{cases} 1/\gamma_1, & i < \nu \;\&\; j \neq 1 \;\&\; k \neq 1, \\ 0, & \text{otherwise} \end{cases}, \quad \gamma_1 := \begin{cases} 2, & i \neq 1 \;\&\; j \neq k, \\ 1, & \text{otherwise}, \end{cases}$$

$$\epsilon_2 := \begin{cases} 1/\gamma_2, & j < \nu \;\&\; i \neq 1 \;\&\; k \neq 1, \\ 0, & \text{otherwise} \end{cases}, \quad \gamma_2 := \begin{cases} 2, & j \neq 1 \;\&\; i \neq k, \\ 1, & \text{otherwise}, \end{cases}$$

$$\epsilon_3 := \begin{cases} 1/\gamma_3, & k < \nu \;\&\; i \neq 1 \;\&\; j \neq 1, \\ 0, & \text{otherwise} \end{cases}, \quad \gamma_3 := \begin{cases} 2, & k \neq 1 \;\&\; i \neq j, \\ 1, & \text{otherwise}. \end{cases}$$

Proof. For $r = 1, 2, 3$ this is an alternative way of formulating Theorem 5 in [18] for $r = 1$, and Theorems 2 and 3. To see this consider first $r = 1$. In [18] it was shown that

$$\begin{aligned}(\boldsymbol{\beta}^T \boldsymbol{u})^3 &= u_1^3 S_1(\boldsymbol{\beta}) + u_1^2 u_2 S_2(\boldsymbol{\beta}) + u_1 u_2^2 S_3(\boldsymbol{\beta}) + u_2^3 S_4(\boldsymbol{\beta}) + u_2^2 u_3 S_5(\boldsymbol{\beta}) \\ &\quad + u_2 u_3^2 S_6(\boldsymbol{\beta}) + u_3^3 S_7(\boldsymbol{\beta}) + u_1 u_3^2 S_8(\boldsymbol{\beta}) + u_1^2 u_3 S_9(\boldsymbol{\beta}) \\ &\quad + u_1 u_2 u_3 \big(S_{10}(\boldsymbol{\beta}) + S_{11}(\boldsymbol{\beta}) + S_{12}(\boldsymbol{\beta})\big),\end{aligned} \tag{40}$$

where $\boldsymbol{\beta} := (\beta_1, \beta_2, \beta_3)$ and $\boldsymbol{u} := (u_1, u_2, u_3)$. For the first 9 (Bernstein) dual functions we find $\epsilon_1 = \epsilon_2 = \epsilon_3 = 0$, $\delta_{0\ell} = 1$, and (39) holds. For the dual function $\rho[1221]$ corresponding to $S_{10} = \triangle_{1221}/3$ we find $(\epsilon_1 u_1 + \epsilon_2 u_2 + \epsilon_3 u_3 + \delta_{0\ell})^{1-\delta_{0\ell}} = u_1$ and (39) gives $\rho_{1221}(\boldsymbol{u}) S^3_{1221}(\boldsymbol{\beta}) = u_1 u_2 u_3 S_{10}(\boldsymbol{\beta})$ as stated in (40). The results for $\rho[2121]$ and $\rho[2211]$ are similar. For $r = 2, 3$ it follows as for $r = 1$ that (39) holds for the Bernstein polynomials in $\mathscr{I}^2_{\text{Bernstein}}$. For $r = 2$ consider for example

$\rho_{2322}(\boldsymbol{u}) = u_1 u_2^2 u_3 \bar{u}_{1,3} \bar{u}_{123}$ and $S_{2322} = 2\triangle_{2332}/3$ in Theorem 2. This is the same as the expressions in (39) since $\epsilon_1 = \epsilon_3 = 1/2$ and $\epsilon_3 = 0$. As an example for $r = 3$, $\rho_{3324}(\boldsymbol{u}) = u_1^2 u_2^2 u_3^2 \bar{u}_{123}^3$ and $S_{3324} = \triangle_{3324}/3$ in Theorem 3 is the same as the expressions in (39) since $\epsilon_1 = \epsilon_2 = 0$ and $\epsilon_3 = 1$. All other cases are verified similarly. □

Acknowledgements We thank the referees for their extremely thorough and careful review and for pointing out a flaw in the first version of the paper.

References

1. Alfeld, P.: A bivariate C^2 Clough-Tocher scheme. Comput. Aided Geom. Des. **1**, 257–267 (1984)
2. Alfeld, P., Schumaker, L.L.: Smooth macro-elements based on Powell-Sabin triangle splits. Adv. Comput. Math. **16**, 29–46 (2002)
3. Cao, J., Chen, Z., Wei, X., Zhang, Y.J.: A finite element framework based on bivariate simplex splines on triangle configurations. Comput. Methods Appl. Mech. Eng. **357**, 112–598 (2019)
4. Ciarlet, P.G.: The Finite Element Method for Elliptic Problems, Classics in Applied Mathematics, vol. 40. Society for Industrial and Applied Mathematics (SIAM), Philadelphia (2002)
5. Clough, R.W., Tocher, J.L.: Finite element stiffness matrices for analysis of plates in bending. In: Proceedings of the Conference on Matrix Methods in Structural Mechanics, pp. 515–545. Wright-Patterson Air Force Base (1965)
6. Cohen, E., Lyche, T., Riesenfeld, R.F.: A B-spline-like basis for the Powell-Sabin 12-split based on simplex splines. Math. Comput. **82**, 1667–1707 (2013)
7. Cohen, E., Riesenfeld, R.F., Elber, G.: Geometric Modeling with Splines. A K Peters Ltd., Natick (2001)
8. Cottrell, J.A., Hughes, T.J.R., Bazilevs, Y.: Isogeometric Analysis: Toward Integration of CAD and FEA. Wiley Publishing (2009)
9. Davydov, O., Yeo, W.P.: Refinable C^2 piecewise quintic polynomials on Powell-Sabin-12 triangulations. J. Comput. Appl. Math. **240**, 62–73 (2013)
10. Dierckx, P.: On calculating normalized Powell-Sabin B-splines. Comput. Aided Geom. Des. **15**, 61–78 (1997)
11. Farin, G.: A modified Clough-Tocher interpolant. Comput. Aided Geom. Des. **2**, 19–27 (1985)
12. Grošelj, J.: A normalized representation of super splines of arbitrary degree on Powell-Sabin triangulations. BIT Numer. Math. **56**, 1257–1280 (2016)
13. Grošelj, J., Speleers, H.: Construction and analysis of cubic Powell-Sabin B-splines. Comput. Aided Geom. Des. **57**, 1–22 (2017)
14. Laghchim-Lahlou, M., Sablonnière, P.: Triangular finite elements of HCT type and class C^ρ. Adv. Comput. Math. **2**, 101–122 (1994)
15. Lai, M.J., Schumaker, L.L.: Macro-elements and stable local bases for splines on Clough-Tocher triangulations. Numerische Mathematik **88**, 105–119 (2001)
16. Lai, M.J., Schumaker, L.L.: Macro-elements and stable local bases for splines on Powell-Sabin triangulations. Math. Comput. **72**, 335–354 (2003)
17. Lai, M.J., Schumaker, L.L.: Spline Functions on Triangulations, Encyclopedia of Mathematics and its Applications, vol. 110. Cambridge University Press, Cambridge (2007)
18. Lyche, T., Merrien, J.L.: Simplex-splines on the Clough-Tocher element. Comput Aided Geom Des **65**, 76–92 (2018)
19. Lyche, T., Muntingh, G.: Stable simplex spline bases for C^3 quintics on the Powell-Sabin 12-split. Constr. Approx **45**, 1–32 (2017)
20. Mann, S.: Cubic precision Clough-Tocher interpolation. Comput. Aided Geom. Des. **16**, 85–88 (1999)

21. Micchelli, C.A.: On a numerically efficient method for computing multivariate B-splines. In: Schempp, W., Zeller, K., (eds.), Multivariate Approximation Theory, International Series of Numerical Mathematics, vol. 51, pp. 211–248. Birkhäuser, Basel-Boston (1979)
22. Powell, M.J.D., Sabin, M.A.: Piecewise quadratic approximations on triangles. ACM Trans. Math. Softw. **3**, 316–325 (1977)
23. Prautzsch, H., Boehm, W., Paluszny, M.: Bézier and B-Spline Techniques. Mathematics and Visualization, Springer, Berlin (2002)
24. Sablonnière, P.: Composite finite elements of class C^k. J. Comput. Appl. Math. **12–13**, 541–550 (1985)
25. Schumaker, L.L., Sorokina, T.: Smooth macro-elements on Powell-Sabin-12 splits. Math. Comput. **75**, 711–726 (2006)
26. Speleers, H.: A normalized basis for quintic Powell-Sabin splines. Comput. Aided Geom. Des. **27**, 438–457 (2010)
27. Speleers, H.: A normalized basis for reduced Clough-Tocher splines. Comput. Aided Geom. Des. **27**, 700–712 (2010)
28. Speleers, H.: Construction of normalized B-splines for a family of smooth spline spaces over Powell-Sabin triangulations. Constr. Approx. **37**, 41–72 (2013)
29. Speleers, H.: A family of smooth quasi-interpolants defined over Powell-Sabin triangulations. Constr. Approx. **41**, 297–324 (2015)
30. Speleers, H.: A new B-spline representation for cubic splines over Powell-Sabin triangulations. Comput. Aided Geom. Des. **37**, 42–56 (2015)
31. Speleers, H., Manni, C.: Optimizing domain parameterization in isogeometric analysis based on Powell-Sabin splines. J. Comput. Appl. Math. **289**, 68–86 (2015)
32. Speleers, H., Manni, C., Pelosi, F., Sampoli, M.L.: Isogeometric analysis with Powell-Sabin splines for advection-diffusion-reaction problems. Comput. Methods Appl. Mech. Eng. **221–222**, 132–148 (2012)
33. Toshniwal, D., Hughes, T.J.R.: Polynomial splines of non-uniform degree on triangulations: Combinatorial bounds on the dimension. Comput. Aided Geom. Des. **75**, 101–763 (2019)
34. Wang, C., Xia, S., Wang, X., Qian, X.: Isogeometric shape optimization on triangulations. Comput. Methods Appl. Mech. Eng. **331**, 585–622 (2018)
35. Wang, R.H., Shi, X.Q.: $S^{\mu}_{\mu+1}$ surface interpolations over triangulations. In: Law, A.G., Wang, C.L. (eds.), Approximation, Optimization and Computing: Theory and Applications, pp. 205–208. Elsevier Science Publishers B.V. (1990)
36. Ženíšek, A.: A general theorem on triangular finite $C^{(m)}$-elements. Rev. Française Automat. Informat. Recherche Opérationnelle Sér. Rouge **8** (1974)

Trimmed Spline Surfaces with Accurate Boundary Control

Florian Martin and Ulrich Reif

Abstract We introduce trimmed NURBS surfaces with accurate boundary control, briefly called ABC-surfaces, as a solution to the notorious problem of constructing watertight or smooth (G^1 and G^2) multi-patch surfaces within the function range of standard CAD/CAM systems and the associated file exchange formats. Our construction is based on the appropriate blend of a base surface, which traces out the intended global shape, and a series of reparametrized ribbons, which dominate the shape near the boundary.

1 Introduction

Trimmed NURBS surfaces are the standard for industrial surface modeling, today. This class of surfaces combines computational efficiency, geometric flexibility, and ease of use for the designer. All current CAD/CAM systems and various file exchange formats support this–and typically only this–method. While the advantages of trimmed NURBS are undisputed, there is one severe and persistent shortcoming that has not found a commonly accepted solution in decades: multi-patch arrangements of trimmed NURBS surfaces are typically discontinuous along common boundary curves of neighboring pieces. The problem is the following: Given a tensor product spline surface $\mathbf{s} : \mathbb{R}^2 \to \mathbb{R}^3$, current CAD systems provide two ways of representing the boundaries of a trimmed NURBS surface. Either one can use a planar spline curve $\gamma : \mathbb{R} \to \mathbb{R}^2$ to bound the domain or one can use a spatial spline curve $\mathbf{c} : \mathbb{R} \to \mathbb{R}^3$ to bound the trace. The first variant leads to the boundary curve $\mathbf{s} \circ \gamma$ in three-space. This is a spline curve, again, but typically, it has high degree and many knots, stemming not only from the breaks of γ but also from all points where the trace of γ intersects a knot line of $\mathbf{s}$. The resulting knot structure is uneven

F. Martin · U. Reif (✉)
Darmstadt University of Technology, Darmstadt, Germany
e-mail: reif@mathematik.tu-darmstadt.de

F. Martin
e-mail: fmartin@mathematik.tu-darmstadt.de

C. Manni and H. Speleers (eds.), *Geometric Challenges in Isogeometric Analysis*, Springer INdAM Series 49, https://doi.org/10.1007/978-3-030-92313-6_6

and hardly predictable.[1] This fact makes it virtually impossible to match $\mathbf{s} \circ \gamma$ with any other surface boundary at reasonable expense. The second variant is equally discouraging. It seems to be impossible to find a surface **s** and a curve **c** such that the trace of **s** contains the trace of **c** exactly. The makeshift for this problem is to accept disjoint traces, and to project **c** onto **s** in order to determine the actual trimming curve. In general, however, this curve is no longer a spline and again not suitable for a watertight match with a neighboring patch.

To avoid the described problem, one can try to find arrangements of patches for the desired shape which do without trimming. This approach leads to methods like geometric continuity [3], singular parametrization [7], or subdivision [4], and also to a variety of constructions for n-sided patches [5, 10]. As a matter of fact, none of these concepts has found widespread acceptance in industrial CAD/CAM applications. It may be speculated that a certain perseverance of designers and executives is responsible for that situation. However, there is also a more solid explanation. The flow of parametric lines is most important for the design process.[2] On one hand, the parametric lines should follow the shape characteristics (e.g., ridges or curvature lines) of the desired surface to obtain convincing results. On the other hand, a 'natural' arrangement of control meshes greatly favors the intuitive character of interactive modeling techniques. Thus, there is little flexibility in the choice of an adequate parametrization, and the corresponding patches must be trimmed at places where neighbors meet. To sum up, the advantages of a well-adapted parametrization seem to be rated higher than the notorious trouble with the trimming issue.

In CAD/CAM applications, the usual solution to the problem of nonmatching trimming curves is to simply accept little gaps in surface representations, and to make them so small that they do not impair the quality of the product to be manufactured. This allows modeling surfaces of adequate quality, but the price to pay is a certain inconvenience of the design process. The challenge is to design patches that are not only fair by themselves but join with neighbors within a typically very small margin.[3] Further, not only positions must be matched, but possibly also surface normals and curvature tensors in the sense of an approximate G^1- or G^2-continuity of the composite surface. For high quality surface design, as requested for instance in the automotive industry, the only way to achieve this is a time-consuming manual process of iterated boundary adaption and fairing.

Besides these practical considerations, there is another issue with the current state of technology. Surface models with disconnected pieces are not well suited for simulation purposes. No matter how small the gaps are, they appear insurmountable[4] for heat conduction, transmission of forces and bending moments, or other physical pro-

[1] Basically, the only case which does not lead to exploding complexity is the special situation where γ parametrizes a straight line parallel to one of the coordinate axes.

[2] This observation is a blatant contradiction to the concepts of differentiable geometry, which rely on the *independence* of geometric properties of the chosen parametrization.

[3] Customary tolerances for traditional production techniques have to be decreased even further for modern 3d printing systems with their incredible attention to detail.

[4] Advanced methods like discontinuous Galerkin may offer a remedy.

cesses. For this reason, multi-patch CAD models are usually converted to watertight meshes by elaborate procedures to prepare them for simulation. Volumetric modeling techniques as requested for Isogeometric Analysis with its potential to save this effort are currently studied. However, they seem to be quite far from a breakthrough for freeform surface design in industrial applications.

From the abundance of material on the design of surfaces of arbitrary topology in general and trimmed NURBS in particular, we want to briefly mention only two recent approaches. In [1], a procedure for converting an assembly of trimmed NURBS patches to a watertight model is suggested. It is based on local remodeling such that trimming curves become parametric lines. Without doubt, this is a reasonable practical procedure, even though some issues with a special kind of singularities appearing there (cf. [2]) and also general fairness aspects are still waiting for a detailed assessment and analysis. Another line of research that has to be discussed in this context is documented in [13]. It is based on ideas similar to those presented here, but there are also significant differences. Before we compare the methods, we briefly explain the basic principles of our approach.

Let $\mathbf{b}$ be a spline surface, called *base*, and $\mathbf{r}_1, \ldots, \mathbf{r}_L$ be a sequence of other spline surfaces, called *ribbons*. Then we define the *ABC-surface*

$$\mathbf{a} := \frac{w \cdot \mathbf{b} + \sum_{\ell=1}^{L} w_\ell \cdot \mathbf{r}_\ell \circ \kappa_\ell}{w + \sum_{\ell=1}^{L} w_\ell}$$

with suitable reparametrizations κ_ℓ and weight functions w and w_ℓ. Figure 1 illustrates the setting. It is instructive, but not necessary, to think of $\mathbf{b}$ as a standard spline surface describing the overall shape to be designed. When trimmed, its boundaries are close to the desired behavior, but do not match exactly. The ribbons $\mathbf{r}_\ell$ are employed to adjust that deviation.

The domain $\Gamma \subset \mathbb{R}^2$ of $\mathbf{a}$ is assumed to be bounded by L smooth segments $\Gamma_1, \ldots, \Gamma_L$ meeting at corners $\sigma_1, \ldots, \sigma_L$. The base $\mathbf{b}$ describes the intended shape globally on Γ, and each ribbon $\mathbf{r}_\ell$ describes that shape locally in a vicinity of the segment Γ_ℓ of the boundary. This is done in such a way that the spline curve $\mathbf{c}_\ell := \mathbf{r}_\ell(\cdot, 0)$

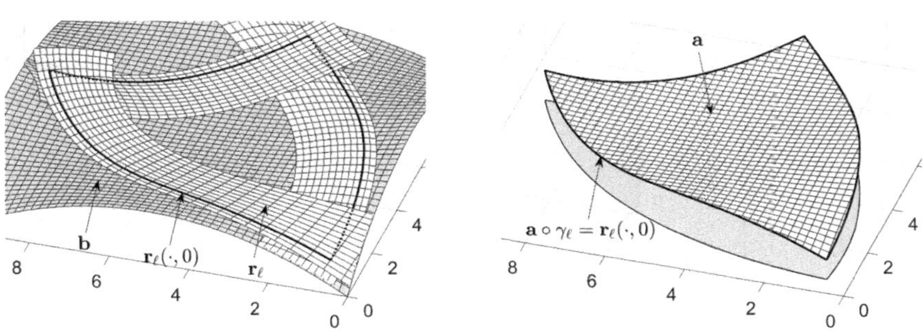

Fig. 1 Base $\mathbf{b}$ and ribbons $\mathbf{r}_\ell$ *(left)*; ABC-surface $\mathbf{a}$ *(right)*

parametrizes the segment $\mathbf{a}(\Gamma_\ell)$ of the boundary of the trace of $\mathbf{a}$. The reparametrizations κ_ℓ are used to reconcile the otherwise unrelated surface representations, and the weights are used for blending. When approaching the boundary, w vanishes and makes the base disappear. Further, near the boundary segment $\Gamma_j \subset \mathbb{R}^2$ corresponding to $\mathbf{c}_j$ all weights w_ℓ except for w_j vanish so that also all ribbons but $\mathbf{r}_j$ are faded out. Hence, this surface dominates the behavior of $\mathbf{a}$ near the boundary in a way that will be made precise below. The *accurate boundary control* of the surface $\mathbf{a}$ by means of ribbons explains the acronym *ABC-surface*.

The basic principle and some technical details of ABC-surfaces were first presented in [6]. Roughly at the same time and independent of our research, another group developed similar ideas [9], which were recently brought to our attention. The formal similarities are obvious, but there are also differences, and one of them is significant.

First, the weighted base $w\mathbf{b}$ does not appear in that form[5] in [9]. Instead, a linear combination $\sum_j b_j \mathbf{p}_j$ of control points $\mathbf{p}_j \in \mathbb{R}^3$ and B-splines b_j is used, where only B-splines with support completely contained in the domain of the surface are permitted so that there is no influence on the shape near the boundary. Accordingly, the term w in the denominator is replaced by $\sum_j b_j$. This can be considered a minor technical detail, but it also reveals a slightly different, though equally valid philosophy behind the construction. We think of the base as a relatively accurate description of the entire surface, and of the ribbons as small, local correctives, used only to adapt $\mathbf{b}$ to the desired boundary behavior. By contrast, the ribbons in [9] are the major building blocks for the shape to be modeled. The summands $b_j \mathbf{p}_j$, if included at all, are only used for variations inside if the shape defined by the ribbons alone does not yet meet expectations.

Second, and this difference is much more important, the surfaces constructed in [9, 13] are *transfinite* in the sense that they do not possess a closed-form representation by means of elementary functions. In particular, these surfaces are not piecewise polynomial or rational and thus cannot be stored in standard file formats, as requested in many applications. The point is that the curves γ_ℓ bounding the domain Γ are defined *explicitly* as splines. Once this is done, there is no way to continue the construction within the spline framework. For instance, it is in general impossible to find splines w_ℓ that vanish on all but one of these boundary curves.[6] In [9, 13], it is suggested to construct reparametrizations and weights as solutions of variational problems with the curves γ_ℓ providing the boundary conditions. This requests a certain computational effort, but visually, the results appear to be excellent. By contrast, in our construction, the boundary curves γ_ℓ are defined *implicitly* as zero levelsets of spline functions. That way, the problem described above disappears. It may be objected that explicit evaluation of the boundary of the domain is now much more

[5] It does appear in a similar form in [12] for the special case of straight boundaries and polynomial ribbons.

[6] The special case of straight boundaries, which admits rational constructions, was discussed in [8, 10–12].

expensive than it was before, but as a matter of fact, this is hardly ever requested in practice as long as the spatial boundary curves $\mathbf{c}_\ell$ are available.

The paper is organized as follows: In the next section, we introduce and analyze the basic principle for the construction of ABC-surfaces in a very general setting. Three theorems concerning the behavior of positions, normals, and curvature tensors at the boundary are provided. They establish the validity of our construction, but also that of the methods presented in the more application-oriented papers [8, 9, 13]. In Sect. 3, we specialize the principle to the spline setting and describe in detail how to construct adequate reparametrizations and weights from given bases and ribbons. Section 4 addresses some important technical issues related to the limitations of current standard formats for file exchange with respect to trimming. Finally, in Sect. 5, we present a few examples of ABC-spline surfaces to demonstrate the potential of our method.

2 Basic Principle

Throughout, boldface letters denote points and functions with values in $\mathbb{R}^3$, while Greek letters denote points and functions with values in $\mathbb{R}^2$. Except for the surface $\mathbf{a}$ to be constructed, all functions appearing below are assumed to be sufficiently smooth. In particular, their derivatives mentioned anywhere (explicitly or implicitly through notions like normals or curvatures) are assumed to exist and to be continuous. Further, to avoid a discussion of degenerate cases, surfaces are always assumed to be immersed, i.e., the partial derivatives are linearly independent everywhere. The following definitions are crucial for the construction of ABC-surfaces:

Definition 1 The function $p : \mathbb{R}^2 \to \mathbb{R}$ *parametrizes* the planar curve $\gamma : [0, 1] \to \mathbb{R}^2$ if $p(\gamma(u)) = u$ for all $u \in [0, 1]$, see Fig. 2 *(left)*. The function $q : \mathbb{R}^2 \to \mathbb{R}$ is an *implicit representation* of the planar curve γ if $q(\gamma(u)) = 0$ for all $u \in [0, 1]$, see

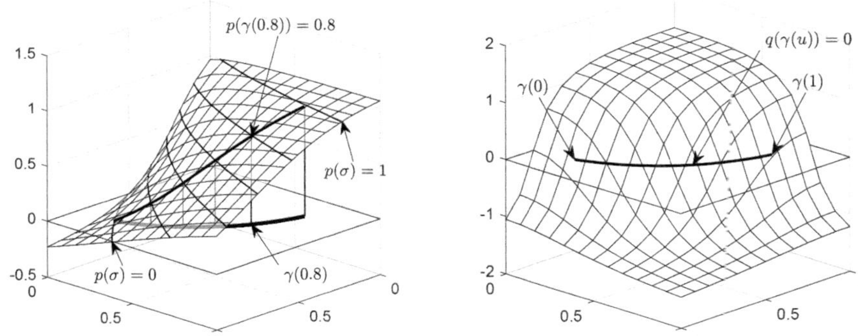

Fig. 2 Parametrizing function p *(left)* and implicit representation q *(right)*

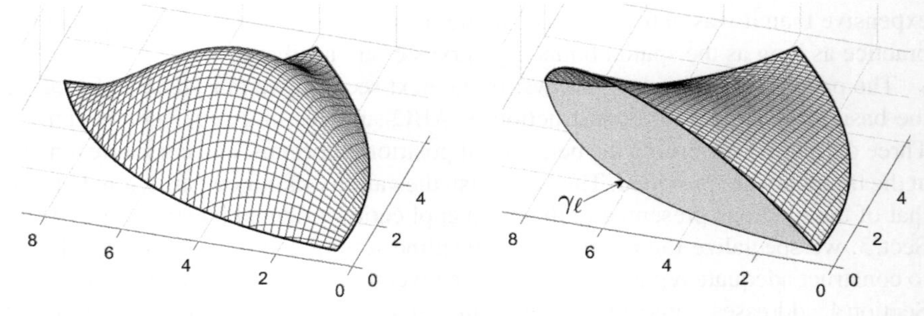

Fig. 3 Weights *w* *(left)* and w_ℓ *(right)*

Fig. 2 *(right)*. The function $\kappa := [p, q] : \mathbb{R}^2 \to \mathbb{R}^2$ *defines* γ if there exists a unique regular curve $\gamma : [0, 1] \to \mathbb{R}^2$ with implicit representation q that is parametrized by p.

Now, we describe the construction of a surface $\mathbf{a}$, whose domain $\Gamma \subset \mathbb{R}^2$ of definition is bounded by a sequence $\gamma_1, \ldots, \gamma_L$ of planar *trimming curves* $\gamma_\ell : [0, 1] \to \mathbb{R}^2$. When concatenated according to

$$\gamma_{\ell-1}(1) = \gamma_\ell(0), \tag{1}$$

these curves form a closed loop without self-intersections. Here and below, the index ℓ always runs from 1 through L and is understood modulo L. Let

$$\sigma_\ell := \gamma_\ell(0) \quad \text{and} \quad \Gamma_\ell := \{\gamma_\ell(u) : 0 \le u < 1\}$$

denote the *corners* and *boundary segments* of Γ, respectively. Formally, the endpoint $\gamma_\ell(1)$ is missing in the definition of Γ_ℓ to obtain a disjoint union $\partial\Gamma = \bigcup_\ell \Gamma_\ell$ for the boundary of Γ. Accordingly, for each boundary point $\beta \in \partial\Gamma$, there exist unique values ℓ and $u \in [0, 1)$ such $\beta = \gamma_\ell(u)$.

The building blocks for the construction of the surface $\mathbf{a}$ are the following:

- Let $\kappa_\ell := [p_\ell, q_\ell] : \mathbb{R}^2 \to \mathbb{R}^2$ be functions that define trimming curves γ_ℓ in the sense of Definition 1 according to the setting described above.
- Let $w : \mathbb{R}^2 \to \mathbb{R}$ be an implicit representation of all trimming curves. That is,

$$w \circ \gamma_j = 0, \quad j = 1, \ldots, L.$$

  Further, it is assumed that w is positive on the interior of Γ, see Fig. 3 *(left)*.
- For each ℓ, let $w_\ell : \mathbb{R}^2 \to \mathbb{R}^2$ be an implicit representation of all trimming curves but γ_ℓ. That is,

$$w_\ell \circ \gamma_j = 0, \quad j = 1, \ldots, \ell - 1, \ell + 1, \ldots, L.$$

Further, it is assumed that w_ℓ is nonnegative on Γ, and that the only zeros of the sum $\sum_\ell w_\ell$ in Γ are the corners, see Fig. 3 *(right)*.
- Let $\mathbf{b} : \mathbb{R}^2 \to \mathbb{R}^3$ be a surface, called the *base*. Unit normal and curvature tensor of $\mathbf{b}$ are denoted by $\mathbf{n}_\mathbf{b}$ and $E_\mathbf{b}$, respectively.
- For $\ell = 1, \ldots, L$, let $\mathbf{r}_\ell : \mathbb{R}^2 \to \mathbb{R}^3$ be surfaces, called *ribbons*. Unit normal and curvature tensor of $\mathbf{r}_\ell$ are denoted by $\mathbf{n}_\ell$ and E_ℓ, respectively.

With these ingredients, we construct a new surface that combines base and ribbons in a specific way:

Definition 2 The *ABC-surface* $\mathbf{a} : \Gamma \to \mathbb{R}^3$ corresponding to the data $\mathbf{b}, \mathbf{r}_\ell, w, w_\ell, \kappa_\ell$, as defined above, is given by

$$\mathbf{a} := \frac{w \cdot \mathbf{b} + \sum_{\ell=1}^{L} w_\ell \cdot \mathbf{r}_\ell \circ \kappa_\ell}{w + \sum_{\ell=1}^{L} w_\ell}. \tag{2}$$

At corners, where the right-hand side is not yet well defined, let

$$\mathbf{a}(\sigma_\ell) := \mathbf{r}_\ell(0,0), \quad \ell = 1, \ldots, L.$$

For the sake of simplicity, this definition is restricted to the case that the domain Γ is simply connected and has always L corners and boundary segments. Generalizations to multiply-connected domains and to disc-like surfaces bounded by one smooth, periodic curve will be discussed in a forthcoming report.

Two basic elements characterize the construction in the definition above. First, the functions κ_ℓ, which define the trimming curves, are used as *reparametrizations* for the ribbons $\mathbf{r}_\ell$ to adapt them to the otherwise unrelated base $\mathbf{b}$. Second, near the boundary segment Γ_j the various implicit representations w and w_ℓ are used as *weights* to fade out the influence of the base $\mathbf{b}$ and all ribbons $\mathbf{r}_\ell$ except for $\mathbf{r}_j$. The ribbon $\mathbf{r}_j$ becomes dominant and determines the shape of $\mathbf{a}$. Thus, the construction provides *accurate boundary control* for $\mathbf{a}$, giving rise to the acronym *ABC-surface*. For a more detailed assessment of this raw idea, we need a further definition:

Definition 3 Let $\mathbf{a}$ be an ABC-surface according to Definition 2. It has *contact order* k at the boundary segment Γ_j if for any $\sigma = \gamma_j(u) \in \Gamma_j$ it holds

$$\bar{w}(\tau) := \frac{w(\sigma + \tau)}{w_j(\sigma + \tau)} = O(|\tau|^{k+1})$$

and, unless $\ell = j$ or $\sigma = \sigma_j$ and $\ell = j - 1$, also

$$\bar{w}_\ell(\tau) := \frac{w_\ell(\sigma + \tau)}{w_j(\sigma + \tau)} = O(|\tau|^{k+1}).$$

Further introducing the notation

$$\bar{\mathbf{r}}_\ell(\tau) := \mathbf{r}_\ell(\kappa_\ell(\sigma + \tau))$$

for the reparametrized ribbons, $\mathbf{a}$ can be written in the form

$$\mathbf{a}(\sigma + \tau) = \frac{\bar{w}(\tau)\mathbf{b}(\sigma + \tau) + \sum_\ell \bar{w}_\ell(\tau)\bar{\mathbf{r}}_\ell(\tau)}{\bar{w}(\tau) + \sum_\ell \bar{w}_\ell(\tau)}. \tag{3}$$

Here and below, the variable σ is always assumed to be fixed and mostly omitted in the notation for the sake of simplicity. Unit normal $\bar{\mathbf{n}}_\ell$ and curvature tensor $\bar{E}_\ell$ of $\bar{\mathbf{r}}_\ell$ coincide with the respective quantities of $\mathbf{r}_\ell$ since these surfaces are related by reparametrization,

$$\bar{\mathbf{n}}_\ell(\tau) = \mathbf{n}_\ell(\kappa_\ell(\sigma + \tau)), \quad \bar{E}_\ell(\tau) = E_\ell(\kappa_\ell(\sigma + \tau)). \tag{4}$$

In the interior of Γ, $\mathbf{a}$ is a blend of smooth surfaces and thus can be expected to be smooth, too. As always, exceptions may occur in case of an incidental loss of regularity. We will not discuss such degenerate cases, but focus on the behavior of $\mathbf{a}$ at the boundary. The following three theorems clarify the situation:

Theorem 1 *(G^0-boundary). Consider an ABC-surface* $\mathbf{a}$ *according to Definition 2 with contact order $k = 0$ at Γ_j. If the ribbons are consistent according to*

$$\mathbf{r}_{j-1}(1, 0) = \mathbf{r}_j(0, 0),$$

then $\mathbf{a}$ *is continuous at Γ_j, and it holds*

$$\mathbf{a}(\gamma_j(u)) = \mathbf{r}_j(u, 0), \quad u \in [0, 1]. \tag{5}$$

In the following proofs, it is convenient to use the abbreviations

$$h(\tau) := \frac{1}{1 + \bar{w}_{j-1}(\tau)} = \frac{w_j(\sigma_j + \tau)}{w_{j-1}(\sigma_j + \tau) + w_j(\sigma_j + \tau)}, \quad h'(\tau) := 1 - h(\tau),$$

to expand $\mathbf{a}$ in a vicinity of the corner σ_j. Since weights are assumed to be nonnegative, it is $h(\tau) \in [0, 1]$. For contact order $k = r - 1$, we obtain

$$\begin{aligned}\mathbf{a}(\sigma_j + \tau) &= \frac{\bar{\mathbf{r}}_j(\tau) + \bar{w}_{j-1}(\tau)\bar{\mathbf{r}}_{j-1}(\tau) + O(|\tau|^r)}{1 + \bar{w}_{j-1}(\tau) + O(|\tau|^r)} \\ &= h(\tau)\bar{\mathbf{r}}_j(\tau) + h'(\tau)\bar{\mathbf{r}}_{j-1}(\tau) + O(|\tau|^r). \end{aligned} \tag{6}$$

Proof Consider the point $\sigma = \gamma_j(u) \in \Gamma_j$ and $\mathbf{a}$ in the form (3). We note that $\bar{\mathbf{r}}_j(\tau) = \mathbf{r}_j(u, 0) + O(|\tau|)$ and, by consistency, $\bar{\mathbf{r}}_{j-1}(\tau) = \mathbf{r}_j(0, 0) + O(|\tau|)$. For $u > 0$, (5) follows from

$$\mathbf{a}(\sigma + \tau) = \frac{\bar{\mathbf{r}}_j(\tau) + O(|\tau|)}{1 + O(|\tau|)} = \bar{\mathbf{r}}_j(\tau) + O(|\tau|) = \mathbf{r}_j(u, 0) + O(|\tau|).$$

For $u = 0$, (6) yields

$$\mathbf{a}(\sigma + \tau) = h(\tau)\bar{\mathbf{r}}_j(\tau) + h'(\tau)\bar{\mathbf{r}}_{j-1}(\tau) + O(|\tau|) = \mathbf{r}_j(0,0) + O(|\tau|).$$

□

In the following, $\partial_1 f, \partial_2 f$ denote the partial derivatives of the bivariate function $f : \mathbb{R}^2 \to \mathbb{R}^n$, respectively. $Df := [\partial_1 f \; \partial_2 f] : \mathbb{R}^2 \to \mathbb{R}^{n\times 2}$ denotes the Jacobian.

Theorem 2 *(G^1-boundary). Consider an ABC-surface* $\mathbf{a}$ *according to Definition 2 with contact order $k = 1$ at Γ_j. If the conditions of Theorem 1 are satisfied, if the unit normals of ribbons are consistent according to*

$$\mathbf{n}_{j-1}(1,0) = \mathbf{n}_j(0,0),$$

and if the regularity condition

$$\operatorname{rank}\big(t D\bar{\mathbf{r}}_j(0) + (1-t) D\bar{\mathbf{r}}_{j-1}(0)\big) = 2 \tag{7}$$

is satisfied for all $t \in [0,1]$, then $\mathbf{a}$ *has a well-defined unit normal* $\mathbf{n}_\mathbf{a}$ *at Γ_j, and it holds*

$$\mathbf{n}_\mathbf{a}(\gamma_j(u)) = \mathbf{n}_j(u,0), \quad u \in [0,1]. \tag{8}$$

Proof Analogous to the previous proof, we obtain for $u > 0$

$$\mathbf{a}(\sigma + \tau) = \frac{\bar{\mathbf{r}}_j(\tau) + O(|\tau|^2)}{1 + O(|\tau|^2)} = \bar{\mathbf{r}}_j(\tau) + O(|\tau|^2).$$

Differing only by second order terms, the surfaces $\mathbf{a}(\sigma + \cdot)$ and $\bar{\mathbf{r}}$ have the same normal at the origin. Hence, using (4), we obtain

$$\mathbf{n}_\mathbf{a}(\sigma) = \bar{\mathbf{n}}_j(0) = \mathbf{n}_j(u,0).$$

Below, we mostly omit not only notating σ, but also the argument τ. For $u = 0$, corresponding to the case $\sigma = \sigma_j$, (6) yields the derivative

$$D\mathbf{a}(\sigma_j + \tau) = hD\bar{\mathbf{r}}_j + h'D\bar{\mathbf{r}}_{j-1} + (\bar{\mathbf{r}}_j - \bar{\mathbf{r}}_{j-1})Dh + O(|\tau|).$$

By consistency of ribbons, it is $\bar{\mathbf{r}}_j - \bar{\mathbf{r}}_{j-1} = O(|\tau|)$ so that

$$D\mathbf{a}(\sigma_j + \tau) = hD\bar{\mathbf{r}}_j(0) + h'D\bar{\mathbf{r}}_{j-1}(0) + O(|\tau|).$$

Let us consider the leading matrix, defined by the first two summands. Since $h \in [0,1]$, it is a convex combination of $D\bar{\mathbf{r}}_j(0)$ and $D\bar{\mathbf{r}}_{j-1}(0)$. By the regularity condition, it has rank 2, and by consistency of normal vectors, both columns are perpendicular to $\mathbf{n}_j(0,0)$. Hence, normalizing the cross product of these columns and passing to the limit $\tau \to 0$ yields $\mathbf{n}_\mathbf{a}(\sigma_j) = \mathbf{n}_j(0,0)$, as stated. Clearly, in general, the derivative $D\mathbf{a}(\sigma_j + \tau)$ itself does not possess a limit for $\tau \to 0$. □

We note that the regularity condition (7) is purely technical and typically satisfied. It rules out a degenerate situation, which can occur in a similar way anywhere in any common surface modeling system for an unfortunate choice of control data.

The crucial conditions for the existence of normal vectors at the boundary are *algebraic* (proper decay of the functions $\bar{w}$ and $\bar{w}_\ell$) and *geometric* (consistency of ribbons and normals). Similar conditions appear in the following theorem concerning curvature tensors. However, and this may be unexpected, also a *parametric* condition (consistency of derivatives of ribbons) has to be met to guarantee existence of a curvature tensor at corners.

Theorem 3 *(G^2-boundary). Consider an ABC-surface* $\mathbf{a}$ *according to Definition 2 with contact order $k = 2$ at Γ_j. If the conditions of Theorems 1 and 2 are satisfied, if the curvature tensors of ribbons are consistent according to*

$$E_{j-1}(1,0) = E_j(0,0),$$

and if the condition

$$D\bar{\mathbf{r}}_{j-1}(0) = D\bar{\mathbf{r}}_j(0) \tag{9}$$

is satisfied, then $\mathbf{a}$ *has a well-defined curvature tensor $E_\mathbf{a}$ at Γ_j, and it holds*

$$E_\mathbf{a}(\gamma_j(u)) = E_j(u,0), \quad u \in [0,1]. \tag{10}$$

Proof As before, we obtain for $u > 0$

$$\mathbf{a}(\sigma + \tau) = \frac{\bar{\mathbf{r}}_j(\tau) + O(|\tau|^3)}{1 + O(|\tau|^3)} = \bar{\mathbf{r}}_j(\tau) + O(|\tau|^3).$$

Differing only by third order terms, the surfaces $\mathbf{a}(\sigma + \cdot)$ and $\bar{\mathbf{r}}$ have the same curvature tensor at the origin. Hence, using (4), we obtain

$$E_\mathbf{a}(\sigma) = E_j(0) = E_j(u,0).$$

Analogous to the previous proof, we obtain for $u = 0$, corresponding to the case $\sigma = \sigma_j$,

$$\mathbf{a}(\sigma_j + \tau) = h\bar{\mathbf{r}}_j + h'\bar{\mathbf{r}}_{j-1} + O(|\tau|^3)$$

and, using (9),

$$D\mathbf{a}(\sigma_j + \tau) = hD\bar{\mathbf{r}}_j(0) + h'D\bar{\mathbf{r}}_{j-1}(0) + O(|\tau|) = D\bar{\mathbf{r}}_j(0) + O(|\tau|).$$

This time, the derivative has a limit at the corner,

$$D\mathbf{a}(\sigma_j) = D\bar{\mathbf{r}}_j(0). \tag{11}$$

The second order partial derivatives are given by

$$\begin{aligned}\partial_{\mu\nu}\mathbf{a}(\sigma_j+\tau) &= \partial_{\mu\nu}h(\bar{\mathbf{r}}_j-\bar{\mathbf{r}}_{j-1})+\partial_\mu h\,(\partial_\nu\bar{\mathbf{r}}_j-\partial_\nu\bar{\mathbf{r}}_{j-1})+\\ &\quad \partial_\nu h\,(\partial_\mu\bar{\mathbf{r}}_j-\partial_\mu\bar{\mathbf{r}}_{j-1})+h\partial_{\mu\nu}\bar{\mathbf{r}}_j+h'\partial_{\mu\nu}\bar{\mathbf{r}}_{j-1}+O(|\tau|)\\ &= h\partial_{\mu\nu}\bar{\mathbf{r}}_j(0)+h'\partial_{\mu\nu}\bar{\mathbf{r}}_{j-1}(0)+O(|\tau|),\end{aligned}$$

where we used that $\bar{\mathbf{r}}_{j-1}(0) = \bar{\mathbf{r}}_j(0)$ and $D\bar{\mathbf{r}}_{j-1}(0) = D\bar{\mathbf{r}}_j(0)$. Further, the curvature tensors of $\bar{\mathbf{r}}_{j-1}$ and $\bar{\mathbf{r}}_j$ coincide at the origin, $\bar{E}_{j-1}(0) = \bar{E}_j(0)$. This implies that also the second fundamental forms coincide,

$$\bar{B}^{\mu\nu}_{j-1}(0) := \langle\bar{\mathbf{n}}_{j-1}(0), \partial_{\mu\nu}\bar{\mathbf{r}}_{j-1}(0)\rangle = \langle\bar{\mathbf{n}}_j(0), \partial_{\mu\nu}\bar{\mathbf{r}}_j(0)\rangle =: \bar{B}^{\mu\nu}_j(0).$$

The second fundamental form of $\mathbf{a}$ is

$$B^{\mu\nu}_{\mathbf{a}}(\sigma_j+\tau) := \langle\mathbf{n}_{\mathbf{a}}(\sigma_j+\tau), \partial_{\mu\nu}\mathbf{a}(\sigma_j+\tau)\rangle.$$

By Theorem 2, the normal vector of $\mathbf{a}$ converges, i.e., $\mathbf{n}_{\mathbf{a}}(\sigma_j+\tau) = \mathbf{n}_{\mathbf{a}}(\sigma_j) + O(|\tau|)$. Hence, we obtain

$$\begin{aligned}B^{\mu\nu}_{\mathbf{a}}(\sigma_j+\tau) &= h\langle\mathbf{n}_{\mathbf{a}}(\sigma_j), \partial_{\mu\nu}\bar{\mathbf{r}}_j(0)\rangle + h'\langle\mathbf{n}_{\mathbf{a}}(\sigma_j), \partial_{\mu\nu}\bar{\mathbf{r}}_{j-1}(0)\rangle + O(|\tau|)\\ &= hB^{\mu\nu}_j(0) + h'B^{\mu\nu}_{j-1}(0) + O(|\tau|) = B^{\mu\nu}_j(0) + O(|\tau|)\end{aligned}$$

and finally, using $\mathbf{n}_{\mathbf{a}}(\sigma_j) = \bar{\mathbf{n}}_j(0) = \bar{\mathbf{n}}_{j-1}(0)$,

$$B^{\mu\nu}_{\mathbf{a}}(\sigma_j) = B^{\mu\nu}_j(0).$$

Together with (11) and (4), this implies coincidence of curvature tensors,

$$E_{\mathbf{a}}(\sigma_j) = \bar{E}_j(0) = E_j(0,0),$$

and the proof is complete. □

The proofs of the three theorems show that G^k-properties on the interior part of the boundary segments ($0 < u < 1$) are easily achieved by requesting proper decay of the functions $\bar{w}$ and $\bar{w}_\ell$. Consistency and regularity conditions are requested only to guarantee good behavior at the corners. Remarkably, condition (9) concerning consistency of the derivatives of reparametrized ribbons shows that the functions κ_ℓ cannot be chosen independently of the ribbons $\mathbf{r}_\ell$. More precisely, the interrelation

$$D\bar{\mathbf{r}}_{j-1}(0) = D\mathbf{r}_{j-1}(1,0)D\kappa_{j-1}(\sigma_j) = D\mathbf{r}_j(0,0)D\kappa_j(\sigma_j) = D\bar{\mathbf{r}}_j(0) \tag{12}$$

must be observed. Assuming that the ribbons are given and the reparametrizations are still unknown, this equation yields a linear constraint on the pair $D\kappa_{j-1}$, $D\kappa_j$ at the corner σ_j. Vice versa, if the reparametrizations are given, the space of admissible ribbons is restricted to a linear subspace.

The following example shows that this condition is in fact crucial. Let

$$\bar{\mathbf{r}}_{j-1}(x, y) = \begin{bmatrix} x \\ y \\ x^2 \end{bmatrix}, \quad \bar{\mathbf{r}}_j(x, y) = \begin{bmatrix} 2x \\ 2y \\ 4x^2 \end{bmatrix}, \quad (x, y) \in [0, 1]^2.$$

Near the origin, the traces of both surfaces coincide so that consistency of point, normal, and curvature tensor is guaranteed. However,

$$D\bar{\mathbf{r}}_{j-1}(0, 0) = \begin{bmatrix} 1 & 0 \\ 0 & 1 \\ 0 & 0 \end{bmatrix} \neq \begin{bmatrix} 2 & 0 \\ 0 & 2 \\ 0 & 0 \end{bmatrix} = D\bar{\mathbf{r}}_j(0, 0).$$

The weight functions $w_{j-1}(x, y) = x^3$, $w_j(x, y) = y^3$ have proper decay at the boundary curves $\gamma_{j-1}(u) = [u, 0, 0]$ and $\gamma_j(u) = [0, u, 0]$. Assuming for simplicity that the base and all other ribbons vanish, we obtain

$$\mathbf{a}(x, y) = \frac{1}{x^3 + y^3} \begin{bmatrix} x(x^3 + 2y^3) \\ y(x^3 + 2y^3) \\ x^2(x^3 + 4y^3) \end{bmatrix}.$$

Skipping the technical details, we note that

$$\lim_{(x,y)\to(0,0)} \mathbf{a}(x, y) = \bar{\mathbf{r}}_{j-1}(0, 0) = \bar{\mathbf{r}}_j(0, 0) = \begin{bmatrix} 0 \\ 0 \\ 0 \end{bmatrix}$$

and

$$\lim_{(x,y)\to(0,0)} \mathbf{n}_{\mathbf{a}}(x, y) = \bar{\mathbf{n}}_{j-1}(0, 0) = \bar{\mathbf{n}}_j(0, 0) = \begin{bmatrix} 0 \\ 0 \\ 1 \end{bmatrix}.$$

However, the curvature tensor $E_{\mathbf{a}}$ is discontinuous at the origin. Figure 4 shows the surface $\mathbf{a}$ *(left)* and the Gaussian curvature $K(r\cos\varphi, r\sin\varphi)$ *(right)* as a function of r. The individual lines correspond to 30 equally spaced angles φ in the interval $[0, \pi/2]$. For $\varphi = 0$ and $\varphi = \pi/2$, corresponding to the trimming curves γ_{j-1} and γ_j, it is $\lim_{r\to 0} K(r, 0) = \lim_{r=0} K(0, r) = 0$, as suggested by the cylindrical shape of $\bar{\mathbf{r}}_{j-1}$ and $\bar{\mathbf{r}}_j$. However, other angles yield non-zero limits.

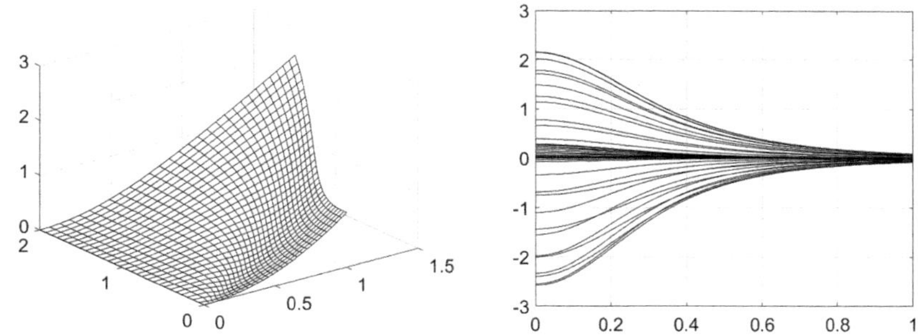

Fig. 4 Surface $\mathbf{a}(x, y)$ obtained by blending inappropriately reparametrized ribbons *(left)* and its Gaussian curvature $K(r \cos\varphi, r \sin\varphi)$ plotted over $r \in (0, 1]$ for 30 different angles $\varphi \in [0, \pi/2]$ *(right)*

3 ABC-Spline Surfaces

In the previous section, the ABC-principle was introduced in great generality. Our main goal, however, is to find a special embodiment of the setup that is fully compatible with the predominant standard of industrial surface modeling: trimmed NURBS.

To this end, we assume that all building blocks of the ABC-surface **a** are tensor product NURBS. Even more, for the sake of simplicity, we consider only purely polynomial, i.e., non-rational splines. Further, it is practical (but not necessary) to use a common set of knot vectors for the base **b**, the weights w, w_ℓ and the reparametrizations κ_ℓ; degrees may be different. By contrast, the knots of the ribbons $\mathbf{r}_\ell$ are completely independent. It is immediately clear that the concatenation, multiplication, summation and division of piecewise polynomial functions yields a piecewise rational function **a**. Denoting the coordinate degree of the bivariate spline s by $\deg s \in \mathbb{N}_0^2$ and its total degree by $|\deg s| \in \mathbb{N}_0$, we obtain

$$\begin{aligned} d &:= \deg(w\mathbf{b}) = \deg w + \deg \mathbf{b} \\ d_\ell &:= \deg(w_\ell \mathbf{r}_\ell \circ \kappa_\ell) = \deg w_\ell + |\deg \mathbf{r}_\ell| \deg \kappa_\ell . \end{aligned}$$

Together, the maximal degree of the rational pieces of **a** equals that of the corresponding numerators, and is given by

$$\deg \mathbf{a} = \max\{d, d_1, \ldots, d_L\}, \tag{13}$$

where the maximum is taken component-wise. More specific results will be presented later on.

The crucial question that has to be addressed now concerns the construction of suitable splines w, w_ℓ, and κ_ℓ. In recent works [9, 13], reparametrizations for domains bounded by curves are obtained as solutions to variational problems with the given trimming curves γ_ℓ as boundary conditions. It can be expected that this yields good visual results, but of course, there does not exist a closed-form solution, and much less a spline fulfilling the condition $\kappa_\ell(\gamma(u)) = [u, 0]$. Similar problems arise when seeking weights vanishing at the boundary or parts thereof.

The solution of the problem to find weights and reparametrizations compatible with the boundary is amazingly simple: There is no need to define the boundary curves γ_ℓ explicitly. It suffices to define them *implicitly* by the functions κ_ℓ according to Definition 1,

$$\gamma_\ell(u) := \{\xi \in \mathbb{R}^2 : \kappa_\ell(\xi) = [u, 0]\}.$$

This setting is justified by the fact that standard CAD systems and also standard file exchange formats like STEP or IGES allow trimming curves to be specified explicitly as a closed loop of spline curves in the parametric domain, or alternatively–and this is the option used here–as a closed loop of spline curves in $\mathbb{R}^3$ bounding the trace of the given NURBS surface. By Theorem 1, the curves

$$[0, 1] \ni u \;\mapsto\; \mathbf{c}_\ell(u) := \mathbf{r}_\ell(u, 0) = \mathbf{a}(\gamma_\ell(u)), \quad \ell = 1, \ldots, L,$$

can serve exactly that purpose. Actually, the planar trimming curves γ_ℓ need not be computed at any processing step of ABC-spline surfaces as long as geometric trimming by spatial curves, as explained in Sect. 4, is available.

Now, we elaborate on methods how to determine suitable reparametrizations κ_ℓ and weights w, w_ℓ under the assumption that the base $\mathbf{b}$ and the ribbons $\mathbf{r}_\ell$ are given.

3.1 Reparametrization

In principle, the functions κ_ℓ can be chosen quite arbitrarily. Basically, only the consistency condition (1) must be satisfied. This means that there exist points $\sigma_\ell \in \mathbb{R}^2$ such that

$$\kappa_\ell(\sigma_\ell) = [0, 0], \quad \kappa_{\ell-1}(\sigma_\ell) = [1, 0]$$

for all ℓ. If continuity of curvature tensors is requested, also condition (12) must be observed, i.e.,

$$D\mathbf{r}_{\ell-1}(1, 0) \cdot D\kappa_{\ell-1}(\sigma_\ell) = D\mathbf{r}_\ell(0, 0) \cdot D\kappa_\ell(\sigma_\ell). \tag{14}$$

Further, to obtain fair surfaces, it is reasonable to request that κ_ℓ reparametrizes $\mathbf{r}_\ell$ in such a way that it is close to the base,

$$\mathbf{r}_\ell \circ \kappa_\ell \approx \mathbf{b}.$$

The simplest way to achieve this (and possibly other desirable properties) is as follows: First, for each ℓ, we consider pairs of parameters $\sigma_\ell, \tau_\ell \in \mathbb{R}^2$, characterized by

$$\mathbf{r}_\ell(\tau_\ell) \approx \mathbf{b}(\sigma_\ell).$$

Collecting a sufficiently large, but finite number of them yields the set K_ℓ. Such pairs $(\sigma_\ell, \tau_\ell) \in K_\ell$ can be found easily by requesting that the line connecting the points $\mathbf{r}_\ell(\tau_\ell)$ and $\mathbf{b}(\sigma_\ell)$ be perpendicular to one of the two surfaces. To address the special role of corners, the set K_ℓ must contain the τ values $\tau_\ell^0 := [0, 0]$ and $\tau_\ell^1 := [1, 0]$. The corresponding σ-values must be consistent in the sense that the corners

$$\sigma_\ell := \sigma_\ell^0 = \sigma_{\ell-1}^1$$

are well-defined. At this and possibly other data points, the function κ_ℓ should not only approximate but interpolate the prescribed values. More precisely, let $K'_\ell \subset K_\ell$ be the subsets of pairs intended for interpolation and $K''_\ell := K_\ell \backslash K'_\ell$ the remainders. In particular, the pairs $(\sigma_\ell^0, \tau_\ell^0)$ and $(\sigma_\ell^1, \tau_\ell^1)$ addressing the corners belong to K'_ℓ. Now, the conditions to be satisfied by κ_ℓ are

$$\begin{aligned} \kappa_\ell(\sigma_\ell) &= \tau_\ell, \quad (\sigma_\ell, \tau_\ell) \in K'_\ell \\ \kappa_\ell(\sigma_\ell) &\approx \tau_\ell, \quad (\sigma_\ell, \tau_\ell) \in K''_\ell. \end{aligned}$$

Standard techniques, including all kinds of least squares methods or smoothing splines can be used to select κ_ℓ accordingly from a given space of splines. If the problem is overconstrained in the sense that no solution exists, the spline space must be enlarged or the conditions must be relaxed. Conversely, if the solution is not unique, the spline space must be reduced or more conditions must be specified.

The interpolation conditions $\kappa_\ell(\sigma_\ell) = [0, 0]$ and $\kappa_{\ell-1}(\sigma_\ell) = [1, 0]$ for the corners yield

$$\gamma_\ell(0) = \gamma_{l-1}(1) = \sigma_\ell,$$

as requested by (1). Still, condition (14) needs to be addressed. When needed, one determines matrices $T_\ell \in \mathbb{R}^{3\times 2}$ such that

$$T_\ell \approx D\mathbf{b}(\sigma_\ell), \quad \mathbf{n}_\ell(0, 0)^{\mathrm{t}} \cdot T_\ell = 0, \quad \ell = 1, \ldots, L.$$

In particular, the columns of T_ℓ span the wanted tangent space at σ_ℓ. Now, κ_ℓ is determined as before, but such that is satisfies the additional conditions

$$\begin{aligned} D\mathbf{r}_\ell(0, 0) \cdot D\kappa_\ell(\sigma_\ell) &= T_\ell \\ D\mathbf{r}_\ell(1, 0) \cdot D\kappa_\ell(\sigma_{\ell+1}) &= T_{\ell+1}. \end{aligned}$$

As the chain rule shows, this is in accordance with $\mathbf{r}_\ell \circ \kappa_\ell \approx \mathbf{b}$, and it also implies (14).

3.2 Weights

Once the reparametrizations $\kappa_\ell = [p_\ell, q_\ell]$ are known, also suitable weights w and w_ℓ can be determined conveniently.

By definition, the second coordinate q_ℓ of κ_ℓ vanishes at γ_ℓ. Assuming that the ABC-surface **a** to be determined should have contact order $k_\ell \geq 0$ at the boundary segment Γ_ℓ, the simplest choice for w is

$$w = q_1^{r_1} \cdots q_L^{r_L}, \tag{15}$$

where the exponents are $r_\ell := k_\ell + 1$. At this point, it has to be made sure that each factor q_ℓ is positive on $\Gamma \setminus \Gamma_\ell$. Additional zeros away from Γ_ℓ are not critical. Increasing some control points of B-splines vanishing at that boundary segment, these zeros can be removed without changing the crucial boundary behavior. However, zeros near the boundary may be persistent so that the whole construction fails. In particular, reentrant corners are likely to cause such a situation. This case requires a special treatment and will be discussed in a forthcoming report.

Omitting one factor yields possible weights

$$w_\ell = q_1^{r_1} \cdots q_{\ell-1}^{r_{\ell-1}} q_{\ell+1}^{r_{\ell+1}} \cdots q_L^{r_L}.$$

Let us check the conditions concerning contact order given in Definition 3. For $\sigma \in \Gamma_j$, we obtain

$$\bar{w}(\tau) = \frac{w(\sigma + \tau)}{w_j(\sigma + \tau)} = q_j^{r_j}(\sigma + \tau) = O(|\tau|^{r_j})$$

since $q_j(\sigma) = 0$. Further,

$$\bar{w}_\ell(\tau) = \frac{w_\ell(\sigma + \tau)}{w_j(\sigma + \tau)} = \frac{q_j^{r_j}(\sigma + \tau)}{q_\ell^{r_\ell}(\sigma + \tau)} = O(|\tau|^{r_j})$$

unless the denominator vanishes at σ, and this is only possible if $\ell = j$ or $\ell = j - 1$ and $\sigma = \sigma_j$. Exactly these cases are excluded in the definition.

The weights suggested above have two drawbacks: First, the influence of each ribbon extends over the whole surface, which might be unwanted. Second, the polynomial degree of such weights is increasing with L. If, for example, $L = 6$, $\deg q_\ell = [3, 3]$, and $r_\ell = 3$ for all ℓ to achieve contact order $k = 2$ everywhere, then $\deg w = \sum_{\ell=1}^{L} r_\ell \deg q_\ell = [54, 54]$. This might cause problems with efficiency of storage and speed of evaluation after converting the surface **a** from its procedural definition (2) to the usual form

$$\mathbf{a} = \frac{\sum_i \omega_i b_i \mathbf{p}_i}{\sum_i \omega_i b_i}$$

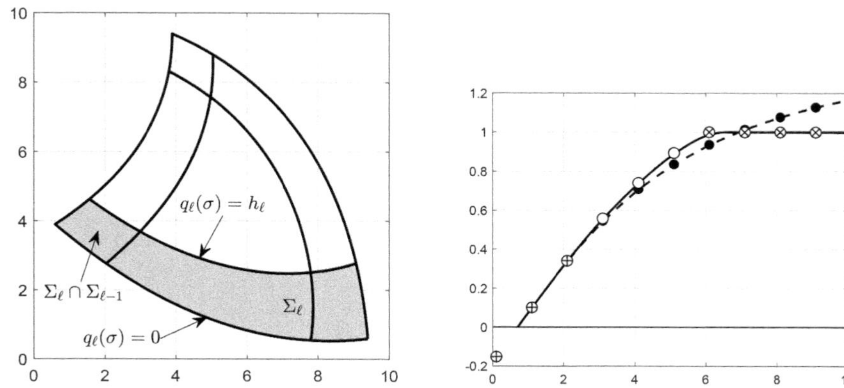

Fig. 5 Boundary stripes *(left)* and univariate plateau generation with original control points •, unchanged control points ⊕, constant control points ⊗, and optimized control points ∘ *(right)*

of a NURBS surface, i.e., the linear combination of control points $\mathbf{p}_i$ by means of B-splines b_i and (constant) weights ω_i. The latter is requested in particular for the export of $\mathbf{a}$ to standard file formats, as addressed in the next section. The following modification of weights away from the boundary of the domain Γ solves both issues:

Denote by

$$\Sigma_\ell := \{\sigma \in \Gamma : q_\ell(\sigma) \leq h_\ell\}$$

the *boundary stripes* of Γ with *width* h_ℓ. The widths $h_\ell > 0$ are assumed to be chosen small enough to make sure that only boundary stripes of neighboring curves intersect, and that these intersections as well as the stripes themselves consist of a single connected piece, see Fig. 5 *(left)*. Now, we derive from q_ℓ a new spline $\bar{q}_\ell$ as follows: First, classify the indices in the B-spline representation $q_\ell = \sum_i q_\ell^i b_i$ according to

$$I_\ell := \{i : q_\ell(\sigma) = 0 \text{ for some } \sigma \in \operatorname{supp} b_i\}$$
$$J_\ell := \{i : q_\ell(\sigma) > h_\ell \text{ for some } \sigma \in \operatorname{supp} b_i\}.$$

That is, B-splines b_i affect the boundary if $i \in I_\ell$, and they affect the part beyond the boundary stripe if $i \in J_\ell$. Second, if the sets I_ℓ and J_ℓ are not disjoint, refine the B-spline representation of q_ℓ by knot insertion and repeat the first step. Third, set

$$\bar{q}_\ell^i := \begin{cases} q_\ell^i & \text{if } i \in I_\ell \\ h_\ell & \text{if } i \in J_\ell. \end{cases} \tag{16}$$

Coefficients with indices neither in I_ℓ nor in J_ℓ are determined such that the resulting spline has reasonable shape, for instance by minimizing a standard fairness functional. Fourth, define the spline $\bar{q}_\ell := \sum_i \bar{q}_\ell^i b_i$. By construction, $\bar{q}_\ell$ is an implicit

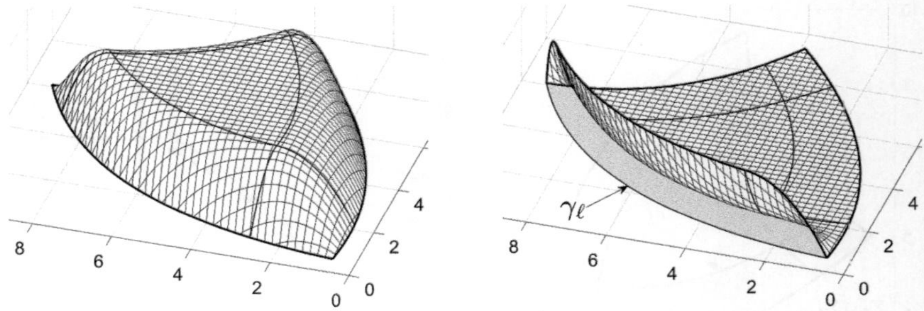

Fig. 6 Weights w *(left)* and w_ℓ *(right)* with plateau

representation of γ_ℓ, but has a *plateau* with constant value h_ℓ beyond the boundary stripe Σ_ℓ. Figure 5 *(right)* illustrates the procedure exemplarily in the univariate case. Replacing (15) by

$$w = \bar{q}_1^{r_1} \cdots \bar{q}_L^{r_L},$$

we obtain an implicit representation of $\partial\Gamma$ with plateau on the interior part of the domain that is not covered by a boundary stripe, see Fig. 6 *(left)*. By construction, the coordinate degree of w is bounded by

$$\deg w \leq \max_\ell \big(r_\ell \deg q_\ell + r_{\ell+1} \deg q_{\ell+1}\big)$$

since at any given point $\sigma \in \Gamma$ at most two factors are non-constant. Notably, the degree does not depend on the number L of boundary segments. In the example above, we have now $\deg q = [18, 18]$. This is still a relatively high degree, but much less than before, and can be considered manageable.

The weights w_ℓ may be modified in a similar spirit. We define the functions

$$\hat{q}_\ell := \bar{q}_{\ell-1}\bar{q}_{\ell+1}$$

and process them exactly like the functions q_ℓ above, but now replacing (16) by

$$\tilde{q}_\ell^i := \begin{cases} \hat{q}_\ell^i & \text{if } i \in I_\ell \\ 0 & \text{if } i \in J_\ell. \end{cases}$$

This way, we produce a function with a zero plateau beyond the boundary stripe. The desired weight is now simply given by

$$w_\ell := \tilde{q}_\ell^{r_\ell},$$

see Fig. 6 *(right)*. The degree

$$\deg w_\ell = r_\ell \max(\deg q_{\ell-1}, \deg q_{\ell+1})$$

is again independent of L.

Let us reconsider the degree of $\mathbf{a}$ for weights w and w_ℓ with plateau, as defined above. For simplicity, we assume that $\deg \mathbf{b} = \deg \kappa_\ell = [n, n]$ and $\deg \mathbf{r}_\ell = [n, m]$, and that all contact orders are equal to $k = r - 1$. Then, using (13),

$$\deg \mathbf{a} = [n, n] \cdot \max\{2r + 1, r + n + m\}.$$

The coordinate degree of the ribbons is set to the unmatched pair $[n, m]$ for the following reason: In the first direction, corresponding to the behavior along the boundary, full flexibility concerning the choice of knots and degrees is provided to model a large variety of shapes. By contrast, in the second direction, corresponding to the behavior transversal to the boundary, it is sufficient to deal with pure polynomials of low degree. For instance, choosing $m = 1$ allows the definition of the boundary curves $\mathbf{r}_\ell(u, 0)$ and the boundary normals $\mathbf{n}_\ell(u, 0)$ along Γ_ℓ. When also the curvature tensors $E_\ell(u, 0)$ shall be prescribed, it suffices to choose $m = 2$. The minimal configuration yielding a G^1-patch and contact order $k = 1$ requests $n = 2$ and $m = 1$, leading to $\deg \mathbf{a} = [10, 10]$. For a G^2-patch and contact order $k = 2$, one chooses at least $n = 3$ and $m = 2$ to obtain $\deg \mathbf{a} = [24, 24]$.

Modeling with surfaces of such a high degree is commonly considered to be impractical. However, this is never requested. The geometric building blocks–base and ribbons–can be assumed to be splines of low degree, and their control points can be manipulated (manually or automatically) as usual. The conversion of the ABC-surface $\mathbf{a}$ to NURBS form

$$\mathbf{a} = \frac{\sum_i \omega_i b_i \mathbf{p}_i}{\sum_i \omega_i b_i},$$

as detailed in the next section, serves only the purpose to prepare it for file export according to industrial standards. It should be considered as a *read-only representation*, to be used for evaluation purposes. Never ever, the control points $\mathbf{p}_i$ should be touched again after being derived from (2).

4 File Export

As a matter of fact, data exchange between industrial CAD/CAM systems is limited to a few completely standardized file formats, like IGES or STEP. The only type of surfaces commonly available therein are trimmed NURBS. Other modeling paradigms, say based on box splines, exponential splines, or subdivision are typically not supported. For the trimming of a NURBS surface $\mathbf{s} : \mathbb{R}^2 \to \mathbb{R}^3$, the following three options can be assumed to be available:

Parametric Trimming. A closed loop of planar spline curves $\gamma_1, \dots, \gamma_L$ without self-intersections is specified. The compact set $\Gamma \subset \mathbb{R}^2$ bounded by these curves is used as the domain of the trimmed surface $\mathbf{s}$.

Geometric Trimming. A closed loop of spatial spline curves $\mathbf{c}_1, \dots, \mathbf{c}_L$ is specified. Since it cannot be assumed that the traces of these curves are contained in the surface, the corresponding planar curves are defined by $\gamma_\ell(u) := \alpha$ where $\alpha \in \mathbb{R}^2$ is chosen such that $\mathbf{s}(\alpha) \approx \mathbf{c}_\ell(u)$ with the help of some projection method. These curves γ_ℓ need not be splines by themselves, but are computed numerically. They are merely assumed to form a closed loop, which is bounding the domain Γ.

Hybrid Trimming. Some segments of the boundary are defined by planar spline curves, and the other ones by spatial spline curves. The planar curves together with the preimages of the projected spatial curves are assumed to form a closed loop, which is bounding the domain Γ.

As we have seen, ABC-spline surfaces are piecewise rational. But nevertheless, in general, they cannot be represented as trimmed NURBS in the requested way. The point is that the summands $\mathbf{r}_\ell \circ \kappa_\ell$ may produce inner boundaries between rational segments that are not axis-aligned in the parameter domain and hence cannot be understood as knot lines. To solve this issue, we present a special construction of the reparametrizations κ_ℓ that leads to a situation where $\mathbf{a}$ can be represented as the union of separate parts, each of which is a genuine trimmed NURBS surface with boundary segments given either geometrically by a 3d spline curve or parametrically by a 2d spline curve.

For a spline s of degree n with knot vector $U = [u^0, \dots, u^N]$, the natural domain of definition is the interval $[u^n, u^{N-n}]$. We assume that s can be evaluated on all of $\mathbb{R}$ by simply prolonging the leftmost polynomial piece from $[u^n, u^{n+1}]$ to $(-\infty, u^{n+1}]$ and the rightmost piece from $[u^{N-n-1}, u^{N-n}]$ to $[u^{N-n-1}, \infty)$. Hence, true break points, where higher order derivatives are discontinuous, appear only at the *inner knots* $u^{n+1}, \dots, u^{N-n-1}$.

Denote the knot vectors of $\mathbf{r}_\ell$ by U_ℓ, V_ℓ. We start with considering the second vector V_ℓ. Critical points $\sigma \subset \mathbb{R}^2$ that are mapped to the knot line $\mathbb{R} \times \{v_\ell^j\}$ for some inner knot $v_\ell^j \in V_\ell$ by the reparametrization $\kappa_\ell = [p_\ell, q_\ell]$ are characterized by

$$q_\ell(\sigma) = v_\ell^j.$$

In general, neither the levelsets of q_ℓ nor their images under $\mathbf{r}_\ell$ are splines. Special constructions for q_ℓ to create such a situation are possible, but impractical. It is much easier and completely sufficient to request that V_ℓ does not possess any inner knots at all. For instance, we may assume $V_\ell = [0, \dots, 0, 1, \dots, 1]$, what is just the knot vector for polynomials in Bézier form. We identified this setting already at the end of the preceding section as a legitimate choice. To sum up, ribbons that are pure polynomials in the second coordinate are able to carry information on location, unit normal, and curvature tensor along the boundary curve, and they do not have breaks in the second coordinate.

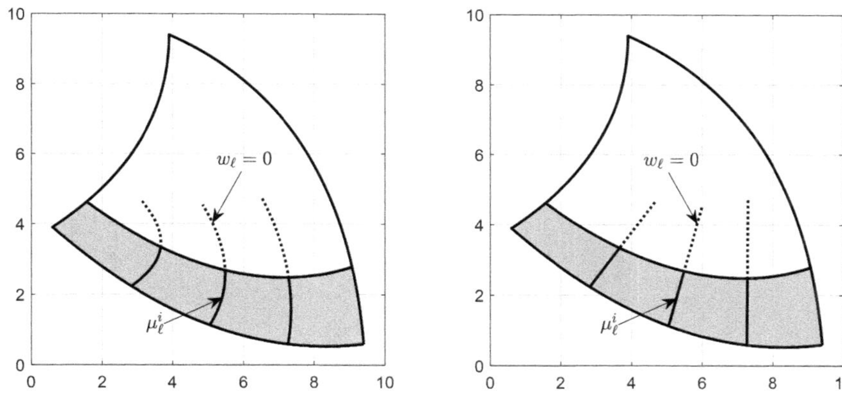

Fig. 7 General *(left)* and straight *(right)* levelsets μ_ℓ^i

The treatment of the first variable is more involved since assuming also this dependence to be purely polynomial would unduly restrict the variety of shapes that can be modeled. Critical points $\sigma \subset \mathbb{R}^2$ that are mapped to the inner knot $u_\ell^i \in U_\ell$ by κ_ℓ are characterized by

$$p_\ell(\sigma) = u_\ell^i.$$

This time, it is appropriate to demand that the critical levelsets of p_ℓ corresponding to inner knots of U_ℓ are linear. More precisely, let

$$\mu_\ell^i := \{\sigma \in \Gamma : p_\ell(\sigma) = u_\ell^i \text{ and } w_\ell(\sigma) > 0\}$$

denote the relevant parts of the levelsets of p_ℓ corresponding to the knots u_ℓ^i. In general, these curves are bent, see Fig. 7 *(left)*, but p_ℓ can be determined such that they become straight lines, see Fig. 7 *(right)*. One may proceed as follows:

First, for each knot $u_\ell^i \in U_\ell$, choose a direction $\delta_\ell^i \in \mathbb{R}^2$ such that

$$\mathbf{r}_\ell(u_\ell^i, t) \approx \mathbf{b}(\gamma_\ell(u_\ell^i) + t\delta_\ell^i)$$

for arguments $\gamma_\ell(u_\ell^i) + t\delta_\ell^i$ within the boundary stripe Σ_ℓ. Denoting by $D\mathbf{b}^+ := (D\mathbf{b}^T D\mathbf{b})^{-1} D\mathbf{b}^T$ the pseudo-inverse of the Jacobian $D\mathbf{b}$, one may set in particular

$$\delta_\ell^i := D\mathbf{b}^+(\gamma_\ell(u_\ell^i)) \cdot \partial_2 \mathbf{r}_\ell(u_\ell^i, 0).$$

Second, define the line segment

$$\mu_\ell^i := \left\{\gamma_\ell^i + t\delta_\ell^i : t \in \mathbb{R}\right\} \cap \Sigma_\ell$$

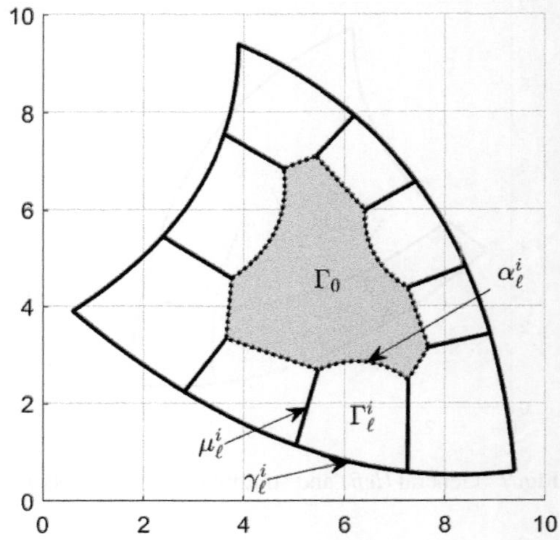

Fig. 8 Partition of the domain Γ into pieces ready for file export

with direction δ_ℓ^i within the boundary stripe Σ_ℓ. Third, choose sufficiently many pairs of points (σ_ℓ, τ_ℓ) of the form

$$\sigma_\ell = \gamma_\ell(u_\ell^i) + t\delta_\ell^i \in \mu_\ell^i, \quad \tau_\ell = (u_\ell^i, t)$$

and add them to the set K_ℓ' of data to be interpolated.

Using this setup, points $\sigma \in \Gamma$ that are mapped to the inner knot u_ℓ^i of $\mathbf{r}_\ell$ by p_ℓ either lie on the straight line μ_ℓ^i or are irrelevant since $w_\ell(\sigma) = 0$.

Assuming that the ribbons $\mathbf{r}_\ell$ are polynomials in the second variable and that the level sets μ_ℓ^i are straight, as described, we can partition the domain Γ into an interior part Γ_0 and boundary parts Γ_ℓ^i by a network of curves, consisting of three different types: First, there are the segments γ_ℓ^i of the trimming curves γ_ℓ corresponding to intervals between consecutive knots in U_ℓ. Second, there are the level sets μ_ℓ^i, and third, there are auxiliary spline curves α_ℓ^i, to be chosen at will, connecting the endpoints of the level sets. Figure 8 illustrates the setting. Restricting the ABC-surface $\mathbf{a}$ to the subdomains Γ_0, Γ_ℓ^i yields pieces $\mathbf{a}_0, \mathbf{a}_\ell^i$, respectively, which are trimmed NURBS surfaces ready for filing with standard data formats. They do not have interior breaks aside from a regular knot grid, and their boundaries are formed by spatial spline curves $\mathbf{a} \circ \gamma_\ell^i$ and planar spline curves $\mu_\ell^i, \alpha_\ell^i$.

5 Examples

The focus of this article is on theory, but still, skipping many technical details, we want to present three examples to illustrate the potential of our approach. They show in turn composite surfaces with G^0, G^1, and G^2 contact according to Theorems 1, 2,

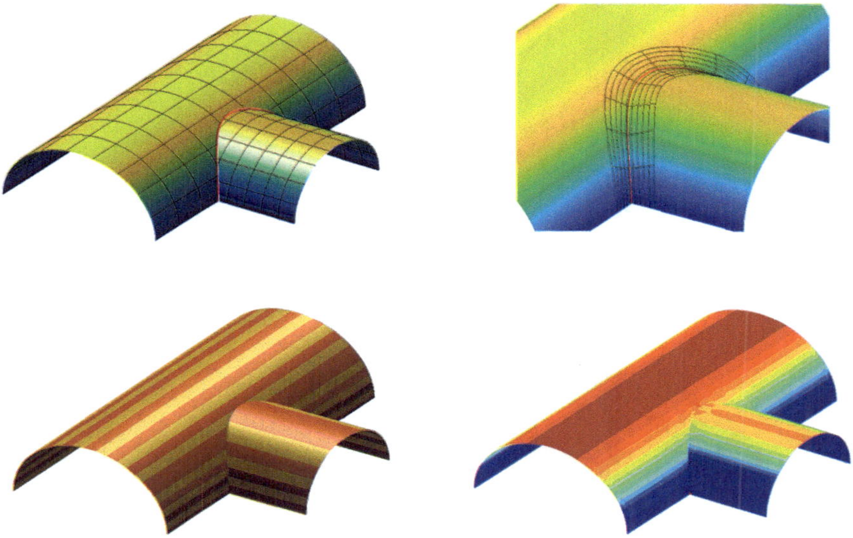

Fig. 9 Two intersecting cylinders—parametric lines *(top left)*, ribbons near intersection curve *(top right)*, isophotes *(bottom left)*, and mean curvature shading *(bottom right)*

and 3, respectively. All surfaces, reparametrizations, and weights are bicubic tensor product splines; all weights are used in the version with plateau.

Figure 9 presents a typical example from solid modeling: two intersecting cylinders that trim each other. The first plot shows two ABC-surfaces $\mathbf{a}^1, \mathbf{a}^2$ together with a few parameter lines, which follow the given shapes in the natural way. The spline representations $\mathbf{b}^1, \mathbf{b}^2$ of the two original cylinders serve as bases for $\mathbf{a}^1, \mathbf{a}^2$, respectively. The second plot shows the pair of ribbons $\mathbf{r}^1, \mathbf{r}^2$ used to model the region near the intersection curve. Both ribbons have 9 interior knots in longitudinal direction. Their parametrizations are adapted to the local situation and chosen such that they share a common curve $\mathbf{c} = \mathbf{r}^1(\cdot, 0) = \mathbf{r}^2(\cdot, 0)$. This curve approximates the exact intersection $\mathbf{b}^1 \cap \mathbf{b}^2$ of the two original cylinders with high accuracy. The watertight connection of ribbons is easy to achieve by coinciding boundary control points since both ribbons are in Bézier form in transversal direction. Theorem 1 says that this property is inherited by the two ABC-surfaces, i.e., they also join seemlessly at $\mathbf{c}$. The third plot shows the impeccable pattern of isophotes, while the fourth one is colored by mean curvature. Only here, one can see that the two surfaces $\mathbf{a}^1, \mathbf{a}^2$ slightly deviate from the perfect cylindrical shape of the original surfaces. This deviation could be made arbitrarily small by choosing finer knot sequences. However, as a matter of fact, it is impossible to do without any deviation since the exact intersection of the original cylinders cannot be represented by a spline curve–neither in geometry space nor in the parametric domain.

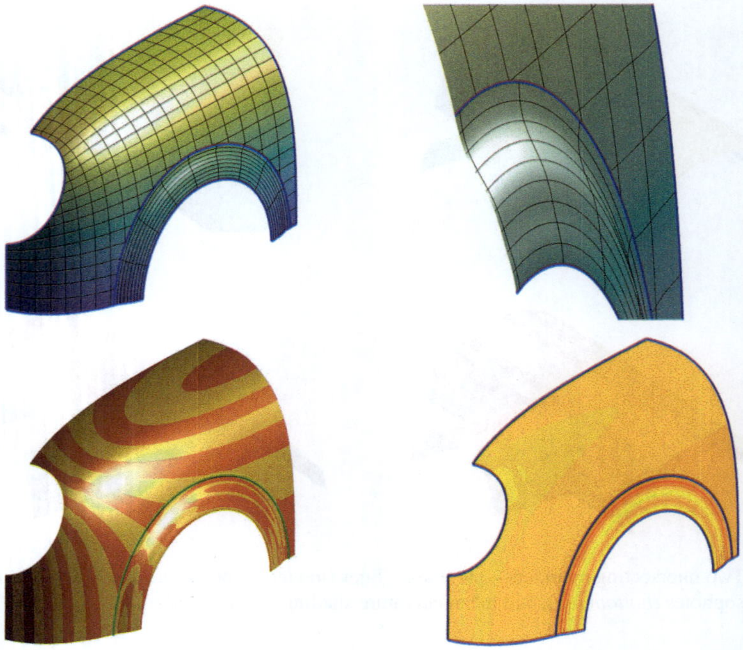

Fig. 10 Fender of a fictional car—whole composite surface *(top left)*, zoom on cross section *(top right)*, isophotes *(bottom left)*, and mean curvature shading *(bottom right)*

Figure 10 shows the fender of a fictional car.[7] The whole geometry consists of a bigger, uniformly curved part and a bulged-out, annular area that can be viewed as part of the wheel case. While it is possible to model the object by a single spline surface, it is much more convenient and natural to assemble it from two pieces. The ring-shaped part is modelled by an untrimmed standard spline surface **s**. The other part is an ABC-surface **a**, whose ribbon along the junction has C^1-contact with **s**, which is easy to achieve. This implies coinciding normals along the common boundary, and hence, by Theorem 2, a G^1 contact between **a** and **s**. The first plot shows the complete composite surface, while the second one features a zoom on a cross-section profile. The third and fourth plot show isophotes (which are continuous) and mean curvature (which is discontinuous), respectively.

Figure 11 presents a classical hole-filling problem. Several standard spline surfaces are surrounding a triangular region, and another surface is sought that closes the gap smoothly. Here, we show that this can be achieved by an ABC-surface with three ribbons, each joining C^2 with the corresponding outer part. By Theorem 3, this guarantees a G^2-contact. The first plot shows the configuration defining the three-

[7] The geometry shown here was kindly provided by Malcolm Sabin, who modeled a similar composite G^2-surface using a completely different, hitherto unpublished technique.

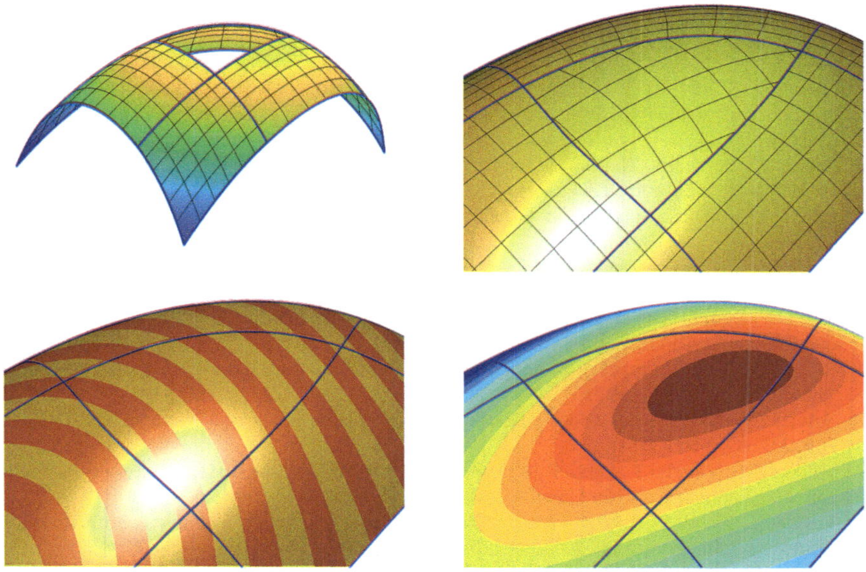

Fig. 11 Filling of a triangular hole—Arrangement of standard patches surrounding the hole *(top left)*, surrounding patches and ABC-filling with some parametric lines *(top right)*, isophotes *(bottom left)*, and mean curvature shading *(bottom right)*

sided hole, and the second one the completed structure together with a few parameter lines. Isophotes and also the distribution of mean curvature, as shown by the third and fourth plot, indicate that the generated shape leaves little room for improvement.

6 Conclusion

All spline-based CAD systems currently in use are unable to generate composite spline surfaces that are watertight along trimming curves. We suggest ABC-surfaces as a solution to this problem. They are a special class of trimmed rational splines that match boundary information given in terms of so-called ribbons exactly, thus enabling the construction of G^k-continuous joints. The framework is relatively simple and can be integrated seamlessly into existing CAD systems and file exchange formats. First experiments show that ABC-surfaces have good shape properties.

Future research will address the treatment of re-entrant corners and applications of the ABC-setup for the approximation of partial differential equations.

Acknowledgements We would like to thank Malcolm Sabin for fruitful discussions and comments.

References

1. Marussig, B., Hughes, T.J.R.: A review of trimming in isogeometric analysis: challenges, data exchange and simulation aspects. Arch. Comput. Methods Eng. **25**, 1059–1127 (2018)
2. Marussig, B., Reif, U.: Surface patches with rounded corners. In preparation
3. Peters, J.: Geometric continuity. In: Farin, G., Hoschek, J., Kim, M.S. (eds.), Handbook of Computer Aided Geometric Design, pp. 193–227. North Holland (2002)
4. Peters, J., Reif, U.: Subdivision surfaces. Series Geometry and Computing, vol. 3. Springer (2008)
5. Pla-Garcia, N., Vigo-Anglada, M., Cotrina-Navau, J.: N-sided patches with B-spline boundaries. Comput. Graph. **30**, 959–970 (2006)
6. Reif, U.: System and method for defining trimmed spline surfaces with accurate boundary control. Patent Application EP19192703.7
7. Reif, U.: A refinable space of smooth spline surfaces of arbitrary topological genus. J. Approx. Theory **90**, 174–199 (1997)
8. Salvi, P., Várady, T., Rockwood, A.P.: Ribbon-based transfinite surfaces. Comput. Aided Geom. Des. **31**, 613–630 (2014)
9. Vaitkus, M., Várady, T., Salvi, P., Sipos, Á.: Multi-sided B-spline surfaces over curved, multi-connected domains (2020). Unpublished manuscript
10. Várady, T., Salvi, P., Karikó, G.: A multi-sided Bézier patch with a simple control structure. Comput. Graph. Forum **35**, 307–317 (2016)
11. Várady, T., Salvi, P., Kovács, I.: Enhancement of a multi-sided Bézier surface representation. Comput. Aided Geom. Des. **55**, 69–83 (2017)
12. Várady, T., Salvi, P., Rockwood, A.P.: Transfinite surface interpolation with interior control. Graph. Models **74**, 311–320 (2012)
13. Várady, T., Salvi, P., Vaitkus, M., Sipos, Á.: Multi-sided Bézier surfaces over curved, multi-connected domains. Comput. Aided Geom. Des. **78**, 101–828 (2020)

Fast Formation and Assembly of Isogeometric Galerkin Matrices for Trimmed Patches

Benjamin Marussig

Abstract This work explores the application of the fast assembly and formation strategy from [9, 17] to trimmed bi-variate parameter spaces. Two concepts for the treatment of basis functions cut by the trimming curve are investigated: one employs a hybrid Gauss-point-based approach, and the other computes discontinuous weighted quadrature rules. The concepts' accuracy and efficiency are examined for the formation of mass matrices and their application to L^2-projection. Significant speed-ups compared to standard element by element finite element formation are observed. There is no clear preference between the concepts proposed. While the discontinuous weighted scheme scales favorably with the degree of the basis, it also requires additional effort for computing the quadrature weights. The hybrid Gauss approach does not have this overhead, which is determined by the complexity of the trimming curve. Hence, it is well-suited for moderate degrees, whereas discontinuous-weighted-quadrature has potential for high degrees, in particular, if the related weights are computed in parallel.

1 Introduction

Isogeometric analysis (IGA) has been introduced to overcome the profound inefficiencies in the conventional interaction of CAD and simulation tools [11, 19]. The ground-breaking idea of IGA is to perform numerical simulations using CAD technology such as (non-uniform rational) B-splines. During the last decade, it has been demonstrated that this paradigm provides several computational benefits. Indeed, IGA outperforms traditional simulations in most academic benchmarks [12, 13, 22, 28] and is nowadays generally recognized as a powerful alternative to the conventional analysis methodology.

The straightforward utilization of high-degree and high-continuity basis functions is an outstanding benefit of IGA. Together with the concept of k-refinement,

B. Marussig (✉)
Graz University of Technology, Graz Center of Computational Engineering (GCCE), Institute of Applied Mechanics, Graz, Austria
e-mail: marussig@tugraz.at

C. Manni and H. Speleers (eds.), *Geometric Challenges in Isogeometric Analysis*, Springer INdAM Series 49, https://doi.org/10.1007/978-3-030-92313-6_7

the resulting analysis features high robustness and accuracy w.r.t. the degrees of freedom employed, see, e.g. [4, 5, 14]. This ability to perform high-order accurate simulations is, however, somewhat limited due to the computational cost, because the formation and assembly of the system matrix gets more involved with increasing polynomial degree p. The state-of-the-art at the core of standard finite element codes is an element-wise assembly. IGA adopted this concept, and the introduction of Bézier extraction [6, 7] has provided a means to map between a smooth spline basis and an element-based representation. Being compatible with conventional analysis routines has played an essential role for the acceptance and dispersion of IGA in the numerical analysis community and allowed a simple integration of this new paradigm into existing simulation software. Yet, the corresponding cost for setting up the system matrix for a C^{p-1}-continuous d-dimensional tensor product B-spline basis is $\mathcal{O}\left(p^{3d}\right)$ per degree of freedom [9]. Consequently, high-order analysis is doomed to be computationally expensive when a conventional matrix formation strategy is employed.

Reduced quadrature rules [3, 16, 20, 21, 36] improve the efficiency of the numerical integration on the element-level. However, the element-wise assembly by itself limits the computation cost to $\mathcal{O}\left(p^{2d}\right)$ [9]. *Sum factorization* [1, 35] approaches this threshold by rearranging the computations to exploit the tensor product structure of the B-spline basis. The resulting cost using element-wise Gaus quadrature is $\mathcal{O}\left(p^{2d+1}\right)$. As shown in [8], sum factorization with Gauss quadrature can even yield a computational complexity of $\mathcal{O}\left(p^{d+2}\right)$, when used globally and not on an element level. *Weighted quadrature* introduced in [9] is a new integration technique that also drops the element perspective. A quadrature rule is set up for each test function by incorporating the test function into the quadrature weights. The outstanding feature of the resulting quadrature is its independence on the degree p for spline with maximal smoothness, i.e., C^{p-1}. The application of weighted quadrature to non-uniform spline with mixed continuity is addressed in [17]. Furthermore, the integration of sum factorization and weighted quadrature into a *row-based assembly* strategy results in the fast formation and assembly approach detailed in [9, 17]. These three ingredients—row assembly, sum factorization, and weighted quadrature—reduce the computational cost to $\mathcal{O}\left(p^{d+1}\right)$.

The treatment of trimmed patches has been denoted as an open challenge for this fast formation and assembly approach at the recent INdAM Workshop on Geometric Challenges in Isogeometric Analysis. Trimming is an essential concept for representing complex geometries with tensor product B-splines. The main component involved is the *trimming curve*, which is specified in the parameter space and restricts the visible part of the spline object to a subregion. The resulting trimmed space consists of interior, exterior, and cut basis functions. The latter introduces several computational challenges as detailed in [30]. For example, the realization of a high-order accurate numerical integration scheme is far from trivial even in the classical finite element setting, see e.g. [15]. Moreover, cut basis functions do not follow the tensor product structure of the basis anymore. In order words, trimming violates an essential property for the fast formation process. Hence, the question arises if a significant reduction of the computational costs is restricted to the non-trimmed case.

This paper provides extensions to the fast formation and assembly approach presented in [9, 17] that allow the analysis of trimmed domains. In particular, the integration over cut basis functions is investigated using either a (i) Gauss quadrature or a (ii) weighted quadrature approach. Their performance and accuracy are compared for the mass matrix formation of trimmed bi-variate spaces with different complexity. Both concepts maintain the optimal approximation order and significantly improve efficiency compared to a standard element based Gaussian assembly. From a conceptional point of view, the extension to tri-variate spaces is straightforward, but the implementation gets more involved due to the increased topological complexity of the cut elements.

The paper is structured as follows: Sect. 2 outlines the fast formation and trimming concepts and highlights the contradiction of their underlying ideas. The proposed approaches to overcome this barrier are presented in Sect. 3 and they are compared by numerical experiments in Sect. 4. The paper closes with concluding remarks.

2 The Discrepancy Between Weighted Quadrature and Trimming

This section provides short introductions to weighted quadrature, sum factorization, and trimmed patches, which are the preliminaries for the proposed approach detailed subsequently. First, the main aspects presented in [9, 17] are recapitulated. Then, the fundamentals of trimmed spaces and the challenge of deriving a fast formation technique for them is outlined.

2.1 *Fast Formation by Weighted Quadrature and Sum Factorization*

Let us recapitulate some essential properties of the basis functions used. A *B-spline* $B_{I,p}$ is described by piecewise polynomial segments of degree p. The continuity between them is specified by the knot vector Ξ, which is a non-decreasing sequence of parametric coordinates $\xi_J \leqslant \xi_{J+1}$. The values of these knots ξ_J define the location where adjacent segments join. The continuity at these breakpoints is C^{p-m}, with m denoting the multiplicity of the corresponding knot value, i.e., $\xi_J = \xi_{J+1} = \cdots = \xi_{J+m-1}$. The knot span S refers to the half-open interval $[\xi_S, \xi_{S+1})$, and if its size is non-zero, it marks an Element. Furthermore, the knot vector Ξ defines an entire set of linearly independent B-splines $\{B_{I,p}\}_{I=0}^{n}$ on the parametric domain Ω. Each $B_{I,p}$ has local support, supp$\{B_{I,p}\}$, specified by the knots $\{\xi_I, \ldots, \xi_{I+p+1}\}$, and each knot span S contains $p+1$ non-zero B-splines. Bi-variate basis functions are obtained by computing the tensor product of uni-variate B-splines B_{I_1,p_1} and B_{I_2,p_2} of degrees

p_1 and p_2, which are defined by separate knot vectors Ξ_1 and Ξ_2 for the parametric directions ξ_1 and ξ_2, respectively. This can be generally expressed as

$$B_I(\boldsymbol{\xi}) = B_{I_1,\ldots,I_d}(\xi_1, \ldots, \xi_d) = \prod_{K=1}^{d} B_{I_K}(\xi_K). \tag{1}$$

In this paper, the focus lies on *weighted quadrature* for the formation of mass matrices. Hence, the aim is to derive an efficient and accurate evaluation of the following uni-variate integral

$$\int_\Omega B_I(\xi) B_J(\xi) c(\xi) d\xi. \tag{2}$$

$B_I(\xi)$ and $B_J(\xi)$ are test and trial functions and $c(\xi)$ is determined by the geometry mapping. In general, numerical quadrature rules $\mathbb{Q}$ are designed to be exact for the case that $c(\xi) = 1$, and they provide quadrature weights w_K, which allow the expression of equation (2) as a sum over corresponding quadrature points x_K, i.e.,

$$\mathbb{Q} = \sum_K B_I(x_K) B_J(x_K) w_K := \int_\Omega B_I(\xi) B_J(\xi) d\xi. \tag{3}$$

The novelty of weighted quadrature is that a quadrature rule $\mathbb{Q}_I$ is designed for each test function B_I by incorporating the test function into the quadrature weights

$$\mathbb{Q}_I = \sum_K B_J(x_K) w_{K,I} := \int_\Omega B_J(\xi) \left(B_I(\xi) d\xi\right). \tag{4}$$

The computation of the weights $w_{K,I}$ requires an adequate distribution of the quadrature points x_K. Here, the procedure proposed in [17] is employed, and the reader is referred to this publication for details. It determines the minimal number of quadrature points for each element of a non-uniform knot vectors with mixed continuity. Subsequently, a uniform distribution within the elements' interior is employed so that no x_K coincides with a knot ξ_I.

Once the position of the quadrature points x_K is fixed, the corresponding weights can be computed by the following system of equations

$$\begin{aligned}
\mathbb{Q}_I\left(B_{J_1}\right) = & \sum_{K \in \mathcal{Q}_I} B_{J_1}(x_K) w_{K,I} && := \int_\Omega B_{J_1}(\xi) \left(B_I(\xi) d\xi\right), \\
\vdots & && \vdots \\
\mathbb{Q}_I\left(B_{J_n}\right) = & \sum_{K \in \mathcal{Q}_I} B_{J_n}(x_K) w_{K,I} && := \int_\Omega B_{J_n}(\xi) \left(B_I(\xi) d\xi\right),
\end{aligned} \tag{5}$$

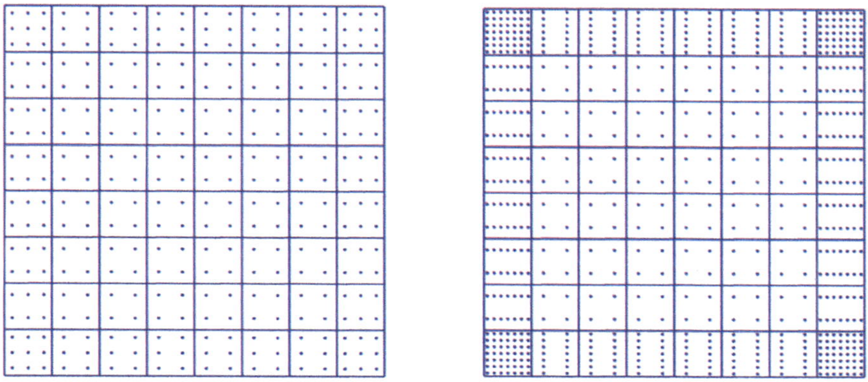

(a) Bi-degree $p = 2$ (b) Bi-degree $p = 6$

Fig. 1 Weighted quadrature points for two bi-variate bases with the same number of elements but with a different polynomial degree

where $J_1, \ldots, J_n$ are the indices of all trial functions whose support overlaps with the one of the current test function, supp$\{B_I\}$, and the index set $\mathcal{Q}_I$ refers to all quadrature points that lie within supp$\{B_I\}$. Solving (5) by QR-factorization yields $w_{K,I}$ for $\forall k \in \mathcal{Q}_I$, and the remaining quadrature weights are set to zero. The solution of (5) maybe not unique, and the system of equations can be weighted to improve positivity and boundedness of the weights as detailed in [17]. In the case of multi-variate tensor product basis functions, quadrature rules $\mathbb{Q}_{I_1}, \ldots, \mathbb{Q}_{I_d}$ are computed for each parametric direction. Figure 1 shows the quadrature layout for two bases of degree 2 and 6, respectively. Note that the number of quadrature points for the inner elements does not change.

Sum factorization takes advantage of the tensor product structure of the quadrature rule and the spline basis by a reordering of the numerical operations such that only uni-variate quadrature rules are required. Consider the computation of an entry of a mass matrix for a bi-variate basis

$$m_{\boldsymbol{I},\boldsymbol{J}} = \int_{\Omega} B_{\boldsymbol{I}}(\boldsymbol{\xi}) B_{\boldsymbol{J}}(\boldsymbol{\xi}) c(\boldsymbol{\xi}) d\boldsymbol{\xi}. \tag{6}$$

Starting from (1), Eq. (6) can be expressed as a recursion of uni-variate integrals

$$m_{\boldsymbol{I},\boldsymbol{J}} = \int_{\Omega_1} B_{I_1}(\xi_1) B_{J_1}(\xi_1) \times \left[\int_{\Omega_2} B_{I_2}(\xi_2) B_{J_2}(\xi_2) c(\xi_1, \xi_2) d\xi_2 \right] d\xi_1. \tag{7}$$

Consequently, it is straightforward to apply the weighted quadrature rules $\mathbb{Q}_{I_1}$, $\mathbb{Q}_{I_2}$ associated with a bi-variate test function $B_{\boldsymbol{I}}$. Finally, it is noted that the equations have to be pulled back to the parametric domain [1, 17] to exploit the structure of Eq. (7).

2.2 Trimmed Patches

Tensor product B-splines allow full control over the continuity and degree of the basis, but they possess an intrinsic four-sided topology limiting their ability to represent arbitrary domains. Trimming provides a remedy to this restriction by defining the valid area Ω^{v} independent from the basis' structure. In particular, curves in the parametric space specify Ω^{v}, and the curve's orientation determines the interior and exterior domain. Usually, these *trimming curves* $\boldsymbol{C}^t$ are represented by B-splines as well [30], but this choice is more or less free.

The presence of $\boldsymbol{C}^t$ divides the basis into three different functions types based on the overlap of the support with the valid domain, i.e., $\mathscr{S}_{\boldsymbol{I}}^{\mathrm{v}} := \mathrm{supp}\{B_{\boldsymbol{I}}\} \cap \overline{\Omega^{\mathrm{v}}}$. That is, a B-spline $B_{\boldsymbol{I}}$ is classified as:

- *Exterior* if $\mathscr{S}_{\boldsymbol{I}}^{\mathrm{v}} = \emptyset$,
- *Interior* if $\mathscr{S}_{\boldsymbol{I}}^{\mathrm{v}} = \mathrm{supp}\{B_{\boldsymbol{I}}\}$,
- *Cut* if $0 < \left|\mathscr{S}_{\boldsymbol{I}}^{\mathrm{v}}\right| < |\mathrm{supp}\{B_{\boldsymbol{I}}\}|$,

where $|\cdot|$ denotes the Lebesgue measure in $\mathbb{R}^d$. Figure 2 illustrated these different classes for a bi-variate cubic basis.

In the context of analysis, exterior basis functions can be neglected from the system of equations, and interior ones can be treated as usual. Cut basis functions, however, induce profound numerical challenges regarding the application of boundary conditions, the conditioning of system matrices, and accurate integration. Addressing these aspects is far from trivial, even in the conventional low-order finite element setting. The loss of the tensor product structure and the requirement of higher-order accuracy complicates the situation further for the fast formation outlined in Sect. 2.1.

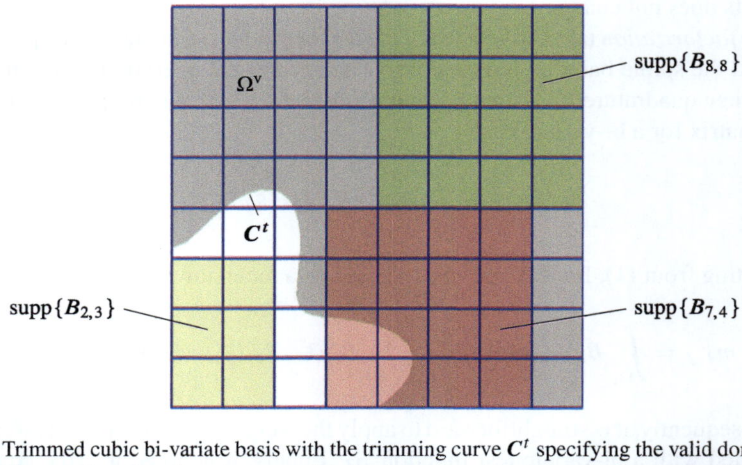

Fig. 2 Trimmed cubic bi-variate basis with the trimming curve $\boldsymbol{C}^t$ specifying the valid domain Ω^{v} (gray). The resulting B-splines types are interior (green), cut (red), or exterior (yellow) based on the overlap of the support, $\mathrm{supp}\{B_{I_1,I_2}\}$, with Ω^{v}

In the following, the focus is on integrating cut basis functions. The application of boundary conditions is not an issue in this paper because only the formation of mass matrices and L^2-projection is considered. The extended B-spline concept is employed to address the conditioning aspect. This procedure is independent of the assembly and formation process, and hence, it is not described here. The interested reader is referred to [18, 29, 31].

3 Integration of Cut Basis Functions

A trimming curve $\boldsymbol{C}^t$ introduces an arbitrarily located jump discontinuity within the parameter space. Simply integrating over this interface or neglecting quadrature points that lie outside of the valid domain Ω^{v} will evidently lead to incorrect results. Consequently, the numerical integration scheme has to account for these arbitrarily located discontinuities. In the case of weighted quadrature, this circumstance affects not only the correct representation of the integration domain $\mathscr{S}_I^{\mathrm{v}}$ of a cut B-spline B_I but also the computation of its quadrature rules $\mathbb{Q}_{I_1}$ and $\mathbb{Q}_{I_2}$. Moreover, sum factorization cannot be applied because $\mathscr{S}_I^{\mathrm{v}}$ does not follow a tensor product structure in general.

However, we can split the domain $\mathscr{S}_I^{\mathrm{v}}$ into a *regular* part $\mathscr{S}_I^{\mathrm{r}}$, which follows the tensor product structure (at least on the element-level), and a *trimmed* part $\mathscr{S}_I^{\mathrm{t}}$, which consists of all elements cut by the trimming curve. The integral over a cut basis function can be written as

$$\int_{\mathscr{S}_I^{\mathrm{v}}} B_I(\boldsymbol{\xi}) d\boldsymbol{\xi} = \int_{\mathscr{S}_I^{\mathrm{r}}} B_I(\boldsymbol{\xi}) d\boldsymbol{\xi} + \int_{\mathscr{S}_I^{\mathrm{t}}} B_I(\boldsymbol{\xi}) d\boldsymbol{\xi}. \tag{8}$$

In the following, the numerical integration of $\mathscr{S}_I^{\mathrm{t}}$ employs a standard element-wise assembly procedure and Sect. 3.1 lists various concepts for the distribution of quadrature points. Furthermore, this work investigates two options for the treatment of the remaining regular part $\mathscr{S}_I^{\mathrm{r}}$: In Sect. 3.2, it is integrated using Gaussian quadrature, while Sect. 3.3 derives a weighted quadrature rule that acknowledges the presence of the trimming curve within the support.

3.1 *Treatment of Cut Elements*

The fast formation and assembly approach opens the path towards efficient higher-order simulation. Hence, it is of utmost importance that the analysis, and therefore the integration of cut elements, is performed with higher-order accuracy. Luckily, there is a substantial body of literature on this topic because integrating over elements cut by an arbitrary interface is a canonical problem in various analysis schemes such as fictitious domain methods, extended finite element approaches, and the simulation with trimmed spline geometries. The proposed concepts can be broadly divided into strategies that (i) set up tailored integration rules or (ii) decompose cut elements into sub-elements, which then employ standard quadrature rules. For the latter type, several works [2, 10, 15, 23–25, 27] have demonstrated that the introduction of curved sub-elements with high degree, whose edges or faces capture the interface, allows for higher-order accurate integration with a modest number of integration points. However, generating these sub-elements is a challenge on its own, and its complexity is determined by the dimension of the domain and the shape of the interface. An alternative is the construction of tailored integration rules for a cut element, see e.g. [32–34]. These strategies usually obtain integration weights by solving a system of moment-fitting equations. Utilizing Lasserre's theorems [26], integrals over the boundaries of the cut elements provide the right-hand side required. Müller et al. [33] showed that this concept achieves higher-order accurate integration over curved three-dimensional domains.

Herein, the decomposition-based approach detailed in [15, 37] is utilized. It represents the valid domain Ω^{v} in the parameter space by a set of higher-order Lagrange elements, which are aligned with the parameter space as well as the trimming curve $\boldsymbol{C}^t$. Furthermore, they have the same polynomial degree as the underlying B-spline basis so that the representation quality of $\boldsymbol{C}^t$ matches the approximation power sought for the analysis. The mappings of these Lagrange elements distribute Gauss quadrature points within cut elements, as shown in Fig. 3. It is worth noting that this particular choice for the integration of cut elements is not essential for the ideas of this work. Any integration scheme could be employed, as long as it allows for higher-order accurate integration of trimmed domains. Therefore, a detailed discussion on this part of the overall integration of cut B-splines is omitted.

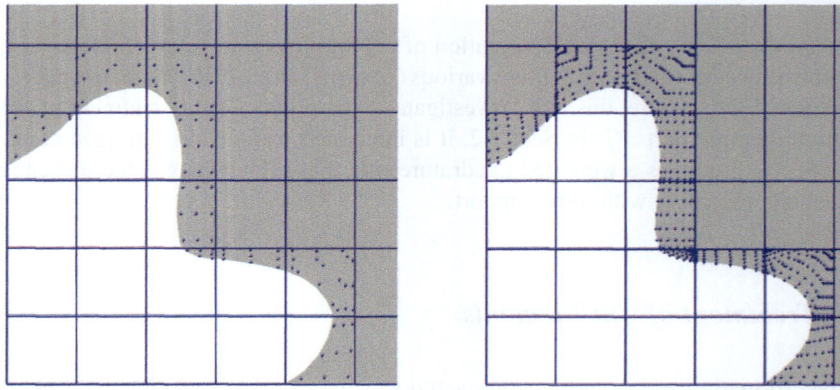

(a) Bi-degree $p = 2$ (b) Bi-degree $p = 6$

Fig. 3 Higher-order accurate distribution of Gaussian quadrature points within cut elements for different degree

3.2 Hybrid Gauss Approach

The perhaps most straightforward way to integrate over the regular portion $\mathscr{S}_I^{\mathrm{r}}$ of a cut basis function is to employ standard Gauss quadrature using a conventional element-wise assembly. In contrast to the cut elements of $\mathscr{S}_I^{\mathrm{t}}$, the evaluations during the formation can exploit the tensor product structure. Hence, it is beneficial to implement separate routines for $\mathscr{S}_I^{\mathrm{r}}$ and $\mathscr{S}_I^{\mathrm{t}}$. As a result, a cut B-spline is treated by a hybrid Gauss-quadrature-based formation concept. Regarding the entire trimmed space, Gauss points are restricted to the vicinity of the trimming curve. Yet, they propagate further into the interior when the degree increases since the supports of the trimmed B-splines increase. Consider the example shown in Fig. 4, all elements are affected in the degree 6 case. Nevertheless, the portion of the computational cost for setting up the entire system of equations decreases with the fineness of the parameter space.

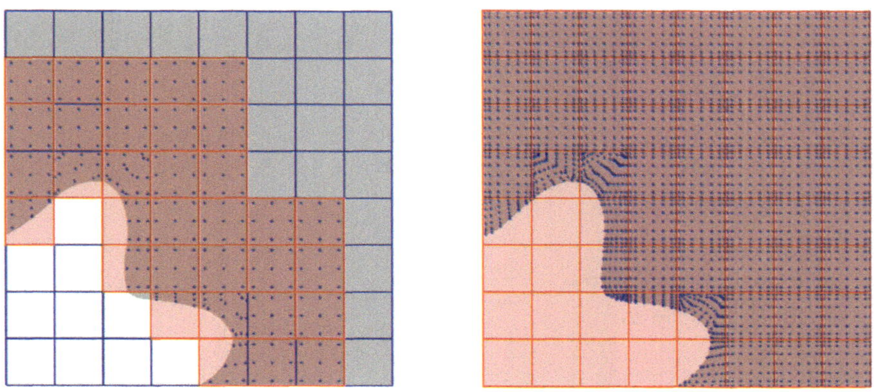

(a) Degree 2 (b) Degree 6

Fig. 4 Gaussian quadrature for the entire valid support $\mathscr{S}^{\mathrm{v}}$ of all cut basis functions. The support of cut basis functions is highlighted in red

Remark 1 Since the elements of $\mathscr{S}_I^{\mathrm{r}}$ follow the tensor product structure, sum factorization with Gauss points $\boldsymbol{x}_K$ may be used by setting the precomputed coefficient $c(\boldsymbol{x}_K)$ of Eq. (6) to zero if $\boldsymbol{x}_K \notin \mathscr{S}_I^{\mathrm{r}}$. This option is, however, not considered here.

3.3 Discontinuous Weighted Quadrature

The domain splitting (8) allows us to extract those elements that follow the tensor product structure. Hence, we can apply sum factorization there, as noted in the previous section. In the context of weighted quadrature, however, we cannot set the coefficients $c(\boldsymbol{x}_K)$ in Eq. (7) to zero if the quadrature point $\boldsymbol{x}_K \notin \mathscr{S}_I^{\mathrm{r}}$. In contrast to Gauss quadrature, the weighted rules take advantage of the continuity between elements. Thus, the zero coefficients introduce artificial discontinuities within the domain of the quadrature, which leads to an error in the evaluation of the integral.

The construction of *discontinuous weighted quadrature* is proposed to resolve this problem. The idea is to incorporate the information on the location ξ^{disc} of these artificial discontinuities into the quadrature weights. As a result, the numerical integration on one side of ξ^{disc} can be treated independently from the other. Each ξ^{disc} marks a knot adjacent to a cut element. Thus, the trimming curve determines the number of ξ^{disc} required. The tools for deriving related quadrature rules are the fact that quadrature is a linear operator and the properties of knot insertion.

Knot insertion denotes the refinement of a B-spline object by adding knots $\tilde{\xi}$ into its knot vector Ξ. This procedure leads to nested spline spaces, and the subdivision matrix $\mathbf{S} : \mathbb{R}^{n+1} \mapsto \mathbb{R}^{\tilde{n}+1}$ encodes the coefficients of the initial coarse representation and the refined one. Considering quadrature weights, we obtain the relation

$$\tilde{w}_i = \sum_{j=0}^{n} \mathbf{S}_{ij}\, w_j, \qquad \text{for } i = 0, \ldots, \tilde{n}, \tag{9}$$

where $\tilde{w}_i$ refers to quadrature weights computed for the refined basis functions. If only one knot is inserted, i.e., $\tilde{\Xi} = \Xi \cup \tilde{\xi}$, where $\tilde{\xi} \in [\xi_S, \xi_{S+1})$, the non-zero entries of $\mathbf{S}$ are determined by

$$\begin{cases} \mathbf{S}(K, K-1) & = 1 - \alpha_K \\ \mathbf{S}(K, K) & = \alpha_K \end{cases}, \qquad \alpha_K = \begin{cases} 1 & K \leqslant S - p \\ \frac{\tilde{\xi} - \xi_K}{\xi_{K+p} - \xi_K} & S - p + 1 \leqslant K \leqslant S \\ 0 & K \geqslant S + 1 \end{cases}. \tag{10}$$

Multiple knots can be inserted by repeating this process, and the multiplication of the individual single-knot matrices yields the overall subdivision matrix.

Using the relations provided by $\mathbf{S}$, the construction of discontinuities weighted quadrature rules for each artificial discontinuities ξ^{disc} goes as follows:

1. Knot insertion at ξ^{disc} so that the uni-variate basis become C^{-1} continuous there, and storing of the corresponding $\mathbf{S}$.
2. Determination of the minimal number of quadrature points n_{min} for the refined basis [17], and addition of new nested quadrature points if necessary.
3. Computation of $\tilde{w}$ for basis functions of the refined basis using Eq. (5).
4. Multiplication of $\tilde{w}$ by $\mathbf{S}^{\mathrm{T}}$ to obtain w for the initial basis.

Figure 5 illustrates these steps for a single cut basis function. Using nested quadrature points in step 2 of the construction allows the reuse of all coefficients set up for the weighted quadrature for interior B-splines during the integration process. The resulting weights w of the initial and nested quadrature points are obtained by a linear combination of weights $\tilde{w}$ that account for the discontinuity at ξ^{disc}. Thus, we can

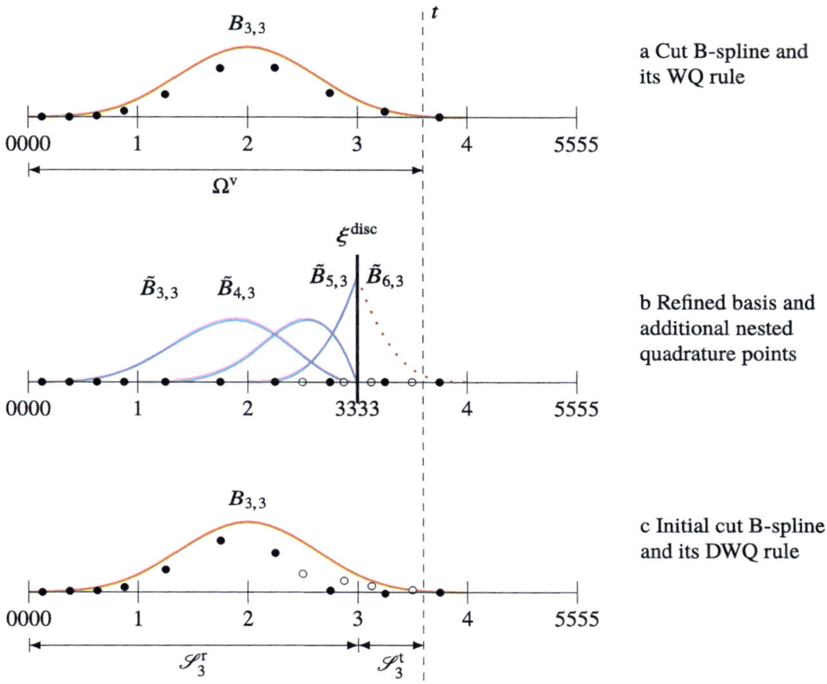

Fig. 5 Discontinuous weighted quadrature (DWQ) for the cubic B-spline $B_{3,3}$ cut at the trimming position t. **a** Conventional weighted quadrature (WQ) points (black dots) for $B_{3,3}$. **b** Quadrature layout with additional quadrature points (white) for WQ of the refined discontinuous $\tilde{B}_{J,3}$ associated to $B_{3,3}$. **c** Linear combination of the refined discontinuous WQ rules to obtain the DWQ for $B_{3,3}$. In (**a**, **c**), the points' height indicates the related weight value

set all coefficients on one side of ξ^{disc} to zero without affecting the integral on the other side.

Remark 2 The nested quadrature points are uniformly distributed between the initial ones. The number of added points per interval is chosen such that the distance between between them gets as large as possible.

In the context of sum factorization, only discontinuous weighted quadrature points within $\mathcal{S}_I^{r}$ are of interest. In a general trimming situation, these points do not cover all elements of $\mathcal{S}_I^{r}$. The remaining parts are integrated by Gaussian quadrature. Consider the example shown in Fig. 6. Note that the artificial discontinuity restricts the Gauss points to the vicinity of the trimming curve, independent of the degree.

Remark 3 Every artificial discontinuity ξ^{disc} increases the number of required quadrature points in the elements adjacent to ξ^{disc}. Hence, the advantage of providing a discontinuous weighted quadrature rule for a cut basis functions decreases with the number of ξ^{disc}. In the present implementation, an individual quadrature is

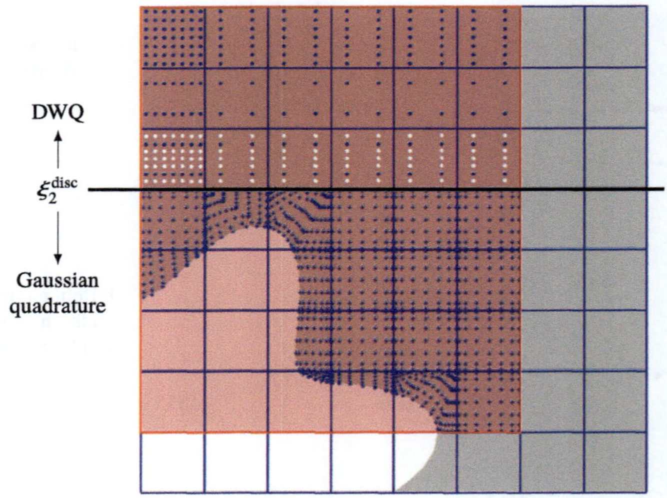

Fig. 6 All quadrature points of a cut B-spline of bi-degree 6: above the artificial discontinuity ξ_2^{disc} discontinuous weighted quadrature (DWQ) is employed, while Gauss points are used below. The white dots mark the nested quadrature points added during the construction

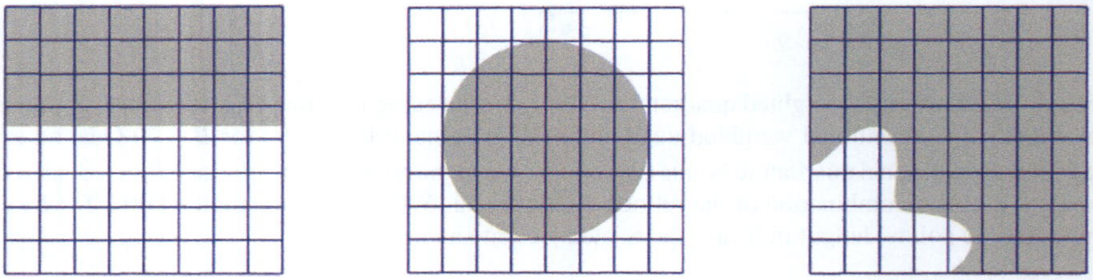

(a) Line (b) Circle (c) Curvy corner cut

Fig. 7 The different trimming cases of the numerical examples

constructed only if not more than one ξ^{disc} for each parametric direction is present in the basis function's support.

4 Numerical Results

The numerical experiments focus on the formation of the mass matrix and L^2-projection for trimmed bi-variate B-splines. Figure 7 illustrates the three investigated trimming cases of different complexity.

In the following, the standard element-wise Gauss procedure provides the reference solutions, and the fast formation and assembly strategies considered are:

- *Weighted quadrature* (Sect. 2.1) for the regular support $\mathscr{S}^{\mathrm{r}}$ of cut and interior B-splines.
- *Hybrid Gauss* (Sect. 3.2) for the regular support $\mathscr{S}^{\mathrm{r}}$ of cut B-splines and weighted quadrature for interior B-splines.
- *Discontinuous weighted quadrature* (Sect. 3.3) for the regular support $\mathscr{S}^{\mathrm{r}}$ of cut B-splines and weighted quadrature for interior B-splines.

Cut elements are integrated by the same element-wise higher-order accurate procedure (Sect. 3.1) in all cases. The related decomposition by Lagrange elements for the distribution of the integration point goes up to degree 6. Thus, this degree marks the upper threshold for the following numerical experiments.

4.1 Mass Matrix Formation

The first numerical example addresses the accuracy of the formation of the mass matrix. Therefore, the line-trim-case (Fig. 7a) is considered for a 10 by 10 element basis. Standard Gauss provides the reference matrix, and the deviation of the other matrices build by the fast concepts is measured in the Euclidean norm. Table 1 summarized the results for various degrees. The results of the other trimming cases are omitted because they lead to the same findings and do not provide further insights.

It is apparent that the direct application of weighted quadrature to cut basis functions is not able to derive the correct mass matrix, whereas the proposed concepts are accurate up to machine precision. Thus, the remaining numerical examples will only employ the hybrid Gauss and discontinuous weighted quadrature approaches.

Table 1 The error of the computed mass matrix for the line-trim-case (Fig. 7a) using different fast formation and assembly concepts

Degree	Hybrid gauss	Disc. weighted quadrature	Weighted quadrature
1	4.68883379e-17	4.69071638e-17	3.33078900e-03
2	2.18894391e-17	2.19723138e-17	3.38923479e-04
3	4.98024622e-17	4.98024919e-17	3.03838755e-04
4	3.80411220e-16	3.84739261e-16	1.78250728e-04
5	1.01851172e-15	1.01851512e-15	1.59974261e-04
6	2.00575179e-15	2.00590425e-15	1.28293909e-04

4.2 L^2-Projection

Here, the accuracy and efficiency of the hybrid Gauss approach and the discontinuous weighted quadrature are investigated for all three trimming cases shown in Fig. 7. One ingredient for the quality of the results is the conditioning of the mass matrix, which may suffer due to the presents of trimmed basis functions. The extended B-spline concept is employed for all simulations to guarantee well-conditioned mass matrices. For details on this approach, the interested reader is referred to [18, 29, 31, 37].

4.2.1 Approximation Quality

L^2-projection is used to fit the trimmed patch to the target function $f(x, y) = \sin(2x)\cos(3y)$, and the quality of the resulting approximation is measured in the relative L^2-error norm $\|\epsilon_{rel}\|_{L^2}$. The convergence results for the line-, circle-, and curvy corner cut trimming cases are illustrated in Figs. 8, 9 and 10, respectively. In all cases, optimal convergence rates and excellent agreement to the reference solutions can be observed for both approaches proposed.

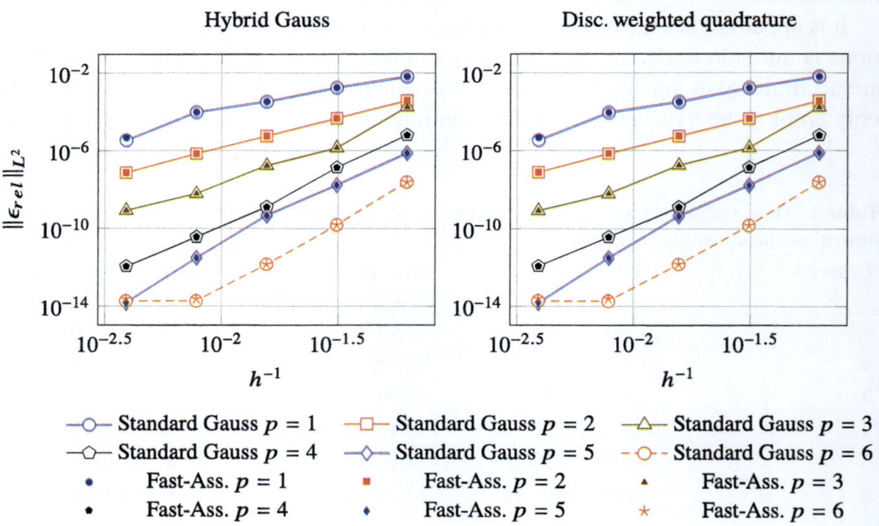

Fig. 8 Line convergence study

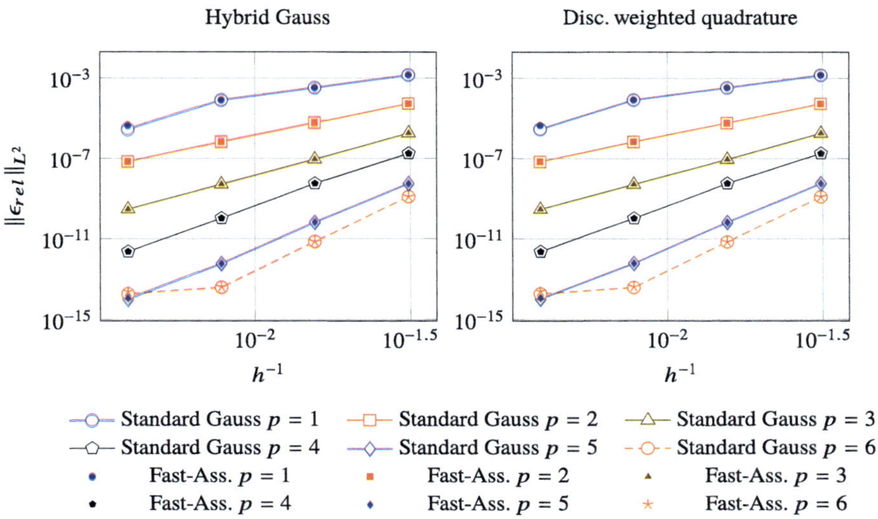

Fig. 9 Circle convergence study

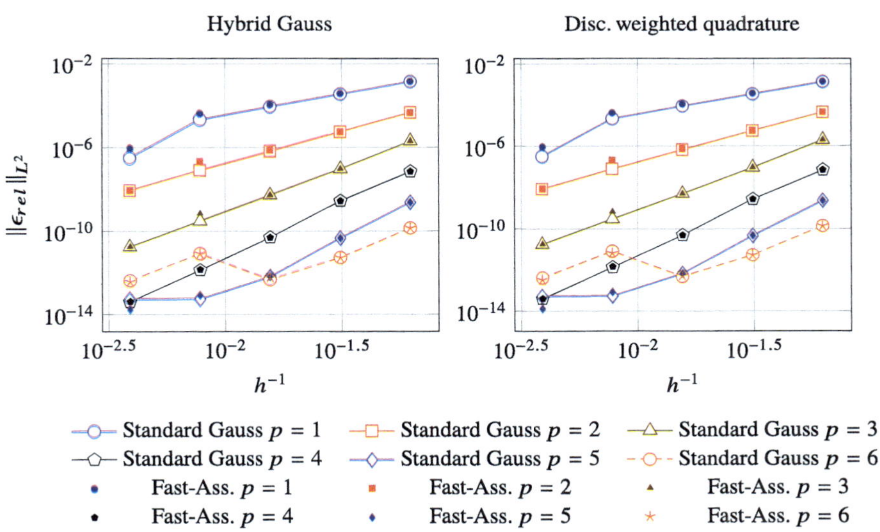

Fig. 10 Curvy corner cut convergence study

4.2.2 Efficiency Comparison

The assessment of the efficiency of the different formation and assembly strategies is based on the timings of the finest discretizations used in the convergence studies. All routines of the proposed concepts have been implemented in an in-house MATLAB® code, which does not utilize any parallelization capabilities. Total timings have been obtained on an Intel® Core™ i7-8700 3.2 GHz processor with 16 GB RAM. The timings have been measured with MATLAB's tic-toc command.

Figure 11 illustrates the total formation and assembly time obtained. The standard Gauss formation is the most expensive concept, as expected. The performance of the other approaches, however, is a bit astonishing since the hybrid Gauss approach

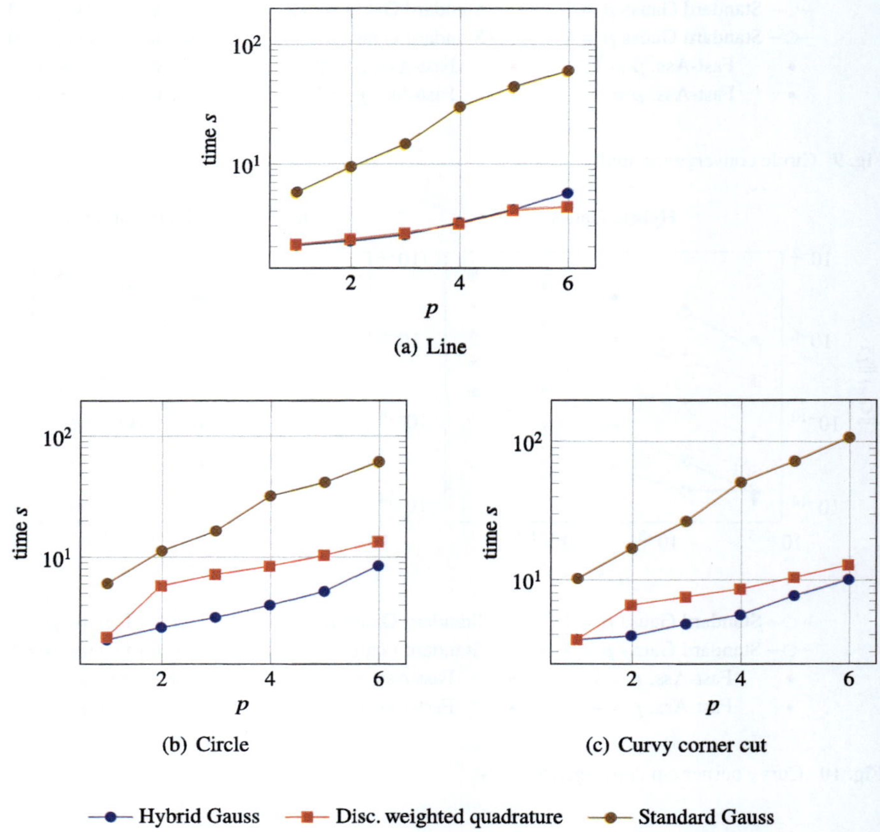

Fig. 11 Total formation and assembly time including (i) the computation of quadrature weights and (ii) the integration of the entire valid support of cut and interior basis functions. Each subfigure corresponds to a trimming case (Fig. 7), while the different graphs refer to the assembly and formation strategy used

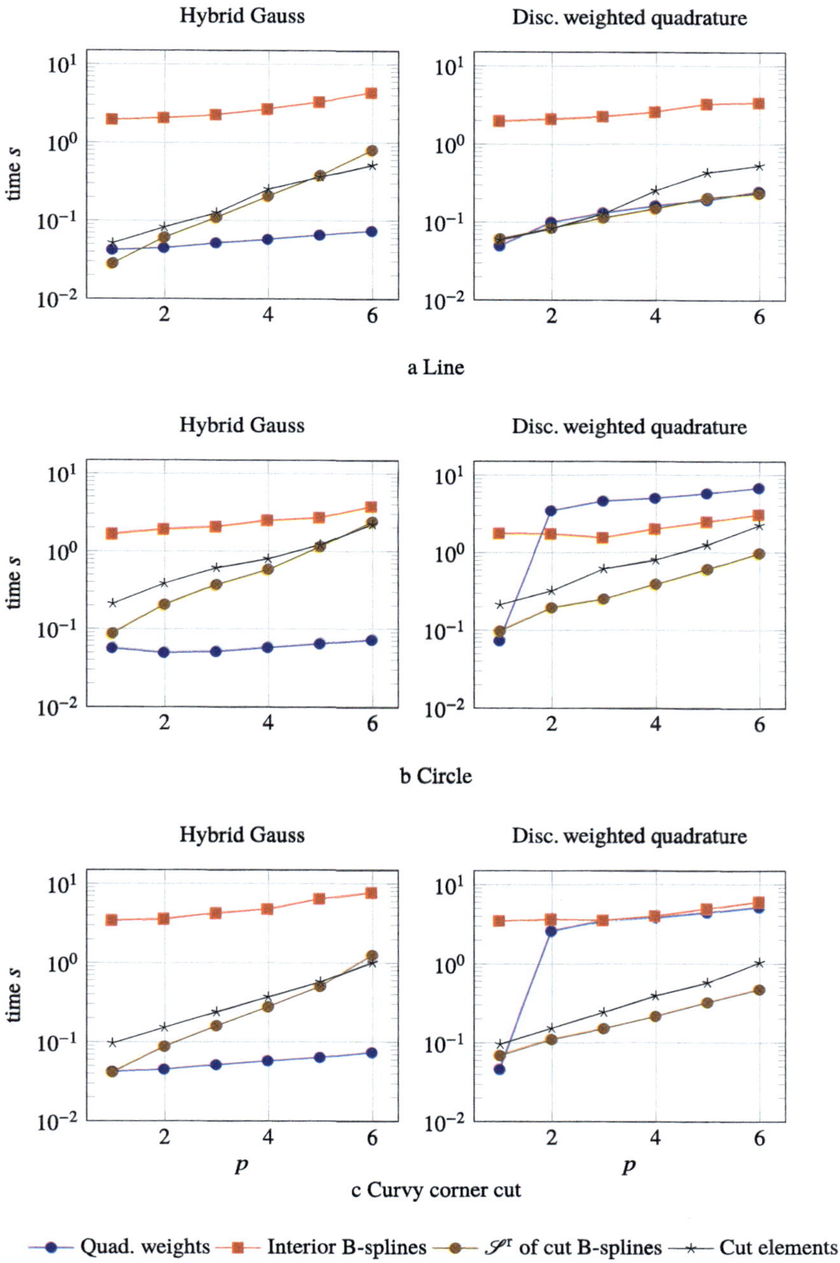

Fig. 12 The various components of the total formation and assembly time for each trimming case (Fig. 7). The graphs refer to the set-up of the quadrature weights (Quad. weights), as well as the integration of the regular (Interior B-splines) and trimmed ($\mathscr{S}^{\mathrm{r}}$ of cut B-splines, Cut elements) B-splines

is more efficient in most situations. In order to gain deeper insight, the various components of the total formation and assembly time are compared for each trimming case in Fig. 12.

These components are (i) the computation of the (normal and discontinuous) weighted quadrature rules, (ii) the formation of interior B-splines, the integration over (iii) the regular part $\mathscr{S}^{\mathrm{r}}$ of cut B-splines and (iv) the remaining portion $\mathscr{S}^{\mathrm{t}}$ covered by cut elements. The components (ii) and (iv) employ the same routines in both approaches. Hence, there are only minor deviations due to the usual time irregularities between two runs of the same code. Looking at the timings related to the regular part $\mathscr{S}^{\mathrm{r}}$ of cut B-splines reveals that the discontinuous weighted quadrature scales better w.r.t. the degree. The computation of the corresponding quadrature weights, however, is more expensive due to the additional expenses for setting up the discontinuous weighted quadrature rules for cut B-splines. This overhead varies with the trimming situation, which determines the number of artificial discontinuous introduced, and therefore, the number of additional quadrature rule computations.

5 Conclusions

It has been demonstrated that the fast assembly and formation strategy presented in [9, 17] cannot be applied directly to trimmed spline spaces. This work examines concepts to overcome this limitation. In particular, B-splines $B_{\boldsymbol{I}}$ cut by the trimming curve $\boldsymbol{C}^t$ require an adaptation of the fast procedure. Therefore, their support is divided into a regular part $\mathscr{S}_{\boldsymbol{I}}^{\mathrm{r}}$ and a trimmed part $\mathscr{S}_{\boldsymbol{I}}^{\mathrm{t}}$. The latter consists of the elements intersected by $\boldsymbol{C}^t$. These elements do not follow the tensor product structure of the underlying spline basis and require tailored integration rules. There are different integration techniques for this task available in the literature. This work employs the approach presented in [15, 37], but any scheme that allows higher-order accurate integration (favorably in an efficient manner) may be used. This property, however, is essential. Otherwise, the integration of the cut elements outweighs the benefits of applying fast assembly and formation of the remaining basis.

The formation of the remaining contributions of a cut B-spline $B_{\boldsymbol{I}}$ involves the integration over its $\mathscr{S}_{\boldsymbol{I}}^{\mathrm{r}}$, i.e., all non-cut elements of $B_{\boldsymbol{I}}$ enclosed by $\boldsymbol{C}^t$. This paper presents two concepts: The first one is a *hybrid Gauss approach* which employs standard Gaussian quadrature. Hence, the formation of a cut B-spline utilizes standard Gauss quadrature in all elements, but the evaluations for $\mathscr{S}_{\boldsymbol{I}}^{\mathrm{r}}$ can be implemented more efficiently. The second concept introduces tailored weighted quadrature rules for a cut B-spline's $\mathscr{S}_{\boldsymbol{I}}^{\mathrm{r}}$. These *discontinuous weighted quadratures* define artificial discontinuous ξ^{disc} between elements for the construction of the quadrature weights. The trimming curve $\boldsymbol{C}^t$ determines the locations of ξ^{disc}. To be precise, they mark the knots adjacent to cut elements. Compared to normal weighted quadrature, additional quadrature points are needed next to ξ^{disc} to fulfill the exactness condition. The resulting discontinuous quadrature rules allow that the coefficients on one side of ξ^{disc} can be set to zero without affecting the contributions of the coefficients on the other side. This property enables a straightforward application of sum factorization to a subregion of a support. Finally, it is emphasized that interior B-splines, i.e., those not cut by $\boldsymbol{C}^t$, are treated by normal weighted quadrature and sum factorization in both approaches.

The approximation quality and efficiency of the proposed approaches have been investigated by the formation of mass matrices and L^2-projection for bi-variate trimmed spline spaces. It has been shown that the hybrid Gauss approach and discontinuous weighted quadrature yield the same accuracy as the reference solutions obtained by standard element-wise formation using Gauss quadrature. Both schemes show significant speedups compared to the reference computations. Using discontinuous weighted quadrature, computing the contribution corresponding to $\mathscr{S}_{\boldsymbol{I}}^{\tau}$ scales better with the degree. However, the computational effort for setting up the additional quadrature rules dominates the total formation and assembly time, especially for complex trimming cases with multiple ξ^{disc}. Overall, the hybrid Gauss approach turned out as the more efficient technique for the majority of the test cases considered. From the numerical experiments, the following conclusions may be drawn: (i) The hybrid Gauss scheme is sufficient for moderate degrees; (ii) discontinuous weighted quadrature rules are beneficial for high degrees such as $p \geqslant 5$, but (iii) the computation of their weights must be performed in parallel.

Future work will focus on this parallelization aspect for the computation of the discontinuous weighted quadrature rules as well as the overall fast assembly and formation scheme. Furthermore, the efficiency of the current hybrid Gauss can be enhanced by utilizing sum factorization. The performance in the case of three-dimensional problems is another topic worth exploring. From a conceptional point of view, the extension of the concepts presented to another parametric dimension is uncomplicated. Considering discontinuous weighted quadrature, adding another dimension merely affects the determination of proper artificial discontinuities. However, the implementation is much more involved due to the increased complexity of the trimming situations. Higher-order accurate integration of trimmed tri-variate splines is already a challenge in the context of traditional finite element setting. Solving this challenge is a prerequisite for applying fast assembly and formation.

References

1. Antolin, P., Buffa, A., Calabrò, F., Martinelli, M., Sangalli, G.: Efficient matrix computation for tensor-product isogeometric analysis: the use of sum factorization. Comput. Methods Appl. Mech. Eng. **285**, 817–828 (2015)
2. Antolin, P., Buffa, A., Martinelli, M.: Isogeometric analysis on V-reps: first results. Comput. Methods Appl. Mech. Eng. **355**, 976–1002 (2019)
3. Auricchio, F., Calabrò, F., Hughes, T.J.R., Reali, A., Sangalli, G.: A simple algorithm for obtaining nearly optimal quadrature rules for NURBS-based isogeometric analysis. Comput. Methods Appl. Mech. Eng. **249–252**, 15–27 (2012)
4. Beer, G., Marussig, B., Duenser, C.: Basis Functions, B-splines, pp. 35–71. Springer International Publishing, Cham (2020)

5. Beirão da Veiga, L., Buffa, A., Rivas, J., Sangalli, G.: Some estimates for h–p–k-refinement in isogeometric analysis. Numerische Mathematik **118**, 271–305 (2011)
6. Borden, M.J., Scott, M.A., Evans, J.A., Hughes, T.J.R.: Isogeometric finite element data structures based on Bézier extraction of NURBS. Int. J. Numer. Methods Eng. **87**, 15–47 (2011)
7. de Borst, R., Chen, L.: The role of Bézier extraction in adaptive isogeometric analysis: local refinement and hierarchical refinement. Int. J. Numer. Methods Eng. **113**, 999–1019 (2018)
8. Bressan, A., Takacs, S.: Sum factorization techniques in isogeometric analysis. Comput. Methods Appl. Mech. Eng. **352**, 437–460 (2019)
9. Calabrò, F., Sangalli, G., Tani, M.: Fast formation of isogeometric Galerkin matrices by weighted quadrature. Comput. Methods Appl. Mech. Eng. **316**, 606–622 (2017)
10. Cheng, K.W., Fries, T.P.: Higher-order XFEM for curved strong and weak discontinuities. Int. J. Numer. Meth. Eng. **82**, 564–590 (2010)
11. Cottrell, J.A., Hughes, T.J.R., Bazilevs, Y.: Isogeometric Analysis: Toward Integration of CAD and FEA. Wiley, Chichester, England (2009)
12. Cottrell, J.A., Hughes, T.J.R., Reali, A.: Studies of refinement and continuity in isogeometric structural analysis. Comput. Methods Appl. Mech. Eng. **196**, 4160–4183 (2007)
13. Cottrell, J.A., Reali, A., Bazilevs, Y., Hughes, T.J.R.: Isogeometric analysis of structural vibrations. Comput. Methods Appl. Mech. Eng. **195**, 5257–5296 (2006)
14. Evans, J.A., Bazilevs, Y., Babuška, I., Hughes, T.J.R.: n-widths, sup-infs, and optimality ratios for the k-version of the isogeometric finite element method. Comput. Methods Appl. Mech. Eng. **198**, 1726–1741 (2009)
15. Fries, T.P., Omerović, S.: Higher-order accurate integration of implicit geometries. Int. J. Numer. Meth. Eng. **106**, 323–371 (2016)
16. Hiemstra, R.R., Calabrò, F., Schillinger, D., Hughes, T.J.R.: Optimal and reduced quadrature rules for tensor product and hierarchically refined splines in isogeometric analysis. Comput. Methods Appl. Mech. Eng. **316**, 966–1004 (2017)
17. Hiemstra, R.R., Sangalli, G., Tani, M., Calabrò, F., Hughes, T.J.R.: Fast formation and assembly of finite element matrices with application to isogeometric linear elasticity. Comput. Methods Appl. Mech. Eng. **355**, 234–260 (2019)
18. Höllig, K.: Finite Element Methods with B-Splines. In: Frontiers in Applied Mathematics, vol. 26. SIAM (2003)
19. Hughes, T.J.R., Cottrell, J.A., Bazilevs, Y.: Isogeometric analysis: CAD, finite elements, NURBS, exact geometry and mesh refinement. Comput. Methods Appl. Mech. Eng. **194**, 4135–4195 (2005)
20. Hughes, T.J.R., Reali, A., Sangalli, G.: Efficient quadrature for NURBS-based isogeometric analysis. Comput. Methods Appl. Mech. Eng. **199**, 301–313 (2010)
21. Johannessen, K.A.: Optimal quadrature for univariate and tensor product splines. Comput. Methods Appl. Mech. Eng. **316**, 84–99 (2017)
22. Kiendl, J., Bletzinger, K.U., Linhard, J., Wüchner, R.: Isogeometric shell analysis with Kirchhoff-Love elements. Comput. Methods Appl. Mech. Eng. **198**, 3902–3914 (2009)
23. Kudela, L.: Highly accurate subcell integration in the context of the finite cell method. Master's thesis, Technical University Munich (2013)
24. Kudela, L., Zander, N., Bog, T., Kollmannsberger, S., Rank, E.: Efficient and accurate numerical quadrature for immersed boundary methods. Adv. Model. Simul. Eng. Sci. **2**, 1–22 (2015)
25. Kudela, L., Zander, N., Kollmannsberger, S., Rank, E.: Smart octrees: accurately integrating discontinuous functions in 3D. Comput. Methods Appl. Mech. Eng. **306**, 406–426 (2016)
26. Lasserre, J.: Integration on a convex polytope. Proc. Am. Math. Soc. **126**, 2433–2441 (1998)
27. Legay, A., Wang, H.W., Belytschko, T.: Strong and weak arbitrary discontinuities in spectral finite elements. Int. J. Numer. Meth. Eng. **64**, 991–1008 (2005)
28. Lipton, S., Evans, J.A., Bazilevs, Y., Elguedj, T., Hughes, T.J.R.: Robustness of isogeometric structural discretizations under severe mesh distortion. Comput. Methods Appl. Mech. Eng. **199**, 357–373 (2010)
29. Marussig, B., Hiemstra, R., Hughes, T.J.R.: Improved conditioning of isogeometric analysis matrices for trimmed geometries. Comput. Methods Appl. Mech. Eng. **334**, 79–110 (2018)

30. Marussig, B., Hughes, T.J.R.: A review of trimming in isogeometric analysis: challenges, data exchange and simulation aspects. Arch. Comput. Methods Eng. **25**, 1059–1127 (2018)
31. Marussig, B., Zechner, J., Beer, G., Fries, T.P.: Stable isogeometric analysis of trimmed geometries. Comput. Methods Appl. Mech. Eng. **316**, 497–521 (2016)
32. Mousavi, S.E., Sukumar, N.: Numerical integration of polynomials and discontinuous functions on irregular convex polygons and polyhedrons. Comput. Mech. **47**, 535–554 (2011)
33. Müller, B., Kummer, F., Oberlack, M.: Highly accurate surface and volume integration on implicit domains by means of moment-fitting. Int. J. Numer. Meth. Eng. **96**, 512–528 (2013)
34. Nagy, A.P., Benson, D.J.: On the numerical integration of trimmed isogeometric elements. Comput. Methods Appl. Mech. Eng. **284**, 165–185 (2015)
35. Orszag, S.A.: Spectral methods for problems in complex geometries. J. Comput. Phys. **37**, 70–92 (1980)
36. Schillinger, D., Hossain, S.J., Hughes, T.J.R.: Reduced Bézier element quadrature rules for quadratic and cubic splines in isogeometric analysis. Comput. Methods Appl. Mech. Eng. **277**, 1–45 (2014)
37. Schöllhammer, D., Marussig, B., Fries, T.P.: A consistent higher-order isogeometric shell formulation (2020). arXiv:2012.11975

Subdivision and G-Spline Hybrid Constructions for High-Quality Geometric and Analysis-Suitable Surfaces

Jörg Peters and Kęstutis Karčiauskas

Abstract This survey of piecewise polynomial surface constructions for filling multi-sided holes in a smooth spline complex focusses on a class of hybrid constructions that, while heterogeneous, combines all the practical advantages of state-of-the-art for modelling and analysis: good shape, easy implementation and simple refinability up a pre-defined level. After reviewing the three ingredients—subdivision, G-spline and guided surfaces—the hybrid is defined to consist essentially of one macro-patch for each of n sectors, leaving just a tiny n-sided central hole to be filled by a G-spline construction. Here tiny means both geometrically small, e.g., two orders of magnitude smaller than pieces of the spline complex, and small in its contribution to engineering analysis, i.e., it is unlikely to require further refinement to express additional geometric detail or resolve a function on the surface, such as the solution of a partial differential equation. Each macro-patch has the local structure of a subdivision surface near, but excluding the central, extraordinary point: all internal transitions are the identity or scale according to contraction speed toward the extraordinary point. Both the number of pieces of the macro-patches and the speed can be chosen application-dependent and adaptively.

1 Introduction

A hybrid construction for filling an n-sided hole in a spline complex consists of two parts: a main body formed by several nested surface rings and a tiny central n-sided cap, see Fig. 1(b, h). Each ring consists of n L-shaped sectors that fit together smoothly at their tips to form a curved annulus, called surface ring. The L-shape also joins smoothly with a smaller L-shaped sector of the next ring, as illustrated in Fig. 2. The surface rings form a nested sequence that contracts rapidly towards the center. Alternatively, one can think of each L-shape as a 4-sided macro-patch with

J. Peters (✉)
University of Florida, Gainesville, FL, USA

K. Karčiauskas
Vilnius University, Vilnius, Lithuania

C. Manni and H. Speleers (eds.), *Geometric Challenges in Isogeometric Analysis*, Springer INdAM Series 49, https://doi.org/10.1007/978-3-030-92313-6_8

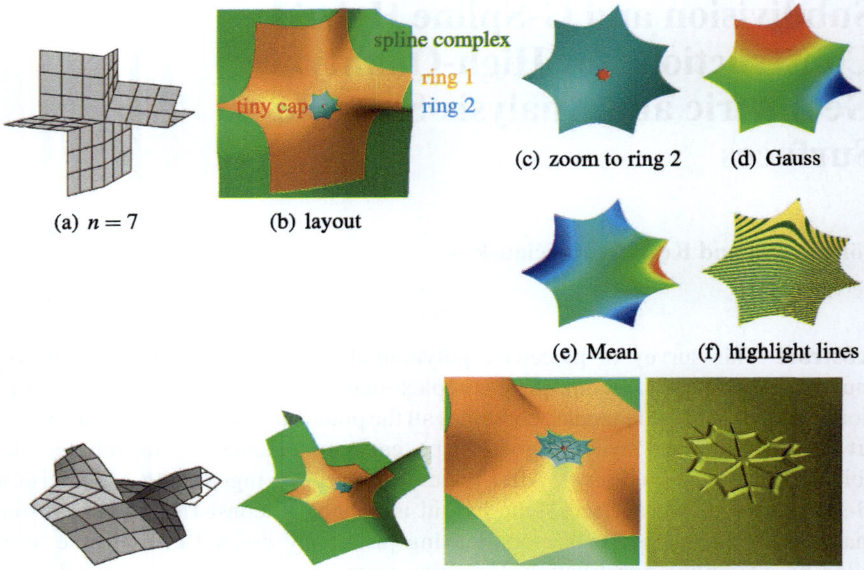

(a) $n = 7$ (b) layout (c) zoom to ring 2 (d) Gauss (e) Mean (f) highlight lines (g) $n = 8$ (h) layout (i) fine geometric edit, magnified

Fig. 1 Examples of high-valence hybrid surfaces. (**a**, **g**) Generalized control net (**c**-net). (**b**, **h**) layout of surface rings with almost invisible tiny cap. (**c**–**f**) Shape interrogation of ring 2 and the tiny cap: (**d**) Gauss curvature, (**e**) mean curvature. (**i**) Localized degrees of freedom used to emboss a pattern [46]

Fig. 2 L-shaped sectors of a macro-patch. The fine lines partitioning each square indicate the bi-quartic (bi-4) BB-net [20]

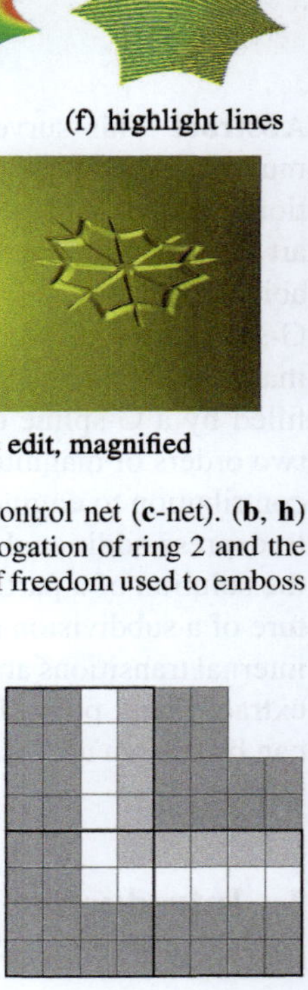

one quadrant punched out to leave space for the next smaller L-shape, i.e., their union is a spline complex with a hole. This is the setup of subdivision algorithms [82] near extraordinary points, except that the process in the hybrid construction stops after a few steps, e.g., 2 or 3 nested surface rings.

The pieces join to form the main body of the surface without change of variables, except possibly scaled derivatives as we adjust the contraction 'speed' of the rings. The main body of the surface is therefore amenable to refinement by knot insertion to increase of degrees of freedom. The same holds for functions on the surface, such as textures.

Only the tiny central n-sided cap is assembled by joining pieces so that derivatives agree after a change of variables (reparameterization), i.e., the cap is G^k continuous. The cap can be one of many in the literature, see Sect. 4. Typically the cap is so small that it needs not be refined. If it needs to be refined, [55] provides a refinable G^2 tiny cap construction.

The hybrid structure seems to just delay the multi-sided hole-filling challenge. The motivation for this seemingly more complex approach is that, if the cap is chosen to be smaller than the maximally anticipated refinement—for geometric modeling or computing on surfaces—then all surface pieces share a simple parameterization for refinement. On the other hand, capping the rings avoids the infinite recursion and central singularity of subdivision surfaces that necessitate smart data structures and special algorithms for proper integration over infinitely many polynomial pieces. By construction a hybrid surface is therefore 'analysis-suitable', in the sense that it offers a uniform increase in degrees of freedom by knot insertion—up to the refinement level of the tiny cap (and further, at some cost in complexity, if the cap is chosen to be refinable).

To ensure that the hybrid constructions are 'geometry-suitable', i.e., yield good shape, a third technique is applied: both the main body and the cap follow a guide surface. A guide surface need not match the surrounding spline complex but is sampled ever closer to the central point to guarantee that subsequent rings and the tiny cap do not freely oscillate but follow a consistent, structurally simpler shape.

Overview. Section 2 classifies constructions for filling n-sided holes and explains requirements for geometric design and engineering analysis. Section 3 reviews subdivision algorithms, Sect. 4 reviews G-splines, and Sect. 5 reviews guided constructions, the three ingredients from which subdivision-and-G-spline-hybrids are built in Sect. 6.

2 Categories of and Requirements for Surfaces with Irregular Configurations

Figure 3(a) proposes a partition of the universe of spline constructions for irregular configurations. The main irregular configurations are illustrated in Fig. 4: star-configurations, multi-sided facets, T-junctions and polar configurations. The partition groups by smoothness, polynomiality and singularity of the surface parameterization. The topmost entry, *trimmed NURBS* surfaces, is the de facto industry standard. Trimming means restricting the domain of surface pieces by curves in the domain. In styling software, polynomial pieces are typically laid down to capture primary shape, cut back where they overlap and blended by fillet transition surfaces. This approach preserves simple, elegant shape in the large but ultimately forces stylists to devote ever more time to ever smaller blends between the primary surfaces [105].

Conforming C^0 surfaces, including curved triangulations, are ubiquitous and sufficient for basic engineering analysis (volume, homogeneous material moments,

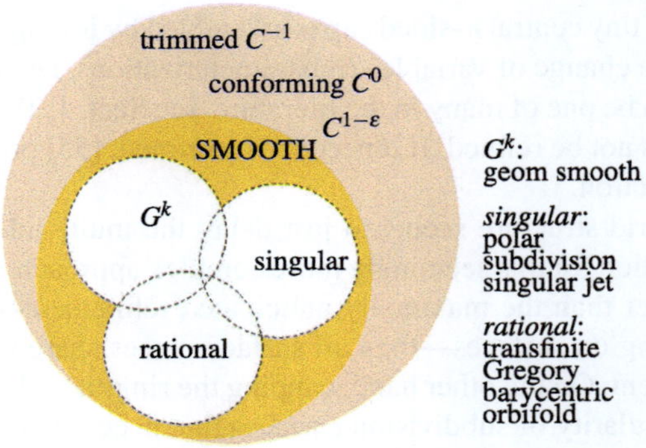

(a) classification of constructions

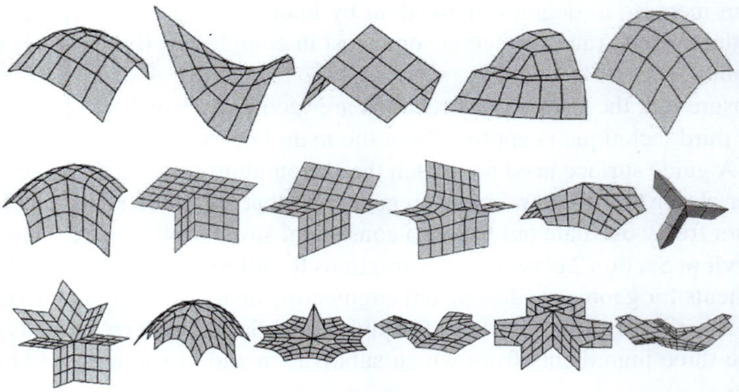

(b) quad-net obstacle course

Fig. 3 (**a**) Categories of surface parameterizations for irregular configurations from [79]. (**b**) Examples of meshes in the quad-net obstacle course [37] that serve as test control nets for the geometric suitability of constructions analogous to the finite element obstacle course for engineering analysis

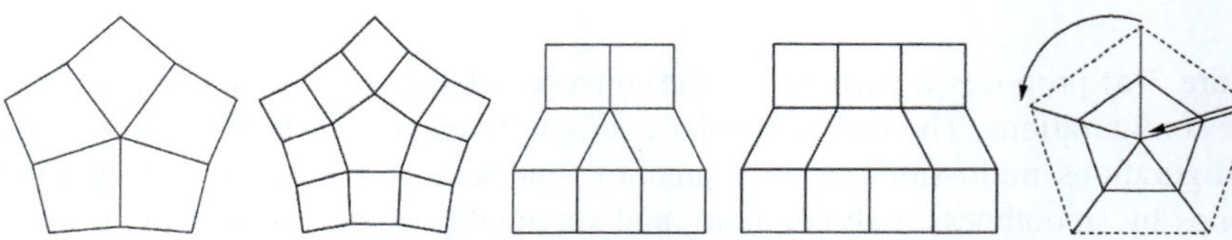

(a) star-net (b) n-gon-net (c) τ_0-net (d) τ_1-net (e) polar-net

Fig. 4 Locally quad-dominant mesh patterns (see [54]). The valence in (**a, b, e**) is $n = 5$. The pentagon in (**d**) is a T_1-gon, the triangle in (**c**) is a T_0-gon

linear elasticity, etc.) and in computer graphics, where visual appearance trumps accurate geometry [28, 64, 101]. Generalized barycentric patches, e.g., [65, 94], and non-4-sided transfinite constructions, e.g., [15, 86, 88, 99, 100], cover multi-sided holes with single, typically C^∞ patches. Due to an underlying very high rational degree, exactly computing their higher-order derivatives is costly. Due to its focus on free-form surfaces, the classification leaves out analysis-suitable functions and their graphs, e.g., [33, 95]. Subdivision surfaces and G^k constructions will be discussed individually in the Sects. 3 and 4.

2.1 Requirements on the Geometry

The quality of a surface is a subjective term. Two tools are commonly used to assess geometric quality: 'curvature profiles' superimpose the curvature function of planar cuts orthogonal to the surface (Fig. 1(d, e) show the less common shading of the whole surface by Gauss and Mean curvature); and 'highlight lines' [6] approximate the effect of parallel arrangement of tube lights in a car show room. Unless explicitly intended as a surface feature, artifacts such as abrupt changes in the distribution of curvature or highlight lines are not wanted because they distort reflections and make the product appear less well designed. When the surfaces generated from the obstacle course quad-meshes (see Fig. 3(b)) avoid artifacts, the construction will be deemed *geometry-suitable*. The automotive design industry also uses the term 'Class A surface' [104] to describe spline surfaces with aesthetic, non-oscillating highlight lines. Class A surfaces satisfy, depending on the application area and contractual agreement, certain hard geometric constraints [2]. Remarkably, they allow a mismatch of normals across curves between two surface pieces up to one tenth of a degree, justifying the surface category $C^{1-\epsilon}$ in Fig. 3(a).

2.2 The Isogeometric Approach

Higher orders of continuity of the solution space are required for correctly solving higher-order differential equations such as the biharmonic equation, Kirchhoff-Love shell formulations [8, 9, 58], the Cahn-Hilliard phase-field model [23] or simplified Navier-Stokes-Korteweg equations. C^1-continuous finite element spaces are needed even when these problems are expressed in the weak formulation [103], e.g., via Galerkin's discretization for solving a partial differential equation [75, 102]. Where the partitions deviate from regular lattices, classical finite elements join typically only C^0. Ignoring the required differentiability between elements and computing piecemeal in the larger C^0 space can give rise to extraneous solutions that do not correspond to physical solutions. For other classes of differential equations, smoothness, while not formally necessary, has been found to improve accuracy, stability or convergence (see, e.g., for contact problems [68, 97]).

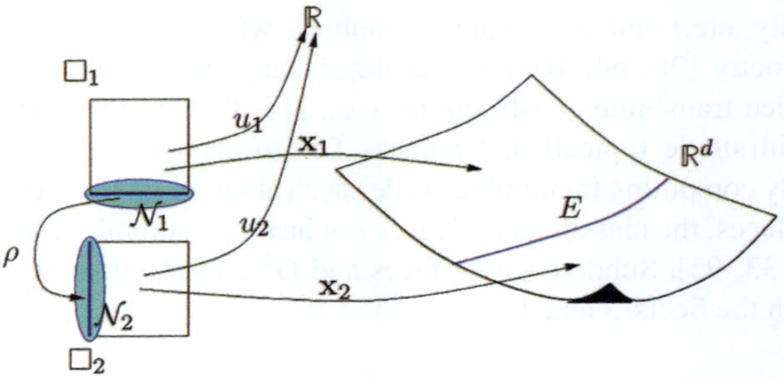

Fig. 5 Atlas **x** with patches $\mathbf{x}_\alpha$ and analysis functions u_α in the formulation of the fundamental theorem: Elements $b_i(\mathbf{x}_\alpha) := u_\alpha \circ \mathbf{x}_\alpha^{-1}$ on manifolds, built with geometric continuity across an interface E with the corresponding edges of the domains $\square_1$ and $\square_2$ related by the change of variable ρ

For a grid-like layout of quadrilateral surface pieces, the practice of using bivariate splines both for modeling the geometry of the domain and for computing functions on this domain goes back at least to the 1980s [13]. To emphasize the use of B-splines for both geometry and engineering analysis, the term iso-geometric analysis (IGA) was coined in the 2005 publication [29]. Publications that developed the isogeometric approach in the 1980s and 1990s used the less specific and less memorable terms 'higher-order', 'isoparametric' or 'finite elements using NURBS' [3, 4, 91, 93].

A fundamental theorem, [26], asserts that generalized splines whose irregular layout captures free-form shape of surfaces, e.g., G-splines, can directly serve to define isogeometric finite elements. The composition of maps (where $\mathbf{x}_\alpha^{-1}$ is the pull back of $\mathbf{x}_\alpha$) that define the finite elements is illustrated in Fig. 5. G-spline implementations in [35, 72–74, 89] have confirmed this fact by solving problems of the classical 'finite element obstacle course' [7].

2.3 Requirements for Analysis: Flexibility-Increasing Refinability

Examples of functions on surfaces are textures in graphics and deformation stress in engineering analysis. Both may need to have increased resolution, even when the surface remains the same. [16] shows that C^2-connected bi-cubics have a sub-optimal approximation order in the presence of extraordinary points and, more generally, that refinement of the domain (h-refinement) in the presence of non-trivial reparameterizations fuses the refined polynomial surface pieces of degree p when the continuity is C^{p-1}. Similarly, in the context of geometric modeling, [80] proved that for bi-3 (bicubic) spline patches the interdependence of partial derivatives forces a minimum separation of the extraordinary points when polynomial pieces are joined G^1 or else they fuse. ([16] uses the term 'locking' for the artificial algebraic stiffness; while evocative this term already has a fixed meaning in the thin-shell community). An alternative term for flexibility-increasing refinability is *analysis-suitability*, hence the title of this paper.

To formalize flexibility-increasing refinability, we follow the exposition of [55] and define smoothness between non-overlapping pieces of manifolds as follows.

Definition 1 (*G^κ Constraints*) Two regular and sufficiently smooth surface pieces $\tilde{\mathbf{f}}, \mathbf{f} : (u, v) \in \mathbb{R}^2 \to \mathbb{R}^d$ that share a boundary curve $\mathbf{e}$ *join G^κ along* $\mathbf{e}$ if there exists a suitably oriented and non-singular reparameterization $\rho : \mathbb{R}^2 \to \mathbb{R}^2$ so that the partial derivatives $\partial^k \tilde{\mathbf{f}}$ and $\partial^k(\mathbf{f} \circ \rho)$, $k = 0, 1, \ldots, \kappa$, agree along $\mathbf{e}$.

Let, according to the focus of this survey, the surface pieces be sufficiently smooth piecewise polynomials, e.g., tensor-product splines, and fix the reparameterization $\rho_\mathbf{e}$ for every edge $\mathbf{e}$. Then the splines joined with these reparameterizations form a space G_ρ linear in the unconstrained polynomial coefficients. That is, any linear combination of elements in G_ρ associated with these free coefficients (set to 1 and all others to 0) is again in G_ρ.

The space G_ρ is (binarily) *refinable* to a space $\dot{G}_{\dot{\rho}}$ in the following sense, see Fig. 6. For each piece $\mathbf{f} \in G_\rho$ restricted to $\square := [0..1]^2$, the space $\dot{G}_{\dot{\rho}}$ has four polynomial pieces $\mathbf{f}^r$, $r = 1, 2, 3, 4$ (wlog. of the same degree as $\mathbf{f}$) defined on the four quarters of $\square$ and joined by the following reparameterizations $\dot{\rho}$:

$$\dot{\rho}_{\mathbf{e}^0}(u, v) = \rho_\mathbf{e}(\frac{u}{2}, \frac{v}{2}), \quad \dot{\rho}_{\mathbf{e}^1}(u, v) = \rho_\mathbf{e}(\frac{1}{2} + \frac{u}{2}, \frac{v}{2}).$$

Then there is a choice of $\mathbf{f}^r$, namely applying de Casteljau's algorithm to $\mathbf{f}$ at $u = v = 1/2$, so that any element $\mathbf{f} \in G_\rho$ can be represented in $\dot{G}_{\dot{\rho}}$. That is, $\dot{G}_{\dot{\rho}}$ refines G_ρ. However, $\dot{G}_{\dot{\rho}}$ is a larger space than G_ρ since many other choices of macro-patches $\mathbf{f}^r$ are allowable. $\dot{G}_{\dot{\rho}}$ can therefore be expected to provide more flexibility than G_ρ. The additional degrees of freedom are new coefficients not constrained by enforcing smoothness.

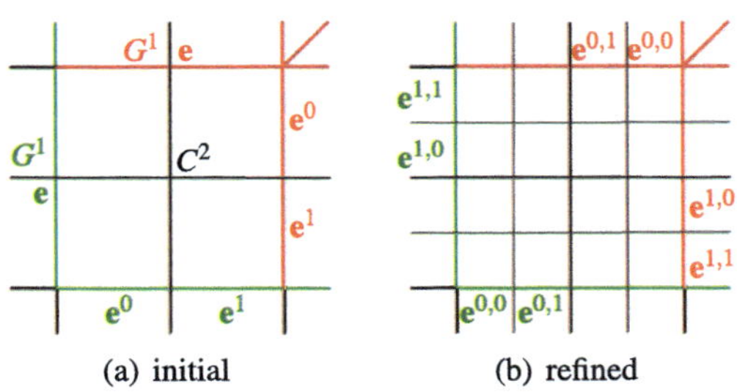

Fig. 6 Refinement of a macro-patch. To be flexibility-increasing, new degrees of freedom must appear both in the interior and along the *red* sector-separating curve boundaries and the *darkgreen* boundaries that connect the cap to the remaining surface. Second superscripts in (**b**) enumerate the (second) splitting of the original edges into parts 0 and 1

Definition 2 (*Flexibility-Increasing Refinable*) A construction is *flexibility-increasing refinable* if, for each domain piece $\square$, $\dot{G}_{\dot{\rho}} \supsetneq G_\rho$ has more degrees of freedom than G_ρ, both along map boundaries and in the interior.

Since all new internal transitions arising from refinement must be parametrically C^2 to reproduce the original polynomial pieces by the finer construction, macro-patches are internally C^k rather than G^k, see the black transitions in Fig. 6 are C^k. Having covered the requirements, we now survey in more detail the three ingredients of hybrid constructions: subdivision, G-spline and guide surfaces.

3 Subdivision Surfaces

More than 40 years ago Catmull and Clark [14] suggested a simple way to smoothly approximate quad meshes by an infinite sequence of nested rings of bi-3 splines; and Doo and Sabin [19] independently generalized bi-2 splines while deriving important mathematical machinery to analyze the resulting surfaces. Following the landmark papers [18, 96] that demonstrated industry support, Catmull-Clark surfaces are nowadays the tool of choice for computer animation. More recently, an efficient set of open source libraries for subdivision on massively parallel CPU and GPU architectures has been agreed upon by Pixar and Microsoft [83] and used by Pixar's proprietary animation system.

Closer mathematical analysis shows that Catmull-Clark surfaces have systemic shape deficiencies that preclude their use in high-end modeling for manufacturing [49, 56]: limits of convex meshes become hyperbolic surfaces, transitions at T-junctions are unduly flat and saddle-like configurations result in undesirable pinched highlight lines. Indeed, distortion of highlight lines for higher valences is a challenge for two members of the category 'singular' in Fig. 3(a): subdivision surfaces and singular jet surfaces. (Only polar surfaces, where one edge is collapsed into a pole, thrive on high valence. Subdivision surfaces can be thought of as collapsing patch size and singular jet surfaces as collapsing the Taylor expansion at a vertex [74, 77, 84, 85, 98].) Indeed, the monograph on the mathematics of subdivision surfaces characterizes subdivision surfaces near irregularities as spline surfaces with singularities [82].

Optimizations of parameters and prescription of the expansion at the extraordinary central point [5, 34, 61, 69] have improved shape outcomes by making the subdivision matrix less sparse. (The subdivision matrix maps control nets surrounding an irregular node to a contracted control net, each of which defines a surface ring, see Fig. 7).

Guided subdivision [38, 50] stabilizes the shape at the cost of a yet denser subdivision matrix. This enables accelerated contraction towards the extraordinary point without noticeable harm to the shape [48]. For example, one high speed contraction

Fig. 7 (*left*) Subdivision surface (from [38]) built from (*right*) a sequence of contracting surface rings

step can shrink the remaining hole by more than two steps of Catmull-Clark subdivision as illustrated in Fig. 11. Accelerated, guided subdivision is a key ingredient of hybrid constructions.

4 G-Splines

While early publications [10, 11, 17, 21, 30, 62, 87] hint at the potential of polynomial pieces joined by a change of variables to fill multi-sided holes in a spline complex, it is arguably Hahn's use of geometric continuity [27] that set the standard for the first generation of G-spline constructions in the 1990s. The focus of the early constructions was on varying the parameterization ρ of Eq. (1) to allow different relations between patch derivatives at irregular non-4-valent points versus regular ones. Expansion by the chain rule of differentiation yields for example G^1 and G^2 constraints in terms of univariate scalar maps $a, b, d, e : u \in \mathbb{R} \to \mathbb{R}$ (partial derivatives of the two coordinates of ρ evaluated on the edge parameterized by $(u, 0)$) and the vector-valued functions $\mathbf{f}, \tilde{\mathbf{f}}$ evaluated at $(u, 0)$:

$$\partial_v \tilde{\mathbf{f}} = a\, \partial_v \mathbf{f} + b\, \partial_u \mathbf{f}, \tag{1}$$

$$\partial^2_{vv} \tilde{\mathbf{f}} = a^2\, \partial^2_{vv} \mathbf{f} + 2a\, b\, \partial_u \partial_v \mathbf{f} + b^2\, \partial^2_{uu} \mathbf{f} + d\, \partial_v \mathbf{f} + e\, \partial_u \mathbf{f}. \tag{2}$$

Since neither $\mathbf{f}$ (the rules for constructing the polynomial pieces) nor ρ (the reparameterization) are known a priori, the constraints define a large non-linear space to explore. Considering ρ free to choose increases the space of possible construction. However all sufficiently smooth constructions have to obey a constraint that arises from the circular arrangement of patches surrounding a point, the vertex enclosure constraint [76]. Moreover, the degree of ρ is bounded by the degree of the surface pieces [76] due to what is now understood to be a syzygy relation, see, e.g., [70]. Moreover, to create a manifold, the composition of n copies of ρ (across each of the sector-separating curves around an interior point) has to form the identity map, up to the degree of smoothness [78].

Choosing ρ to Hermite interpolate and so separate the computation at either end of an edge between two patches, simplifies solving the system of equations in the polynomial coefficients of Eqs. (1) and (2) but results in many unconstrained coefficients: Hahn and Gregory's early G^2 construction used pieces of degree 18 in u and in v (bi-18) [24] and [57, 106] of degree bi-9. Modern G-splines include G^2 constructions of degree as low as bi-5. Throughout the 1990s work focused on optimizing ρ to minimize the polynomial degree of the surface. Although compatible with the NURBS standard adopted by the manufacturing industries, the shape of early G-spline constructions was often worse than that of subdivision surfaces.

The first decade of this millennium saw improved shape, e.g., [63, 66] still of degree bi-7. Also the distinction between smoothness in the large, assessed via highlight lines, and infinitesimal smoothness, measured as matching derivatives came into focus: a formally only G^1 construction of degree bi-5 or bi-4 [42, 49] showed better curvature distribution than higher degree curvature continuous constructions; and, remarkably, $C^{1-\epsilon}$ construction in [41] trades a slight mismatch in the normal (still within the bounds of class A surfaces) for good highlight line distributions over the multi-sided patch constructed from a finite number of bi-3 patches; the quality of the highlight line distributions could not be matched by the subdivision surfaces of the time (except of the guided variety, see, e.g., [47] and Sect. 5). Additionally, the degree of formally G^2 surfaces with a single patch per sector was reduced to bi-6 [44]. and to degree bi-5 for 2×2 macro-patches [40]. Special scaffold- and sphere-like configurations even allow for curvature continuous bi-4 constructions [43].

A next step forward for G-splines was to allow not only irregular points but also T-junctions as part of a generalized B-spline control net for free-form modeling [36]. T-junctions occur where two quads on one side meet one facet on the other and serve in polyhedral modeling to start or stop quad-strips and so increase or decrease the number of free points to be set. T-junctions prominently feature in quad-dominant remeshing (see, e.g., [1, 31, 60, 90]) where they allow to side-step the otherwise stringent global quad-meshing constraints (see, e.g., [12, 32, 71]). Also popular in this context are (isolated) triangles that merge and so reduce the number of quad-strips, see Fig. 8. The novelty is that all mesh nodes act as coefficients of linear combinations of piecewise polynomials, i.e., as B-spline-like control points.

While G-splines for multi-sided holes or generalized subdivision can, in principle, convert quad-dominant meshes with T-junctions into smooth surfaces, they do not preserve the two preferred directions and so cause visible shape artifacts. Hierarchical and T-splines such as [22, 59, 92] need to carefully coordinate knot intervals to admit meshes with T-gons as control nets. For many meshes a globally consistent choice of intervals is impossible [36]. That is, these approaches excel at refining tensor-product patches, but may not be able to produce a smooth surface from a given polyhedral mesh including T-junctions.

The G^1 constructions [51–53] differ in their polynomial degree, their flexibility-increasing refinability and how close their τ-nets can be placed to each other and

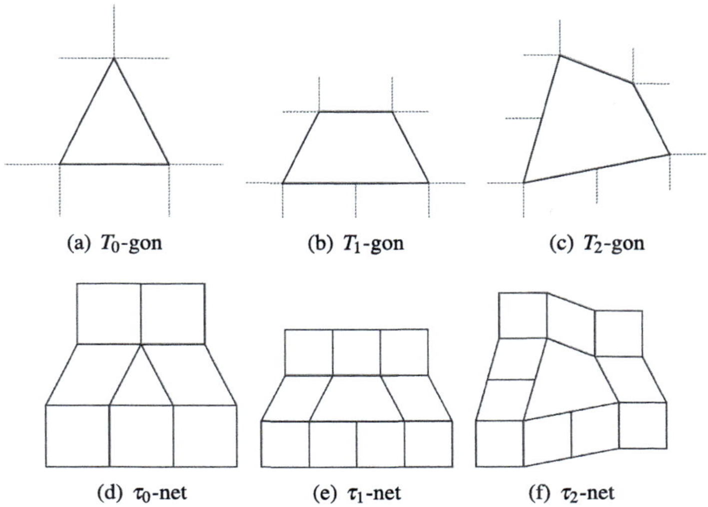

Fig. 8 T-gons and τ-configurations. The subscript counts the number of T-junctions

to irregular nodes. Built from pieces of degree 3×5, [51] is suitable for modeling class A surfaces. And by complementing the existing G-spline constructions for multi-sided facets, splines for the τ-configurations of Fig. 8 allow interpreting locally quad-dominant meshes as spline control nets.

Since the majority of high-end surface constructions now are G-splines, their flexibility-increasing refinability has come under scrutiny. The detailed analysis of flexibility-increasing-refinement of [42] showed that, unlike for tensor-product splines, the resulting unconstrained new coefficients (degrees of freedom) are not convenient geometric handles due to their irregular distribution and support [45]. Indeed the full characterization of G^2 flexibility-increasing refinability in [55] proves that multi-sided refinement of G-splines can not be flexibility-increasing when the

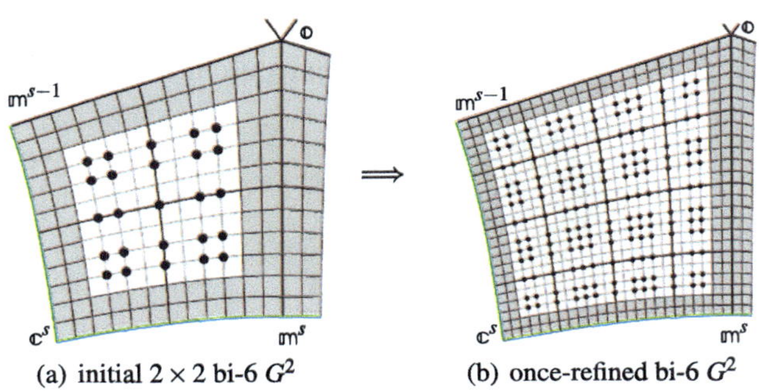

Fig. 9 Internal degrees of freedom of refinement of [55] marked as •

construction uses bi-5 macro-patches, regardless of the number of $N \times N$ pieces. Conversely, [55] exhibits a bi-6 construction with 2×2 pieces per sector that is G^2 flexibility-increasing refinable, see Fig. 9. (A bi-4 publication, currently under review, presents parallel results for the construction of multi-sided G^1 surfaces).

5 Guided Surfaces

Guided Subdivision [38] is an effective tool to overcome the defects of standard subdivision algorithms. Moreover, unlike standard subdivision, guided subdivision can easily define curvature continuous subdivision surfaces with controllable polynomial reproduction at the limit point. Figure 10 illustrates the underlying principle: the control mesh (a) defines a guide surface (b) and the Hermite prolongation of the surrounding regular ring (c). Since the guide surface and the surrounding surface frame do not fit together as illustrated in (d), the subdivision step retains the outer part of the ring that fits its predecessor ring and determines its new inner part by sampling the guide. The result is shown in (e). We note that the guide surface can have a different structure, smoothness and polynomial degree than the final surface. For example, it can consist of 3-sided patches, whereas the final surface consists of 4-sided patches. Remarkably the overall process is linear and stationary and can so be interpreted as subdivision with large stencil, i.e., with a denser subdivision matrix than Catmull-Clark subdivision. While the rules become more complex, the mathematical analysis of the limit becomes much simpler—and the shape is far better [49].

Moreover, since guided subdivision stabilizes the shape, an additional technique can be leveraged: *accelerated* guided subdivision [81, Sect. 5], [48] widens the

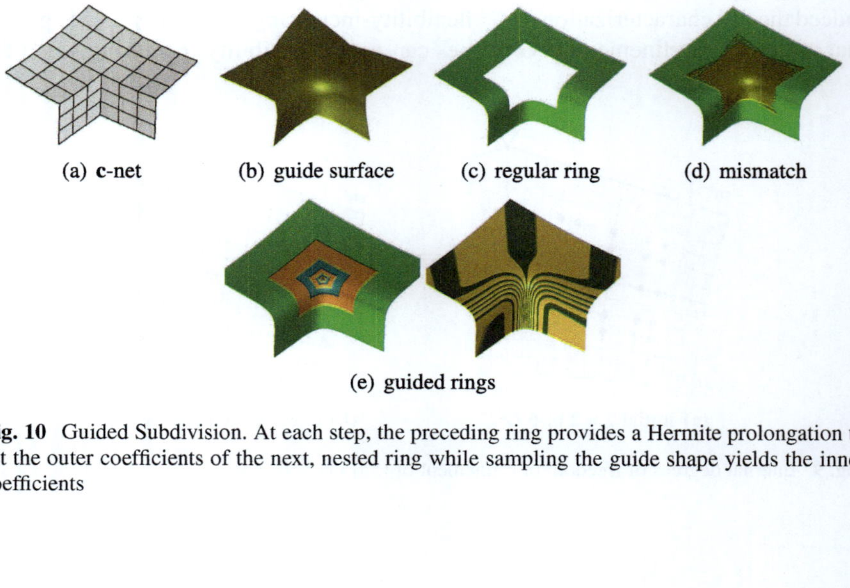

Fig. 10 Guided Subdivision. At each step, the preceding ring provides a Hermite prolongation to set the outer coefficients of the next, nested ring while sampling the guide shape yields the inner coefficients

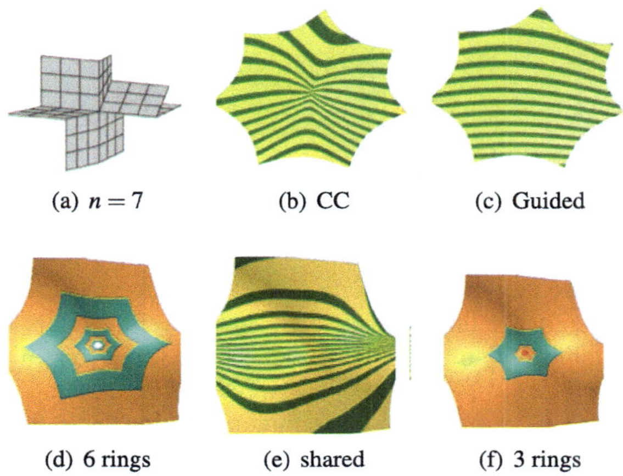

Fig. 11 Guided Subdivision and Accelerated Guided Subdivision. The highlight line distributions for input **c**-net (**a**) for (**b**) Catmull-Clark subdivision and (**c**) Guided Subdivision. For a different input $n = 6$, surfaces (**d**) and (**f**) have visually the same highlight line distribution shown in (**e**)

surface rings and so achieves a more rapid contraction of the remaining gap, see Fig. 11. When the surface rings and final cap follow a guiding shape, good highlight line distributions are obtained for the shape obstacle course Fig. 3(a), already when the accelerated sequence of C^k-joined surface rings is G^1 completed after 2 or 3 rings, see Fig. 11(e, f).

We note that guides can equally well be applied to finite constructions in order to harness excess free parameters. In hybrid surfaces, the same guide surface is applied both to accelerated subdivision rings and the final tiny cap.

6 Hybrid Surfaces

While guided subdivision improves shape, it does not address the problem of infinite recursion. Stopping the recursion after a few steps and filling the remaining hole with a triangulated multi-sided facet leads to noticeable flaws in the highlight line rendering. A better solution is to fit a multi-sided G^1-cap. Since the resulting surface consists of a fixed number of surface pieces, it is industry compatible. To combine the best features of subdivision and geometrically continuous surface constructions, we observe that in practice, design and analysis can often predict a maximal level of refinement. When the maximal anticipated refinement level at the irregularity suffices, the cap need not be refined and refinement in the surrounding finite coarser surface requires only standard spline knot insertion [46]. If the anticipated level does not suffice, it is easy to reconstruct with additional surface rings—or once can use a more complex geometrically smooth flexibility-increasing-refinable construction such as Fig. 9.

The earliest hybrid construction, [41], addressed whether bi-cubic surfaces can be class A. The construction, of a main body and cap of degree bi-3, is formally only

$n = 5$ 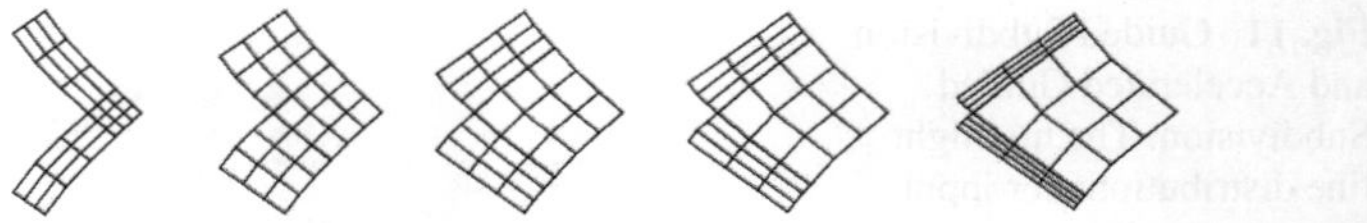

Fig. 12 A sector of the characteristic map for $n = 5$ and speeds $\frac{1}{4}, \frac{1}{2}, \frac{2}{3}, \frac{3}{4}, \frac{7}{8}$ from [39]

C^0, with empirically less than a 0.1° normal mismatch between the surrounding surface and the main body. The default uses just three pieces per L-shape and one piece for the limit. The results, summarized visually in [41, Fig. 1] show the hybrid to noticeably improve on methods that yield a formally smoother result [14, 25, 67]. The focus of [47] is on improving the shape of subdivision surfaces by constructing both a C^2 subdivision algorithm generating surfaces of polynomial degree bi-6. An appendix adds G^1 parameterized tiny cap. The full concept of a hybrid construction, including acceleration to obtain few pieces, is realized in [46] (see Fig. 1). In [48] the degree of the C^2 main body is reduced to bi-5. In [49], see Fig. 2, the degree is bi-4 using macro-patches. The approaches use guide surfaces built from n three-sided patches with high smoothness at the central point. Finally [55] proves that the central caps of [46, 48, 49] are not flexibility-increasing refinable and presents a bi-6 tiny cap that both completes the surface G^2 to be is flexibility-increasing refinable. For practical purposes the latter may be more than needed: G^1 suffices up to fourth order differential operators and good geometric shape can be guaranteed by tiny G^1 caps whose refinement is much simpler, pointing to [48] as the method of choice.

The key ingredient of hybrid constructions is the guide surface and the sampling, via characteristic maps Fig. 12, of Hermite data along the sequences of boundary curves of the patches. Presented in BB-form these Hermite data are called tensor-borders. For details on the characteristic maps see [48]. The rapid contraction of the guided rings means that the resulting macro-patches consist of few pieces per sector, e.g., of seven pieces: three for each of two polynomial L-shapes and one for the central G-spline patch. This yields smooth surfaces consisting of a finite number of pieces whose formulas derive linearly from the input control net, hence can be implemented, for fixed contraction speed and valence, as a single matrix multiplication applied to the local control net. In practice, for [48], one splits the matrix into a 21×31 matrix for creating a sector of the guide, a 21×21 matrix for the de Casteljau steps that re-represent the guide for a contracted domain, a 27×21 matrix defining one sector of the prolongation of the L-shape and 36×21 matrix for one sector of the tiny cap.

7 Conclusion

In order to advertise a useful class of piecewise polynomial surface constructions for filling multi-sided holes in a smooth spline complex, we surveyed subdivision, G-spline and guided surfaces. Combining accelerated guided subdivision with a tiny

G-spline cap yields a number of advantages for modelling and analysis: good shape, easy implementation and simple refinability up to a pre-defined level.

References

1. Alliez, P., Cohen-Steiner, D., Devillers, O., Lévy, B., Desbrun, M.: Anisotropic polygonal remeshing. ACM Trans. Graph. **22**, 485–493 (2003)
2. Andjelic, Z.: What is the definition of a class A surface? https://grabcad.com/questions/what-is-definition-of-class-a-surface. Accessed May 2015
3. Au, F.T.K., Cheung, Y.K.: Spline finite elements for beam and plate. Comput. Struct. **37**, 717–729 (1990)
4. Au, F.T.K., Cheung, Y.K.: Isoparametric spline finite strip for plane structures. Comput. Struct. **48**, 22–32 (1993)
5. Augsdoerfer, U.H., Dodgson, N.A., Sabin, M.A.: Removing polar rendering artifacts in subdivision surfaces. J. Graph., GPU, Game Tools **14**, 61–76 (2009)
6. Beier, K.-P., Chen, Y.: Highlight-line algorithm for realtime surface-quality assessment. Comput. Aided Des. **26**, 268–277 (1994)
7. Belytschko, T., Stolarski, H., Liu, W.K., Carpenter, N., Ong, J.S.J.: Stress projection for membrane and shear locking in shell finite elements. Comput. Methods Appl. Mech. Eng. **51**, 221–258 (1985)
8. Benson, D.J., Bazilevs, Y., Hsu, M.-C., Hughes, T.J.R.: A large deformation, rotation-free, isogeometric shell. Comput. Methods Appl. Mech. Eng. **200**, 1367–1378 (2011)
9. Benson, D.J., Hartmann, S., Bazilevs, Y., Hsu, M.-C., Hughes, T.J.R.: Blended iso-geometric shells. Comput. Methods Appl. Mech. Eng. **255**, 133–146 (2013)
10. Bézier, P.E.: Essai de Définition Numérique des Courbes et des Surfaces Experimentales. PhD thesis, Université Pierre et Marie Curie (1977)
11. Boehm, W.: Smooth curves and surfaces. In: Farin, G. (ed.) Geometric Modeling: Algorithms and New Trends, pp. 175–184. SIAM, Philadelphia (1987)
12. Bommes, D., Campen, M., Ebke, H.-C., Alliez, P., Kobbelt, L.: Integer-grid maps for reliable quad meshing. ACM Trans. Graph. **32**, 98 (2013)
13. Braibant, V., Fleury, C.: Shape optimal design using B-splines. Comput. Methods Appl. Mech. Eng. **44**, 247–267 (1984)
14. Catmull, E., Clark, J.: Recursively generated B-spline surfaces on arbitrary topological meshes. Comput. Aided Des. **10**, 350–355 (1978)
15. Charrot, P., Gregory, J.: A pentagonal surface patch for computer aided geometric design. Comput. Aided Geometr. Design **1**, 87–94 (1984)
16. Collin, A., Sangalli, G., Takacs, T.: Analysis-suitable G^1 multi-patch parametrizations for C^1 isogeometric spaces. Comput. Aided Geometr. Design **47**, 93–113 (2016)
17. DeRose, T.: Geometric continuity: a parametrization independent measure of continuity for computer aided design. Ph.D. thesis, UC Berkeley, California (1985)
18. DeRose, T., Kass, M., Truong, T.: Subdivision surfaces in character animation. In: Proceedings of the ACM Conference on Computer Graphics (SIGGRAPH-98), pp. 85–94. ACM Press, New York (1998)
19. Doo, D., Sabin, M.: Behaviour of recursive division surfaces near extraordinary points. Comput. Aided Des. **10**, 356–360 (1978)
20. Farin, G.: Curves and Surfaces for Computer Aided Geometric Design: A Practical Guide. Academic (1988)
21. Garrity, T., Warren, J.: Geometric continuity. Comput. Aided Geometr. Design **8**, 51–66 (1991)
22. Giannelli, C., Jüttler, B., Speleers, H.: THB-splines: the truncated basis for hierarchical splines. Comput. Aided Geometr. Design **29**, 485–498 (2012)

23. Gomez, H., Calo, V.M., Bazilevs, Y., Hughes, T.J.R.: Isogeometric analysis of the Cahn-Hilliard phase-field model. Comput. Methods Appl. Mech. Eng. **197**, 4333–4352 (2008)
24. Gregory, J.A., Hahn, J.M.: A C^2 polygonal surface patch. Comput. Aided Geometr. Design **6**, 69–75 (1989)
25. Gregory, J.A., Zhou, J.: Filling polygonal holes with bicubic patches. Comput. Aided Geometr. Design **11**, 391–410 (1994)
26. Groisser, D., Peters, J.: Matched G^k-constructions always yield C^k-continuous isogeometric elements. Comput. Aided Geometr. Design **34**, 67–72 (2015)
27. Hahn, J.M.: Geometric continuous patch complexes. Comput. Aided Geometr. Design **6**, 55–67 (1989)
28. Hettinga, G.J., Kosinka, J.: Phong tessellation and PN polygons for polygonal models. In: Peytavie, A., Bosch, C., (eds.), EG 2017 - Short Papers. The Eurographics Association (2017)
29. Hughes, T.J.R., Cottrell, J.A., Bazilevs, Y.: Isogeometric analysis: CAD, finite elements, NURBS, exact geometry and mesh refinement. Comput. Methods Appl. Mech. Eng. **194**, 4135–4195 (2005)
30. Kahmann, J.: Continuity of curvature between adjacent Bézier patches. In: Barnhill, R.E., Böhm, W. (eds.) Surfaces in CAGD, pp. 65–75. North-Holland Publishing Company, Amsterdam (1983)
31. Jakob, W., Tarini, M., Panozzo, D., Sorkine-Hornung, O.: Instant field-aligned meshes. ACM Trans. Graph. **34**, 189 (2015)
32. Kälberer, F., Nieser, M., Polthier, K.: QuadCover - surface parameterization using branched coverings. Comput. Graph. Forum **26**, 375–384 (2007)
33. Kapl, M., Vitrih, V., Jüttler, B., Birner, K.: Isogeometric analysis with geometrically continuous functions on two-patch geometries. Comput. Math. Appl. **70**, 1518–1538 (2015)
34. Karčiauskas, K., Myles, A., Peters, J.: A C^2 polar jet subdivision. In: Scheffer, A., Polthier, K.,(eds.), Geometry Processing 2006, pp. 173–180. Eurographics Association (2006)
35. Karčiauskas, K., Nguyen, T., Peters, J.: Generalizing bicubic splines for modelling and IGA with irregular layout. Comput. Aided Design **70**, 23–35 (2016)
36. Karčiauskas, K., Panozzo, D., Peters, J.: T-junctions in spline surfaces. ACM Trans. Graph. **36**, 170 (2017)
37. Karčiauskas, K., Peters, J.: Quad-net obstacle course. http://www.cise.ufl.edu/research/SurfLab/shape_gallery.shtml. Accessed June 2020
38. Karčiauskas, K., Peters, J.: Concentric tesselation maps and curvature continuous guided surfaces. Comput. Aided Geometr. Design **24**, 99–111 (2007)
39. Karčiauskas, K., Peters, J.: Adjustable speed surface subdivision. Comput. Aided Geometr. Design **26**, 962–969 (2009)
40. Karčiauskas, K., Peters, J.: Biquintic G^2 surfaces via functionals. Comput. Aided Geometr. Design **33**, 17–29 (2015)
41. Karčiauskas, K., Peters, J.: Can bi-cubic surfaces be class A? Comput. Graph. Forum **34**, 229–238 (2015)
42. Karčiauskas, K., Peters, J.: Improved shape for multi-surface blends. Graph. Models **82**, 87–98 (2015)
43. Karčiauskas, K., Peters, J.: Curvature continuous bi-4 constructions for scaffold- and sphere-like surfaces. Comput. Aided Des. **78**, 48–59 (2016)
44. Karčiauskas, K., Peters, J.: Minimal bi-6 G^2 completion of bicubic spline surfaces. Comput. Aided Geometr. Design **41**, 10–22 (2016)
45. Karčiauskas, K., Peters, J.: Refinable G^1 functions on G^1 free-form surfaces. Comput. Aided Geometr. Design **54**, 61–73 (2017)
46. Karčiauskas, K., Peters, J.: Fair free-form surfaces that are almost everywhere parametrically C^2. J. Comput. Appl. Math. **349**, 470–481 (2019)
47. Karčiauskas, K., Peters, J.: A new class of guided C^2 subdivision surfaces combining good shape with nested refinement. Comput. Graph. Forum **37**, 84–95 (2018)
48. Karčiauskas, K., Peters, J.: Rapidly contracting subdivision yields finite, effectively C^2 surfaces. Comput. Graph. **74**, 182–190 (2018)

49. Karčiauskas, K., Peters, J.: Refinable bi-quartics for design and analysis. Comput. Aided Des. **102**, 204–214 (2018)
50. Karčiauskas, K., Peters, J.: Curvature-bounded guided subdivision: Biquartics versus bicubics. Comput.-Aided Design **114**, 122–132 (2019)
51. Karčiauskas, K., Peters, J.: High quality refinable G-splines for locally quad-dominant meshes with T-gons. Comput. Graph. Forum **38**, 151–161 (2019)
52. Karčiauskas, K., Peters, J.: Localized G-splines for quad & T-gon meshes. Comput. Aided Geometr. Design **71**, 244–254 (2019)
53. Karčiauskas, K., Peters, J.: Refinable smooth surfaces for locally quad-dominant meshes with T-gons. Comput. Graph. **82**, 193–202 (2019)
54. Karčiauskas, K., Peters, J.: Low degree splines for locally quad-dominant meshes. Comput. Aided Geometr. Design **83**, 1–12 (2020)
55. Karčiauskas, K., Peters, J.: A sharp degree bound on G^2-refinable multi-sided surfaces. Comput. Aided Des. **125**, 102867 (2020)
56. Karciauskas, K., Peters, J., Reif, U.: Shape characterization of subdivision surfaces - case studies. Comput. Aided Geometr. Design **21**, 601–614 (2004)
57. Kiciak, P.: Spline surfaces of arbitrary topology with continuous curvature and optimized shape. Comput. Aided Des. **45**, 154–167 (2013)
58. Kiendl, J., Bletzinger, K.-U., Linhard, J., Wüchner, R.: Isogeometric shell analysis with Kirchhoff-Love elements. Comput. Methods Appl. Mech. Eng. **198**, 3902–3914 (2009)
59. Kraft, R.: Adaptive und linear unabhängige Multilevel B-Splines und ihre Anwendungen. Ph.D. thesis, University of Stuttgart (1998)
60. Lai, Y.-K., Kobbelt, L., Hu, S.-M.: An incremental approach to feature aligned quad dominant remeshing. In: Haines, E., McGuire, M., (eds.), Symposium on Solid and Physical Modeling, pp. 137–145. ACM (2008)
61. Li, X., Finnigan, G.T., Sederberg, T.W.: G^1 non-uniform Catmull-Clark surfaces. ACM Trans. Graph. **35**, 135 (2016)
62. Liu, D.: A geometric condition for smoothness between adjacent Bézier surface patches. Acta Math. Appl. Sin. **9**, 432–442 (1986)
63. Loop, C.: Second order smoothness over extraordinary vertices. In: Eurographics Symposium on Geometry Processing, pp. 169–178 (2004)
64. Loop, C., Schaefer, S.: Approximating Catmull-Clark subdivision surfaces with bicubic patches. ACM Trans. Graph. **27**, 8 (2008)
65. Loop, C.T., DeRose, T.D.: A multisided generalization of Bézier surfaces. ACM Trans. Graph. **8**, 204–234 (1989)
66. Loop, C.T., Schaefer, S.: G^2 tensor product splines over extraordinary vertices. Comput. Graph. Forum **27**, 1373–1382 (2008)
67. Loop, C.T., Schaefer, S., Ni, T., Castaño, I.: Approximating subdivision surfaces with Gregory patches for hardware tessellation. ACM Trans. Graph. **28**, 151 (2009)
68. De Lorenzis, L., Temizer, I., Wriggers, P., Zavarise, G.: A large deformation frictional contact formulation using NURBS-based isogeometric analysis. Int. J. Numer. Meth. Eng. **87**, 1278–1300 (2011)
69. Ma, Y., Ma, W.: Subdivision schemes with optimal bounded curvature near extraordinary vertices. Comput. Graph. Forum **37**, 455–467 (2018)
70. Mourrain, B., Vidunas, R., Villamizar, N.: Geometrically continuous splines for surfaces of arbitrary topology. Comput. Aided Geometr. Design **45**, 108–133 (2016)
71. Myles, A., Pietroni, N., Zorin, D.: Robust field-aligned global parametrization. ACM Trans. Graph. **33**, 135 (2014)
72. Nguyen, T., Karčiauskas, K., Peters, J.: A comparative study of several classical, discrete differential and isogeometric methods for solving Poisson's equation on the disk. Axioms **3**, 280–299 (2014)
73. Nguyen, T., Karčiauskas, K., Peters, J.: C^1 finite elements on non-tensor-product 2d and 3d manifolds. Appl. Math. Comput. **272**, 148–158 (2016)

74. Nguyen, T., Peters, J.: Refinable C^1 spline elements for irregular quad layout. Comput. Aided Geometr. Design **43**, 123–130 (2016)
75. Encyclopedia of Mathematics. Galerkin method. https://encyclopediaofmath.org/wiki/Galerkin_method. Accessed Dec 2020
76. Peters, J.: Fitting Smooth Parametric Surfaces to 3D Data. Ph.D. thesis, University of Wisconsin (1990)
77. Peters, J.: Parametrizing singularly to enclose vertices by a smooth parametric surface. In: MacKay, S., Kidd, E.M., (eds.), Graphics Interface '91, Calgary, Alberta, 3–7 June 1991: proceedings, pp. 1–7. Canadian Information Processing Society (1991)
78. Peters, J.: A characterization of connecting maps as roots of the identity. In: Laurent, P.-J., Le Méhauté, A., Schumaker, L.L., (eds.), Curves and Surfaces in Geometric Design, pp. 369–376 (1994)
79. Peters, J.: Splines for meshes with irregularities. SMAI J. Comput. Math. **S5**, 161–183 (2019)
80. Peters, J., Fan, J.: On the complexity of smooth spline surfaces from quad meshes. Comput. Aided Geometr. Design **27**, 96–105 (2009)
81. Peters, J., Karčiauskas, K.: An introduction to guided and polar surfacing. In: æhlen, M.D., et al., (eds.), Mathematical Methods for Curves and Surfaces. Lecture Notes in Computer Science, vol. 5862, pp. 299–315. Springer (2010)
82. Peters, J., Reif, U.: Subdivision Surfaces. Geometry and Computing, vol. 3. Springer, New York (2008)
83. Pixar. http://graphics.pixar.com/opensubdiv/docs/intro.html. Accessed June 2020
84. Reif, U.: Neue Aspekte in der Theorie der Freiformflächen beliebiger Topologie. Ph.D. thesis, University of Stuttgart (1993)
85. Reif, U.: A refinable space of smooth spline surfaces of arbitrary topological genus. J. Approx. Theory **90**, 174–199 (1997)
86. Sabin, M.: Transfinite surface interpolation. In: Mullineux, G., (ed.), The Mathematics of Surfaces VI, pp. 517–534. Clarendon Press (1996)
87. Sabin, M.A.: The use of piecewise forms for the numerical representation of shape. Ph.D. thesis, Computer and Automation Institute (1977)
88. Salvi, P., Várady, T.: Multi-sided Bézier surfaces over concave polygonal domains. Comput. Graph. **74**, 56–65 (2018)
89. Sangalli, G., Takacs, T., Vázquez, R.: Unstructured spline spaces for isogeometric analysis based on spline manifolds. Comput. Aided Geometr. Design **47**, 61–82 (2016)
90. Schertler, N., Tarini, M., Jakob, W., Kazhdan, M., Gumhold, S., Panozzo, D.: Field-aligned online surface reconstruction. ACM Trans. Graph. **36**, 77 (2017)
91. Schramm, U., Pilkey, W.D.: The coupling of geometric descriptions and finite elements using NURBS - a study in shape optimization. Finite Elem. Anal. Des. **340**, 11–34 (1993)
92. Sederberg, T.W., Zheng, J., Bakenov, A., Nasri, A.: T-splines and T-NURCCs. ACM Trans. Graph. **22**, 477–484 (2003)
93. Shyy, Y.K., Fleury, C., Izadpanah, K.: Shape optimal design using higher-order elements. Comput. Methods Appl. Mech. Eng. **71**, 99–116 (1988)
94. Smith, J., Schaefer, S.: Selective degree elevation for multi-sided Bézier patches. Comput. Graph. Forum **34**, 609–615 (2015)
95. Speleers, H., Manni, C., Pelosi, F., Sampoli, M.L.: Isogeometric analysis with Powell-Sabin splines for advection-diffusion-reaction problems. Comput. Methods Appl. Mech. Eng. **221**, 132–148 (2012)
96. Stam, J.: Exact evaluation of Catmull-Clark subdivision surfaces at arbitrary parameter values. In: Proceedings of SIGGRAPH 1998, pp. 395–404. ACM Press (1998)
97. Temizer, I., Wriggers, P., Hughes, T.J.R.: Contact treatment in isogeometric analysis with NURBS. Comput. Methods Appl. Mech. Eng. **200**, 1100–1112 (2011)
98. Toshniwal, D., Speleers, H., Hughes, T.J.R.: Smooth cubic spline spaces on unstructured quadrilateral meshes with particular emphasis on extraordinary points: Geometric design and isogeometric analysis considerations. Comput. Methods Appl. Mech. Eng. **327**, 411–458 (2017)

99. Várady, T., Rockwood, A.P., Salvi, P.: Transfinite surface interpolation over irregular n-sided domains. Comput. Aided Des. **43**, 1330–1340 (2011)
100. Várady, T., Salvi, P., Karikó, G.: A multi-sided Bézier patch with a simple control structure. Comput. Graph. Forum **35**, 307–317 (2016)
101. Vlachos, A., Peters, J., Boyd, C., Mitchell, J.L.: Curved PN triangles. In: 2001 Symposium on Interactive 3D Graphics, pp. 159–166. ACM Press (2001)
102. Wikipedia. Galerkin method. https://en.wikipedia.org/wiki/Galerkin_method. Accessed Dec 2020
103. Wikipedia. Weak formulation. https://en.wikipedia.org/wiki/Weak_formulation. Accessed Dec 2020
104. Wikipedia. Class A surface. http://en.wikipedia.org/wiki/Class_A_surface. Accessed June 2020
105. Yamada, Y.: Clay Modeling: Techniques for Giving Three-dimensional Form to Idea. Nissan Design Center, Kaneko Enterprises (1997)
106. Ye, X.: Curvature continuous interpolation of curve meshes. Comput. Aided Geometr. Design **14**, 169–190 (1997)

Meshing as the Choice of Basis Functions for Finite Element Analysis

Malcolm A. Sabin

Abstract Finite Element Analysis is seen as a particular technique for solving problems in Field Analysis in which some quantity which varies with position in a domain satisfies a Partial Differential Equation (PDE). Traditionally the approach to finite element analysis has been first to partition the domain into elements, sharing nodes. Each element then has a shape function associated with each node, and the shape functions associated with a given node together form a basis function. The coefficients of these basis functions are the field values at their nodes, and so the basis functions are interpolating. Gradually it became evident that some of these assumptions were unnecessary. This paper traces the development of the technology, and leads to the conclusion that in order to achieve significantly higher performance in our field analysis we need to regard the partitioning as a side-effect of the choice of basis functions instead of being the first step. Explicitly choosing the basis functions can give very significant enhancements in the cost/accuracy pareto tradeoff, which is of particular importance when the analyses are embedded in optimisation.

1 Content

The remainder of this paper contains

- the criteria which have driven the development of field analysis (Sect. 2);
- the context placing the techniques in their generality (Sect. 3);
- specific issues relating to the criteria (Sect. 4);
- quad and hex meshing techniques (Sect. 5);
- situations where there is no good hex mesh (Sect. 6);
- ways in which focussing on the basis functions rather than the mesh can overcome these and give significant performance advantages (Sect. 7).

It is neither a survey of splines (which have a strong rôle to play) nor of all meshing techniques nor of analysis techniques for different fields. Some, like electrostatics

M. A. Sabin (✉)
Ely, UK
e-mail: malcolm.sabin@btinternet.com

C. Manni and H. Speleers (eds.), *Geometric Challenges in Isogeometric Analysis*,
Springer INdAM Series 49, https://doi.org/10.1007/978-3-030-92313-6_9

and thermal analysis, have very similar PDEs to elasticity. Others, like supersonic aerodynamics and electrodynamics, are very different and are not addressed. They will need their own understanding.

It does contain personal views, opinions and conjectures, and therefore should be read critically.

2 What is Analysis About, and What Are the Real Criteria?

Analysis is about predicting the performance of an object being designed from its shape, material properties and intended environment, and there are three criteria in a practical context.

1. **Acceptable Cost**
   If the analysis will cost more than can be afforded, or even more than the cost of a more conservative design, then it is not worth doing. If the cost is only a tiny fraction of the available budget, then the exact cost does not matter.
2. **Adequate Accuracy**
   If the analysis will not provide better accuracy than needed then it is not worth doing. If the accuracy is much better than necessary, then the exact precision does not matter.
   The performance of various bases is often measured by the rate of convergence, typically at the limit of very fine mesh. In fact this is almost irrelevant, because there is no point in aiming for an error much smaller than the uncertainties in the material properties and the fabrication tolerances. The relevant errors are those near the top of the $e(h)$ curve[1] shown in Fig. 1, where it has not yet settled down into a definite convergence rate.
   In the optimisation context smaller errors might be important if incoherence between optimisation iterations spoils the optimiser's estimates of "downhill". This would not be relevant if design velocity ideas [1] were used in the optimiser.
3. **Timeliness**
   The results of the analysis have to be available before the product design is frozen and manufacture starts. The market does not wait for the ideal product.

The balance of importance between the first two criteria has varied over the decades. When the Finite Element Method (**FE** or **FEM**) first emerged in the late 1950s as a practical tool, computing was really expensive and the computers only had the capacity for small problems. The trade-off between cost and accuracy was all that mattered. This led to the accepted wisdom as shown in Fig. 2 (left) that quad and hex meshing were better than triangle and tet meshing.[2] The book [11] describes the state of this art very readably.

---

[1] e denotes the error, and h the distance between freedoms.

[2] This might have been because only low degree elements were available, or because the triangular meshes used were not even approximately regular. I suspect that box-spline elements in centroidal

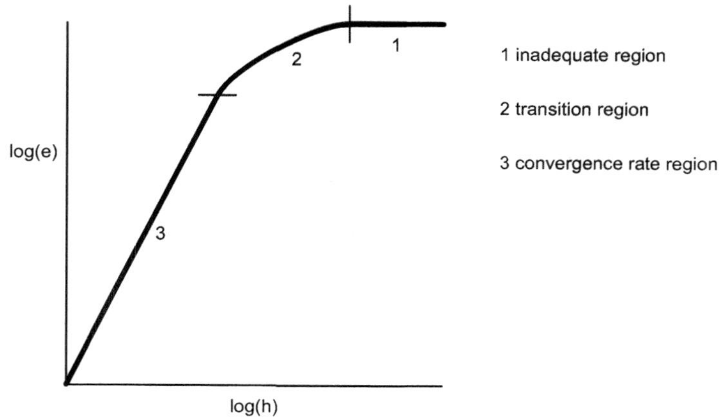

Fig. 1 A stylised $e(h)$ curve. It typically has three regions, –1, a region where the density is totally inadequate to represent the target function, and so the error is controlled by the feature sizes in that function, –2, a transition region which depends in its shape on the target function and the detail of the basis functions used and –3 a convergence rate region whose slope gives the **rate of convergence**. Publications usually stress the convergence rate for extremely tiny errors. The region of interest for practical analysis is probably near the junction of regions 2 and 3 where the rate of convergence is relevant, but not necessarily dominant. This plot is totally relevant to the cost/error plots of Fig. 2. e maps directly, and h is tightly linked to the cost, depending on the solution techniques used and the dimensionality of the problem

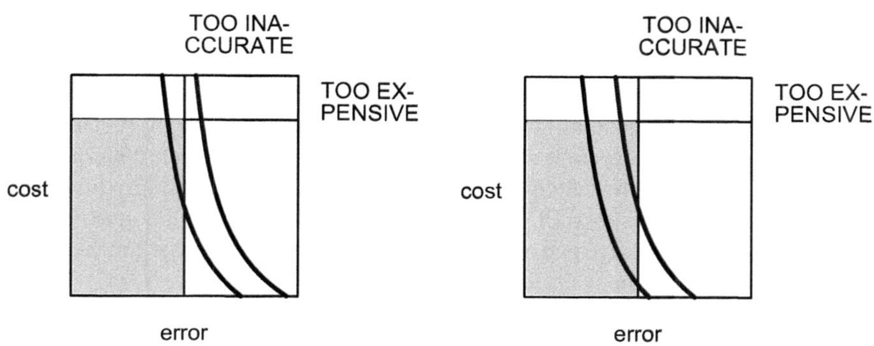

Fig. 2 These figures show schematic cost/error graphs for quad meshes (left bold curve) and triangular meshes (right bold curve), on sharply graded log scales in both directions in about 1970 (left), and in about 1990 (right). Technology cutting the top right corner of the shaded box is relevant and accessible. The graphs for expense have moved down (not across) between the two graphs because the cost scale has seriously reduced in that 20 years

As computer speeds and capacities increased, the thresholds on speed and accuracy became less restricting, and the length of time to prepare good quad and hex meshes became critical, which led to the widespread adoption of triangle and tet meshing for which reliable automation became available in the 1980s, see Fig. 2 (right).

We now want to include the analysis inside the optimisation loop. This demands that the meshing should be automatic, but it also makes the speed/accuracy trade-off important again.

3 Generality

Abstract FE [22] covers Boundary Elements and Cartesian Grids. It also covers all linear elliptic PDEs. In this paper the language of FE and elastostatics (stress analysis) is used but its message is not limited to them.

We cover this generality by looking at this topic as **Field Analysis**,[3] and (within the wide range of physics covered by linear elliptic PDEs) we address it by representing the field by

$$F(P) = \sum_i F_i \phi_i(P), \tag{1}$$

where P is the position of a point within the domain, $F(P)$ is the value of the field there, the F_i are **coefficients** which are the same kind of quantity as the field (it might be scalar for temperature, but is a displacement vector for stress analysis, and might be an arbitrary kind of object in principle[4]) and the $\phi_i(P)$ are scalar **basis functions** which can be evaluated (along with their derivatives) at any value of P. This form is also used in shape representation of curves and surfaces, which underpins the definition of **isoparametric elements**. We can think of this (and of approximation in general) as field analysis for a PDE of degree zero. Shape representation experience gives the very clear message that any variation from a smoothly flowing mesh causes serious deficiencies in the resulting shape.

We also call the F_i the **freedoms**, and computing their values is the essence of what the computation in the analysis does. The set of the ϕ_i is called the **basis**. Different bases can span the same space of representable fields.

---

voronoi triangulations would have many of the advantages of tensor product B-spline elements in regular quad grid meshes. However, in 3D there are no regular tetrahedron meshes: the semi-regular ones which do exist (e.g. the body centred cubic lattices) are all based on rectangular lattices and their automatic generation is still essentially the hex meshing problem.

[3] Bold font indicates that a term is being defined by its context.

[4] Not quite: the coefficients must lie in a vector field so that linear combinations can be formed, but differences (like displacements) will satisfy this criterion.

3.1 The Patch Test

There are also constraints on the ϕ_i:

$$\exists\{f_i\},\ \forall P,\ \sum_i f_i\phi_i(P) = 1, \tag{2}$$

$$\exists\{F_i\},\ \forall P,\ \sum_i F_i\phi_i(P) = P. \tag{3}$$

These constraints were discovered early in the development of FE, and stem from the fact that if the boundary constraints are consistent with a state of constant strain within the domain, then that has to be the solution.

In the surface definition context the same criteria give the conditions for shape invariance under solid body transformations.

In equations (2) and (3), the sets which must exist are frequently assumed to be the entire sets of coefficients, but this is over-constraining.

3.2 Computation

If the field is represented in this way, the computational approach is to set up a linear system

$$\forall j,\ \sum_i S_{ij}F_i = R_j, \tag{4}$$

which is solved for the F_i by various techniques of linear algebra based on matrix computations. R is a vector called the **right hand side**, which represents the context of the analysis—in particular the boundary conditions (for example loads and constraints) which normally force the field to be non-zero. There can often be multiple right-hand-sides, covering a range of load cases.

The entries in the **stiffness matrix** S_{ij} are computed from properties of the ϕ_i. Three options are discussed in the following.

3.2.1 Collocation

Here the equations to be solved set the residual of the PDE to zero at as many points as there are freedoms. This gives a square matrix which is sparse, better conditioned if the collocation points are well distributed.

3.2.2 Least Squares Collocation

Here the number of collocation points is rather higher (probably by a factor of four for 2D and eight for 3D) and the residuals are minimised in a least squares sense. The matrix is square, symmetric and sparse (but with a wider band than pure collocation). It is the product of a non-square matrix, like that of pure collocation, with its transpose, and so the condition number is squared.

3.2.3 Weak Form

Here calculus of variations is used to minimise the integral of an energy over the domain. Numerical integration is used, but the quantity integrated uses derivatives of one degree lower than for both pure collocation and least squares collocation. It can be compared with least squares collocation with totally dense collocation points, and it has a correspondingly high condition number.

The weak form is the standard assumption, but if higher derivatives of the basis functions can be evaluated, the others might have advantages.

3.3 Competitors

These are competitors in the elasticity context. There are other approaches (such as Finite Volumes for trans- and super-sonic aerodynamics, Transmission Line Method for the extreme electrodynamics found in analysis of lightning strikes) for other field problems.

3.3.1 Boundary Elements

Here the basis functions are solutions of the PDE. Because, at least for elliptic PDEs, the solution is continuous in all derivatives wherever no loads or constraints are applied (for convenience here called the **boundary**), each basis function covers the whole domain which is therefore not partitioned. However, in order to make a finite set of variables in the linear system, the boundary is approximated instead. The original approaches, described in [4], used basis functions with a singularity at one point, positioned on or just outside the boundary. A more recent approach described in [2, 15] is to take the convolution of such functions with a B-spline representation of the boundary, which neatly removes the singularity.[5]

[5] In fact, [4, page 39] does already introduce the idea of convolution, but only with classic high degree piecewise representation of the boundary.

3.3.2 Cartesian Grid Elements

If, instead of trying to match the partitioning to the boundary, we just apply standard piecewise polynomials over a regular grid in domain space, it is easy to achieve a reasonably well conditioned basis. The boundary conditions are no longer just nodal values on the boundary, but they can be articulated in a way which is almost trivial for entry into the linear system. An early example of this was the commercial code later called "Scan'n'Solve", described in [20].

There are two main variants – the embedded boundary and the immersed boundary. In one of these the material and its strain energy is considered only within the domain: in the other, material (and its strain energy) is assumed to cover the entire region covered by those basis functions which overlap the domain. The latter will still give the same results (who cares if an aerodynamic analysis says that there is flow of air inside the wing?) and may make the formulation simpler to code.

3.3.3 Isogeometric Analysis

As the analogy with shape representations became more and more evident, and as the desire strengthened for stronger integration with the source of shape information (the CAD systems) to improve work-flow procedures, the idea arose of using the basis functions in the surface representations (piecewise polynomials called B-splines, or their rational extensions called NURBS) also as the basis functions for analysis. This was presented in [6].

There were in fact three ideas here:

(i) Using the raw functions from the CAD system representation of the boundary. This argument was in fact totally flawed, because actual CAD system representations of real shapes and components are utterly fragmented, with sliver faces and trimming with mismatches at the boundaries of the pieces. The spatial frequencies of the CAD representation have no reason to reflect the spatial frequencies present in the field we are trying to solve for, and they do not.

(ii) B-splines are actually very good basis functions. This is a good argument.

(iii) If the partitioning resulting from B-spline basis functions is used in an optimisation it will be easy to export the result back to a CAD system for further detailing thus enhancing the overall work-flow of the entire design/manufacture process. This is another good argument, though a little ahead of its time.

4 Evaluating the Criteria

We need to consider the sources of the quantities involved in our criteria.

4.1 Computing Cost

In this section n denotes the number of freedoms, b the bandwidth of the stiffness matrix, and h the typical distance between the places where freedoms can be assigned (the **nodes**), so that $n \propto h^{-d}$ where d is the dimensionality of the domain.

The costs depend on the relationship of the linear system to the problem and on the numerical technique to be used for the solution of the linear system. We can separate the number of floating point operations needed into the two main stages of the computation, formulation of S and solution. In each case we can relate the cost to the number of freedoms, the number of non-zero freedoms in each row and the bandwidth of S, depending on the technique to be used to solve the system.

4.1.1 Formulation

The formulation cost depends on the number of non-zero entries in the stiffness matrix and on exactly how those entries are computed. In principle every entry can be computed independently of every other, and even the integration over different elements can be computed independently. On big problems, there is therefore scope for massively more parallelism than our largest computers are likely to offer in the immediate future.

Higher degree spline basis functions, with their larger support, will have more overlaps and so this cost increases with the square of the degree in 2D and the cube in 3D. In fact after elimination of the internal freedoms of separate high degree elements the situation is much the same.

4.1.2 Solution

If explicit solution (for example, by LDL^T factorisation) is used, then the cost of solution is proportional to nb^2, which is ideal for one-dimensional analysis (b is a small constant), tolerable for two-dimensional analysis (b is proportional to the width of the domain, while n is proportional to its area, thus giving a cost of $O(h^{-4})$), but a serious problem for three-dimensional domains, (where b is proportional to the cross-sectional area[6] and n to the volume, giving a cost of $O(h^{-7})$).

Iterative solutions include Relaxation, which is slow for 1D because the effects of boundary conditions have to propagate many times through the domain, Preconditioned Conjugate Gradient, and Multigrid, which relaxes low spatial frequencies

[6] Minimisation of b by permutation of the sequence of coefficients is actually a Graph Theory problem, with methods such as Reverse Cuthill-McKee being developed quite early. There is a sensible heuristic based on the idea of an elimination wavefront spreading through the domain. If this starts at one end of a long thin object it is kept small in volume as it works its way to the other. Clearly this is not appropriate for strongly branched objects, where substructuring would be appropriate.

efficiently on a coarse grid and high ones efficiently on a denser grid. Adaptive solution, where higher frequencies are introduced only where an error estimator indicates that they are needed, can be regarded as an extreme of Multigrid.

4.1.3 Field Recovery

For the analyst to be able to understand the result of the analysis, it is necessary to evaluate the field (and its derivatives such as peak stress) at a reasonably dense set of points in the domain. This just evaluates equation [1] at those points. There is nothing special about nodal values which are evaluated in exactly the same way. This cost is not considered further here.

4.2 Error

There are five important sources of error as discussed in the following.

4.2.1 Floating Point Rounding Errors

These are outside the scope of this note.

4.2.2 Approximation of the Geometry

The shape analysed is the shape covered by the mesh. If isoparametric elements do not match this shape exactly there will be some error involved. This is one of the arguments for Isogeometric Analysis, where the basis functions used to define the shape are also used for the analysis. This is seldom practical for predefined objects, and this source of error is likely to be small compared with manufacturing tolerances.

4.2.3 Approximate Integration

When the weak form is used, the entries in the stiffness matrix are integrals of products of derivatives of the basis functions over the domain. These integrals are typically computed by Gaussian integration over the individual elements in the intersection of the supports of ϕ_i and ϕ_j. Gaussian integration is only accurate when the integrand is polynomial of a chosen degree. Isoparametric elements, where both the position field and the field of interest are defined by equations of the form [1] over a parametric domain, require the inclusion of the Jacobian of the position field, which is not polynomial.

If collocation is used (and more specifically least squares collocation), the solution is less sensitive to exact positioning of the collocation points, but this issue is not normally considered to be significant if the mesh density is sufficient for the other issues.

4.2.4 Inadequate Mesh Density

In order to represent a field accurately, there must be enough freedoms to match the spatial frequencies present in that field. Shannon's theorem [19], that a waveform cannot be reconstructed from less than two samples per cycle in the highest frequency present in the signal is equally applicable to spatial variations. This explains the shape of the $e(h)$ curves at their very top, see Fig. 1. The $e(h)$ curve is a critical one, because while e is the error as in the cost/error concerns of the original criteria, h dominates the computing cost, often at a very high power.

4.2.5 Basis Function Shape

If the basis functions themselves have high frequency components, then these are introduced in the analysis as artifacts driven by the lower frequency basis functions, and then need higher frequency basis functions to remove them again. In fact all basis functions of finite support will have some high frequency components, and these decay as a power of frequency which depends on the degree and continuity of the basis functions. A deep understanding of the relationship will someday be achieved, but it is beyond the scope of this paper. Independent or weakly coupled basis functions[7] of degree d have a rate of decay (and hence a convergence rate) proportional to $d + 1$. Spline elements of continuity $d - 1$ also have a rate of decay of $d + 1$. It can be conjectured that the equality of these two effects is because the solution process finds solutions in the independent case that are very close to the spline solution. However, the number of freedoms in each coordinate direction is higher by a factor of d for the independent basis.

4.3 *Timeliness*

At the point where a candidate design is stable enough to be analysed, there are two tasks which need to be carried out prior to the computation. The first is the identification of the relevant load cases. This is essentially a part of the design process, linked to the design intent, and will be a manual process for as long as design itself is done by humans. It can be kept outside any optimisation loop. The second is the meshing, which traditionally has been regarded as the partitioning of the domain

[7] I.e. classic high degree elements.

into elements. Partitioning into triangles or tetrahedra can be done automatically by well-established algorithms, but the error/cost tradeoff is believed to be worse than for quad/hex elements. For this reason, robust automatic hex meshing has been regarded as the "Holy Grail", even though it is only part of the story.

5 Quad and Hex Meshing

Before elaborating on the known approaches to quad and hex meshing, it is important to understand first the issues.

5.1 Issues

To fill a rectangle evenly with quads is trivial. The challenges arise from

- *the shape of the domain*: The rule of the FE meshing "game" is that, for ease of application of boundary conditions and integration, edges of elements should be aligned with the boundary. Cartesian grid analysis does not have this problem, but that is a separate approach.
- *the variation of the mesh density*: If we know that the solution field will have higher spatial frequencies in one region than in another we require **grading** of the mesh to match the local densities to the required spatial frequencies.

5.1.1 Shape of Domain

If a 2D domain is not four-sided, then there have to exist vertices where other than four quad elements meet. By analogy with subdivision surfaces I call these **EVs** (Extraordinary Vertices).[8] Near such places it is harder to find smooth basis functions, and so we would like to minimise their effect.

The number and type of EVs needed is determined by the number and type of the corners of the domain. If we define **convex corners** as being those places on the boundary where a good mesh will have one element, **concave corners** as being those where it will have three elements,[9] and **crack-tip corners** as those where it will have four, see Fig. 3, the following algorithm determines the "net" number of EVs.

(i) Give each convex corner a score of $+1$, each concave corner -1, and each crack-tip corner -2.

[8] This phrase was coined by Steve Coons.

[9] Such corners really need basis functions with fractional power behaviour, but that is not usually considered by meshing software.

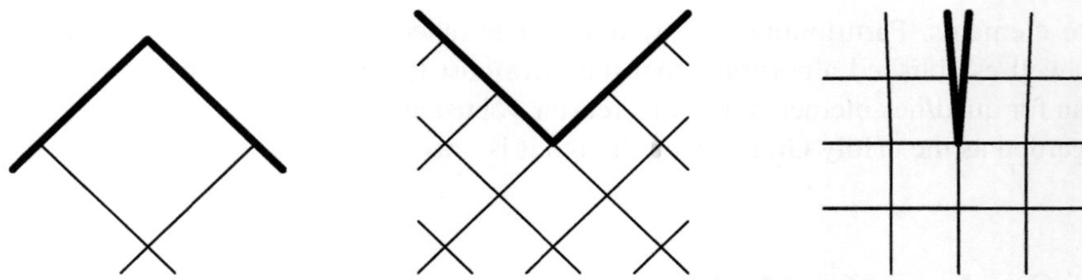

Fig. 3 Various convexities of corners. Left: convex, middle: concave, right: crack-tip

(ii) For each loop of the boundary of the domain, add up the scores of its corners and subtract 4.
(iii) Subtract the scores of any inner loops from the outer loop.

This gives a number which needs to be matched by EV scores which are given by the valency minus 4. Trivalent EVs provide negative scores and those with a valency of 5 or more a positive one, and so the minimum number will be achieved if the EVs used all have scores of the same sign. The fact that a 6-valent EV has twice the score of a 5-valent one is linked to the fact that it is equivalent to two 5-valent EVs in terms of the overall flow of the mesh. However, the local distortion grows worse as the valency increases, and so unless there are important reasons (such as symmetry) for using valencies higher than 5 they should be avoided.

The 3D analog is clear. Here the analog of an EV is a chain of **EEs** (Extraordinary Edges), where a non-standard number of solid elements meet. They can emerge through the boundary of the domain in places where they look like EVs on the faces. They can also form closed loops inside the domain, or they can collide with other chains. If we limit the EEs to valencies of 3 and 5, such collisions can only happen in a small number of possible configurations which, although they were articulated earlier [17], are well-explained in [13].

5.1.2 Variation of Mesh Density

A local subproblem is characterised by the numbers of element edges along each side of the local domain. We call these numbers **division numbers**. The exact positioning of the nodes along the edges is an aspect of mesh smoothing and is ignored here. There is a parity constraint: if a 2D region is to be filled with quads, the total number of element edges around the boundary of the domain needs to be even. Hex meshing has a parity constraint on every face of the object.

Obviously, if the domain is tapered, a regular grid will give a variation of density from one place to another. We would need to vary this if we wanted evenly sized elements, and equally, if the boundary conditions imply denser mesh in one part of the boundary than another we might need a non-regular mesh.

The classical approach is to use patterns like the "Patran Zoom", which puts two extra edges along one side of a quad domain, as shown in Fig. 4 (top). However,

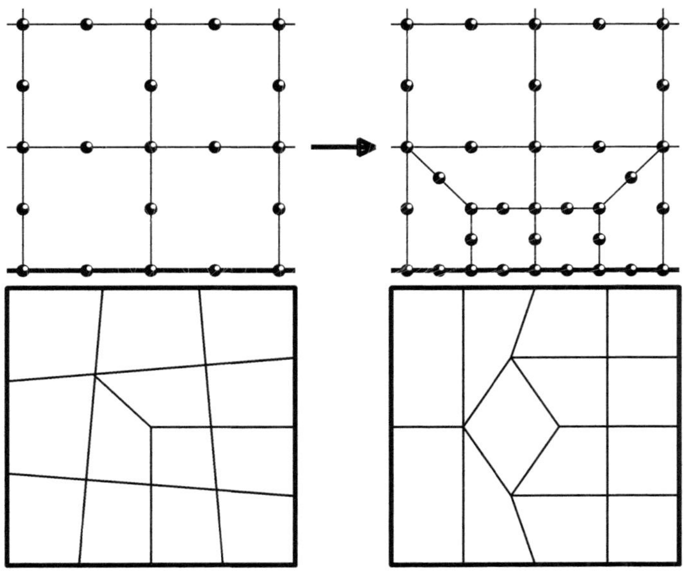

Fig. 4 Top: The PATRAN ZOOM: Bullets show the nodes of serendipity biquadratic elements.. Bottom: Dipoles achieving more flexible variation of mesh density

this means the introduction of extra EVs, with their distortion of the mesh and unsmoothing of the field. There is some generality in the use of pairs of EVs for this purpose. (Once the minimum number of EVs is identified for matching boundary orientation, additional EVs can only be added in matching + and − pairs. I call such a pair a **dipole**. It is analogous to a **dislocation** in crystallography.)

The number which is needed is easily assessed.[10] In a quad (sub)region of the mesh, identify the differences between the division numbers on opposite sides of the domain. There are two pairs of opposite sides and so we get a vector (the **transition vector**) with two integer components, see Fig. 4 (bottom). This vector gives the displacement vector (in mesh units) between a positive and a negative EV.

From the point of view of division numbers it makes no difference where in the domain this is positioned, though the positioning does influence the total number of elements. If a lot of variation of density is required it can be achieved either by a single large dipole (if it will fit into the domain) or by many small ones.

Incidentally, if a face already has an EV, adding a dipole with one end cancelling out the original EV, the effect is essentially that of moving it, so that the division numbers on such a face provide a way of positioning the EV.

[10] The description here is over-simplified, though it is an adequate starting point. In a mesh with a lot of dipoles a band of elements can be turned by passing through another dipole, so that it contributes to the wrong division number. Sorting this out properly would be a good box-ticking exercise for an appropriate student.

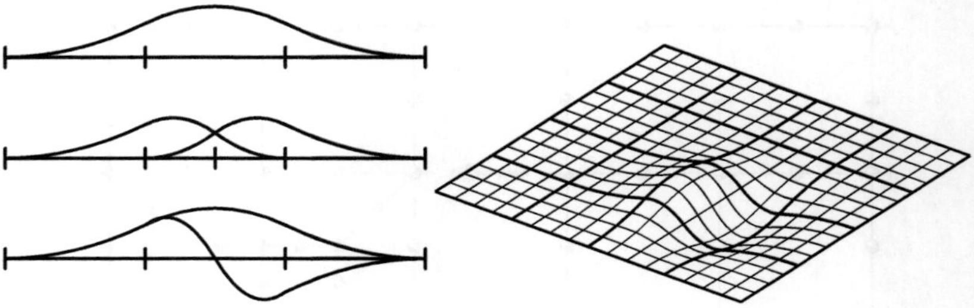

Fig. 5 Left: An extra basis function from splitting a quadratic B-spline. Right: An extra basis function from splitting a biquadratic B-spline

That is the classic approach. It is bad for performance of the analysis, because it introduces EVs which in turn upset the quality of the element formulations. This is particularly unfortunate if the density is being increased locally in order to get higher density in places where the error is still too large. This increase is called **adaptivity**. The problem is that adding dipoles gives a mesh whose basis functions do not span the space originally spanned by the coarse mesh. This means that there is no guarantee that the refined mesh will give a smaller error that the coarse one. It fails the **nesting** property.

The spline basis function approach gives a far better approach. Any individual tensor product bivariate B-spline basis function can be split into two in either direction. The sum of the two is equal to the original (thus preserving summation to unity) and the difference can be used as a single new basis function (with a vector coefficient rather than a control value). Such a biquadratic difference function is illustrated in Fig. 5. The idea is similar to T-splines but is far, far simpler and applies easily in 3D [18], where use of combinations of extraordinary edges gets really complicated.

Extra density can therefore be created exactly where it is needed with an exactly nested sequence of bases.

In fact this effect can be achieved with current commercial FE software by using **multi-point constraints** (MPCs), as shown in Fig. 6, but it fits more naturally into software designed from the outset to exploit B-spline elements.

5.2 *Known Approaches to Quad and Hex Meshing*

There are at least five known approaches. They are described here in the quad-meshing context with comment about the hex-meshing equivalent where relevant. Within each approach there are many possible variations in the detail.

The first three of these focus on the mesh, the last two on identifying and placing irregularities, letting the mesh in regions without irregularity just flow.

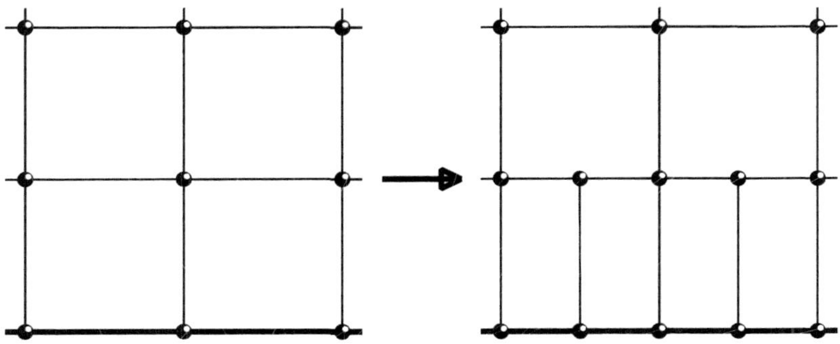

Fig. 6 MPC-based addition of extra freedoms. The T-junction nodes are constrained to be compatible with the element on the coarser side

5.2.1 Inside-Out

This method is described by George [10].

Place a regular grid of quads so that it covers the domain. Delete those quads which overlap the boundary and join the newly exposed edges to the boundary of the domain, making new quad elements. See Fig. 7.

There are, of course, variants which move vertices lying just inside or just outside the domain on to the domain boundary, but this variant guarantees quad elements.

To get better shape into the elements some mesh smoothing will need to be applied. That is not controversial. However, this method does result typically in more than the minimum number of EVs, and they are positioned near the boundary, which is where they do most harm to the accuracy of the solution. Remember that in a solution of an elliptic PDE all extrema (e.g. places where the stress may be high enough to be critical) lie on the boundary.

The trivariate analogue does not seem to play many tricks on us, although joining up to the boundary is less trivial in its small print.

Fig. 7 Inside-out quad meshing

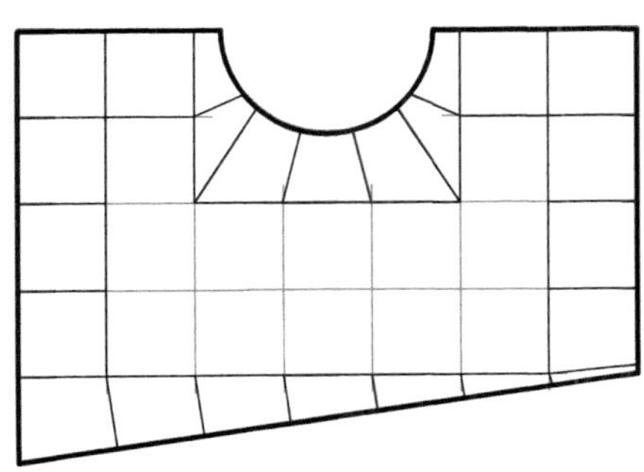

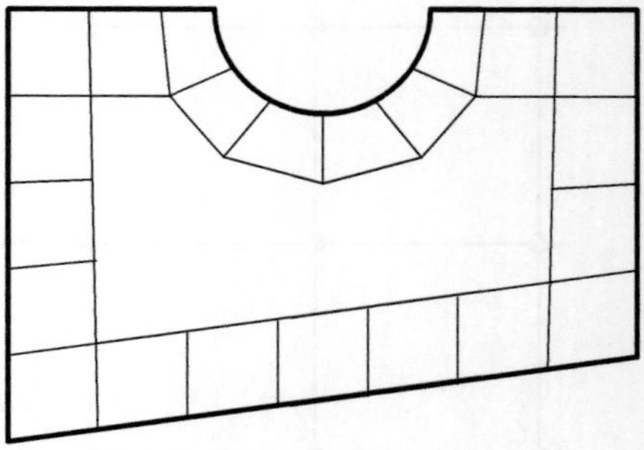

Fig. 8 Outside-in quad meshing

5.2.2 Outside-In

This method forms the basis of the Paving algorithm [3], which is regarded as a standard method for quad meshing, offered in many commercial mesh generators.

Create a layer of well-formed quad elements immediately inside the domain boundary, leaving a hole which will be smaller than the original domain and also have fewer element edges. Then just apply the same process to the hole until the number of edges around it drops to zero. See Fig. 8.

The boundary of the hole is typically less smooth than the original, and EVs emerge fairly naturally from this as corners appear or disappear. The main problem is that opposite parts of the boundary of the hole can approach each other as successive layers of elements are added, and the collision needs to be managed so that if the hole splits into two each half satisfies the parity condition.

This can be managed in the bivariate context, but in the trivariate context the patterns on the approaching faces can be totally different from each other, and fitting hex elements to join them up is an enormous problem. Although the Plastering algorithm [21] often works, the mesh quality where this happens is very poor indeed. It is only acceptable because this is where the field has very little happening.

5.2.3 Conversion from a Triangle/Tet Mesh

This method is described by Remacle [16].

Make a triangle mesh with an even number of triangles. Join them together in pairs to make quads. Then smooth the result.

Looking at examples in the papers there are places where there are unnecessary pairs of EVs spoiling the smooth flow of the mesh. This might be attributed to the need for speed in the meshing process causing the pairing decisions to be taken as locally as possible. The 3D analog joins small collections of tets to form hexes. It is not claimed that this approach can always produce an all-hex mesh.

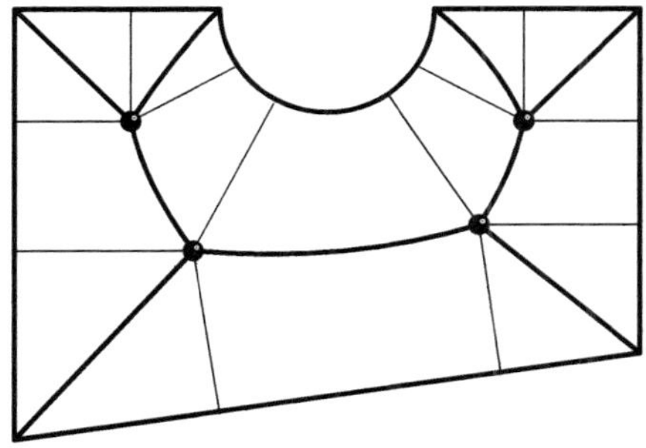

Fig. 9 EVs at medial vertices. The bold lines are the edges of the medial axis, the fainter ones partitioning lines

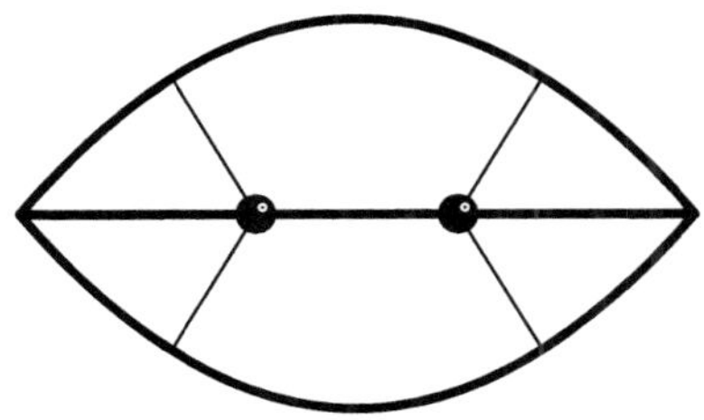

Fig. 10 EVs on medial edges at transition angles

5.2.4 Explicit Positioning of EVs

There are two current approaches to this, with variants within each. The first uses the **medial axis**, which, of a 2D shape, is the locus of the centres of maximal circles (touching the boundary in at least two places, called the **supports**).[11] This was first suggested by a PhD student who, after graduating, joined IBM, who promptly patented his method, thus discouraging its exploration by anybody else.

Join each edge in the medial axis of the domain to its support boundaries. This splits the domain into rectangles which can be meshed regularly. See Fig. 9.

This puts EVs at the vertices of the medial axis of the domain. These EVs have six adjacent elements and so they cause more severe distortion of the mesh than 5-valent ones would. There is a lot of special casing, so that a medial vertex which has two pieces of medial axis joining it to convex corners of the domain need not get an EV at all. The 3D equivalent was not explored.

More recently Fogg [7] suggested placing EVs not at the medial vertices, but at places on the medial edges where the support angle passes through thresholds. This gives well-separated EVs, and neatly covers the situation of a two-sided domain where the medial axis does not have any vertices. See Fig. 10.

The second approach to explicit positioning of EVs is based on **frame fields**. A good example is the Quadcover method [12] but it has been well-explored recently.

[11] The 3D analogue is the **medial object**, the locus of centres of spheres touching the boundary in at least three places.

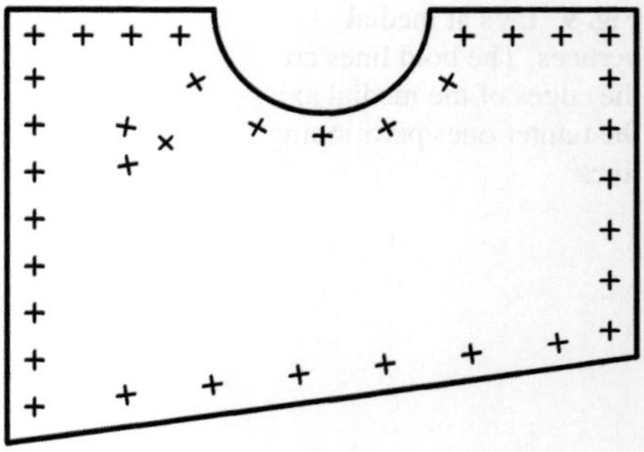

Fig. 11 EVs at singularities of the frame field. The + symbols indicated the local orientation

Interpolate a field of element orientations between the known orientations on the boundary. The field is locally flat except at singularities at which the EVs are placed. See Fig. 11.

There is an implementation problem that tracing element edges in this field can lead to spiralling. It may be that interpolating not a simple frame field but an element metric tensor might solve this, but I am not aware of work in this direction.

The really beautiful and elegant result used in Quadcover itself is based on the fact that if the rays of a frame are represented by complex numbers, then it is only necessary to interpolate the fourth power (which is the same for all four rays) and then take the fourth root to get the four rays at a point.

The problem arises in 3D that these chains of extraordinary edges themselves need to be aligned with the frames, so the theory is much deeper than in the 2D case. This issue is well addressed in [13].

5.2.5 Separation of EVs

A promising alternative is not even to try to position the singularities at first, but instead merely to separate them by partitioning the complete domain into simpler pieces. This idea is not new. It is mentioned and applied by Mukherjee [14] but not claimed as new even then. It is also an obvious application of the ideas in [5]. However, the medial object makes it easier to apply.

Use the medial axes of the medial faces of the domain to identify cutting surfaces which a good mesh will have collections of element faces on. See Fig. 12.

In the trivariate case, the cuts are surfaces which naturally partition the domain into pieces which match the classes of collisions of extraordinary edges. Indeed except for a couple of cases collisions can be avoided completely, giving a very smoothly flowing mesh. The exact positions of the extraordinary edges are then determined by the mesh density requirements. Spiralling is avoided.

This subsection should be read with caution. Only the 2D case has been implemented, and although the 3D analog appears clear, there may be some nasty surprises in store.

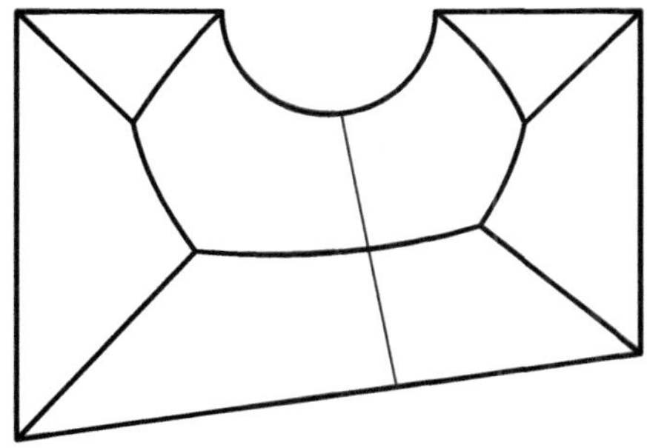

Fig. 12 Separation of EVs by lines cutting the medial edges. Bold lines are the boundary and medial axis of the shape: the thinner line is a potential place to separate the domain into simpler pieces. Such cuts form a foliation of the domain

6 BUT There Exist Solid Shapes for Which No Good Hex Mesh Exists

Here are a few counterexamples for any hex meshing algorithm.

6.1 The Pyramid

Consider a pyramid with four equilateral triangular faces on a square base in Fig. 13 (left). Each face has an obvious quad mesh, which in a nice sparse setting will have four quads on the base and three on each of the triangular sides. Clearly this object will need at least six hex elements. However, even though there is a theorem which says that any polyhedral solid which has quad mesh on all its faces will contain a hex mesh matching those quad meshes, the smallest number of hexes in such a mesh is significantly larger than six, and therefore the average volume of each such element will be a lot smaller than the cube of the edge length of the quads. There must be hex elements inside which have very poor aspect ratios.

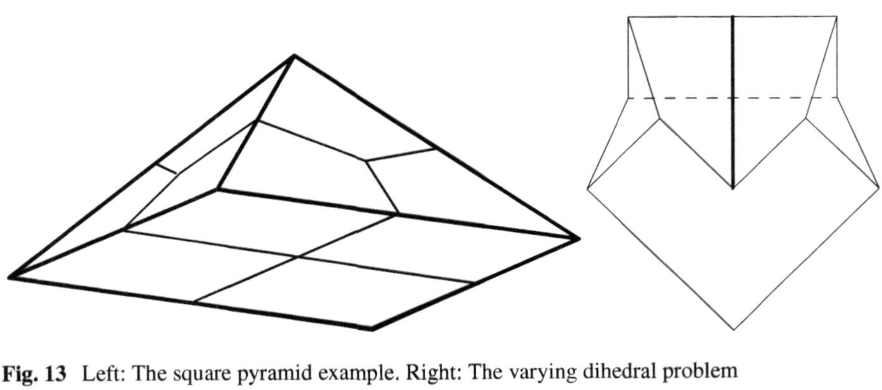

Fig. 13 Left: The square pyramid example. Right: The varying dihedral problem

6.2 *The Varying Dihedral Angle Along an Edge*

Consider the way in which an engine nacelle is attached to the wing of an airliner. There is a pylon the surface of which is almost perpendicular to the nacelle surface in the middle of its section, but is almost tangential to the nacelle surface at the front. When analysing the airflow around this configuration, we require that edge to contain two hex elements at the front, but just one further aft. This problem is illustrated in Fig. 13 (right).

6.3 *The Leadscrew*

A numerically controlled machine tool has the position of the cutter controlled by **leadscrews**, the rotation of each of which moves one of the axes. A very schematic leadscrew is illustrated in Fig. 14.

The thread cross section is typically a very squarish shape into which hex elements fit nicely. The question is *"how does the mesh flow in the central core of the screw?"*.

6.4 *The Slab with Different Features on Opposite Faces*

In Fig. 15 (left) the body has sharply tapered trapezoidal faces which vary significantly between the front and back of the object. Any topological extrusion will give very severe distortion of the mesh. Indeed, in more extreme examples there may not even be any 1:1 correspondences between front faces and back ones on which extrusion can be based. However, the object is thin enough that we want only a small number of layers of elements between front and back, and we certainly do not want mesh features running perpendicular to the view of that figure.

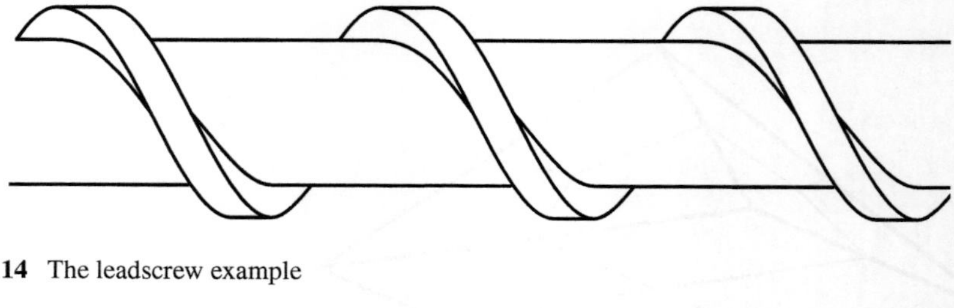

Fig. 14 The leadscrew example

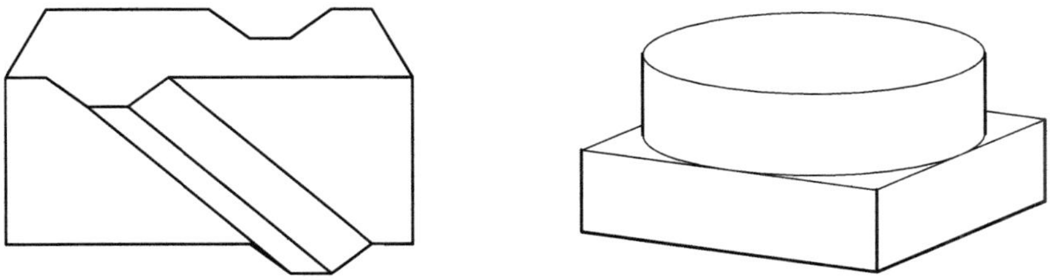

Fig. 15 Left: The slab with different features on opposite faces. Right: An example where edges meet tangentially with a tiny angle on one face

6.5 The Block with a Tangential Cylinder on Top

The first problem in Fig. 15 (right) is that the flat face on top of the block needs badly shaped elements to fit into its sharp corners. The second is that you do not want to extrude the mesh with its irregularity all the way down to the bottom of the block, but this is essentially the same issue as in Fig. 15 (left).

7 MAGIC!

The section illustrates two ways in which the *"think about basis functions, not partitioning of the domain"* approach can give serious advantage.

7.1 Transition to the Boundary

For the quadratic PDE which underpins a lot of applications, there are some known solutions. One, widely used as a reference point for boundary elements, is the point load solution. Another provided Fourier with the motivation for inventing Fourier analysis [8]. (A more accessible reference is [9], pages 209–210.) He was studying thermal analysis and found that if the temperature along one edge of a rectangular block varied sinusoidally, the variation at a parallel slice inside the block had the same phase and frequency but an amplitude which varied negative-exponentially with depth and the decay rate was the spatial frequency of the edge conditions. High frequencies decay very quickly, and so the density of basis functions in the interior can be much lower than on the boundary. Figure 17 has fewer than half the number of freedoms of Fig. 16, but will have about the same error, because the local density of freedoms need only match the spatial frequencies in the field which have magnitude comparable to the tolerated error.

Here the advantage is that the total number of freedoms in a bulky 3D object is proportional only to the surface area (in units of one element face) instead of to the

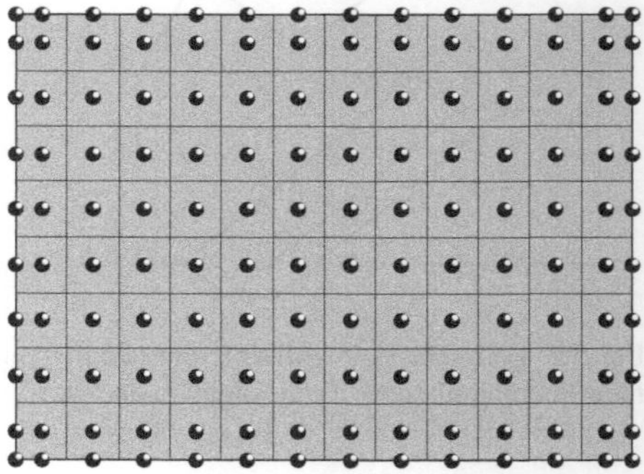

Fig. 16 A regular dense grid. The dots indicate the "centres" of biquadratic B-spline basis functions; the lines the resulting partitioning of the domain into polynomial pieces

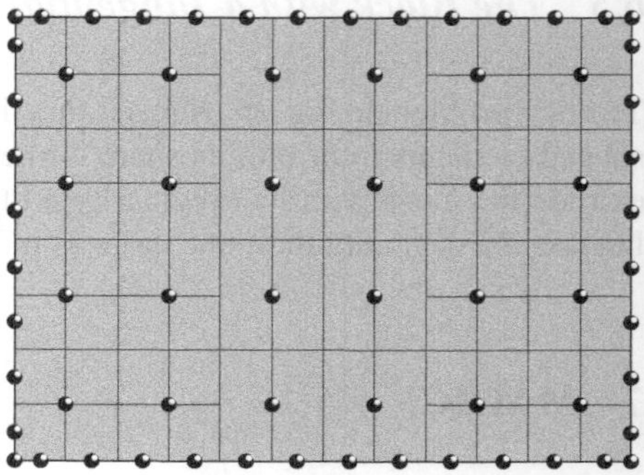

Fig. 17 A sparser grid with the same expected error

volume of the object (in units of one element volume). A $1000 \times 1000 \times 1000$ block would need only about 10^7 freedoms, not the 10^9 of a regular mesh. Even for $O(n)$ solvers a factor of 100 is worth a lot! Run it overnight instead of taking three weeks, or on your desktop computer instead of on the big parallel server.

The small print is the detail of what to do at corners and edges of a 2D domain and at the edges of a solid, but the potential gains make thinking about this worth while.

7.2 *Fairly Thin Regions with Different Patterns on the Opposite Faces*

"Fairly thin" means not thin enough for shell elements to be appropriate.

Figure 18 shows a 2D mesh whose resolution varies from one end to the other. This set of basis functions works because the functions are all tensor products with the common horizontal factors. The sum of a vertical column is therefore an extrusion of its horizontal factor. It does not matter how it is split vertically.

The same principle works to even more dramatic effect in 3D meshing. Using an estimate of how the field will vary between the faces to choose a number of layers

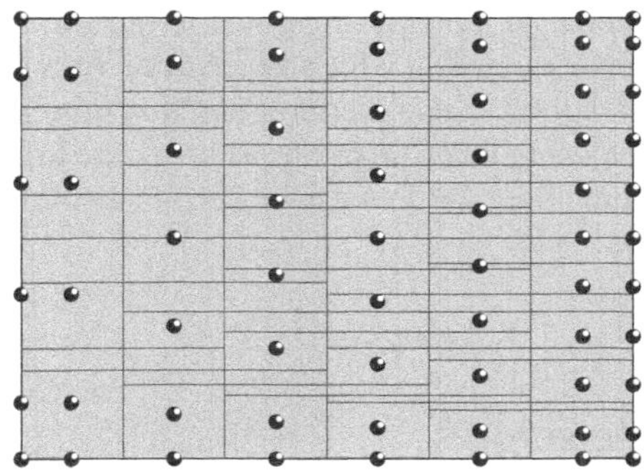

Fig. 18 Tapered density

of control points, and divide the domain between the incompatible faces into an appropriate number of layers. Each layer of control points can have its own pattern of quad mesh independent of all the others, chosen to match the spatial frequencies of its layer. Adaptivity still works because each basis function is still a tensor product B-spline.

There is again some small print about edges and corners of the domain. While this is not trivial it is not prohibitive either.

This provides an effective approach, not just to Fig. 15 but also to the leadscrew example of Fig. 14.

The downside of this approach is that the building of the stiffness matrix for weak form solution will require considerable housekeeping. The collocation approaches will be simpler, though still non-trivial. The fragmentation of the medial faces is a good diagnostic tool to find when the problem is serious.

7.3 Other Opportunities

Two techniques widely used in pre-FE stress analysis are **abstraction** which replaces parts of the object of low dimensionality explicitly by low-dimension models which can be analysed faster and **stress concentration factors** which replace small features (such as tiny holes) by rules for increasing the local peak stresses. Both of these fit fairly naturally, though non-trivially, into the ‘choose basis functions and adapt’ approach.

8 Conclusions

Many of the original assumptions made when finite elements was first introduced as an approach to the analysis of structures have proved to be unnecessary. The particular assumption that the first step needs to be a partitioning of the domain into elements seems particularly ingrained in the street wisdom of structural engineers, and it is this assumption which needs to be reversed. The partitioning is a side effect of the choice of basis functions, and it is of totally secondary importance. If we switch to choosing basis functions as being the primary choice, letting the partitioning being something we do not even need to display, very significant gains in performance can be made.

References

1. Agarwal, D., Robinson, T.T., Armstrong, C.G., Marques, S., Vasilopoulos, I., Meyer, M.: Parametric design velocity computation for CAD-based design optimization using adjoint methods. Eng. Comput. **34**, 225–239 (2018)
2. Belibassakis, K.A., Gerostathis, T.P., Kostas, K.V., Politis, C.G., Kaklis, P.D., Ginnis, A.I., Feurer, C.: A BEM isogeometric metod for the ship wave-resistance problem. Ocean Eng. **60**, 53–67 (2013)
3. Blacker, T.D., Stephenson, M.: Paving: a new approach to automated quadrilateral mesh generation. Int. J. Numer. Methods Eng. **32**, 811–847 (1991)
4. Brebbia, C.A., Walker, S.: Boundary Element Techniques in Engineering. Butterworth and Co Ltd (1980)
5. Campen, M., Bommes, D., Kobbelt, L.: Dual loops meshing. ACM Trans. Graph. **31**, 1–11 (2012)
6. Cottrell, J.A., Hughes, T.J.R., Bazilevs, Y.: Isogeometric Analysis. Wiley Ltd, New York (2009)
7. Fogg, H.J.: Automatic generation of multiblock decompositions for meshing. Ph.D. thesis, Queens University, Belfast (2014)
8. Fourier, J.B.J.: Sur la Théorie Analytique de la Chaleur. Firmin Didot (1829)
9. Fourier, J.B.J.: Analytic Theory of Heat. Cambridge Library Collection - Mathematics. Cambridge University Press, Cambridge (2009)
10. George, P.L.: Automatic Mesh Generation: Applications to Finite Element Methods. Wiley Inc, USA (1992)
11. Irons, B., Shrive, N.: Finite Element Primer. Ellis Horwood Ltd (1983)
12. Kaelberer, F., Nieser, M., Polthier, K.: QuadCover - surface parameterization using branched coverings. Comput. Graph. Forum **26**, 375–384 (2007)
13. Liu, H., Zhang, P., Chien, E., Solomon, J., Bommes, D.: Singularity-constrained octahedral fields for hexahedral meshing. ACM Trans. Graph. **37**, 1–17 (2018)
14. Mukherjee, N.: CSALF-Q: A bricolage algorithm for anisotropic quad mesh generation. In: Proceedings of 20th International Meshing Round Table, pp. 489–511 (2011)
15. Palitis, C., Ginnis, A.J., Kaklis, P.D., Belibassikis, K., Feurer, C.: An isogeometric BEM for exterior potential-flow problems in the plane. In: 2009 SIAM/ACM Joint Conference on Geometric and Physical Modelling, pp. 349–354. ACM (2009)
16. Remacle, J.F., Lambrechts, J., Seny, B., Marchandise, E., Johnen, A., Geuzainet, C.: Blossom-Quad: A non-uniform quadrilateral mesh generator using a minimum-cost perfect-matching algorithm. Int. J. Numer. Meth. Eng. **89**, 1102–1119 (2012)
17. Sabin, M.A.: Use of B-spline basis functions in finite element analysis. Ph.D. thesis, Leeds University (1997)
18. Sabin, M.A.: Adaptivity with B-spline elements. In: Mathematical Methods for Curves and Surfaces. Lecture Notes in Computer Science, vol. 10521, pp. 218–232. Springer (2016)
19. Shannon, C.E., Weaver, W.: A Mathematical Theory of Communication. University of Illinois Press, USA (1963)

20. Shapiro, V., Tsukanov, I.: The architecture of SAGE-a meshfree system based on RFM. Eng. Comput. **18**, 295–311 (2002)
21. Stephenson, M.B., Canann, S.A., Blacker, T.D.: Plastering: a new approach to automated, 3D hexahedral mesh generation. In: 33rd Structures, Structural Dynamics and Materials Conference, pp. 662–670 (1992)
22. Strang, G., Fix, G.: An Analysis of the Finite Element Method. Prentice-Hall (1973)

Scattered Data Approximation by LR B-Spline Surfaces: A Study on Refinement Strategies for Efficient Approximation

Vibeke Skytt and Tor Dokken

Abstract Locally refined B-spline (LRB) surfaces provide a representation that is well suited to scattered data approximation. When a data set has local details in some areas and is largely smooth elsewhere, LR B-splines allow the spatial distribution of degrees of freedom to follow the variations of the data set. An LRB surface approximating a data set is refined in areas where the accuracy does not meet a required tolerance. In this paper we address, in a systematic study, different LRB refinement strategies and polynomial degrees for surface approximation. We study their influence on the resulting data volume and accuracy when applied to geospatial data sets with different structural behaviour. The relative performance of the refinement strategies is reasonably coherent for the different data sets and this paper concludes with some recommendations. An overall evaluation indicates that bi-quadratic LRB are preferable for the use cases tested, and that the strategies we denote as "full span" have the overall best performance.

1 Introduction

Tensor-product B-spline surfaces are a mature and standardized geometry representation that has been known at least since the 1970s. The first uses of Tensor-product B-splines were in Computer Aided Design (CAD). In Isogeometric Analysis (IgA) [6], B-splines replace the traditional shape functions used in Finite Element Analysis (FEA). A guide to splines can be found in [9]. Tensor-product spline surfaces have very good numerical properties, but lack local refinement of the spline space. In recent years several approaches have been proposed for local refinement of spline surfaces including T-splines [3, 20], Truncated Hierarchical B-splines (THB) [11], and LR B-splines (LRB) [2, 7]. We will, in this paper, focus on the use of LRB surfaces for approximation of scattered data.

V. Skytt · T. Dokken (✉)
SINTEF, Oslo, Norway
e-mail: tor.dokken@sintef.no

V. Skytt
e-mail: vibeke.skytt@sintef.no

C. Manni and H. Speleers (eds.), *Geometric Challenges in Isogeometric Analysis*,
Springer INdAM Series 49, https://doi.org/10.1007/978-3-030-92313-6_10

We perform a systematic study on the effects of different strategies for local refinement of LRB surfaces for the approximation of geospatial point clouds. Section 2 provides an overview of relevant locally refined spline methods and outlines the algorithmic approach we employ for scattered data approximation. Section 3 gives a brief overview over previously published local refinement strategies for bi-variate splines and discusses the concept of a good refinement strategy for approximation of large point clouds. A set of candidate strategies are defined in Sect. 4. Section 5 presents five data sets and the corresponding approximation results for the selected refinement strategies along with analysis of the result related to each data set. In Sect. 6 the case specific analysis is summarized to provide a unified understanding, and finally in Sect. 7 some conclusions are drawn.

2 Background

Section 2.1 presents locally refined splines in general and LRB in particular. We also explain why we focus on LRB rather than T-splines and THB. We then turn to scattered data approximation in Sect. 2.2.

2.1 Locally Refined Splines

The lack of local refinement for tensor-product splines provides severe restrictions for IgA as well as for scattered data approximation. Using tensor product B-splines will in most cases introduce significantly more degrees of freedom than actually needed. This makes the data volume grow and restricts the size of problems that can be addressed. Two basic approaches exist for building local spline surfaces:

- Refinement in the mesh of vertices/control points: the approach used for T-splines.
- Refinement in the parameter domain: the approach used for THB and LRB.

For simulation and approximation purposes, it is convenient to span the locally refined spline space by a set of linearly independent functions. T-spline, LRB and THB spaces are all spanned by functions that are composed from tensor products of univariate B-splines. It is also attractive to have quasi interpolants for hierarchical spaces [22].

2.1.1 T-Splines

T-splines were introduced by Sederberg et. al. [20] to enhance the modelling flexibility in CAD design. The starting point of the T-spline T-mesh is the regular grid of control points in a tensor product B-splines surface. New control points in the T-mesh are added on axis parallel lines between existing control points according to a set of T-spline rules. Each control point in the T-mesh corresponds to a tensor product B-spline. The knot vectors of each such B-spline are identified by traversing the T-mesh starting from the control point and going outwards, in all four axis parallel directions, until a degree dependent number of meshlines are intersected. In the bi-cubic case this traversing stops after two lines in the T-mesh are intersected in each of the four directions. The new control points are used to model local details in a preexisting surface. Frequently, the B-splines spanning the T-spline space have to be scaled to form a partition of unity.

The most general version of T-splines possesses neither nested spline spaces nor a guarantee for linear independence [5] of B-splines. In IgA, linear independence is important and additional rules were added to the T-spline creation algorithm to ensure it, giving rise to Analysis Suitable T-splines [3].

2.1.2 Truncated Hierarchical B-Splines

Hierarchical B-splines (HB), introduced by Forsey and Bartels [10], are based on a dyadic sequence of grids determined by scaled lattices over which uniform spline spaces are defined. HB provide nested spline spaces spanned by tensor product B-splines, but do not form a partition of unity and they are not linearly independent. How to select B-splines that gives linearly independent HB was solved in [15]. To provide partition of unity for HB, THB [11] were introduced. THB allow non-uniform spline spaces, are linearly independent and reduce the support of the basis functions. The basis functions of THB are made by truncating tensor product B-splines with tensor product B-splines from finer refinement levels. That is, they are made by eliminating the contribution corresponding to the subset of the finer B-splines included in the hierarchical basis from the representation of such coarser B-splines. The truncated B-splines can be described as a sum of scaled tensor product B-splines from finer levels.

2.1.3 Locally Refined B-Splines

An LRB surface is a piecewise polynomial or piecewise rational polynomial surface defined on an LR-mesh. An LR-mesh is a locally refined mesh made by applying a sequence of refinements starting from a tensor-product mesh. LRB are algorithmically defined throughout the refinement process of the mesh. An LRB surface is defined as

$$F(u, v) = \sum_{i=1}^{L} P_i s_i R_{i,p_1,p_2}(u, v) = \sum_{i=1}^{L} P_i N_{i,p_1,p_2}(u, v), \tag{1}$$

where $P_i, i = 1, \dots, L$ are the surface coefficients, and s_i are scaling factors introduced to ensure partition of unity of the resulting collection of tensor product B-

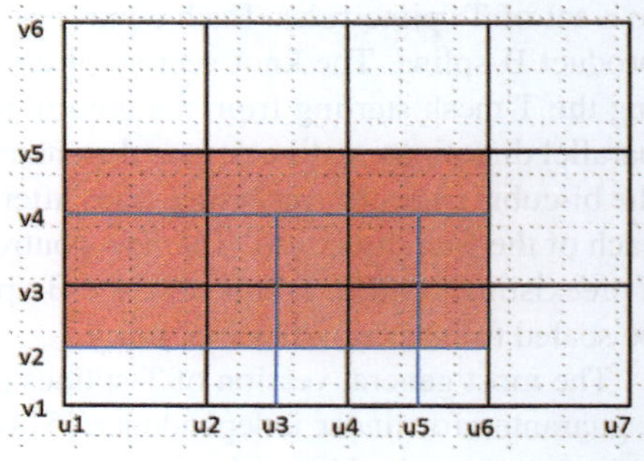

Fig. 1 LR-mesh. The support of one tensor product B-spline visualized as a red pattern. Initial knotlines are shown as black lines, the inserted knotline segments are blue

splines. The tensor product B-splines R_{i,p_1,p_2} are of bi-degree (p_1, p_2) defined on knot vectors of lengths $p_1 + 2$ and $p_2 + 2$ on the parametric domain in the u and v directions respectively. N_{i,p_1,p_2} are the tensor product B-splines multiplied with their scaling factor.

The initial tensor-product mesh corresponding to the LR-mesh shown in Fig. 1 is given by the knot vectors $[u_1, u_1, u_1, u_2, u_4, u_6, u_7, u_7, u_7]$ and $[v_1, v_1, v_1, v_3, v_5, v_6, v_6, v_6]$. It corresponds to a polynomial spline surface of degree two in both parameter directions with multiple knots in the end parameters. The LR-mesh is constructed by first inserting knots at v_2 and v_4 covering a part of the surface domain, and next inserting knots at u_3 and u_5. The B-spline with support shown in red has the knot mesh $[u_1, u_2, u_4, u_6] \times [v_2, v_3, v_4, v_5]$. Some lines of the LR-mesh intersecting the support of such B-spline do not correspond to knotlines of its knot mesh as they do not traverse the support completely.

The procedure for refining an LRB surface is the following:

1. Add a new line segment that triggers the refinement of at least one existing tensor-product B-spline. It can be an extension of an existing line segment, can be independent of existing line segments, or increase the multiplicity of an existing the line segment. Thus, a line segment going from (u_5, v_4) to (u_5, v_5) is a legal choice.
2. Subdivide all tensor product B-splines with support that is completely traversed by the new line/extended line.
3. The tensor product B-splines are required to have minimal support. This means that all line segments traversing the support of a tensor product B-spline are required to be knotline segments of the B-spline taking the multiplicity of line segments into account. After inserting a new line segment and performing the subdivision in Step 2., there might still be tensor product B-splines that do not have minimal support with respect to the LR-mesh. Consequently all such B-splines must be refined. This process is continued until all tensor product B-splines have minimal support.

LRB surface refinement is described in detail in [14]. If more than one new knotline segment is defined simultaneously, the refinement process is applied to one segment at the time.

The LRB construction results in a sequence of nested spline spaces. LRB are non-negative and have compact support. The scaling factors s_i are computed during the refinement process to ensure partition of unity. LRB are not guaranteed to be linearly independent, but a dependency relation can be detected and resolved by dedicated knot insertions. Linear dependency can only occur in a situation with overloaded LR B-splines. An element is overloaded if it belongs to the support of more LRBs than necessary to span the polynomial space over the element. An LR B-spline is overloaded if all elements in its support are overloaded. Overloading can be detected by the peeling algorithm [7]. Patrizi and Dokken address configurations that can lead to a linear dependency relation in [18]. In a situation with local linear independence over all polynomial elements all weights, s_i $i = 1, \ldots, L$, will be equal to one [2].

2.1.4 Why the Focus on LRB?

THB refinement requires that a refinement region is defined that is wide enough to contain the support of at least one B-spline on the refined level. LRB is more flexible as a single mesh line segment can be inserted thus allowing more targeted refinements. Comparing T-splines and LRB refinement is not so simple, as T-splines refine in the coefficient T-mesh, while LRB refines in the LR-mesh. The LR-mesh corresponds to the extended T-mesh of T-spline. T-spline refinement has to be performed between adjacent control points connected by an axis parallel line in the T-mesh, thus relating to two B-splines. We focus on LRB refinement as refinement is allowed as long as the support of one B-spline is split. It is thus less restrictive.

2.2 *Scattered Data Approximation*

The aim is to approximate a scattered data point cloud by an LRB surface. We focus on data sets with projectable points that can be parameterized by their xy-coordinates leaving the z-coordinate to be approximated by a height function. Also non-projectable points being parameterized by some appropriate method can be handled by the general algorithm used in this paper.

Algorithm 1 gives an overview of the iterative approximation algorithm. The starting point is a tensor-product B-spline surface defining an initial spline space.

The focus of this article is to investigate the term "Refine the surface where needed" of Algorithm 1. The elements of the surface mesh will, during the computation, keep track of the data points situated within their domain as well as some accuracy statistics. This includes maximum and average distances between the surface and the points in the element, and the number of points with a distance to the surface that is larger than the specified tolerance. This information provides a basis for selecting where and how new degrees of freedom shall be added.

Two surface approximation methods are applied during the iterative algorithm: Least squares approximation with a smoothing term and multi-level B-spline approx-

Data: Point cloud, maximum number of iterations, tolerance
Result: Approximating surface, information on approximation accuracy
Generate initial surface;
Compute accuracy;
while *there exists points with larger distance than the given tolerance and the maximum number of iterations is not reached* **do**
    Refine the surface where needed;
    Perform approximation in the current spline space;
    Compute accuracy;
end

Algorithm 1: Iterative algorithm for LRB surface generation

imation (MBA). Let $\mathbf{x} = (x_k, y_k, z_k)$, $k = 1, \ldots, K$ be the projectable point cloud we want to approximate. Least squares approximation with a smoothing term is a global method where a minimization functional

$$\min_{\mathbf{P}}[\alpha_1 J(F(x, y)) + \alpha_2 \sum_{k=1}^{K}(F(x_k, y_k) - z_k)^2]$$

is differentiated leading to a sparse, linear equation system. Linear independence of the B-splines is a prerequisite for a non-singular equation system. $F(x, y)$, as defined in Eq. 1, denotes the LRB height function we want to obtain and $\mathbf{P} = P_1, \ldots P_L$ are the coefficients of this function. The solution to this functional results in the best possible approximation with the given degrees of freedom in a least squares sense. The actual smoothing term $J(F(x, y))$ is described in [21] and other possible smoothing terms can be found in [17]. Our focus is on the approximation. As the input points may or may not represent a smooth surface, the weight α_1 on smoothing must be kept low. Still the term is important to handle areas in the surface domain without points.

The multi-level B-spline approximation algorithm (MBA) was introduced by Lee et. al. in [16] for scattered data interpolation, and is described in detail [23]. It is an iterative, local method, where the coefficients are calculated individually. The result is a hierarchical structure of tensor-product B-spline surfaces.

Assume that N_i is a B-spline of the residual surface for which we want to calculate the coefficient Q_i,

$$Q_i = \frac{\sum_{c=1}^{C} N_i(x_c, y_c)^2 \phi_{i,c}}{\sum_{c=1}^{C} N_i(x_c, y_c)^2}. \tag{2}$$

Here the coefficients $\phi_{i,c}$ are calculated by a local surface approximation of the residuals, $\mathbf{x}_c = (x_c, y_c, z_c)$, $c = 1, \ldots, C$, that are inside the support of N_i. All B-splines that have a support that overlaps any of these residuals take part in the approximation. The local approximation is formulated as an under-determined equation system

$$z_c = \sum_{h=1}^{H_c} \phi_{h,c} N_{h,c}(x_c, y_c), \quad c = 1, \ldots, C.$$

The equation for residual c only involves the B-splines that are none zero at (x_c, y_c). Consequently which B-splines, $N_{h,c}$, $h = 1, \dots, H_c$, $c = 1, \dots, C$, are involved in the equation is dependent on the residual addressed in the equation. The solution selected for the under-determined system is

$$\phi_{h,c} = \frac{N_{h,c}(x_c, y_c)z_c}{\sum_{k=1}^{H_c} N_{k,c}(x_c, y_c)^2}, \quad h = 1, \dots, H_c, \quad c = 1, \dots C.$$

From the above coefficients we select $\phi_{i,c}$, $c = 1, \dots, C$, to be used in (2). This equation is obtained by minimizing with respect to Q_i the error

$$e(Q_i) = \sum_{c=1}^{C}(Q_i N_i(x_c, y_c) - \phi_{i,c} N_i(x_c, y_c))^2.$$

In the LRB setting, the residual surface is incorporated in the expression for the approximating surface avoiding a hierarchical representation. The MBA method does not require linearly independent LRB to find a solution.

Both approximation methods are in some sense minimizing the average distance between the point cloud and the surface. Thus, neither the maximum distance nor the number of points outside the tolerance can be expected to decrease monotonically. The average distance will in general be steadily reduced, but temporary stagnation may occur in particular in the context of outliers or if the elements have been refined to the extent that we model noise. In general we expect the input point clouds to have high data sizes and a varying degree of smoothness over the domain of the point cloud, which makes the property of local refinement essential.

3 Refinement Strategies and Success Criteria

LRB surfaces are appropriate for approximation of large scattered data sets due to the ability of increasing the data size in the areas where more degrees of freedom are required while keeping the data size low in other areas. However, it is not obvious how to decide where new mesh line segments should be inserted. There is a wide range of possibilities in selecting these lines, and we will investigate the effect of different choices.

Previous studies on refinement strategies for LRB have focused on the use of LRB in isogeometric analysis or refinement strategies that ensure local or global linear independence. Local linear independence ensures that a minimum amount of the B-spline supports overlap each element of the mesh. Global linear independence implies that a linear equation system originating from a least squares approximation or finite element analysis is non-singular.

Johannessen et. al. give an introduction to LRB in the context of isogeometric analysis in [14] where a detailed description of the refinement basics is provided and the effect on a number of test cases is investigated for the three following refinement strategies:

Full span: Given an element selected for refinement, all B-splines with a support that overlaps this element are refined by adding a line traversing the element.

Minimum span: The shortest possible line overlapping a chosen element that splits the support of at least one B-spline, is selected. Several candidate B-splines can exist and even if additional selection criteria are added, the candidate may not be unique. The selected line is not necessarily symmetric with respect to the element.

Structured mesh: Choose a B-spline and refine all knot intervals in this B-spline.

The sensitivity towards the different choices of refinement strategies were in [14] found to be moderate. However, the authors also investigated the effect of different knot multiplicity and found that meshes with low knot multiplicity tended to give less error in the computation compared to meshes with higher multiplicity for the same number of degrees of freedom. In [12] the structured mesh approach is analyzed theoretically and numerically in a set of test cases.

Bressan and Jüttler [4] look at refinement of LRB surfaces from the perspective of local linear independence and present a mesh construction where this property is proved. Patrizi et. al. [19] propose a practical refinement strategy where local linear independence is ensured. The strategy consists of a modified structured mesh refinement where some of the splits are prolonged so the refined mesh satisfies the so called non-nested support property, see [19] for a definition.

Other authors have focused on refinement of hierarchical B-spline surfaces or T-spline surfaces. Bracco et al. define two classes of admissible meshes for hierarchical B-splines and compare them for use in isogeometric analysis in [1]. A comparison between two refinement strategies applied to hierarchical B-splines and T-splines in the context of IgA is investigated in [13].

The applications of scattered data approximation and isogeometric analysis have some fundamental differences. In the context of IgA, the refined meshes normally belong to an intermediate stage in the computations. Thus, the possibly large locally refined spline surfaces are not kept. Moreover, the need for refinement in an adaptive isogeometric analysis computation is typically concentrated in localized areas. One of the main motivations for approximating scattered data with a locally refined surface is the need for a compact representation of data with a locally non-smooth behaviour. Areas where extra degrees of freedom are required for an accurate approximation, may be scattered around in the entire domain of the data set.

We restrict ourselves to polynomial LRB surfaces of bi-degree one, two and three and place all new knotline segments in the middle of existing knot intervals. With this configuration, no linear dependence relation has been encountered.

If there exist data points within an element with a distance to the surface larger than the tolerance, there is either a need for more degrees of freedom corresponding to the element, or the required accuracy cannot be met by a smooth surface due to outliers or a lack of smoothness in the data points. Any B-spline whose support overlaps this element will give new degrees of freedom to the element, if refined. This implies that there is a choice of how new degrees of freedom are defined. It is also clear that all choices are not equally good.

We define some criteria for a good refinement strategy:

1. Best possible accuracy with the minimum degrees of freedom.
2. If supported by the data set, it should be possible to adapt the surface to the point cloud within the prescribed tolerance.
3. Avoid a premature stop where the requested accuracy is not reached and the maximum number of iteration steps is not performed.
4. A steady improvement in accuracy when adding new degrees of freedom.
5. Keep the execution time low.
6. Keep the memory consumption low.
7. Affected elements should preferably be split in the middle if a new mesh line is positioned in the middle of some element.
8. Linear independence or local linear independence of B-splines.

We will mainly focus on the first five criteria. Critera 5 and 6 are to some extent linked and also dependent on the previous ones. A lean surface will lead to lower memory consumption and the part of the execution time spent in surface refinement is connected to the number of coefficients. The execution time also depends on the number of steps applied in the iterative algorithm and whether or not the knot insertion at late iteration steps is focused in a few areas or spread out in the entire surface domain.

The importance of linear independence is related to the use of the resulting surface. The approximation algorithm outlined in Algorithm 1 does not rely on linear independence. We will, thus, not focus on the last criterion.

We can now formulate two empirical rules for a good refinement strategy. They are taken into account when the strategies to be tested are defined in the next section:

Rule 1: A gradual introduction of new degrees of freedom gives better approximation efficiency.

Rule 2: An improved accuracy can be blocked by a failure to identify one or more B-splines that need to be refined. A disproportionate high number of B-spline supports overlapping an element is not desirable.

An important term in this discussion is **approximation efficiency**. It is defined as the number of resolved points divided by the number of surface coefficients for a particular refinement strategy and a particular iteration level. This figure will, together with other criteria, be used to evaluate the success of one refinement strategy compared to others.

4 Applied Refinement Strategies

4.1 Main Categories

The selected data sets are approximated using a variety of refinement strategies. They belong to four main categories: **Full span**, **Minimum span** and **Structured mesh** as described in Sect. 3 and an additional category named **Restricted mesh**. In the full span and minimum span strategies an element is selected for refinement while in the structured and restricted mesh strategies the refinement process is started from a selected B-spline.

In the full span strategy all B-splines with a support that overlaps the selected element are refined. The new mesh line splits the element in half in the direction of refinement and it is sized to split the support of all associated B-splines.

In the minimum span strategy one B-spline with a support overlapping the element, is selected. We will test three different minimum span strategies. They differ in how the B-spline to be refined is chosen. The selection criteria are: the candidate B-spline with **largest** support; the B-spline that has the highest number of **unresolved points** compared to the total number of points in the support; and, a **combination** of the two where the criteria are given equal weights. Several B-splines may fit the selected criterion equally well. In case of doubt, the most centered B-spline with respect to the initial element is selected. If there is still no unique B-spline satisfying the criterion, one candidate is randomly chosen.

In the structured mesh strategy, the selected B-spline is refined in the middle of all knot intervals. This is similar to the refinement strategy for hierarchical B-splines. In the restricted mesh strategy, refinement is performed in knot intervals where the corresponding elements contain significant points outside the tolerance with respect to number and distance. Let the elements of one knot interval in one parameter direction be denoted an element strip. The distances between the points and the surface that exceed the tolerance are accumulated for all such strips and scaled with respect to the element sizes. Furthermore the element strip measures are scaled with the position of the strip in the B-spline thereby prioritizing the middlemost knot intervals. Knot intervals with strips that have a measure larger than the average strip or larger than half the maximum strip measure are refined. In addition, knot intervals that exceed the average knot interval in the B-spline support by a factor of three are refined. This implies that refinement can be performed in one or more knot intervals in one or two parameter directions. The minimum span strategies and the restricted mesh strategy are similar in the respect that they are both restrictive, but differ in how the refinement selection is made. The effect of the strategies differ as can be seen in Sect. 5.

4.2 Restrictions to the Introduction of New Knots

We want to test the empirical rule that a gradual introduction of new degrees of freedom will lead to a lean final surface with good accuracy. In strategies with refinement in **alternating parameter directions** the surfaces are refined in the first parameter direction at odd iterations levels and in the second parameter direction at even levels. Thus, the number of potential new knot line segments is limited compared to the case for strategies that perform refinement in **both parameter directions** at each level.

A threshold can also be applied to restrict the number of new mesh line segments at each step, possibly in addition to alternating parameter directions and/or restrictions inherent in the strategy itself. Thresholds can imply that only the elements or B-splines with the most significant approximation errors, either with respect to distance or number of unresolved points, are selected for refinement. The threshold factor is set globally for each iteration level. In the restricted mesh strategy, thresholds can be used to reduce the number of knot intervals that are refined in the selected B-spline. A refinement strategy can be combined with zero, one or two different types of thresholds as described in Sect. 4.4.

4.3 Modifications to Particularly Restricted Strategies

The restricted mesh strategy may fail to capture elements with significant points outside the tolerance belt. This relates to situations where the corresponding element strips of the B-spline support otherwise have few unresolved points, and refinement in the associated knot interval is not performed. The situation occurs typically at the border of the point set. To ensure refinement in such elements, an element extension to the strategy is applied. Elements with a significant approximation error that are not already split, are identified at each iteration level and trigger an additional refinement related to the full span strategy. However, refinement may be performed in one parameter direction only, even if the main strategy imposes refinement in both parameter directions.

In addition to the restricted mesh strategy, the various minimum span strategies can lead to few new lines being inserted at each iteration level, in particular if the strategy is combined with a threshold. We investigate the effect of combining these strategies with a full span strategy as this is less restrictive in the introduction of new mesh line segments. Line segments added by the full span strategy will most often be longer than line segments added by the minimum span strategy. The two strategies, which are combined, are applied with refinement in the same number of parameter directions. The switch between strategies is performed when the fraction between newly resolved points in the last iteration and the number of new coefficients drops below 0.1.

4.4 Components for Refinement Strategies

Each refinement strategy and each combination of strategy and threshold is given a unique label. Each label is a compositions of letters. The following explains the meaning of such letters:

Main category: Full span (F), Minimum span (M), Structured mesh (S) or Restricted mesh (R).
Element extension to restricted mesh strategy Select significant elements left out in the B-spline refinement (L).
Minimum span selection criterion: B-spline with largest support overlapping the element (l), B-spline with a support that overlaps the element with most unresolved points (u) or a combination (c).
Parameter direction: Refine in both (B) or alternating (A) parameter directions at each iteration level.
Threshold: With respect to distance (td), the number of unresolved points in an element (tn) or the number of unresolved points in a strip of elements corresponding to a B-spline knot interval (tk).

Let *maxdist* denote the maximum distance between the surface and the point cloud and let *avdist* be the average distance in out-of-tolerance points globally. Let further *nmb* be the number of points situated in an element, *nmb_out* the number of unresolved points in the element and *avdist2* the average distance in these unresolved points. We can now define the threshold factors as:

td = *(tolerance + maxdist + avdist)/3*. An element or a B-spline is selected for refinement if the maximum distance between the points in the support of this B-spline or element and the surface exceeds the threshold.
tn = $factor \times min(wgt) + (1\text{-}factor) \times average(wgt)$ where $wgt = nmb\_out + scale \times avdist2$ in each element with unresolved points. Early in the computation *nmb_out* is large and this term will dominate the formula for *wgt*. Later the distance in the points will dominate. The *scale* is introduced to make the expression independent of the distance unit. It is set to one in the examples in this article. The weight scores are used to rank the elements with respect to importance. The *factor* depends on the range between the minimum and maximum score. If this range is small then the importance of the elements cannot be distinguished and all elements lead to refinement. If the range is large compared to the size of this score, only the highest ranked elements trigger refinement.
tk = $max(average(nmb\_out), 0.01 \times average(nmb))$ for all elements. The tk threshold is only applied to the restricted mesh strategy. A knot interval in a selected B-spline is refined only if the number of out-of-tolerance points in the corresponding element strip exceeds this threshold factor.

The tk threshold is very strong as the number of points outside the tolerance in an element strip is compared to the number of points in the entire B-spline support. In general, the threshold types and associated factors are set experimentally. The aim is to get an impression of how different approaches influence the approximation process. All threshold factors are reduced with successive iterations.

4.5 *Composition of Refinement Strategies for the Tests*

The refinement strategies are applied to a number of test cases. Each strategy is tested with and without threshold and with different choices in the number of parameter directions to refine at each iteration step. Table 1 shows how the strategy labels are composed. Here the type specifies the strategy category as full span, minimum span, structured mesh or reduced mesh. The sub type applies to the minimum span strategy only. If several strategies are applied during the iterations, a second category (type2) is specified. The two strategies can be applied simultaneously as in the case for the element extended restricted mesh strategy, or one after the other. Two strategies or thresholds that are applied to the same iteration level are combined with + in the label, and with/if they are used at different levels. An extensive set of combinations of strategies and threshold types are tested, but there are still possible combinations that are not applied. We believe, however, that the results give a broad basis for drawing conclusions on the suitability of the various strategies.

Table 1 Composition of strategy labels

Type	Sub type	Type2	Dir	Threshold	E.g.	Comment
F			A or B	None, td or tn	FA tn	Full span
M	l or u or c		A or B	None, td or tn	MlB	Minimum span
M	l or c	F	A or B	tn	Mc/FA tn	Combine minimum and full span
S			A or B	None or td	SB td	Structured mesh
R			A or B	None, td, tk or td and tk	RA td+k	Reduced mesh
R		F	A or B	tk and tn	R/FB tk/n	Combine reduced mesh and full span
R		L	A or B	None, td, tk or td and tk	R+LA tk	Element extended reduced mesh

5 Refinement Study

Five geospatial data sets are used in the study. All data sets have points that are projectable on the xy-plane, but are different in size and properties. The number of points varies from 71 888 to 131 160 220. The regularity distribution varies from completely regular to scanlines where the distances between scans are very large compared to the distances between points in the scans. The smoothness also varies from data set to data set. Two data sets contain known outlier points.

The point clouds are approximated using various combinations of refinement strategies and thresholds. After defining a tensor-product spline surface using the least squares method the approximation is continued with the least squares method applied to an LRB surface before turning to the MBA. The swap is performed after five iteration steps or earlier. The least squares method involves solving a linear equation system. This is done by an iterative method, which may struggle with convergence when the point distribution is very uneven. In such situation, the MBA method is a better choice. Moreover, an early switch is made if the approximation accuracy is not improved.

All data sets, but one, are tested with a tolerance of 0.5 m. The number of iterations varies according to properties of the data set, and to some extent with regard to whether alternating parameter directions are applied. Varying the tolerance and the number of iterations for each data set could bring additional information to the range of refinement strategies, but fall outside the scope of this article. The study is run using bi-degrees (1,1), (2,2) and (3,3).

The information related to each data set includes measurements of execution time. The recorded time includes all aspects of the approximation, but excludes reading and writing to file. The computations are performed on a stationary desktop with 64 GB of DDR4-2666 RAM. It has a i9-9900K CPU with 8 cores and 16 threads, but a single core implementation is used in the experiments. Approximation results are collected at the final stage of the computations and at an intermediate stage. This stage will be defined according to the properties of each data set.

The presentation of each data set follows the same pattern. First the data set is introduced, then the range in execution time and number of surface coefficients is presented, and finally some selected details. A table is also presented, showing results from a number of strategies: strategies that have low execution time; strategies that produce few coefficients; strategies that have an overall good behaviour; strategies that have the highest number of unresolved points at the end of the computations; and some additional strategies to complement the picture or unveil differences between strategies.

For each strategy the following information is given: iteration level at the two stages; the number of points with a distance larger than the tolerance; the corresponding number of coefficients; the maximum distance; and the execution time. Distances are given in meters. Refinement strategies having the best overall results at the intermediate or final stage are highlighted using bold font and also single best results for some criteria is shown in bold font. The worst result for a given criterion is highlighted using italic font. Graphs providing more continuous accuracy information will complement the tabular entries.

Similar tabulated results on the accuracy of all refinement strategies and combinations with threshold types will be made available in a separate report along with more analysis of the results.

5.1 Banc du Four

5.1.1 Data Set

Figure 2a shows a sub sea data set representing sand dunes, (b) zooms in on a detail. The data is obtained from the Banc de Four outside the coast of Brittany. The data set consists of 5 054 827 well distributed points. The elevation range is [−92.76, −54.84] and the standard deviation is 8.08. Sand dunes have a relatively smooth shape and the point cloud does not contain outliers. The approximating surface, Fig. 2c, has an accuracy of 0.2 m and is approximated using refinement strategy RB tk and biquadratic splines. The approximation is in this case performed using least squares approximation for two iteration steps before turning to MBA.

5.1.2 Experience Setup and Performance Ranges

We use a tolerance of 0.2 m for this data set. The approximation algorithm is allowed to run for 40 iterations. The intermediate stage is defined as: *At least 99.9% of the points, are closer to the surface than the tolerance and the maximum distance between the point cloud and the surface is less than 0.5 m. This implies that at most 5 055 points are further away from the surface than the tolerance.*

(a) (b) (c)

Fig. 2 **a** Outline of point cloud. **b** A detail. **c** Corresponding LRB surface, the surface is trimmed to adapt to the point cloud domain

Table 2 Computation time and number of coefficients. Performance ranges and associated strategies for polynomial degrees one to three

Deg	Min t.	Strategy	Max t.	Strategy	Min cf	Strategy	Max cf	Strategy
1	0m21s	SB	1m32s	MuA td	42 503	R+LA tk	91 050	SB
2	0m33s	FB, SB	2m58s	MuA td	39 366	R+LA tk	111 992	SB
3	1m6s	FB	5m46s	MuB td	45 826	RB tk	162 972	SB

The data set is well adapted for approximation with LRB surfaces and most combinations of refinement strategies and threshold types converge within the prescribed number of iterations. Exceptions exist, some cases are referenced in Table 3. Moreover, the execution time and final number of coefficients differ. Table 2 presents the minimum and maximum execution time at the end of the computations and the final number of coefficients together with the corresponding refinement strategies. The structured mesh strategy appears both as the strategy with the lowest execution time and the highest number of coefficients. The minimum number of coefficients are found for restricted mesh strategies while a minimum span strategy with threshold, MuB td, has the highest computational time. This strategy also fails to converge completely. In the bi-linear, bi-quadratic and bi-cubic cases respectively 3, 6 and 7 points remain with a larger distance to the surface than 0.5 m.

The execution time and number of coefficients increase with increasing polynomial degree. Some strategies converge well for the bi-quadratic and bi-cubic cases while they fail in the bi-linear case. One example is FA tn, see Table 3.

5.1.3 Selected Refinement Strategies

Table 3 shows the result for a set of selected strategies including the best performing ones. Some strategy combinations provide both a low execution time and a lean final surface. This is the case for FA, in particular when combined with threshold tn for the bi-quadratic and bi-cubic cases. FB has lower computation times, but results in surfaces with more coefficients. Refinement in both parameter directions at every iteration level tends to result in fewer iterations, and to some extent in a lower execution time. An exaggerated introduction of new mesh lines sometimes leads to fast convergence, but for many cases also results in a high number of coefficients. This is the case for the structured mesh strategy SB. Adding a threshold reduces only to a limited extent the number of surface coefficients, see SB td in the bi-linear and bi-cubic cases.

The restricted span strategy RB tk yields good results both at the intermediate and final stage even though it has higher computation time than the fastest strategies and a high number of iterations between the intermediate and final stage. RA tk also performs well at the intermediate stage, but does not converge within the prescribed number of iterations. Restricted mesh strategies tend to perform better at the inter-

Table 3 Accuracy results of some selected strategies at the intermediate and final stage

Refine	Intermediate stage					Final results				
Strategy	It	Pt out	No coef	Max	Time	It	Pt out	No coef	Max	Time
Bi-linear										
FB	7	104	62 291	0.258	**0m18s**	10	0	62 532	0.2	0m22s
RA tk	14	4 385	**36 540**	0.324	0m32s	40	4	42 338	0.255	1m5s
R+LA tk	14	2 640	38 681	0.324	0m31s	33	0	**42 503**	0.2	0m55s
SB	7	26	*90 733*	0.25	0m20s	8	0	*91 050*	0.2	**0m21s**
FA tn	**13**	**3 866**	**40 472**	**0.465**	**0m27s**	40	26	47 047	0.24	0m57s
McB	7	651	47 442	0.353	**0m18s**	**15**	**0**	**49 212**	**0.2**	**0m27s**
RA	13	449	46 733	0.443	0m29s	**23**	**0**	**47 430**	**0.2**	**0m40s**
MuA tn	18	3 922	43 218	0.494	0m40s	40	*73*	56 661	0.332	1m18s
MuA	13	2 649	41 503	0.456	0m27s	40	22	46 411	0.305	1m
FA	13	434	47 970	0.443	0m26s	23	0	48 588	0.2	0m37s
SB td	8	2 104	75 076	0.417	0m24s	11	0	83 505	0.2	0m29s
SA	13	240	68 128	0.44	0m27s	17	0	68 891	0.2	0m32s
RB tk	7	2 539	45 394	0.324	0m21s	28	0	48 636	0.2	0m45s
Bi-quadratic										
FB	7	21	65 475	0.225	**0m31s**	8	0	65 523	0.2	**0m33s**
RA tk	15	4 435	**30 855**	0.449	1m10s	40	178	39 799	0.288	2m18s
R+LA tk	13	2 444	35 935	0.426	1m	31	0	**39 336**	0.2	1m39s
SB	7	0	*111 992*	0.2	0m33s	7	0	*111 992*	0.2	**0m33s**
RB tk	**7**	**2 501**	**39 029**	**0.486**	**0m37s**	32	0	44 414	0.2	1m34s
McA	13	1 263	40 328	0.367	0m54s	**20**	**0**	**41 952**	**0.2**	**1m9s**
RA td+k	25	2 481	31 889	0.4	*1m46s*	40	*289*	37 994	*0.309s*	2m25s
FA	13	258	50 220	0.366	0m48s	19	0	50 629	0.2	1m
FA td	16	1 470	40 180	0.306	1m10s	21	0	41 751	0.2	1m20s
FA tn	14	1 335	41 305	0.411	0m53s	24	0	43 067	0.2	1m12s
MuB	7	659	47 732	0.323	0m36s	30	3	50 860	0.216	1m27s
MuA	13	1 725	39 933	0.4	0m55s	40	7	45 762	0.224	1m57s
R+LB tk	7	1 326	41 928	0.441	0m37s	14	0	44 232	0.2	0m51s
Bi-cubic										
FB	7	23	81 895	0.281	**0m59s**	9	0	82 064	0.199	**1m6s**
RA tk	24	2 544	**33 015**	0.379	3m47s	40	296	39 339	0.275	5m20s
RB tk	**8**	**3 674**	**38 581**	**0.463**	**1m25s**	40	0	**45 826**	0.2	3m58s
R+LB tk	7	1 678	51 047	0.499	1m16s	**13**	**0**	**55 706**	**0.2**	**1m41s**
R+LA td	13	1 648	48 549	0.497	1m42s	**18**	**0**	**51 488**	**0.2**	**2m3s**
RA td+k	29	3 058	34 625	0.409	*4m20s*	40	*604*	40 256	0.281	5m41s
SB	7	17	*161 779*	0.248	1m4s	9	0	*162 972*	0.199	1m13s
SB td	8	223	137 571	0.274	1m19s	10	0	142 166	0.2	1m28s
FA	13	406	60 346	0.443	1m32s	17	0	61 084	0.2	1m46s
FA tn	13	1 327	51 977	0.496	1m36s	23	0	54 377	0.2	2m10s
R+LA tk	14	1 098	46 508	0.381	2m9s	23	0	48 120	0.2	2m45s
MuA	13	2 270	49 786	0.455	2m4s	40	32	59 356	0.253	4m12s

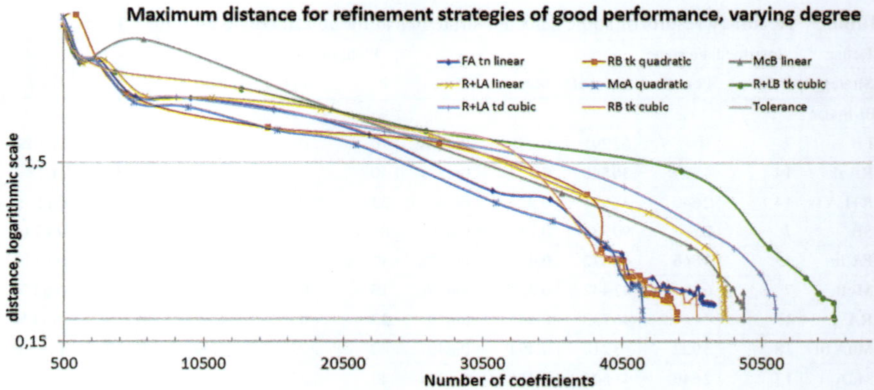

Fig. 3 Maximum distance for selected well performing refinement strategies, all degrees. The tolerance is shown as the horizontal line just above the value 0.15

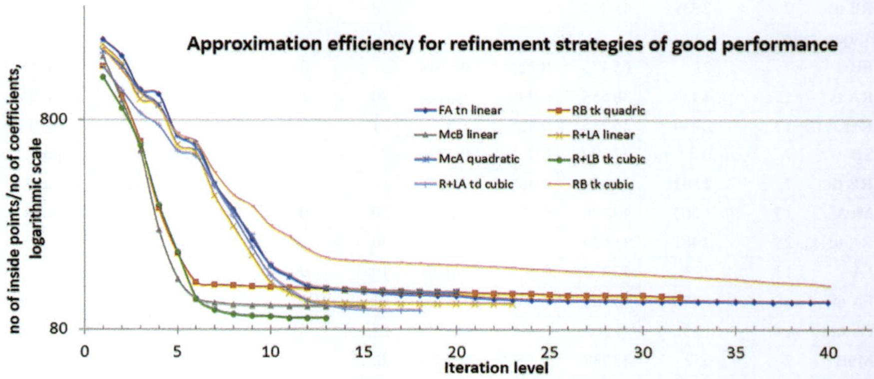

Fig. 4 Approximation efficiency for selected well performing refinement strategies, all degrees. A high efficiency is preferred, but must be coupled with the actual convergence of the strategy

mediate than the final stage, in particular when a threshold is involved. However, extending the method with refinement based on elements, e.g., R+LA tk, speeds up the convergence in the latter part of the computation.

Minimal span strategies do not always converge. In particular MuA struggles also without a threshold. Note also that MuB quadratic has unresolved points at the end of the computations without reaching the maximum number of iterations. One or more critical mesh line segments are not inserted due to restrictions in the strategy.

A continuous picture of the performance of the selected best strategies at the intermediate and final stage is given in Figs. 3 and 4. Figure 3 relates the number of coefficients to the number of points within the tolerance for a further reduced set of strategies. RB tk bi-cubic has high approximation efficiency throughout the computations and finishes with relatively few coefficients despite a large increase in

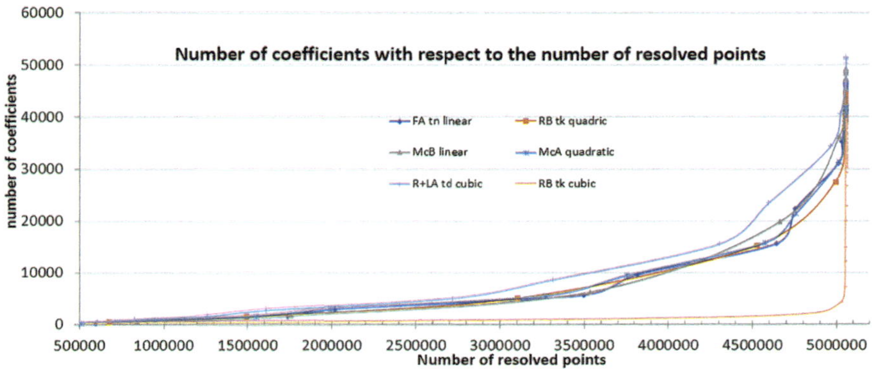

Fig. 5 Number of coefficients with respect to the number of points within the tolerance

number after the stage where almost all points are resolved. However, the execution time is more than the double of FB bi-cubic. RB tk with bi-quadratic splines has a relatively poor performance in the first part of the computation, but reduces the maximum distance quickly after about 40 000 coefficients and has a good approximation efficiency after 12 iterations.

The relative performance of the strategies with respect to approximation efficiency is, with the exception of RB tk bi-cubic, mainly retained throughout the computations. Typically the approximation efficiency is lower for strategies that refine in both parameter directions simultaneously. The increase in the number of coefficients is rapid and the corresponding decrease in the number of unresolved points is not comparatively high. The ranking of the strategies with respect to the maximum distance varies throughout the iterations. For instance the minimum span strategy McB with bi-linear splines increases the maximum distance early in the computation. It continues with a high distance until it drops below the distance of several other strategies in the last part of the computation.

Minimal span strategy McA tn with bi-quadratic splines has the best performance throughout the computation. The maximum distance is constant in the lower part of the group and the approximation efficiency lies in the group with highest efficiency if RB tk bi-cubic is kept out. Furthermore, it finishes well within the allowed number of iterations and with an acceptable computation time, see Table 3.

The two bi-cubic strategies in Fig. 5 can be found at both ends of the spectrum. The strategies are both of the reduced mesh type, but differ with respect to the number of parameter directions to refine at each level and threshold. The latter difference is probably the main source of the difference in behaviour. In general a threshold with respect to the number of out-of-tolerance points will lead to fewer coefficients with regard to the number of points resolved. Figures 6 and 7 compare respectively the evolution of the maximum distance between the surface and the point cloud, and the approximation efficiency for the various full span refinement strategies in the bi-quadratic case.

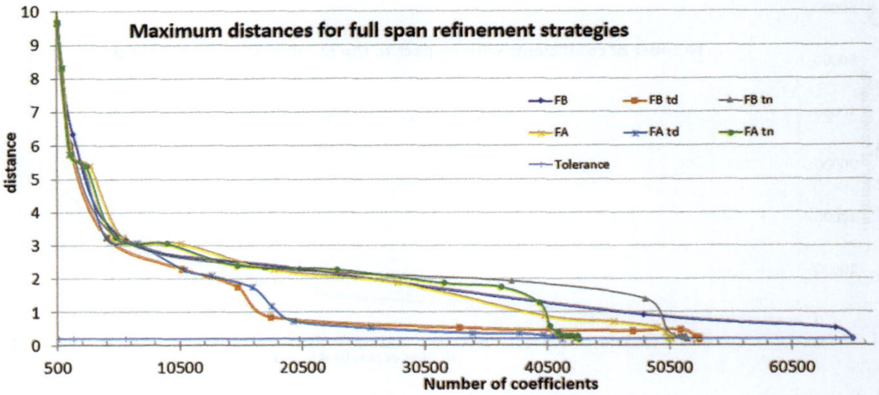

Fig. 6 Maximum distance for full span strategies with and without threshold, bi-quadratic

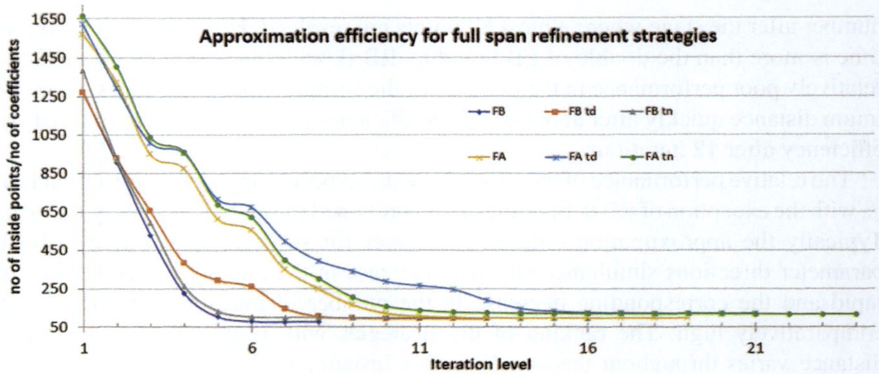

Fig. 7 Approximation efficiency for full span strategies with and without threshold, bi-quadratic

We see in both Figures that the graphs of the refinement strategies fall into two groups, however, the groups are different. The maximum distance (Fig. 6) is reduced more quickly for strategies that are combined with the distance based threshold type td. The evolution for other strategies diverge at a later stage of the computations. Strategies with alternating parameter directions finish with fewer coefficients than strategies that refine in both parameter directions at every iteration step. Figure 7 shows that strategies with alternating parameter directions have higher approximation efficiency than the others. Within each group, the strategy that is combined with threshold type td has the highest efficiency at each iteration level, but needs also a high number of iterations to converge. This threshold may be beneficial if the computation is stopped at an early stage, but not necessarily if a sufficient number of iterations to reach convergence are allowed.

5.2 *Gaustatoppen*

5.2.1 Data Set

Figure 8a shows a completely regular point set containing 490 000 points. The corresponding surface in (b) is bi-linear and created with refinement strategy FA and a tolerance of 0.5 m. Least squares approximation is used for the first five iteration steps, then MBA is used. The points are extracted from a sparse raster approximating measurement data from the mountain area Gaustatoppen in Norway. The elevation range is [216.6, 1877.3] measured in meters and the standard deviation is 370.542. The data set covers an area of 6 507 times 6 507 m^2. The data set is regular, but very sparse. It has already been processed and thinned considerably. Every point carries a lot of information, thus an accurate approximation with a lean surface is not feasible. However, the data set can still be used to distinguish between different refinement strategies, and in particular evaluate the strategies with respect to robustness.

5.2.2 Experience Setup and Performance Ranges

The Gaustatoppen terrain is much more demanding than in the previous example. The area has one high mountain peak and a valley with a river. There are also several other peaks and valleys in the neighbourhood. We specify a tolerance of 0.5 m and a maximum number of iterations of 40. The intermediate stage is defined as the iteration level where *the maximum distance between the point cloud and the surface is less than two meters, 99% of the points are resolved meaning that the number of points where the tolerance is not satisfied does not exceed 4 900, and the average distance in out-of-tolerance points is less than one meter.*

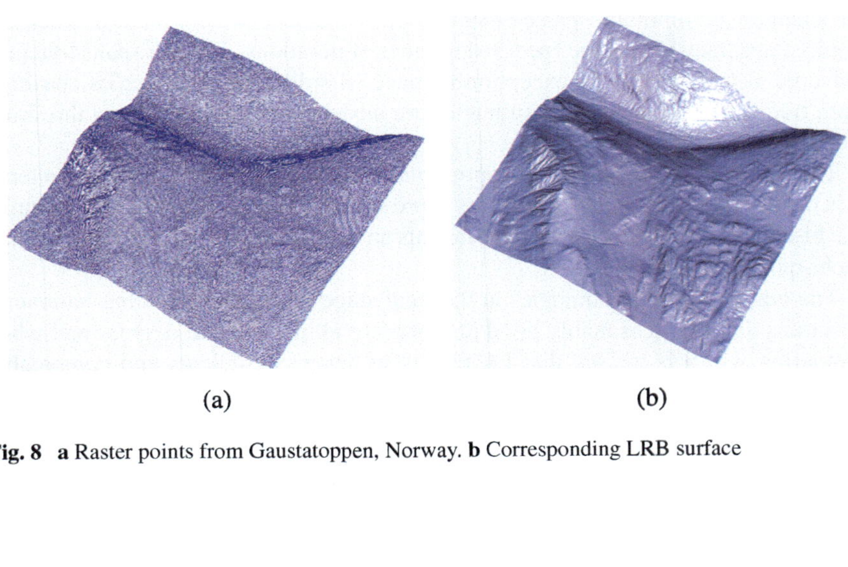

(a) (b)

Fig. 8 **a** Raster points from Gaustatoppen, Norway. **b** Corresponding LRB surface

Table 4 Computation time and number of coefficients. Performance ranges and associated strategies for polynomial degrees one to three

Deg	Min t.	Strategy	Max t.	Strategy	Min cf	Strategy	Max cf	Strategy
1	1m26s	FA	29m40s	MuB td	270 613	R+LA tk	1 227 085	SB
2	0m31s	FA	1h23m28s	MuB td	195 532	RA tk	814 295	SB
3	0m52s	SB	3h13m30s	MuB td	233 482	RA td+k	757 012	MuB td

Table 4 presents the ranges in computational time and number of surface coefficients for the bi-linear, bi-quadratic and bi-cubic cases. The range in execution time is wide for this data set. A high number of surface coefficients and some distributions of new mesh line segments cause the time spent in maintaining the data structure of the surface to increase very rapidly during the iteration. The highest execution times occur for MuB td strategy. The minimum execution times are obtained for the full span strategy with alternating parameter directions and the structured mesh strategy with refinement in both directions. The latter strategy occurs also as the one that results in most coefficients, and opposite to the pattern found in the other test cases bi-linear splines lead to the highest number of coefficients. The surfaces with fewest coefficients are obtained for various restricted mesh strategies with threshold.

5.2.3 Selected Refinement Strategies

Table 5 presents selected results for the Gaustatoppen example. Several refinement strategies fail to resolve all points and also to reach the intermediate stage. The required maximum distance of less than two meters is most demanding.

Bi-linear approximation struggles if a threshold is applied. A restraint selection of elements to refine leads to convergence failure also without additional thresholds. In particular, most minimum span element based refinement strategies fail to reach the required accuracy within the specified number of iterations. The full span strategies converge in most cases, the exception is FA td. B-spline based strategies converge more frequently, but also these strategies are unstable, especially when a threshold is applied.

The bi-quadratic and bi-cubic approximation have similar convergence patterns as in the bi-linear case, but less extreme. More strategies reach the intermediate stage. The bi-cubic case results in more coefficients and shows higher execution times than the bi-quadratic case.

The best performing strategies at the final stage refine in alternating parameter directions. FA gets reasonably good numbers for all polynomial degrees and when comparing with FB we see that FA results in fewer coefficients and comparable execution times. The same relation can be seen for RA tk and RB tk degree two and three.

Table 5 Accuracy results of some selected strategies at the intermediate and final stage

Refine	Intermediate stage					Final results				
Strategy	It	Pt out	No coef	Max	Time	It	Pt out	No coef	Max	Time
Bi-linear										
RA tk	39	1	**265 050**	1.788	1m47s	40	1	265 054	1.786	1m54s
R+LA tk	31	18	270 465	1.667	1m29s	39	0	**270 613**	0.5	1m51s
R+LA	**30**	**15**	**265 776**	**1.73**	**1m13s**	40	1	265 933	1.279	1m40s
FA	38	0	271 450	0.5	1m26s	**38**	**0**	**271 450**	**0.5**	**1m26s**
FA tn	37	2	274 141	1.588	1m38s	39	0	274 150	0.5	1m39s
MuA td						40	*2481*	429 270	*6.293*	14m20s
MuB td						40	499	595 629	4.522	*29m40s*
McA tn						40	564	325 784	3.515	4m34s
SB	13	636	*1 230 874*	1.729	6m27s	31	0	*1 272 085*	0.5	10m18s
FB	20	57	493 999	1.804	1m56s	29	0	494 933	0.5	2m37s
RA td+k						40	146	315 269	5.539	3m9s
eFA td						40	31	310631	3.999	2m40s
Bi-quadratic										
FB tn	6	2 942	214 016	1.452	**0m13s**	18	0	247 540	0.5	0m49s
FA	12	838	210 068	1.159	0m17s	18	0	213 774	0.5	**0m31s**
FA tn	13	3 127	181 930	1.18	0m20s	**22**	**0**	**198 013**	**0.5**	**0m41s**
RA tk	**13**	**4 006**	**171 341**	**1.681**	**0m24s**	**22**	**0**	**195 532**	**0.5**	**0m53s**
R+LA td+k	12	1 672	193 022	1.545	0m21s	**19**	**0**	**199 654**	**0.5**	**0m37s**
MuA td						40	*2 125*	429 427	*5.506*	29m
MuB td						40	950	660 319	4.387	*1h23m28s*
SB	6	575	442 656	1.452	0m25s	40	18	*814 295*	1.577	8m1s
FB	6	669	293 921	1.452	0m17s	18	0	305 672	0.5	0m54s
R+LA tk	12	2 018	190 310	1.349	0m23s	20	0	201 934	0.5	0m45s
RB tk	13	17	234 394	1.979	0m41s	25	0	235 108	0.5	1m19s
McA tn	15	2 101	194 789	1.642	0m43s	40	3	211 543	2.527	2m20s
Mc/FA tn	15	2 101	194 789	1.642	0m43s	29	0	211 336	0.5	1m40s
RA td+k	20	1 729	188 638	1.132	0m48s	31	0	200 427	0.5	1m24s
eFA td	19	3 844	193 220	0.95	0m42s	40	0	220 853	0.5	1m50s
Bi-cubic										
MlB tn	7	1 953	261 332	1.762	**0m21s**	18	0	292 258	0.5	2m5s
RB	7	152	354 459	0.848	0m41s	9	0	355 993	0.499	**0m52s**
RA td+k	22	2 238	204 748	1.081	1m29s	35	0	**233 482**	0.5	2m52s
RB tk	**7**	**4 797**	**193 719**	**1.908**	**0m30s**	13	0	261 432	0.5	1m19s
FA tn	13	2 687	213 500	1.451	0m32s	**21**	**0**	**233 552**	**0.5**	**1m7s**
MuA td						40	*2 145*	466 088	*4.325*	1h3m25s
MuB td						40	1 135	*757 012*	2.716	*3h13m30s*
FB	7	92	381 781	0.777	0m41s	10	0	382 654	0.5	0m57s
FA	13	566	259 657	1.273	0m36s	19	0	262 633	0.5	1m1s
R+LA td+k	13	870	238 837	1.44	0m40s	20	0	243 719	0.5	1m10s
RA tk	16	1 469	207 691	1.053	1m1s	27	0	224 220	0.5	2m1s
SB	7	39	*570 533*	0.641	1m15s	18	0	579 502	0.5	2m51s
eFA td	20	3 251	238 037	0.971	1m28s	30	0	271 902	0.5	2m46s

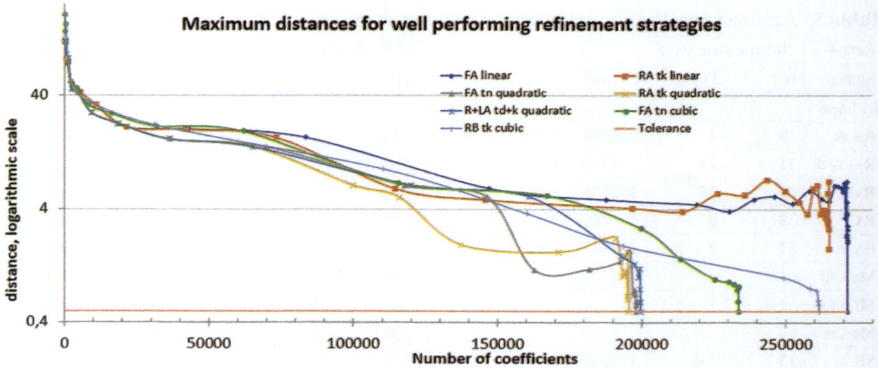

Fig. 9 Maximum distance for selected well performing refinement strategies, all degrees

The restricted mesh strategies in general perform well although there are some exceptions in the bi-linear case, e.g. RA td+k. This strategy does well in the bi-quadratic and bi-cubic case although the execution times are not among the best. Introducing the element extension to these strategies does not make a big difference.

The full span strategy FA has good results, in particular at the final stage and combined with threshold tn. The minimum span strategies Mu, and to some degree Mc, struggle with convergence and do not always reach the intermediate stage. The minimum and full span combination Mc/FA tn has the same results as McA tn at the intermediate stage, but ensures convergence at the final stage for bi-quadratic splines. In Fig. 11 we address combined strategies.

The structured mesh strategies lead to many surfaces coefficients and do not have, in contrast to the case for the Banc du Four data set, particularly low computation times. We will, based on the results for the two first data sets, omit further testing of the structured mesh and the minimum span strategies with sub-type u. The first approach produces surfaces with too much data and the second is too restrictive.

Figures 9 and 10 provide more information on the behaviour of the selected best strategies throughout the entire computations. Figure 11 shows the approximation efficiency for some combined refinement strategies. Figure 9 shows that the maximum distance decreases steadily for a while before it for most strategies starts to oscillate. The oscillations are strongest in the bi-linear case and least profound for bi-cubic splines. It is no correspondence between lack of oscillation and few coefficients at the final stage.

The maximum distance is, over most of the computations, lowest with respect to the number of coefficients for RA tk bi-quadratic. This strategy also has the highest approximation efficiency, see Fig. 10. The best quadratic strategies have the least overall number of coefficients in the final surfaces and the highest approximation efficiency. RB tk bi-cubic has a steep descent in approximation efficiency, but ends up at the same level as several of the other strategies and the resulting number of coefficients (261 462) is quite competitive compared to other bi-cubic strategies.

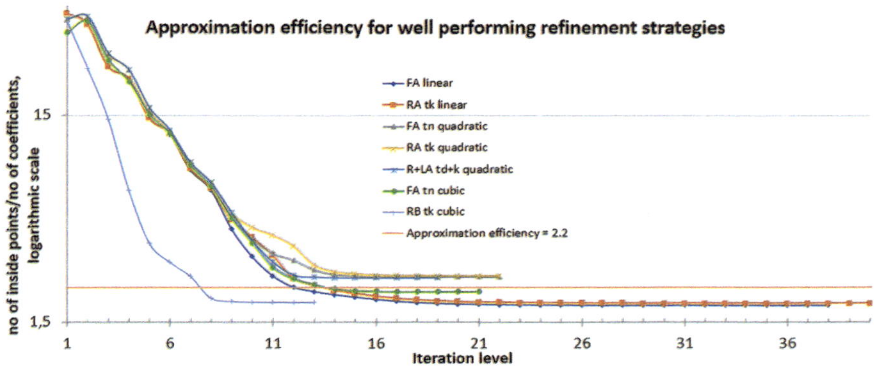

Fig. 10 Approximation efficiency for selected well performing refinement strategies, all degrees

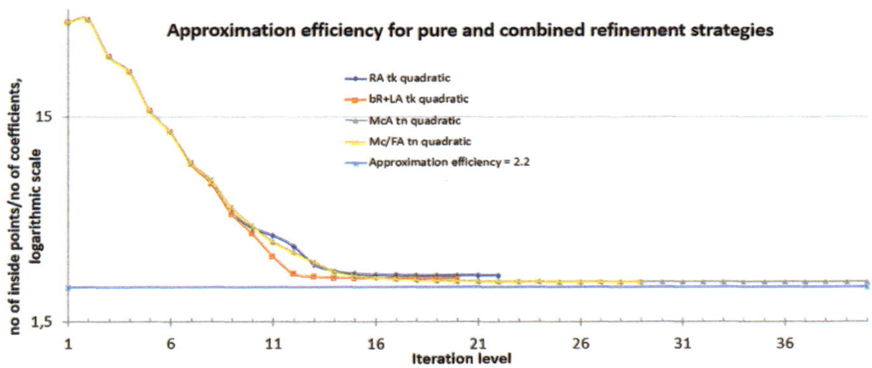

Fig. 11 Approximation efficiency for selected combined refinement strategies with threshold type tk or tn, bi-quadratic splines

Figure 11 compares the approximation efficiency of some minimum span strategies with the corresponding combined strategy for the middle and last part of the computation. We see that RA tk quadratic has the best efficiency. This strategy can also be found in Fig. 10 and can thus support a comparison between the two graphs along with the line marking an approximation efficiency of 2.2. R+LA tk finishes second best, but has a lower efficiency in some part of the iterations. McA tn does not converge completely alone, but does so when combined with a full span strategy. The B-spline based strategies covered here have better execution times than the minimum span strategies.

5.3 *Scan Lines*

5.3.1 Data Set

Figure 12a shows a small data set of 71 888 points. The data set is obtained from the English channel and the points are organized in scan lines with varying distances. A detail can be seen in (b). The point set contains a number of outlier points situated off the scan lines. These are inconsistent with the general trend of the data. The regular points represent a smooth part of the sea bed. The data set covers about 23 km^2, the elevation range is $[-47.8, -12.1]$ m and the standard deviation is 2.408. The corresponding surface (c) is created with an accuracy of 0.5 m and is approximating also the outlier points closely. The surface is bi-cubic and the applied refinement strategy is FA. Least squares approximation is performed for two iterations before the algorithm switches to MBA.

5.3.2 Experience Setup and Performance Ranges

The approximation is run with a tolerance of 0.5 m and the maximum number of iterations is 30 and 40 for refinement strategies refining in both parameter directions at every iteration level and with alternating parameter directions, respectively. *An intermediate stage is defined when 99% of the points are within the resolution. This corresponds to 719 unresolved points*. Due to the outliers, no condition is put on the maximum distance at the intermediate stage.

The current test case contains few points and this is reflected in fast execution times also for the strategies with the poorest performance as can be seen in Table 6. FB is the fastest strategy while minimum span strategies are the slowest and do not always converge within the given number of iterations.

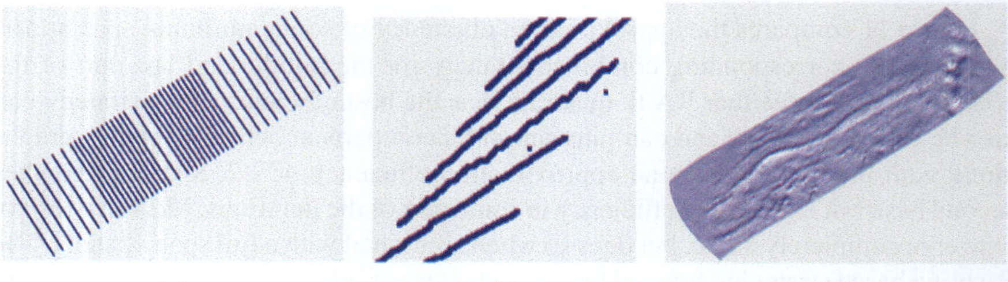

(a) (b) (c)

Fig. 12 **a** Sub sea scan line points, English channel. **b** Detail emphasizing outliers. **c** Corresponding surface

Table 6 Computation time and number of coefficients. Performance ranges and associated strategies for polynomial degrees one to three

Deg	Min time	Strategy	Max time	Strategy	Min cf	Strategy	Max cf	Strategy
1	0m1.3s	FB	0m4.6s	McA tn	10593	FA td	11978	RA
2	0m3.1s	FB	0m12.4s	MlA tn	16453	FA td	23264	R+LB tk
3	0m8.9s	FB	0m35.6s	McA	25331	FA tn	35386	RB td+k

5.3.3 Selected Refinement Strategies

Table 7 presents the results from a set of refinement strategies at the intermediate stage and final stage. Several minimum and full span strategies obtain the same lowest computational time at the intermediate stage for bi-linear splines. The increases in iteration level and execution time from the intermediate to the final stage for these strategies are systematically higher for the minimum span strategies than for full span.

The combination strategy Mc/FA tn bi-linear performs identically to McA tn at the intermediate stage, but converges in contrast to the pure minimum strategy at the final stage. Similar pairs can be found for McB tn and Mc/FB tn in the bi-linear case, and McA tn and Mc/FA tn in both the bi-quadratic and bi-cubic cases. The combined strategy performs in all cases better than the pure one. Combinations including the restricted mesh strategy also perform better than the pure restricted mesh strategy Here the element extension in R+LA tk is better than R/FA tk/n in the bi-cubic case. R/FA tk/n needs more iterations than specified to converge.

The full span strategy yields good results both with and without a threshold and FA is superior to FB when it comes to the number of coefficients in the surface. FB delivers the lowest computational time.

The strategies that perform best at the intermediate stage are not always those that finish with the best result. Figures 13 and 14 give a more continuous picture of the behaviour of the selected strategies. Approximation with bi-cubic splines results in most coefficients, and RA tk and R/FA tk/n have low approximation efficiency in the later part of the iterations. Early in the iterations these two strategies have a good efficiency. Bi-linear approximation gives the least number of coefficients. McA tn bi-linear does not converge in isolation, but changing to FA tn when the convergence speed decreases results in convergence and a moderate number of coefficients. Most strategies maintain a high maximum distance in the start before it starts decreasing and in particular RA bi-linear has a very rapid decrease.

The maximum distance for FA td quadratic decreases fast and the approximation efficiency is high for most of the iterations. In the bi-quadratic case FA td maintains its lead on FA tn throughout the computation despite a high number of iterations, while in the bi-cubic case FA td does not finish with all points resolved, see Table 7. This pattern with a fast decrease in maximum distance, high approximation efficiency

Table 7 Accuracy results of some selected strategies at the intermediate and final stage. * Converges, but needs more than the specified iterations to finish

Refine	Intermediate stage					Final results				
Strategy	It	Pt out	No coef	Max	Time	It	Pt out	No coef	Max	Time
Bi-linear										
FB	5	587	11 001	25.235	**0m0.5s**	10	0	14 678	0.5	**0m1.3s**
FB tn	6	604	7 574	17.377	**0m0.5s**	13	0	14 106	0.5	0m1.5s
MIB	6	617	6 884	18.22	**0m0.5s**	15	0	13 347	0.5	0m2.1s
McB	5	709	8 792	25 249	**0m0.5s**	14	0	13 185	0.5	0m1.9s
McB tn	6	658	6 483	23.884	**0m0.5s**	30	3	13 173	1.138	0m4s
Mc/FB tn	6	658	6 483	23.884	**0m0.5s**	30	0	13 983	0.5	0m2s
FA td	24	513	7 411	1.358	0m1.2s	36	0	**10 593**	0.5	0m2.4s
McA tn	**13**	**672**	**6 313**	**19.893**	**0m0.9s**	40	*12*	12 010	1.337	*0m4.6s*
Mc/FA tn	**13**	**672**	**6 313**	**19.893**	**0m0.9s**	26	0	11 836	0.5	0m2.8s
RA	10	583	9 170	25.178	0m0.7s	**18**	**0**	**11 978**	**0.5**	**0m1.6s**
FA	10	521	9 270	25.178	0m0.6s	18	0	11 728	0.5	0m1.4s
FA tn	12	620	6 532	17.213	0m0.7s	25	0	11 534	0.5	0m2.1s
Bi-quadratic										
FB	5	606	14 422	23.663	**0m1s**	11	0	21 058	0.5	**0m3.1s**
FB tn	6	603	10 074	18.220	**0m1s**	15	0	20 902	0.5	0m4s
RB td+k	22	667	**8 235**	13.404	0m2.5s	30	0	22 898	0.5	0m6.5s
FA tn	**12**	**664**	**8 848**	**18.153**	**0m1.3s**	26	0	18 101	0.5	0m5.3s
FA td	24	432	10 969	1.644	0m2.6s	**35**	**0**	**16 453**	**0.5**	**0m5.6s**
MIB td	23	655	10 071	2.354	0m3.1s	30	*52**	22 802	*1.308*	0m8.2s
McA tn	13	682	8 785	22.079	0m2s	36	0	18 914	0.5	0m11.5s
Mc/FA tn	13	646	8 365	22.594	0m2s	26	0	18 248	0.5	0m6.5s
RA tk	15	656	9 554	23.685	0m2.1s	30	0	19 661	0.5	0m7.7s
R/FA tk/n	15	656	9 554	23.685	0m2.1s	29	0	18 855	0.5	0m6.7s
R+LA tk	12	533	13 448	18.159	0m2.2s	22	0	20 035	0.5	0m6.1s
FA	10	644	12 709	23.641	0m1.3s	21	0	18 050	0.5	0m4.1s
Bi-cubic										
FB	6	434	22 420	19.046	0m3s	13	0	29 908	0.499	**0m8.9s**
FB tn	6	614	12 387	18.919	**0m1.8s**	15	0	28 715	0.499	0m9s
RA tk	**22**	**712**	**9 994**	**23.505**	**0m5.1s**	40	2*	30 046	0.576	0m22.9s
R/FA tk/n	**22**	**712**	**9 994**	**23.505**	**0m5.1s**	40	2*	27 328	0.514	0m19.8s
FA tn	**12**	**635**	**11 231**	**18.964**	**0m2.5s**	**24**	**0**	**25 331**	**0.5**	**0m10.4s**
R+LA tk	12	679	13 997	23.908	0m3.6s	**24**	**0**	**23 396**	**0.5**	**0m15.9s**
RA td+k	34	696	**9 955**	17.748	0m7s	40	*289**	23 321	*9.952*	0m12.1s
FA	12	448	20 017	18.966	0m3.6s	21	0	25 926	0.5	0m9.7s
FA td	24	459	13 281	1.405	0m5.3s	40	1*	23 216	0.797	0m14.6s
McB tn	7	618	11 816	16.395	0m2.9s	23	0	29 385	0.5	0m23.7s
McA tn	14	587	12 898	17.01	0m4.8s	38	0	27 803	0.499	0m32s
Mc/FA tn	14	587	12 898	17.01	0m4.8s	26	0	27 002	0.5	0m16.6s

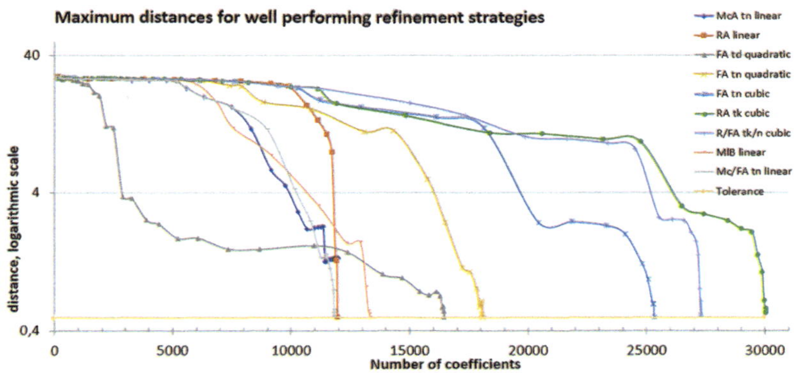

Fig. 13 The scanline data set. Maximum distance for selected well performing refinement strategies, all degrees

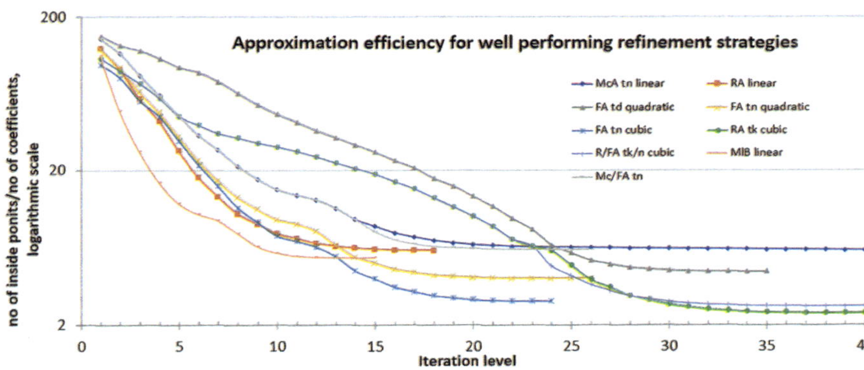

Fig. 14 Approximation efficiency for selected well performing refinement strategies, all degrees. Mc/FA tn has polynomial degree one

and the need for many iterations to converge can be found also for other polynomial degrees and refinement strategies when this threshold is applied.

The scan line data set is, despite the outlier points, approximated with good accuracy at the cost of surface smoothness. An alternative stop criteria for configurations where particular points have a severe lack of convergence should be considered.

5.4 Søre Sunnmøre, Sea Bed

5.4.1 Data Set

The point cloud, Fig. 15a, consists of 11 150 110 points obtained from a 881 100 m^2 area at Søre Sunnmøre in Norway. In (b) the points are scaled with a factor of 10,

(a) (b) (c)

Fig. 15 **a** Outline of point set from Søre Sunnmøre, Norway. **b** Detail of scaled point cloud. **c** Corresponding LRB surface

zoomed in and seen from the right corner to show some features. It is a sub sea data set in shallow waters, the depth ranges from –27.94 to –0.55 m. The point cloud is dense with some holes representing land. It has some outliers, although less extreme than for the scanline data set. At one position, two points with the same xy-coordinates and differing depth with a distance of 2.38 m are given. Thus, the theoretical smallest obtainable maximum distance between the point cloud and the surface is 1.19 m. In contrast to the previous example, the outliers are situated close to other points in the cloud. The point cloud has a standard deviation of 2.52 m. The surface in (c) is bi-quadratic and created with strategy FA td. The approximation algorithm switches from the least squares method to MBA after five iterations.

5.4.2 Experience Setup and Performance Ranges

Despite it being impossible to reach this resolution, we apply a tolerance of 0.5 m and run with a maximum number of iterations of 30 for refinement strategies refining in both parameter directions and 40 for strategies with alternating parameter directions. *The intermediate stage is where 99.0% of the points are within the tolerance, meaning that at most 11 150 points have a larger distance to the surface than 0.5 m.* In most cases, the iteration runs the maximum allowed number of steps.

Table 8 presents the minimum and maximum execution times and the lowest and highest numbers of coefficients in the final surfaces for the bi-linear, bi-quadratic and bi-cubic cases. We see that the ranges between best and worst are large, both with respect to time and size and for all polynomial degrees. The largest differences can be found internal to each polynomial degree. Bi-linear splines lead in general to lower execution times and less coefficients, while the bi-cubic splines give results at the other end of the scale.

Due to outlier points, none of the refinement strategies converge with all points within the resolution. The minimum number of out-of-tolerance points obtained is 469 for RA and RA td with bi-linear splines. We regard all results with less than

Table 8 Computation time and number of coefficients. Performance ranges and associated strategies for polynomial degrees one to three

Deg	Min t.	Strategy	Max t.	Strategy	Min cf	Strategy	Max cf	Strategy
1	8m47s	FA	29m26s	Mc/FB tn	282 422	FA	486 106	RB tk
2	15m8s	FA	1h9m59s	McB td	384 488	FA	785 442	RB td+k
3	36m7s	FA	3h13m39s	MIB td	550 034	FA	1 237 194	RB tk

480 points outside the tolerance and a maximum distance of 1.19 m as having an accepted accuracy. The least possible maximum distance is reached by all strategies except bi-quadratic McA td, which finishes with a maximum distance of 1.191 m.

5.4.3 Selected Refinement Strategies

Table 9 presents the results of some strategies at the intermediate and final stage. Strategies with refinement in alternating parameter directions have, for this data set, lower computation time than corresponding strategies that refine in both directions. A complete convergence is not possible in this case, and the extra number of iterations in the case of alternating directions does not account for the additional number of possible knot insertions when refinement is performed in both directions. The least number of coefficients is obtained with alternating parameter directions.

At the final stage, the full span strategy without threshold delivers the best results for all polynomial degrees and evaluation categories. This result does not change when observing the situation at the stage where the computation is regarded as converged. Variations of the full span strategy dominate also at the intermediate stage.

Several minimum span strategies finish with more than 480 points outside the tolerance, Table 9 lists some occurrences. This occurs for strategies with alternating parameter directions in particular. Minimum span strategies and some restricted mesh strategies, both with threshold, are the strategies that have the highest computation time and result in the surfaces with most coefficients.

Restricted mesh refinement strategies behave very similar with and without element extension, see RA tk and R+LA tk in Table 9. The final number of coefficients and the execution time are much higher than for corresponding full span strategies. Combining RA tk with FA tn gives better results for the bi-linear and bi-quadratic cases although it is not compatible with the pure full span strategy.

Most strategies continue several iterations after the stage where the number of unresolved points and the maximum distance are regarded as satisfactory. The number of coefficients is increased, in some cases considerably, during these iterations. The differences are typically largest for the strategies that refine in both parameter directions and in particular for the restricted mesh strategies. For instance the number of coefficients for RB increases by 122 418 during the last 13 iterations in the

Table 9 Accuracy results of some selected strategies at the intermediate and final stage

Refine	Intermediate stage					Final results				
Strategy	It	Pt out	No coef	Max	Time	It	Pt out	No coef	Max	Time
Bi-linear										
FA	19	9183	220761	2.177	2m52s	**40**	**473**	**282422**	**1.19**	**8m47s**
FA td	**19**	**11093**	**208656**	**2.177**	**2m54s**	40	473	284022	1.19	9m15s
FA tn	**21**	**9967**	**205036**	**1.895**	**4m27s**	40	471	286189	1.19	12m21s
McA tn	22	10423	**201216**	1.679	7m32s	40	636	299240	1.19	24m28s
McA	20	9087	212104	2.118	5m39s	40	*642*	289739	1.19	19m23s
RA	19	9037	262744	2.027	4m6s	40	469	350282	1.19	14m2s
RA tk	20	9561	261547	2.0	5m9s	40	472	363197	1.19	16m12s
RA td+k	20	10571	251670	2.0	4m47s	40	471	362604	1.19	16m39s
R/FA tk/n	21	9937	214600	1.887	4m9s	40	472	289134	1.19	11m17s
R+LA tk	19	10930	255163	2.027	5m21s	40	471	364468	1.19	17m33s
FB	10	6313	302982	2.118	2m59s	28	473	382159	1.19	11m8s
RB	10	7052	*348745*	2.085	4m15s	30	473	470721	1.19	18m12s
Mc/FA tn	22	8776	227991	1.602	7m28s	40	471	300277	1.19	16m42s
Bi-quadratic										
FA	19	8756	298113	2.05	**5m2s**	**40**	**471**	**384488**	**1.19**	**15m8s**
FA tn	**20**	**11070**	**247881**	**1.717**	**5m57s**	40	471	390876	1.19	19m38s
MlA tn	22	10351	303734	2.101	16m4s	40	*595*	502258	1.19	1h1m38s
RA tk	21	10996	345175	2.533	7m37s	40	479	614831	1.19	32m48s
R/FA tk/n	22	10151	283088	2.5	7m1s	40	473	410131	1.19	18m31s
R+LA tk	19	11132	390101	2.05	9m57s	40	474	619375	1.19	40m4s
FA td	19	10585	284391	2.05	5m22s	40	472	386719	1.19	15m59s
McA tn	22	9256	277757	2.315	16m11s	40	560	436114	1.19	56m20s
Mc/FA tn	21	8760	292773	1.926	11m49s	40	475	419197	1.19	31m20s
McB tn	11	10903	258016	1.554	10m9s	30	474	504463	1.19	1h6m51s
FB	10	5687	399843	1.7	5m6s	28	475	496494	1.19	18m34s
FB td	11	7065	368518	1.393	5m53s	30	473	504523	1.19	24m16s
RB tk	11	7895	473178	2.523	8m34s	30	472	783590	1.19	48m8s
McA td	21	8583	301687	1.805	16m43s	40	559	425780	1.191	49m55s
Bi-cubic										
FB	10	7738	523751	1.844	**11m45s**	30	474	693975	1.19	45m34s
FA	20	7560	438026	1.816	13m45s	**40**	473	**550034**	**1.19**	**36m7s**
FA tn	**21**	**9047**	**375301**	**1.779**	**15m22s**	40	477	558671	1.10	46m50s
McA tn	22	10802	**357114**	2.673	34m51s	40	725	643960	1.19	2h40m17s
MlA tn	23	9890	444465	2.462	41m57s	40	*742*	740829	1.19	2h44m42s
RB tk	12	10219	613700	2.667	16m26s	30	474	*1237194*	1.19	2h21m27s
Mc/FA tn	21	9307	385652	1.779	21m2s	40	475	576446	1.19	1h0m32s
FA td	20	8759	424838	1.816	14m48s	40	475	554411	1.19	40m55s
RA	20	7182	657945	1.813	24m10s	40	474	880517	1.19	1h16m2s
FB tn	11	8760	410358	1.376	11m57s	30	474	683441	1.19	58m36s
RA tk	24	10028	527792	2.482	19m50s	40	481	946074	1.19	1h20m10s
R+LA tk	20	9823	601843	1.813	27m3s	40	474	917541	1.19	1h35m30s
R/FA tk/n	25	9019	413007	1.813	16m54s	40	510	585346	1.19	42m18s

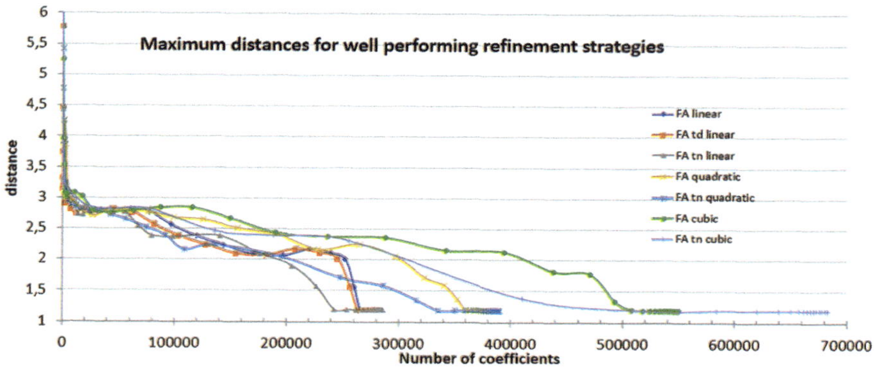

Fig. 16 Maximum distance for selected well performing refinement strategies, all degrees

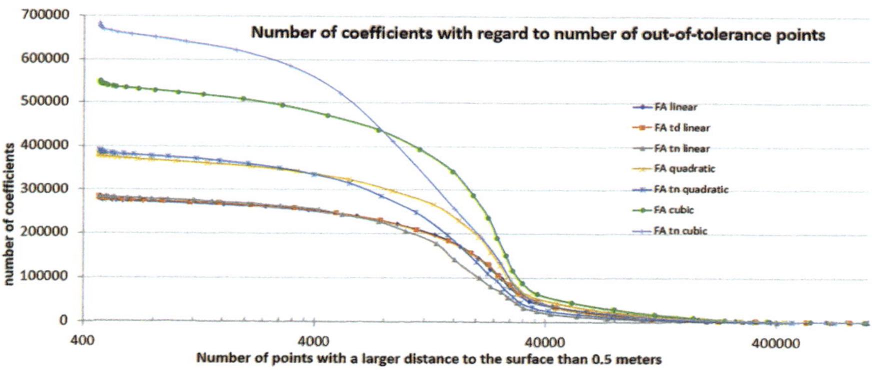

Fig. 17 Number of coefficients with regard to the number of unresolved points

bi-cubic case. Other examples are bi-quadratic FA tn that reaches 478 unresolved points at level 37 and increases the number of coefficients with 2 379 in the last three iterations, and bi-linear FB that goes from 475 to 473 unresolved points and gets 8 015 in new coefficients the last eight iterations. A more precise criterion for stopping the iterations should be applied to avoid unnecessary data size and execution time.

Figures 16 and 17 give a more continuous picture of the performance of the strategies selected as best at the intermediate and final stage. Bi-cubic splines have the slowest decrease in the maximum distance and most coefficients, and this behaviour can be seen throughout the approximation process. Cases with bi-linear splines are found at the other end of the spectrum, but FA tn bi-quadratic is compatible with the bi-linear cases during most of the process.

The full span strategies give the best results for this data set as can be seen from Table 9. However, some combinations of minimum span strategies and threshold have good scores at the intermediate stage. Combinations of a minimum span strategy

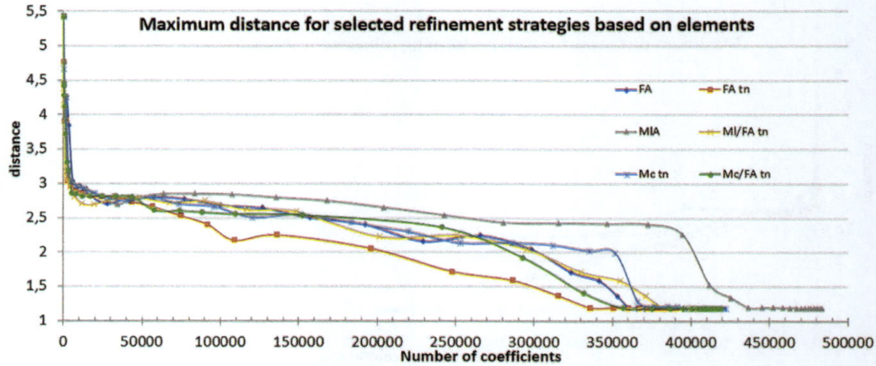

Fig. 18 Maximum distance for selected alternating refinement strategies, bi-quadratic splines

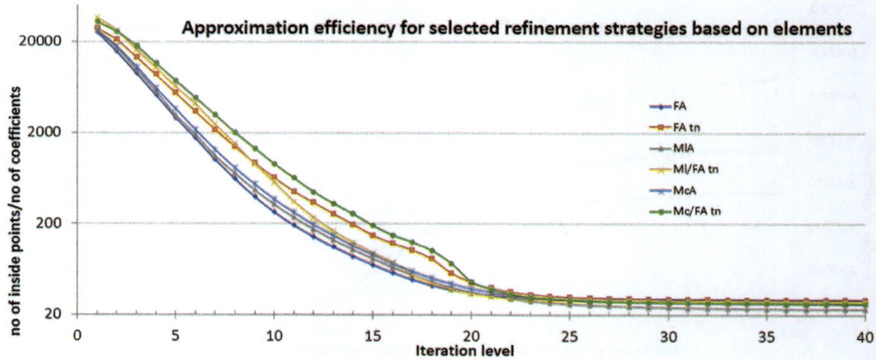

Fig. 19 Approximation efficiency, bi-quadratic splines

and a full span strategy turn in several cases the accuracy result from not accepted to accepted. These combinations are only tested with threshold type tn. Figures 18 and 19 compare the performance of two full span strategies with two minimum span strategies, MlA and McA, and combinations between minimum and full span strategies. The maximum distance decreases most rapidly for FA tn and stays low during the computation. This strategy also has a high approximation efficiency. FA has a better total score at the finishing stage, but for most of the computations the threshold improve the performance. Also the minimum span strategy Mc/FA tn has a higher approximation efficiency during large parts of the computation. Combining this strategy with the full span strategy leads to convergence with an acceptable number of coefficients, see Table 9, but the execution time is higher than for full span strategies also at the intermediate stage.

5.5 *Sea Bed, Large Point Cloud*

5.5.1 Data Set

The final data set contains 131 160 220 well distributed points acquired from the English Channel. The point cloud is relatively well behaved and represents an area of mostly smooth sea bed, with some areas of reduced smoothness. Approximation with an LRB surface should, thus be well adapted to the properties of the data set. The data set covers about 285 km^2 of shallow waters including a few meters above sea level. The elevation ranges from –32.46 to 3.26 m and the standard deviation is 8.04. The outline of the point cloud, a scaled version of the point cloud to visualize some features and the surface created with strategy R+LA and bi-quadratic splines are shown in Fig. 20a–c. The approximation uses the least squares method for three iterations, then turns to MBA due to lack of convergence while solving the linear equation system from the least squares approximation.

5.5.2 Experience Setup and Performance Ranges

We apply a tolerance of 0.5 m and we let the maximum number of iterations be 60 for alternating directions and 40 if we refine in both parameter directions at each level. *An intermediate stage is defined where either 99.9% of the points are within the tolerance (this equals 131 160 points with a larger distance to the surface than 0.5 m), or the maximum distance is less than two meters.*

The current point cloud has two complicating factors, the number of data points and the fact that some points stand out as untypical compared to other points in the neighbourhood. The maximum distance is for all combinations of refinement strategies and thresholds quickly reduced to about five meters, then the distance is kept at this level for a number of iterations before it decreases again, the effect can be observed in Fig. 21 for some strategies. Most strategies converge within the specified number of iterations, a few fail to reach the specified accuracy and some strategies have a very slow reduction in the number of points outside the tolerance for the last iterations, a few examples can be found in Table 10.

(a) (b) (c)

Fig. 20 **a** Outline of sea bed point cloud, **b** point cloud scaled by a factor of 20 and seen from the left, **c** LRB surface created from the full point cloud

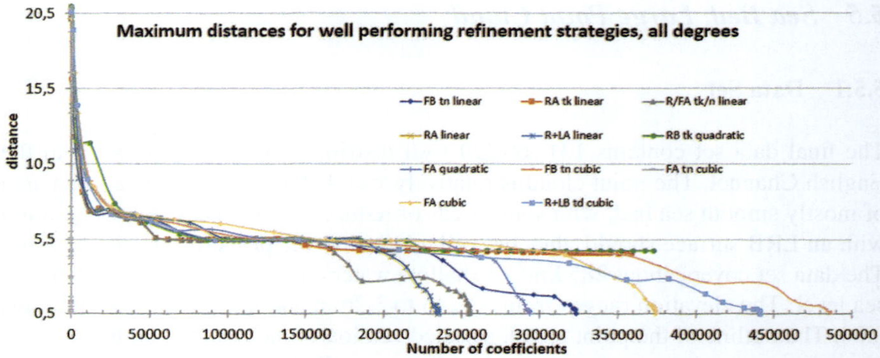

Fig. 21 Maximum distance for selected well performing refinement strategies, all degrees

Table 10 Computation time and number of coefficients. Performance ranges and associated strategies for polynomial degrees one to three

Deg	Min t.	Strategy	Max t.	Strategy	Min cf	Strategy	Max cf	Strategy
1	16m31s	RB	1h16m34s	RA td+k	235 033	RA	440 917	MIB td
2	30m39s	FB	2h31m52s	MIB td	291 753	R+LA td	729 432	MIB td
3	56m59s	FB	37h44m5s	RA tk	374 490	FA	860 910	MIB td

The point cloud really distinguishes the strategies both with respect to the number of coefficients, the execution time and whether or not the strategy converges.

Table 10 shows the ranges in execution time and number of coefficients in the final surface for the various polynomial degrees applied. The difference between the best and worst result is tremendous especially with regard to time. In the quadratic case the minimum span strategy MIB td results in the surface with the maximum number of coefficients and the highest execution time. In the linear and cubic case, restricted mesh strategies have the poorest performance with regard to time. The lowest computational time is, for both bi-degree two and three, obtained for FB.

5.5.3 Selected Refinement Strategies

Table 11 summarizes the results of some selected strategies. The strategies with the least number of coefficients tend also to have a relatively low computational time. Bi-cubic splines at the intermediate stage deviate from this trend, but the strategy resulting in the least number of coefficients still has an execution time in the lower range also in this case, see Table 10. Strategies that refine in both parameter directions simultaneously tend to have the lowest computational time, while strategies with alternating directions tend to result in the least number of coefficients. The difference between the iteration levels at the intermediate and final stage differs between the

Table 11 Selected strategies at the intermediate and final stage

Refine	Intermediate stage					Final results				
Strategy	It	Pt out	No coef	Max	Time	It	Pt out	No coef	Max	Time
Bi-linear										
FB	8	42 841	223 855	5.412	**9m45s**	29	0	334 788	0.5	19m59s
RB	8	45 344	213 499	5.473	10m11s	19	0	325 787	0.5	**16m31s**
RA tk	**16**	**119 540**	**76 349**	**5.438**	**17m45s**	60	1 658	359 131	4.566	1h11m40s
R/FA tk/n	**16**	**119 540**	**76 349**	**5.438**	**17m41s**	47	0	255 247	0.5	40m32s
R+LB tk	**8**	**79 170**	**113 003**	**5.441**	**11m43s**	40	10	306 673	0.772	34m39s
RA	15	77 436	144 195	5.359	14m40s	**37**	**0**	**235 033**	**0.5**	**25m11s**
R+LA	15	73 579	145 284	5.359	14m36s	**38**	**0**	**235 264**	**0.5**	**25m23s**
FA td	21	107 275	146 793	4.679	23m36s	43	0	280 494	0.5	37m44s
FA tn	15	123 366	88 414	5.482	16m44s	45	0	254 364	0.5	35m12s
RB td+k	16	126 755	147 874	5.436	20m2s	40	*3 986*	362 681	4.643	51m4s
Bi-quadratic										
FB	8	46 144	274 288	5.392	**16m8s**	22	0	412 811	0.5	**30m39s**
R+LA td	15	117 818	100 311	5.402	31m21s	42	0	**291 753**	0.5	1h2m7s
RB tk	**9**	**130 965**	**97 052**	**5.449**	**22m1s**	40	13 811	373 079	4.703	1h31m20s
FA	15	77 438	180 018	5.399	26m18s	**36**	**0**	**293 853**	**0.5**	**44m47s**
FA tn	15	115 818	101 492	5.432	30m15s	46	0	301 098	0.5	1h5m40s
R+LA	15	79 275	175 492	5.398	27m56s	41	0	294 369	0.5	50m24s
R+LA td+k	16	83 659	112 329	5.508	34m22s	55	0	292 697	0.5	1h18m9s
RB td+k	18	104 638	176 639	5.412	59m24s	40	*16 365*	437 539	5.302	1h59m10s
Bi-cubic										
FB	8	71 175	315 679	5.318	**31m26s**	21	0	518 726	0.5	**56m59s**
McA tn	16	126 520	**125 691**	5.374	1h10m34s	60	13	452 751	0.598	4h2m38s
Mc/FA tn	16	126 520	**125 691**	5.374	1h10m50s	48	0	435 790	0.5	2h59m40s
FB tn	**8**	**106 573**	**174 903**	**5.338**	**34m27s**	26	0	486 144	0.5	1h24m34s
FA tn	**16**	**84 792**	**145 020**	**5.482**	**58m53s**	45	0	375 612	0.5	2h7m22s
R+LB td	**8**	**130 334**	**145 641**	**5.488**	**35m11s**	22	0	441 405	0.5	1h11m29s
FA	15	116 928	205 512	5.435	48m57s	**33**	**0**	**374 490**	**0.5**	**1h24m20s**
RA	15	130 559	193 111	5.438	54m41s	34	0	387 211	0.5	1h32m48s
R+LA	15	121 854	197 582	5.438	54m7s	33	0	386 891	0.5	1h28m17s
RB td+k	20	118 312	245 578	5.326	*7h42m3s*	40	*20 740*	550 574	4.823	10h57m56s

strategies, but is in general high. This indicates slow convergence in the latter part of the computation.

Various versions of the full span strategy stand out. Also some versions of the restricted mesh strategy show good results at the intermediate stage, but fail in several cases to converge in the end. Strategies with a very restrictive introduction of new degrees of freedom can yield good results early in the computations, but may fail

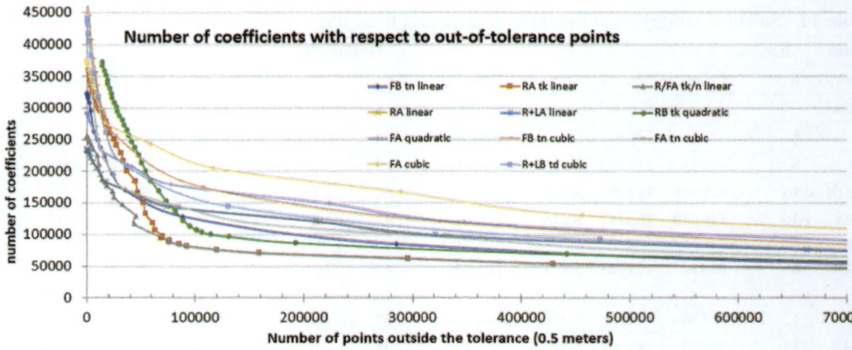

Fig. 22 Number of coefficients with respect to number of outside points, final stages, all degrees

to insert crucial new mesh line segments in the later stages and finish with a large number of coefficients, high execution times and no convergence. RB td+k is a very restrictive strategy that struggles already at the intermediate stage and finishes with poor results for all polynomial degrees.

Table 11 includes some strategies that perform reasonably well without being amongst the best. In the bi-linear case we compare some full span strategies with alternating parameter directions and different thresholds. Threshold tn performs best, which matches the normal threshold pattern. In the bi-quadratic case we see that the threshold slows down the convergence for FA and the final number of coefficients is only slightly lower. Applying a threshold has little effect for the element extended restricted mesh strategy R+LA, but the version without a threshold performs slightly faster. In the bi-cubic case we can compare the restricted mesh with and without element extension. The difference is small for this data set.

Figure 21 shows the continuous development of the maximum distance between the point cloud and the surface throughout the computation for some selected methods. The distance quickly reduces to about five meters, is maintained at that level for several iterations while the number coefficients increases. Then the distance drops again for most strategies. The exceptions are two restricted mesh strategies, RA tk bi-linear and RB tk bi-quadratic. In the linear case, also the combination of this restricted mesh strategy with a full span strategy is shown and results in few coefficients. The corresponding numbers can be found in Table 10.

The bi-linear surfaces have the least number of coefficients, then come the bi-quadratic surfaces and finally the bi-cubic ones. The difference in the number of coefficients is large even for strategies with the same polynomial degree.

Figure 22 shows the number of coefficients given the number of out-of-tolerance points for the same strategies. Only the last part of the computation is shown and the development is from right to left. The number of coefficients increases rapidly towards the end of the computation while the decrease in the number of points outside the tolerance is slow. The ranking of the strategies is stable until less than 100 000 points (less than 0.1% of the total number of points) are unresolved; then there is a shift. This indicates that whether or not we want to continue the computation to a complete convergence must be taken into account when selecting the refinement strategy.

6 Summary of Results

This section provides a summary of the results for the individual data sets. The full span strategy and the structured mesh strategy are very clearly defined. All other strategies have an element of choice and tuning included in them. Even for the minimum span strategies Ml there might be more than one candidate B-spline of the same size.

Bi-cubic approximation almost always results in more coefficients in the resulting surface and higher execution times than corresponding strategies for lower order polynomial degrees. The pattern regarding the relative performance of the strategies for a given data set is quite similar in the bi-quadratic and the bi-cubic case. The number of surface coefficients and execution time is often low in the bi-linear case, but strategies with restrictions on the introduction of new mesh lines are more vulnerable with respect to convergence for this degree.

Full span strategies have the most consistent behaviour throughout the data sets and tend to have the lowest execution times. The resulting surfaces do not always have the smallest number of coefficients, but the results are always among the best.

Minimum span strategies have a high risk of lack of convergence. The effect is most pronounced for bi-linear splines and reinforced if thresholds are applied. Minimum span strategies perform better in the first part of the computation than for the final iterations. The combined strategy for selecting the B-spline to subdivide performs better than focusing either on the size of the B-spline support or the number of out-of-tolerance points in the support.

The execution time for the structured mesh strategy is low, but the number of coefficients is very high for the tested versions of the strategy, which is omitted for further testing after two data sets.

The restricted mesh strategy does not have a consistent behaviour for all data sets. It often has a high approximation efficiency in the first and middle part of the computation, but may need a back-up strategy to converge completely. Applying a threshold most often reduces the number of surface coefficients, but increases the risk of not converging. All variants of the restricted mesh strategy lead to a high number of coefficients for the Søre Sunnmøre data set.

Strategies with alternating parameter directions tend to lead to fewer coefficients in the final surface. The execution time is often lowest when refinement is performed in both parameter directions simultaneously, but the difference is moderate.

Applying a threshold in general reduces the number of surface coefficients and increases the computation time, but is not always effective. The selected B-splines are not always those that lead to the highest improvement in accuracy when refined. The effect of applying a threshold depends on the type. Thresholding with respect to distance (td), tends to reduce the maximum distance quite rapidly, but has little effect on the number of resolved points. These strategies will often result in more coefficients than strategies without a threshold. This threshold type in general has a more positive effect on the B-spline based strategies than on the element based ones. A threshold with respect to the number of out-of-tolerance points (tn and tk) often leads to good approximation efficiency, that is, it leads to few coefficients with respect to the number of resolved points. The effect on the final surface size varies, the surface can be more lean when applying a threshold, but this is not always the case. It would probably be beneficial if applied thresholds are released more rapidly than the case in this study.

Two partly conflicting rules were formulated in Sect. 3. Both are confirmed. A reduced pace in the introduction of new knots results in leaner surfaces, while a too restrictive approach has the opposite result.

The convergence is typically rapid in the start of the approximation before it slows down drastically towards the end. The cost of resolving the last data points is high both with regard to data size and time consumption.

7 Conclusion

We have performed a study to find out how different strategies for knot selection in adaptive refinement of LRB surfaces perform in the context of scattered data approximation. The tests on the refinement strategies are applied to five different geospatial data sets with a large variety of sizes and properties. It is not the case that one size fits all. The various strategies perform differently on different data sets and the performance of each refinement strategy varies throughout the computations. The purpose of approximating a point cloud with an LRB surface is not always to achieve an accurate approximation. It can also be to represent the smooth part of the data points with a smooth surface in order to analyze the residuals. A very accurate approximation of noisy data is also unattractive. The recommendation of the refinement strategy to select depends on the goals for the approximation.

Bi-cubic splines should only be used if a surface with a continuous second derivative is required. A bi-linear surface is a good choice if no smoothness is needed. However, in general, we recommend the use of bi-quadratic splines.

If the aim is to create very accurate surfaces with respect to the point data, the full span strategy with alternating parameter directions should be applied. A threshold with respect to the number of points outside the given tolerance should probably be considered, but there is room for improvement regarding the actual composition. Thresholding with respect to the maximum distance should not be applied in this case.

If the aim is to get a smallest possible surface after a restricted number of iterations or when applying a stop criterion different from complete convergence, also other strategies are applicable. The restricted mesh strategy performs well for four out of five test cases. Minimum span strategies are more seldom selected as the best choice, but the results for the best versions are good at the intermediate stage and quite consistent with regard to the various data sets for bi-degree two and three. In both cases a switch to the full span strategy should be applied when the convergence slows down. The full span strategy is applicable also in this case.

Refinement in alternating parameter directions is recommended also for early finalization of the computation. Threshold with respect to distance should be used only if the most important aim is to reduce the maximum distance rapidly with less focus on the number of out-of-tolerance points. A threshold with respect to the number of points is recommended.

The applied study is extensive. More combinations are possible, but it is questionable whether or not they will provide more insight. The criteria used to define thresholds and restricted strategies have room for improvement.

The study focuses on the number of surface coefficients, execution time and approximation accuracy. The structure of the polynomial mesh is not considered, although it may be of importance. Structured and restricted mesh strategies give a more orderly mesh than the full and minimum span strategies. Furthermore, an analysis of the level of the accuracy threshold and the best criterion to stop the iteration would complement this study in a good way. A good refinement strategy should not break down when pushed to the limits. However, the most accurate surface does not necessarily give the best representation of the terrain or sea bed as the quality of the point clouds may vary.

Acknowledgements This work has been supported by the Norwegian research council under grant number 270922. The data sets are provided by the University of Brest, the UK Hydrographic Office and the Norwegian map authorities.

References

1. Bracco, C., Giannelli, C., Vázquez, R.: Refinement algorithms for adaptive isogeometric methods with hierarchical splines. Axioms **7**, 43 (2018)
2. Bressan, A.: Some properties of LR-splines. Comput. Aided Geometr. Design **30**, 778–794 (2013)
3. Bressan, A., Buffa, A., Sangalli, G.: Characterization of analysis-suitable T-splines. Comput. Aided Geometr. Design **39**, 17–49 (2015)
4. Bressan, A., Jüttler, B.: A hierarchical construction of LR meshes in 2D. Comput. Aided Geometr. Design **37**, 9–24 (2015)
5. Buffa, A., Cho, D., Sangalli, G.: Linear independence of the T-spline blending functions associated with some particular T-meshes. Comput. Methods Appl. Mech. Eng. **199**, 1437–1445 (2010)
6. Cottrell, J.A., Hughes, T.J.R., Bazilevs, Y.: Isogeometric Analysis: Toward Integration of CAD and FEA. Wiley, Chichester (2009)
7. Dokken, T., Pettersen, K.F., Lyche, T.: Polynomial splines over locally refined box-partitions. Comput. Aided Geometr. Design **30**, 331–356 (2013)
8. Dokken, T., Skytt, V., Barrowclough, O.: Trivariate spline representations for computer aided design and additive manufacturing. Comput. Math. Appl. **78**, 2168–2182 (2019)
9. Farin, G.: Curves and Surfaces for CAGD, A Practical Guide, 5th edn. Morgan Kaufmann Publishers (1999)

10. Forsey, D.R., Bartels, R.H.: Hierarchical B-spline refinement. ACM SIGGRAPH Comput. Graph. **4**, 205–212 (1988)
11. Giannelli, C., Speleers, H., Jüttler, B.: THB-splines: the truncated basis for hierarchical splines. Comput. Aided Geometr. Design **29**, 485–498 (2012)
12. Gu, J., Yu, T., Lich, L.V., Nguyen, T.-T., Bui, T.Q.: Adaptive multi-patch isogeometric analysis based on locally refined B-splines. Comput. Methods Appl. Mech. Eng. **339**, 704–738 (2018)
13. Hennig, P., Kästner, M., Morgenstern, P., Peterseim, D.: Adaptive mesh refinement strategies in isogeometric analysis - a computational comparison. Comput. Methods Appl. Mech. Eng. **316**, 424–448 (2017)
14. Johannessen, K.A., Kvamsdal, T., Dokken, T.: Isogeometry analysis using LR B-splines. Comput. Methods Appl. Mech. Eng. **269**, 471–514 (2014)
15. Kraft, R.: Adaptive and bi-linearly independent multi level B-splines. In: Le Méhauté, A., Rabut, C., Schumaker, L.L. (eds.) Surface Fitting and Multiresolution Methods, pp. 209–218. Vanderbilt University Press, Nashville (1997)
16. Lee, S., Wolberg, G., Shin, S.Y.: Scattered data interpolation with multilevel B-splines. IEEE Trans. Vis. Comput. Graph. **3**, 229–244 (1997)
17. Nowacki, H., Westgaard, G., Heinemann, J.: Creation of fair surfaces based on higher order smoothness measures with interpolation constraints. In: Nowacki, H., Kaklis, P.D. (eds.), Creating Fair and Shape-Preserving Curves and Surfaces, pp. 141–162. B. G. Teubner Stuttart, Leipzig (1998)
18. Patrizi, F., Dokken, T.: Linear dependence of bivariate minimal support and locally refined B-splines over LR-meshes. Comput. Aided Geometr. Design **77**, 101803 (2020)
19. Patrizi, F., Manni, C., Pelosi, F., Speleers, H.: Adaptive refinement with locally linearly independent LR B-splines: theory and applications. Comput. Methods Appl. Mech. Eng. **369**, 113230 (2020)
20. Sederberg, T.W., Zheng, J., Bakenov, A., Nasri, A.: T-splines and T-NURCCs. ACM Trans. Graph. **22**, 477–484 (2003)
21. Skytt, V., Barrowclough, O., Dokken, T.: Locally refined spline surfaces for representation of terrain data. Comput. Graph. **49**, 58–68 (2015)
22. Speleers, H., Manni, C.: Effortless quasi-interpolation in hierarchical spaces. Numer. Math. **132**, 155–184 (2016)
23. Zhang, W., Tang, Z., Li, J.: Adaptive hierarchical B-spline surface approximation of large-scale scattered data. In: Proceedings Pacific Graphics '98, pp. 8–16. Singapore (1998)

A Block ILUT Smoother for Multipatch Geometries in Isogeometric Analysis

Roel Tielen, Matthias Möller, and Kees Vuik

Abstract Since its introduction in [20], Isogeometric Analysis (IgA) has established itself as a viable alternative to the Finite Element Method (FEM). Solving the resulting linear systems of equations efficiently remains, however, challenging when high-order B-spline basis functions of order $p > 1$ are adopted for approximation. The use of Incomplete LU (ILU) type factorizations, like ILU(k) or ILUT, as a preconditioner within a Krylov method or as a smoother within a multigrid method is very effective, but costly [37]. In this paper, we investigate the use of a block ILUT smoother within a p-multigrid method, where the coarse grid correction is obtained at $p = 1$, and compare it to a global ILUT smoother in case of multipatch geometries. A spectral analysis indicates that the use of the block ILUT smoother improves the overall convergence rate of the resulting p-multigrid method. Numerical results, obtained for a variety of two dimensional benchmark problems, illustrate the potential of this block ILUT smoother for multipatch geometries.

1 Introduction

Isogeometric Analysis [20] has established itself as a viable alternative to the Finite Element Method (FEM). The use of high-order B-spline basis functions allows for an accurate description of curved geometries, thereby bridging the gap between Computer Aided Design (CAD) and Finite Element Analysis (FEA). Furthermore, a higher accuracy per degree of freedom can be obtained with high-order B-spline basis functions compared to standard FEM [21].

Solving the resulting linear systems of equations efficiently remains, however, a challenging task. The condition number of the resulting system matrices grows

R. Tielen (✉) · M. Möller · K. Vuik
Delft University of Technology, Delft, The Netherlands
e-mail: r.p.w.m.tielen@tudelft.nl

M. Möller
e-mail: m.moller@tudelft.nl

K. Vuik
e-mail: c.vuik@tudelft.nl

C. Manni and H. Speleers (eds.), *Geometric Challenges in Isogeometric Analysis*, Springer INdAM Series 49, https://doi.org/10.1007/978-3-030-92313-6_11

exponentially with the approximation order p [12], making the use of (standard) iterative solvers not straightforward. Over the years, special preconditioners have been developed [3, 19, 34] to obtain convergence rates independent of the approximation order p. A viable alternative are multigrid methods [5, 15], which can be designed to be of optimal complexity for elliptic problems. Multigrid methods combine the use of a smoother (e.g. Gauss-Seidel) and a correction obtained on coarser levels. Typically, this correction is obtained based on a coarser discretization, leading to so-called h-multigrid methods. The use of h-multigrid methods in IgA has been investigated in detail, see for example [10, 12, 17, 18, 27, 32].

Alternatively, for high-order methods, the correction can be determined at a low-order level as well, resulting in p-multigrid methods [11, 16, 22, 23, 31]. Recently, the use of p-multigrid methods has been investigated in the context of Isogeometric Analysis [35–38]. Throughout this paper, p-multigrid methods will be adopted as a stand-alone solver or as a preconditioner within a Krylov solver. More precisely, a two-level method is considered, where the second level (i.e. the "coarse" level) corresponds to a discretization of order $p = 1$. Furthermore, k-refinement is applied throughout this paper, implying that the considered spaces are non-nested.

In recent years, Incomplete LU (ILU) factorizations have gained increasing popularity within Isogeometric Analysis. As a preconditioner, ILU(k) has shown to be very efficient [9]. When applied as a smoother in multigrid methods [37], convergence rates independent of p have been observed. A drawback of ILU factorizations are, however, the high setup costs, making their use as a smoother or preconditioner within a solver costly [37].

In this paper, we focus on the use of ILUT factorizations as a smoother within multigrid methods. In particular, we propose the use of a block ILUT smoother in case of multipatch geometries. The smoother is based on a splitting of the domain in multiple patches and has been applied successfully in the context of domain decomposition methods for reservoir simulation problems [26].

This paper is organized as follows. In Sect. 2, we present a model problem that will be considered throughout this paper. Section 3 presents the adopted p-multigrid method. The global and block ILUT smoother considered in this paper are presented in Sect. 4. The performance of both ILUT smoothers is compared for several two dimensional benchmark problems in Sect. 5. Furthermore, both smoothers are applied to a challenging benchmark, a Yeti footprint consisting of 21 patches, in Sect. 6. Finally, conclusions are drawn in Sect. 7.

2 Model Problem

As a model problem, we consider the convection-diffusion-reaction (CDR) equation on a connected, Lipschitz domain $\Omega \subset \mathbb{R}^d$ (where d is the spatial dimension):

$$- \nabla \cdot (\mathbf{D} \nabla u) + \mathbf{v} \cdot \nabla u + R u = f, \quad \text{on } \Omega. \tag{1}$$

Here, $\mathbf{D}$ denotes the diffusion tensor, $\mathbf{v}$ a divergence-free velocity field and R a reaction term. Furthermore, we have $f \in L^2(\Omega)$ and $u = 0$ on the boundary $\partial\Omega$. Let $\mathcal{V} = H_0^1(\Omega)$ be defined as the Sobolev space $H^1(\Omega)$ with functions that vanish on $\partial\Omega$. The variational form of the CDR-equation is obtained by multiplying Eq. (1) with a test function $v \in \mathcal{V}$, integrating over the domain Ω and applying integration by parts to the second-order term:

$$\int_{\Omega} (\mathbf{D}\nabla u) \cdot \nabla v + (\mathbf{v} \cdot \nabla u)v + Ruv \, d\Omega = \int_{\Omega} fv \, d\Omega \quad \forall v \in \mathcal{V}. \tag{2}$$

The physical domain Ω is then parameterized by a geometry function that describes an invertible mapping connecting the parameter domain $\Omega_0 = (0, 1)^d$ with the physical domain Ω:

$$\mathbf{F} : \Omega_0 \to \Omega, \qquad \mathbf{F}(\boldsymbol{\xi}) = \mathbf{x} \in \Omega \quad \forall \boldsymbol{\xi} \in \Omega_0. \tag{3}$$

Many physical domains of practical relevance cannot be modelled by a single geometry function, e.g. a channel with one or more obstacles. In that case, Ω is divided into a collection of non-overlapping subdomains $\Omega^{(k)}$ such that

$$\overline{\Omega} = \bigcup_{k=1}^{K} \overline{\Omega}^{(k)} \tag{4}$$

and we refer to Ω as a multipatch geometry consisting of K patches. For each $\Omega^{(k)}$, a geometry function $\mathbf{F}^{(k)}$ is then defined to parameterize each subdomain individually.

$$\mathbf{F}^{(k)} : \Omega_0 \to \Omega^{(k)}, \qquad \mathbf{F}^{(k)}(\boldsymbol{\xi}) = \mathbf{x} \in \Omega^{(k)} \quad \forall \boldsymbol{\xi} \in \Omega_0. \tag{5}$$

Figure 1 shows a multipatch geometry, consisting of 4 patches. For the spatial discretization, multivariate B-spline basis functions are adopted which are constructed as tensor product of univariate B-spline basis functions. The latter ones are defined on the parameter domain (0, 1) and are uniquely determined by their underlying knot vector $\Xi = \{\xi_1, \xi_2, \dots, \xi_{N+p}, \xi_{N+p+1}\}$, consisting of a sequence of non-decreasing knots $\xi_i \in (0, 1)$. Here, N denotes the number of basis functions and p is their polynomial order. B-spline basis functions are defined recursively by the Cox-de Boor formula [4].

The resulting B-spline basis functions $\phi_{i,p}$, $(i = 1, \dots, N)$, are non-zero on the interval $[\xi_i, \xi_{i+p+1})$, implying a compact support that increases with p. As a consequence, system matrices resulting from the substitution of B-spline basis functions into Galerkin formulations of differential equations are sparse, although the number of non-zero entries increases with p. Furthermore, at every knot ξ_i the basis functions are C^{p-m_i}-continuous, where m_i denotes the multiplicity of knot ξ_i. Finally, the basis functions possess the partition of unity property and are non-negative:

$$\sum_{i=1}^{N} \phi_{i,p}(\xi) = 1 \quad \forall \xi \in [\xi_1, \xi_{n+p+1}], \tag{6}$$

$$\phi_{i,p}(\xi) \geq 0 \quad \forall \xi \in [\xi_1, \xi_{n+p+1}]. \tag{7}$$

As a consequence, mass matrices can be lumped within the variational formulation.

Throughout this paper, B-spline basis functions are considered based on an open uniform knot vector with knot span size h, that is, the first and last knots are repeated $p+1$ times. Therefore, the basis functions considered are C^{p-1} continuous and interpolatory only at the two end points.

Multivariate B-spline basis functions $\Phi_{i,p}$ ($i = 1, \ldots, N_{\text{dof}}$) are then defined as the tensor product of univariate basis functions of order p. Let

$$\mathscr{V}_{h,p} := \text{span}\left\{\Phi_{i,p} \circ \mathbf{F}^{-1}\right\}_{i=1,\ldots,N_{\text{dof}}}. \tag{8}$$

Then, the Galerkin formulation of (2) becomes: Find $u_{h,p} \in \mathscr{V}_{h,p}$ such that

$$a(u_{h,p}, v_{h,p}) = (f, v_{h,p}) \qquad \forall v_{h,p} \in \mathscr{V}_{h,p}. \tag{9}$$

where

$$a(u,v) = \int_{\Omega} (\mathbf{D}\nabla u) \cdot \nabla v + (\mathbf{v} \cdot \nabla u)v + Ruv \, \mathrm{d}\Omega, \qquad (f,v) = \int_{\Omega} fv \, \mathrm{d}\Omega. \tag{10}$$

The discretized problem can be written as a linear system

$$\mathbf{A}_{h,p}\mathbf{u}_{h,p} = \mathbf{f}_{h,p}, \tag{11}$$

where $\mathbf{A}_{h,p}$ denotes the system matrix. For a more detailed description of the spatial discretization in Isogeometric Analysis, we refer to [20].

Figure 1 illustrates a multipatch geometry consisting of 4 patches $\Omega^{(k)}$, $k = 1, \ldots, 4$, and the resulting block structure of the system matrix $\mathbf{A}_{h,p}$. The first 4 diagonal blocks are associated with the interior degrees of freedom on Ω_i, $i = 1, \ldots, 4$, while the right-bottom block, shown in grey, denotes the degrees of freedom at the interface Γ. Finally, the off-diagonal blocks denote the coupling between degrees of freedom at the interior and the interface, resulting in an arrowhead matrix. It should be noted that this structure is related to a particular numbering of the degrees of freedom. Note that this block structure is common within domain decomposition (DD) methods [33], which decompose the global domain Ω into subproblems and solve small boundary value problems per subdomain individually using data at the interface as boundary conditions. A coarse problem is then typically solved to exchange information about the solution between the subdomains. When non-overlapping subdomains are considered, the resulting block structure is similar to the structure shown in Fig. 1. The definition of the block ILUT smoother considered in this paper is based on this structure and will be discussed in detail in Section 4.

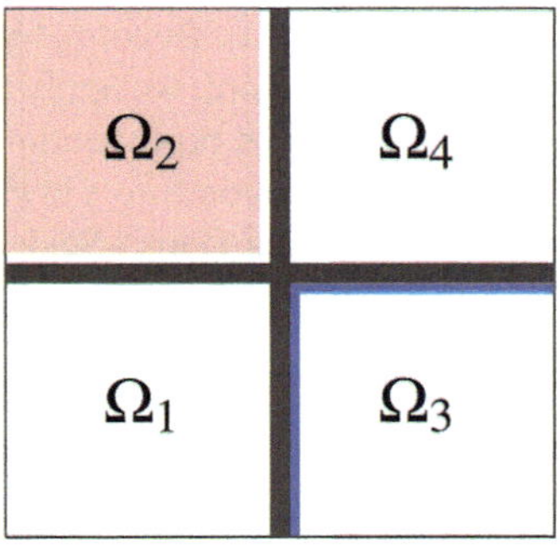

$$\begin{bmatrix} \mathbf{A}_{11} & 0 & 0 & 0 & \mathbf{A}_{\Gamma 1} \\ 0 & \mathbf{A}_{22} & 0 & 0 & \mathbf{A}_{\Gamma 2} \\ 0 & 0 & \mathbf{A}_{33} & 0 & \mathbf{A}_{\Gamma 3} \\ 0 & 0 & 0 & \mathbf{A}_{44} & \mathbf{A}_{\Gamma 4} \\ \mathbf{A}_{1\Gamma} & \mathbf{A}_{2\Gamma} & \mathbf{A}_{3\Gamma} & \mathbf{A}_{4\Gamma} & \mathbf{A}_{\Gamma\Gamma} \end{bmatrix}$$

Fig. 1 A multipatch geometry, consisting of 4 patches and the resulting block structure of the system matrix. The block associated to the interior degrees of freedom of Ω_2 is highlighted in red, while the coupling between the interface Γ and Ω_3 is highlighted in blue. Furthermore, the degrees of freedom at Γ result in the gray block

3 p-Multigrid Method

In order to solve the linear system from Eq. (11), we consider the p-multigrid method presented in [37]. With a p-multigrid method, a correction is obtained at level $p = 1$ by a direct projection from the high-order level. Starting from the high-order problem, the following steps are performed:

1. Apply a fixed number ν_1 of presmoothing steps to the initial guess $\mathbf{u}_{h,p}^{(0,0)}$:

$$\mathbf{u}_{h,p}^{(0,m)} = \mathbf{u}_{h,p}^{(0,m-1)} + \mathscr{S}_{h,p}\left(\mathbf{f}_{h,p} - \mathbf{A}_{h,p}\mathbf{u}_{h,p}^{(0,m-1)}\right), \quad m = 1, \ldots, \nu_1, \tag{12}$$

   where $\mathscr{S}_{h,p}$ is a smoothing operator applied to the high-order problem.
2. Determine the residual at level p and project it onto the space $\mathscr{V}_{h,1}$ using the restriction operator $\mathscr{I}_p^1$:

$$\mathbf{r}_{h,1} = \mathscr{I}_p^1\left(\mathbf{f}_{h,p} - \mathbf{A}_{h,p}\mathbf{u}_{h,p}^{(0,\nu_1)}\right). \tag{13}$$

3. Solve the residual equation to determine the coarse grid error:

$$\mathbf{A}_{h,1}\mathbf{e}_{h,1} = \mathbf{r}_{h,1}. \tag{14}$$

4. Project the error $\mathbf{e}_{h,1}$ onto the space $\mathscr{V}_{h,p}$ using the prolongation operator $\mathscr{I}_1^p$ and update $\mathbf{u}_{h,p}^{(0,\nu_1)}$:

$$\mathbf{u}_{h,p}^{(0,\nu_1)} := \mathbf{u}_{h,p}^{(0,\nu_1)} + \mathscr{I}_1^p\left(\mathbf{e}_{h,1}\right). \tag{15}$$

5. Apply ν_2 postsmoothing steps of the form (12) to obtain $\mathbf{u}_{h,p}^{(0,\nu_1+\nu_2)} =: \mathbf{u}_{h,p}^{(1,0)}$.

Note that, in contrast to h-multigrid methods, the number of degrees of freedom for the residual equation from Eq. (14) can still be relatively large. Therefore, we apply a single W-cycle of a standard h-multigrid method [15], using canonical prolongation and weighted restriction, to solve Eq. (14) approximately. This corresponds to a low-order Lagrange discretization, and hence, h-multigrid methods (using Gauss-Seidel as a smoother) are known to be both efficient and cheap [15, 39]. The resulting p-multigrid is shown in Fig. 2.

Instead of applying a single W-cycle, the h-multigrid method could be applied until the residual is reduced by a fixed factor as well. By doing so, the h-multigrid method basically becomes a solver, leaving the high-order errors to the (more expensive) ILUT smoother. Numerical results show, however, that the total number of p-multigrid iterations is not affected by this for the benchmarks considered in this paper. Therefore, only a single W-cycle is applied at level $p = 1$ throughout the remainder of this paper.

To project the error to $\mathscr{V}_{h,p}$ and the residual to $\mathscr{V}_{h,1}$, prolongation and restriction operators based on an L_2 projection are adopted which have been used extensively in the literature [6, 7, 30]. The prolongation operator $\mathscr{I}_1^p : \mathscr{V}_{h,1} \to \mathscr{V}_{h,p}$ is given by

$$\mathscr{I}_1^p(\mathbf{v}_1) = (\mathbf{M}_p)^{-1}\mathbf{P}_1^p\, \mathbf{v}_1. \tag{16}$$

Here, the mass matrix $\mathbf{M}_p$ and transfer matrix $\mathbf{P}_1^p$ are defined as follows:

$$(\mathbf{M}_p)_{(i,j)} := \int_\Omega \Phi_{i,p}\Phi_{j,p}\, \mathrm{d}\Omega, \qquad (\mathbf{P}_1^p)_{(i,j)} := \int_\Omega \Phi_{i,p}\Phi_{j,1}\, \mathrm{d}\Omega. \tag{17}$$

The restriction operator $\mathscr{I}_p^1 : \mathscr{V}_{h,p} \to \mathscr{V}_{h,1}$ is defined by

$$\mathscr{I}_p^1(\mathbf{v}_p) = (\mathbf{M}_1)^{-1}\mathbf{P}_p^1\, \mathbf{v}_p. \tag{18}$$

To prevent the explicit solution of a linear system of equations for each projection step, the consistent mass matrix $\mathbf{M}$ in both transfer operators is replaced by its lumped counterpart $\mathbf{M}^L$ by applying row-sum lumping:

$$\mathbf{M}^L_{(i,i)} = \sum_{j=1}^{N_{\text{dof}}} \mathbf{M}_{(i,j)}. \tag{19}$$

Note that, row-sum lumping can be applied within the variational formulation, due to the partition of unity and non-negativity of the B-spline basis functions. Alternatively, the mass matrix could be inverted by exploiting the tensor product structure, cf. [13]. Numerical results, not presented in this paper, indicate that the use of a lumped mass matrix does not influence the overall convergence or accuracy of the p-multigrid method. Note that this choice of prolongation and restriction operator leads to a non-symmetric multigrid operator even for symmetric PDE problems. As a consequence, the resulting p-multigrid method can only be applied as a preconditioner to a Krylov

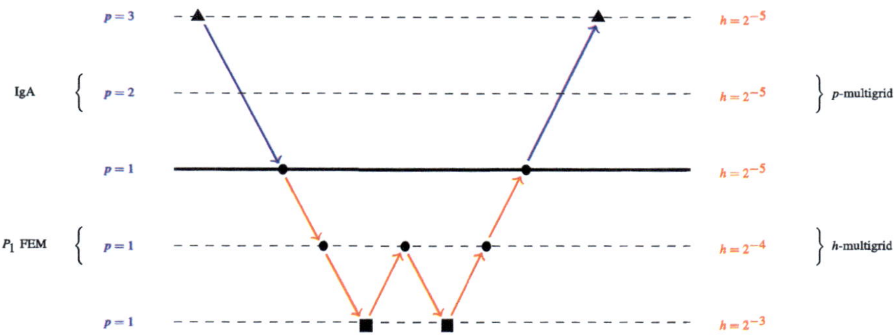

Fig. 2 Illustration of the p-multigrid method [37]. At $p = 1$, Gauss-Seidel is always adopted as a smoother (•), whereas at the high-order level different smoothers can be applied (▲). At the coarsest level, a direct solver is applied to solve the residual Eq. (■)

solver suited for non-symmetric matrices like the stabilized Bi-Conjugate Gradient (Bi-CGSTAB) method. Choosing the prolongation and restriction operator as the transpose of each other would restore symmetry. However, numerical experiments, not presented in this paper, indicate that his choice leads to a less robust p-multigrid method [37].

Different choices can be made with respect to the smoother. The use of Gauss-Seidel or (damped) Jacobi as a smoother at level p leads to convergence rates that depend significantly on the approximation order p [35]. Alternative smoothers have been developed in recent year to overcome this shortcoming [10, 18, 27, 32]. In particular, the use of ILUT factorizations has shown to be very effective. However, the costs of setting up the smoother (i.e. performing the ILUT factorization) are relatively high compared to other smoothers. A computationally efficient alternative consists in using a *block* ILUT smoother, which lends itself to parallelization. Therefore, we consider a *block* ILUT smoother in this paper and compare it to a global ILUT smoother.

4 (Block) ILUT Smoother

The global ILUT smoother is based on an Incomplete LU factorization with dual Threshold strategy [28]. The factorization is based on two parameters, the fill factor f and threshold τ that specifies that all smaller entries of the factorization are dropped. Furthermore, the maximum number of nonzero entries in each row is f times the average number of nonzero entries in a row of the original operator. Throughout this paper, we choose $f = 1$ and $\tau = 10^{-13}$ so that the number of nonzero entries in the ILUT factorization is comparable to the number of nonzero entries of the original operator $\mathbf{A}_{h,p}$.

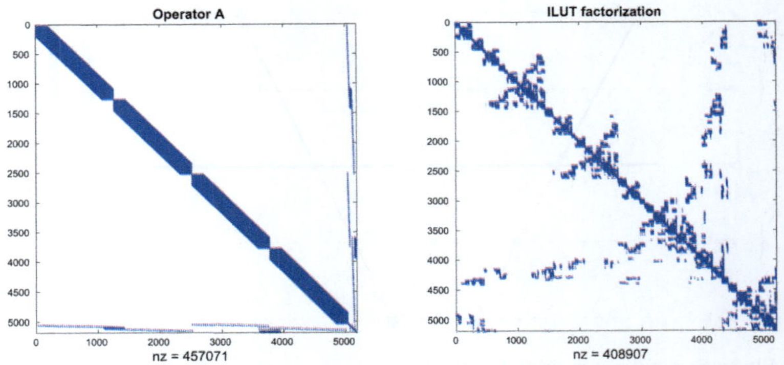

Fig. 3 Sparsity pattern of $\mathbf{A}_{h,p}$ (left) and $\mathbf{L}_{h,p} + \mathbf{U}_{h,p}$ (right) for $p = 4$, $h = 2^{-5}$ and 4 patches. The number of nonzero (nz) entries for both operators is provided as well

A typical sparsity pattern of the system matrix for a multipatch geometry is shown in Fig. 3 (left). The sparsity pattern of the resulting (global) ILUT factorization plotted as $\mathbf{L}_{h,p} + \mathbf{U}_{h,p}$ is shown in Fig. 3 (right). Here, we consider a discretization with $p = 4$ and $h = 2^{-5}$, consisting of 4 patches. Since an approximate minimum degree (AMD) ordering [1] is applied during the ILUT factorization to reduce the fill-in, sparsity patterns differ significantly. Note that, indeed, $\mathbf{A}_{h,p}$ and $\mathbf{L}_{h,p} + \mathbf{U}_{h,p}$ have a comparable number of nonzero entries.

An efficient implementation of ILUT is available in the Eigen library [14] based on [28, 29]. Once the factorization is obtained when setting up the multigrid solver, a single smoothing step at multigrid cycle i is applied as follows:

$$\mathbf{u}_{h,p}^{(i,n+1)} = \mathbf{u}_{h,p}^{(i,n)} + (\mathbf{L}_{h,p}\mathbf{U}_{h,p})^{-1}(\mathbf{f}_{h,p} - \mathbf{A}_{h,p}\mathbf{u}_{h,p}^{(i,n)}), \tag{20}$$

$$= \mathbf{u}_{h,p}^{(i,n)} + \mathbf{U}_{h,p}^{-1}\mathbf{L}_{h,p}^{-1}(\mathbf{f}_{h,p} - \mathbf{A}_{h,p}\mathbf{u}_{h,p}^{(i,n)}), \tag{21}$$

where the two matrix inversions in Eq. (21) amount to forward and backward substitutions.

In the following, we consider a block ILUT smoother as an alternative. This smoother is based on the observation that, in case of a multipatch geometry consisting of K patches, the resulting system matrix can be written as follows (see Fig. 1):

$$\mathbf{A} = \begin{bmatrix} \mathbf{A}_{11} & & \mathbf{0} & \mathbf{A}_{\Gamma 1} \\ & \ddots & & \vdots \\ \mathbf{0} & & \mathbf{A}_{KK} & \mathbf{A}_{\Gamma K} \\ \mathbf{A}_{1\Gamma} & \cdots & \mathbf{A}_{K\Gamma} & \mathbf{A}_{\Gamma\Gamma} \end{bmatrix} \tag{22}$$

Note that, in the remainder of this section, we drop the h and p as subscripts to improve readability.

In [26] a preconditioner $\mathscr{P}$ is defined consisting of a coarse part $\mathscr{P}_C$ and fine part $\mathscr{P}_F$ in a multiplicative way. That is, the iteration matrix of the resulting preconditioner is obtained by multiplying the iteration matrix of both preconditioners:

$$\mathbf{I} - \mathscr{P}\mathbf{A} = (\mathbf{I} - \mathscr{P}_C\mathbf{A})(\mathbf{I} - \mathscr{P}_F\mathbf{A}), \tag{23}$$

$$= \mathbf{I} - \mathscr{P}_F\mathbf{A} - \mathscr{P}_C\mathbf{A} + \mathscr{P}_F\mathbf{A}\mathscr{P}_C\mathbf{A} \tag{24}$$

and hence:

$$\mathscr{P} = \mathscr{P}_F + \mathscr{P}_C - \mathscr{P}_F\mathbf{A}\mathscr{P}_C. \tag{25}$$

The coarse part consists of a subdomain deflation approach [25] and takes care of the low frequency components. However, it is known from multigrid that the coarse grid correction decreases the low frequency components of the error. We therefore only adopt the fine part of the preconditioner (i.e. $\mathscr{P}_F$) in our block ILUT smoother and ignore $\mathscr{P}_C$. Hence, we have $\mathscr{P} = \mathscr{P}_F$ and will use the linear iteration with respect to $\mathscr{P}$ as our smoother.

Based on the considerations above, the smoother $\mathscr{P}$ is then defined by observing that $\mathbf{A}$ can be written as follows:

$$\mathbf{A} = \mathbf{L}\mathbf{U} = \begin{bmatrix} \mathbf{L}_1 & & & \\ & \ddots & & \\ & & \mathbf{L}_K & \\ \mathbf{B}_1 & \cdots & \mathbf{B}_K & \mathbf{I} \end{bmatrix} \begin{bmatrix} \mathbf{U}_1 & & & \mathbf{C}_1 \\ & \ddots & & \vdots \\ & & \mathbf{U}_K & \mathbf{C}_K \\ & & & \mathbf{S} \end{bmatrix}. \tag{26}$$

Here, $\mathbf{A}_{ii} = \mathbf{L}_i\mathbf{U}_i$ is based on a complete LU factorization. Furthermore, we have $\mathbf{B}_i = \mathbf{A}_{i\Gamma}\mathbf{U}_i^{-1}$ and $\mathbf{C}_i = \mathbf{L}_i^{-1}\mathbf{A}_{\Gamma i}$. Finally, we have $\mathbf{S} = \mathbf{A}_{\Gamma\Gamma} - \sum_{i=1}^{K}\mathbf{B}_i\mathbf{C}_i$. These choices indeed imply that $\mathbf{A} = \mathbf{L}\mathbf{U}$.

A smoother can now be obtained by replacing the LU factorizations by ILUT factorizations:

$$\mathbf{A} \approx \tilde{\mathbf{L}}\tilde{\mathbf{U}} = \begin{bmatrix} \tilde{\mathbf{L}}_1 & & & \\ & \ddots & & \\ & & \tilde{\mathbf{L}}_K & \\ \tilde{\mathbf{B}}_1 & \cdots & \tilde{\mathbf{B}}_K & \mathbf{I} \end{bmatrix} \begin{bmatrix} \tilde{\mathbf{U}}_1 & & & \tilde{\mathbf{C}}_1 \\ & \ddots & & \vdots \\ & & \tilde{\mathbf{U}}_K & \tilde{\mathbf{C}}_K \\ & & & \tilde{\mathbf{S}} \end{bmatrix}. \tag{27}$$

Here, $\mathbf{A}_{ii} \overset{(!)}{\approx} \tilde{\mathbf{L}}_i\tilde{\mathbf{U}}_i$ and all other blocks ($\tilde{\mathbf{B}}_i = \mathbf{A}_{i\Gamma}\tilde{\mathbf{U}}_i^{-1}$, $\tilde{\mathbf{C}}_i = \tilde{\mathbf{L}}_i^{-1}\mathbf{A}_{\Gamma i}$ and $\tilde{\mathbf{S}}$) are based on $\tilde{\mathbf{L}}_i$ and $\tilde{\mathbf{U}}_i$. Figure 4 depicts the sparsity pattern of the operator $\mathbf{A}_{h,p}$ and the block ILUT factorization $\mathbf{L}_{h,p} + \mathbf{U}_{h,p}$ for $p = 4$, $h = 2^{-4}$ and 4 patches. Note that, compared to the global ILUT factorization, the factorizations are obtained within each block. As a consequence, all non-zero entries of the global matrix $\mathbf{L}_{h,p} + \mathbf{U}_{h,p}$ stay within these blocks.

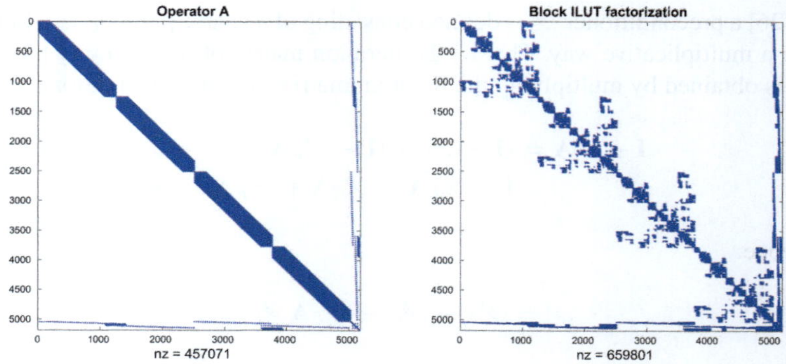

Fig. 4 Sparsity pattern of $\mathbf{A}_{h,p}$ (left) and $\mathbf{L}_{h,p} + \mathbf{U}_{h,p}$ (right) for $p = 4$, $h = 2^{-5}$ and 4 patches. The number of nonzero (nz) entries for both operators is provided as well

Computational Costs

In this subsection, the computational costs of the p-multigrid method adopting both smoothers are discussed. First, we consider the assembly costs of the considered operators after which both set-up and application costs of the block ILUT and global ILUT smoother are presented.

Assuming an element-based assembly loop with standard Gauss-quadrature, the assembly costs at level p for the stiffness matrix and transfer matrix are $\mathcal{O}(N_{\text{dof}}\, p^{3d})$ floating point operations (flops) [9]. Note that the stiffness matrix has to be assembled both at the high-order level and level $p = 1$. Furthermore, the (variationally) lumped mass matrices have to be assembled for the prolongation and restriction operator, at the cost of $\mathcal{O}(N_{\text{dof}})$ flops. It was shown in [37] that the assembly costs form a significant part of the total computational costs within the p-multigrid method. Alternative (and more efficient) assembly techniques exist [2, 8, 24], but will not be explored in this paper.

At the low-order level, Gauss-Seidel is applied as a smoother, which costs $\mathcal{O}(N_{\text{dof}})$ flops per smoothing step and has no set-up costs. The costs of setting up the block ILUT and global ILUT smoother at the high-order level differ significantly. The global ILUT factorization of $\mathbf{A}_{h,p}$ costs $\mathcal{O}(N_{\text{dof}}\, p^{2d})$ flops, provided that $f = 1$ and $\tau = 10^{-13}$ (i.e. no fill-in is allowed). Under these assumptions, applying the global ILUT smoother costs $\mathcal{O}(N_{\text{dof}}\, p^{d})$ flops. Setting up the block ILUT smoother consists of different steps. Here, we assume that the total number of degrees of freedom N_{dof} is given by $N_{\text{dof}} = K N_{\text{patch}} + N_{\text{interface}}$, where N_{patch} is the number of degrees of freedom associated to the interior of a single patch and $N_{\text{interface}}$ denotes the total number of degrees of freedom associated to the interface. In general, N_{patch} is significantly higher compared to $N_{\text{interface}}$.

First, an ILUT factorization is determined for all patches, which costs $\mathcal{O}(N_{\text{patch}}\, p^{2d})$ for each patch. The matrices $\tilde{\mathbf{B}}_i$, $\tilde{\mathbf{C}}_i$ $(i = 1, \ldots, K)$ are obtained by solving the following systems, respectively, with multiple matrices:

$$\mathbf{U}_i^\top \mathbf{B}_i^\top = \mathbf{A}_{i\Gamma}^\top, \qquad \mathbf{L}_i \mathbf{C}_i = \mathbf{A}_{\Gamma i}$$

Note that these solves only require forward substitutions and can therefore be performed efficiently and in parallel at the cost of $\mathscr{O}(N_{\text{patch}} p^d)$. Finally, $\tilde{\mathbf{S}}$ can be determined at the costs of $\mathscr{O}(N_{\text{interface}}^3)$, assuming the worst-case scenario of a full matrix $\mathbf{S}$. Applying the block ILUT smoother $\mathscr{O}(N_{\text{patch}} p^d)$ flops, assuming the smoother is applied in parallel.

The analysis of the computational costs show that the efficiency of the block ILUT smoother depends heavily on an efficient parallel implementation of this smoother. This paper, however, places the focus on the analysis of the above block ILUT smoother when applied to multipatch IgA discretizations and not on its efficient parallel implementation. Future research will focus on a parallel and distributed implementation of the considered block ILUT smoother.

5 Numerical Results

In this section, we apply both the global ILUT and the block ILUT smoother within a p-multigrid method for a variety of benchmarks. In particular, the spectral properties of the resulting p-multigrid are analyzed for different values of the mesh width h, the approximation order p and a different number of patches. Furthermore, the number of iterations to reach convergence are determined adopting both smoothers for the benchmarks described in Sect. 5.1.

5.1 Benchmarks

To analyze the block ILUT smoother throughout this paper, we consider the following two dimensional benchmarks:

Benchmark 1. Let Ω be the unit square, i.e. $\Omega = [0, 1]^2$, so that the geometry function is the identity and a simple shrinking of Ω_0 to the subdomain $\Omega^{(k)}$ in case of a multipatch discretization. The CDR equation is considered, where the exact solution u is given by $u(x, y) = \sin(\pi x)\sin(\pi y)$ and the coefficients are chosen as follows:

$$\mathbf{D} = \begin{bmatrix} 1.2 & -0.7 \\ -0.4 & 0.9 \end{bmatrix}, \qquad \mathbf{v} = \begin{bmatrix} 0.4 \\ -0.2 \end{bmatrix}, \qquad R = 0.3.$$

Benchmark 2. Let Ω be the quarter annulus with an inner and outer radius of 1 and 2, respectively. Here, Poisson's equation is considered, where the exact solution u is given by $u(x, y) = -(x^2 + y^2 - 1)(x^2 + y^2 - 4)xy^2$:

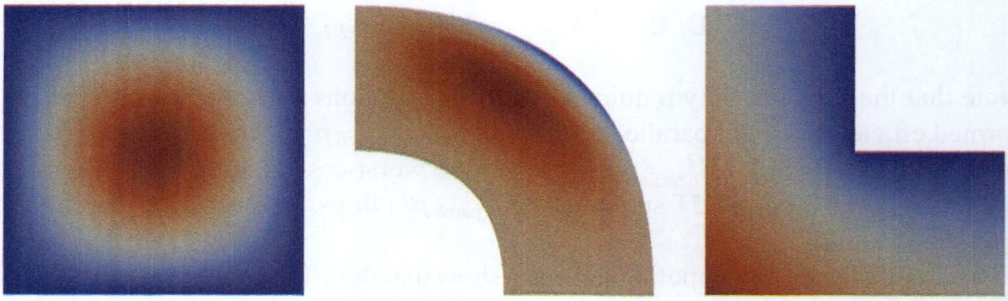

Fig. 5 Exact solutions on the three domains of the considered benchmarks

$$\mathbf{D} = \begin{bmatrix} 1 & 0 \\ 0 & 1 \end{bmatrix}, \quad \mathbf{v} = \begin{bmatrix} 0 \\ 0 \end{bmatrix}, \quad R = 0.$$

Benchmark 3. Let $\Omega = \{[-1, 1] \times [-1, 1]\} \backslash \{[0, 1] \times [0, 1]\}$ be an L-shaped domain. As with the previous benchmark, Poisson's equation is considered, where the exact solution u is given by

$$u(x, y) = \begin{cases} \sqrt[3]{x^2 + y^2} \sin\left(\frac{2\text{atan2}(y,x)-\pi}{3}\right) & \text{if } y > 0 \\ \sqrt[3]{x^2 + y^2} \sin\left(\frac{2\text{atan2}(y,x)+3\pi}{3}\right) & \text{if } y < 0 \end{cases},$$

where atan2 is the 2-argument arc tangent function. The right-hand side is chosen according to the exact solution.

For the first two benchmarks, homogeneous Dirichlet boundary conditions are applied on the entire boundary $\partial\Omega$, while for the third benchmark inhomogeneous Dirichlet boundary conditions are used given by the exact solution. Figure 5 shows the exact solution for all three benchmarks. All three geometries can be described by a single patch which is converted into a multipatch form by splitting Ω uniformly in both directions. As a consequence, the number of patches K is given by 4^S, where S denotes the number of splittings.

5.2 Spectral Analysis

For any multigrid method, the asymptotic convergence rate is determined by the spectral radius (the largest eigenvalue in absolute value) of the iteration matrix. This matrix can easily be obtained numerically following the procedure described in [39]. Table 1 shows the spectral radius obtained for the first benchmark for different values of p and K with the global ILUT and block ILUT smoother for $h = 2^{-5}$. As the spectral radii are obtained numerically, we use a relatively coarse grid.

For both smoothers, the spectral radius increases when the number of patches is increased. For higher values of p, the spectral radius decreases when block ILUT

Table 1 Asymptotic convergence rate of the p-multigrid adopting a global ILUT and block ILUT smoother for different values of the approximation order p and number of patches K

	$p = 2$		$p = 3$		$p = 4$	
# patches K	Global	Block	Global	Block	Global	Block
4	0.094	0.015	0.090	0.011	0.062	0.005
16	0.156	0.019	0.187	0.015	0.146	0.005
64	0.275	0.039	0.374	0.067	**3.659**	0.058

is applied as a smoother. With global ILUT, the dependence on p is more erratic. In particular, the resulting p-multigrid method is diverging for one of the configurations, indicated by a spectral radius larger than 1. Note that, for all configurations, the spectral radius obtained with the block ILUT smoother is significantly lower compared to the one obtained with the global ILUT smoother. As a consequence, the p-multigrid method (using block ILUT as a smoother) is expected to show superior convergence behaviour.

Figure 6 shows the spectrum of the iteration matrix for the p-multigrid method ($h = 2^{-5}$) adopting the global and block ILUT smoother. For the global ILUT smoother, the increased spectral radius for a higher number of patches can be related to single eigenvalues positioned further away from the origin. As a consequence, the use of an outer Krylov method is expected to mitigate this dependency on the number of patches. This has been shown and verified numerically in literature for the global ILUT solver, see [36]. Recall that our specific choice of the prolongation and restriction operator leads to a non-symmetric p-multigrid method, making the use of a Krylov method suited for non-symmetric systems, like Bi-CGSTAB, necessary. For the block ILUT smoother, the spectra are more clustered around the origin, resulting in a lower spectral radius. Furthermore, the dependency on the number of patches is hardly visible in the spectra, although it is visible in Table 1. Therefore, the use of an outer Krylov solver is not expected to significantly decrease the dependency on the number of patches when block ILUT is applied as a smoother.

5.3 *Iteration Numbers*

Based on the spectral analysis in the previous subsection, the use of the block ILUT smoother is expected to show improved iteration numbers compared to the p-multigrid method equipped with the global ILUT smoother. In this section, the number of p-multigrid cycles needed to achieve convergence is determined using both smoothers. As a stopping criterion, a reduction of the relative residual by 10^8 is chosen. We consider a random vector as an initial guess, where each entry is sampled from a uniform distribution on the interval $[-1, 1]$ using the same seed.

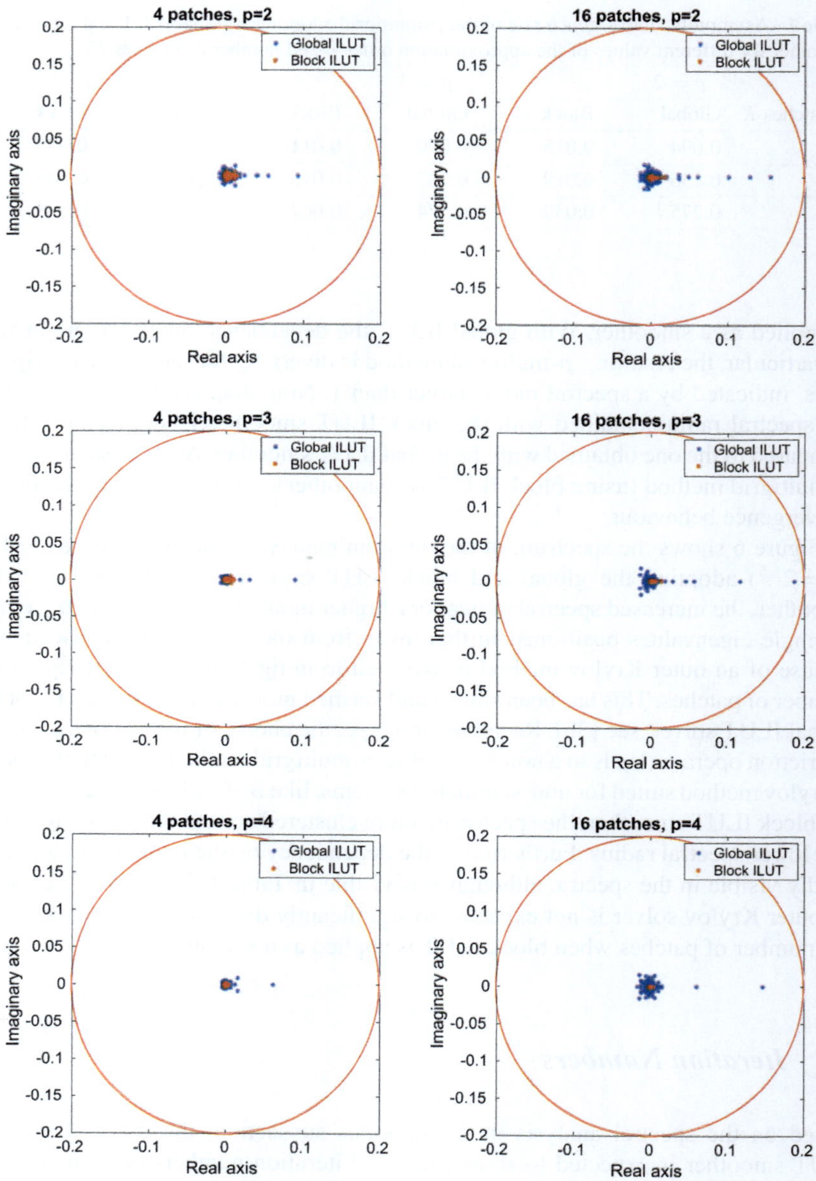

Fig. 6 Spectra of the iteration matrix for the first benchmark obtained with p-multigrid ($h = 2^{-5}$). Here, global ILUT and block ILUT are applied as a smoother

Table 2 shows the number of multigrid cycles needed to achieve convergence for the considered benchmarks with block ILUT as a smoother. The number of iterations needed to achieve convergence with block ILUT as a smoother is, in general, independent of h. For two configurations, however, the p-multigrid method is diverging (indicated by $-$). A small dependence on the number of patches is visible on the coarse mesh, but this dependency becomes negligible for smaller values of h. For a fixed number of patches and higher values of p, the number of iterations slightly decreases. This behavior has been observed before in literature (See [37]) and is related to the fact that the fill-in allowed increases with p. It should be noted, however, that the extra fill-in significantly increases the set-up times of the smoother for higher values of p.

The number of p-multigrid cycles needed to reach convergence with the global ILUT smoother are shown in Table 3. In general, the iteration numbers are (more or less) independent of h and p. However, a dependency on the number of patches can be observed, in particular for coarser meshes. Furthermore, the number of iterations needed to reach convergence is higher for all configurations compared to the use of block ILUT as a smoother.

As the number of p-multigrid cycles when adopting (block) ILUT as a smoother is very low, the use of (block) ILUT as a stand-alone solver has been investigated as well. Table 4 shows the number of iterations needed for the second benchmark when both block ILUT and global ILUT are applied as a solver. For all configurations, the number of iterations is significantly higher compared to the number of p-multigrid cycles (see Tables 2 and 3). Furthermore, a dependency on the mesh width h can be

Table 2 Number of p-multigrid V-cycles needed to achieve convergence using block ILUT as a smoother

	$p=2$			$p=3$			$p=4$			$p=5$		
	# patches			# patches			# patches			# patches		
	4	16	64	4	16	64	4	16	64	4	16	64
(a) CDR-equation on the unit square												
$h=2^{-5}$	4	4	7	3	3	5	2	3	5	2	2	4
$h=2^{-6}$	4	4	5	3	3	4	3	3	4	3	3	3
$h=2^{-7}$	4	4	4	3	3	3	3	3	3	4	3	3
(b) Poisson's equation on a quarter annulus												
$h=2^{-5}$	3	4	4	3	3	4	2	2	4	2	2	–
$h=2^{-6}$	3	3	4	3	3	4	3	3	3	3	3	3
$h=2^{-7}$	3	3	3	3	3	3	3	3	3	–	6	3
(c) Poisson's equation on an L-shaped domain												
$h=2^{-5}$	3	3	4	2	3	4	2	2	3	2	2	2
$h=2^{-6}$	3	3	3	3	3	3	2	2	3	2	2	2
$h=2^{-7}$	3	3	3	2	3	3	2	2	3	2	2	3

Table 3 Number of p-multigrid cycles needed to achieve convergence using global ILUT as a smoother

	$p=2$			$p=3$			$p=4$			$p=5$		
	# patches			# patches			# patches			# patches		
	4	16	64	4	16	64	4	16	64	4	16	64
(a) CDR-equation on the unit square												
$h=2^{-5}$	6	8	11	6	9	15	6	8	15	5	7	14
$h=2^{-6}$	6	7	8	6	8	10	7	9	13	7	8	13
$h=2^{-7}$	6	6	7	6	7	8	7	7	10	6	8	12
(b) Poisson's equation on a quarter annulus												
$h=2^{-5}$	5	7	9	5	7	11	4	6	–	4	6	–
$h=2^{-6}$	5	5	7	5	7	10	6	7	11	5	7	10
$h=2^{-7}$	5	5	5	5	6	8	5	6	10	5	7	11
(c) Poisson's equation on an L-shaped domain												
$h=2^{-5}$	6	7	11	5	8	13	5	6	11	4	5	–
$h=2^{-6}$	6	6	8	6	8	10	5	8	12	5	7	10
$h=2^{-7}$	7	7	8	6	7	8	5	6	10	5	7	12

Table 4 Number of iterations needed to achieve convergence for Poisson's equation on a quarter annulus. Here, block ILUT and global ILUT are used as a solver

	$p=2$			$p=3$			$p=4$			$p=5$		
	# patches			# patches			# patches			# patches		
	4	16	64	4	16	64	4	16	64	4	16	64
(a) Block ILUT												
$h=2^{-5}$	28	22	16	14	10	14	8	6	10	4	4	–
$h=2^{-6}$	112	90	68	48	40	26	24	18	12	14	10	8
$h=2^{-7}$	406	386	312	178	162	130	86	64	54	48	38	26
(b) Global ILUT												
$h=2^{-5}$	56	64	100	22	30	56	14	16	–	8	12	–
$h=2^{-6}$	220	218	234	84	84	100	40	42	56	24	26	36
$h=2^{-7}$	710	700	782	276	302	296	138	138	148	80	80	76

observed, making the use of (block) ILUT as a solver even less efficient for smaller values of h.

Alternatively, a single p-multigrid cycle can be applied as a preconditioner within a Bi-CGSTAB method. Table 5 shows the number of Bi-CGSTAB iterations needed to achieve convergence for the second benchmark, Poisson's equation on the quarter annulus. Compared to the use of p-multigrid as a solver, the number of Bi-CGSTAB iterations is only slightly lower when block ILUT is applied as a smoother. However, for the global ILUT smoother, a significant reduction of the iteration numbers can be

Table 5 Number of Bi-CGSTAB iterations needed to achieve convergence for the second benchmark. Here, a single p-multigrid cycle is applied as a preconditioner using block ILUT and global ILUT as smoother

	$p = 2$			$p = 3$			$p = 4$			$p = 5$		
	# patches			# patches			# patches			# patches		
	4	16	64	4	16	64	4	16	64	4	16	64
(a) Block ILUT smoother												
$h = 2^{-5}$	2	2	2	2	2	2	1	1	2	1	1	50
$h = 2^{-6}$	2	2	2	2	2	2	2	2	2	2	2	2
$h = 2^{-7}$	2	2	2	2	2	2	2	2	2	34	3	2
(b) Global ILUT smoother												
$h = 2^{-5}$	2	3	3	2	3	4	2	3	13	2	3	353
$h = 2^{-6}$	2	2	3	3	3	4	2	3	4	2	3	3
$h = 2^{-7}$	2	2	3	2	3	3	2	3	3	2	3	4

achieved, as expected from the spectral analysis. Note that the use of a Krylov solver restores stability for configurations that diverged before, but lead to a relatively high number of iterations.

6 Application: Yeti Footprint

In the previous section, we applied the proposed block ILUT smoother on benchmarks defined on geometries consisting of a different number of patches. In general, the application of both smoothers within a p-multigrid method resulted in iteration numbers independent of h and p. A small dependency on the number of patches was noticeable, however, in particular on coarser meshes.

In this section, the block ILUT and global ILUT smoother are applied on a more challenging geometry. We consider Poisson's equation on a two dimensional Yeti footprint, where the exact solution is given by $u(x, t) = \sin(5\pi x)\sin(5\pi y)$:

$$\begin{aligned} -\Delta u &= 50\pi^2 \sin(5\pi x)\sin(5\pi y), && (x, y) \in \Omega, \\ u &= 0, && (x, y) \in \partial\Omega. \end{aligned}$$

This benchmark has been considered in literature in the context of parallel multigrid methods for IgA, see [17]. Figure 7 shows the Yeti footprint multipatch geometry (left) and its exact solution (right) consisting of 21 patches.

Table 6 shows the number of multigrid cycles needed to reach convergence for the p-multigrid method adopting both smoothers and the aforementioned convergence criterion. For all configurations, the use of block ILUT leads to a lower number of cycles compared to the use of global ILUT. Furthermore, the number of iterations

Fig. 7 Underlying multipatch geometry (left) and the exact solution (right) of the Yeti footprint

Table 6 Number of multigrid cycles needed to reach convergence for the Yeti footprint for a p-multigrid method with global and block ILUT as a smoother

	$p = 2$		$p = 3$		$p = 4$		$p = 5$	
	Global	Block	Global	Block	Global	Block	Global	Block
$h = 2^{-3}$	5	4	4	2	4	2	4	2
$h = 2^{-4}$	8	4	5	3	5	3	4	2
$h = 2^{-5}$	8	4	6	3	5	3	5	3

needed to reach convergence is, in general, lower for higher values of the approximation order p. Results are, to a large extent, comparable with the ones presented in the previous section. A small h-dependence can be observed, however, for smaller values of p when applying the global ILUT smoother.

7 Conclusions

In this paper, we presented a block ILUT smoother for p-multigrid methods in Isogeometric Analysis for multipatch geometries. The block ILUT smoother is based on a preconditioner, which has been developed in the context of domain decomposition methods [26]. The key idea is to make use of the block structure of the resulting system matrix when setting up the smoother. As a consequence, the ILUT factorizations are only obtained for smaller blocks, where the number of blocks depends on the number of patches describing the geometry.

A spectral analysis has shown that this smoother is more effective when applied within a p-multigrid method compared to a global ILUT smoother. Numerical results, obtained for a variety of two dimensional benchmarks, indeed showed that the number of iterations needed to achieve convergence is lower for all considered configurations when adopting the block ILUT smoother. The presented block ILUT smoother lends itself for a parallel implementation, which will be the focus of future research. In particular, focus will lie on the comparison with the global ILUT smoother in terms of computational efficiency.

References

1. Amestoy, P.R., Davis, T.A., Duff, I.S.: An approximate minimum degree ordering algorithm. SIAM J. Matrix Anal. Appl. **17**, 886–905 (1996)
2. Antolin, P., Buffa, A., Calabro, F., Martinelli, M., Sangalli, G.: Efficient matrix computation for tensor-product isogeometric analysis: The use of sum factorization. Comput. Methods Appl. Mech. Eng. **285**, 817–828 (2015)
3. Beirao da Veiga, L., Cho, D., Pavarino, L.F., Scacchi, S.: Overlapping Schwarz methods for isogeometric analysis. SIAM J. Numer. Anal. **50**, 1394–1416 (2012)
4. de Boor, C.: A Practical Guide to Splines, 1st edn. Springer, New York (1978)
5. Brandt, A.: Multi-level adaptive solutions to boundary-value problems. Math. Comput. **31**, 333–390 (1977)
6. Brenner, S.C., Scott, L.R.: The Mathematical Theory of Finite Element Methods. Texts in Applied Mathematics. Springer, New York (1994)
7. Briggs, W.L., Henson, V.E., McCormick, S.F.: A Multigrid Tutorial, 2nd edn. SIAM, Philadelphia (2000)
8. Calabro, F., Sangalli, G., Tani, M.: Fast formation of isogeometric Galerkin matrices by weighted quadrature. Comput. Methods Appl. Mech. Eng. **316**, 606–622 (2017)
9. Collier, N., Dalcin, L., Pardo, D., Calo, V.M.: The cost of continuity: performance of iterative solvers on Isogeometric Finite Elements. SIAM J. Sci. Comput. **35**, 767–784 (2013)
10. Donatelli, M., Garoni, C., Manni, C., Capizzano, S., Speleers, H.: Symbol-based multigrid methods for Galerkin B-spline isogeometric analysis. SIAM J. Numer. Anal. **55**, 31–62 (2017)
11. Fidkowski, K.J., Oliver, T.A., LU, J., Darmofal, D.L.: p-Multigrid solution of high-order discontinuous Galerkin discretizations of the compressible Navier-Stokes equations. J. Comput. Phys. **207**, 92–113 (2005)
12. Gahalaut, K.P.S., Kraus, J.K., Tomar, S.K.: Multigrid methods for isogeometric discretizations. Comput. Methods Appl. Mech. Eng. **253**, 413–425 (2013)
13. Gao, L., Calo, V.: Fast isogeometric solvers for explicit dynamics. Comput. Methods Appl. Mech. Eng. **274**, 19–41 (2014)
14. Guennebaud, G., Jacob, B.: Eigen v3 (2010). http://eigen.tuxfamily.org
15. Hackbush, W.: Multi-Grid Methods and Applications. Springer, Berlin (1985)
16. Helenbrook, B., Mavriplis, D., Atkins, H.: Analysis of p-multigrid for continuous and discontinuous finite element discretizations. In: 16th AIAA Computational Fluid Dynamics Conference, Fluid Dynamics and Co-located Conferences (2003)
17. Hofer, C., Takacs, S.: A parallel multigrid solver for multi-patch isogeometric analysis. In: Apel, T., Langer, U., Meyer, A., Steinbach, O., (eds.), Advanced Finite Element Methods with Applications. FEM 2017. Lecture Notes in Computational Science and Engineering, vol. 128, pp. 205–219. Springer, Cham (2019)
18. Hofreither, C., Takacs, S., Zulehner, W.: A robust multigrid method for isogeometric analysis in two dimensions using boundary correction. Comput. Methods Appl. Mech. Eng. **316**, 22–42 (2017)

19. Hofreither, C., Takacs, S.: Robust multigrid for isogeometric analysis based on stable splittings of spline spaces. SIAM J. Numer. Anal. **55**, 2004–2024 (2017)
20. Hughes, T.J.R., Cottrell, J.A., Bazilevs, Y.: Isogeometric analysis: CAD, finite elements, NURBS, exact geometry and mesh refinement. Comput. Methods Appl. Mech. Eng. **194**, 4135–4195 (2005)
21. Hughes, T.J.R., Reali, A., Sangalli, G.: Duality and unified analysis of discrete approximations in structural dynamics and wave propagation: comparison of p-method finite elements with k-method NURBS. Comput. Methods Appl. Mech. Eng. **197**(49–50), 4104–4124 (2008)
22. Luo, H., Baum, J.D., Löhner, R.: A p-multigrid discontinuous Galerkin method for the Euler equations on unstructured grids. J. Comput. Phys. **211**, 767–783 (2006)
23. Luo, H., Baum, J.D., Löhner, R.: Fast p-multigrid discontinuous Galerkin method for compressible flows at all speeds. AIAA J. **46**, 635–652 (2008)
24. Mantzaflaris, A., Jüttler, B., Khoromskij, B.N., Langer, U.: Low rank tensor methods in Galerkin-based isogeometric analysis. Comput. Methods Appl. Mech. Eng. **316**, 1062–1085 (2017)
25. Nicolaides, R.A.: Deflation of conjugate gradients with applications to boundary value problems. SIAM J. Numer. Anal. **24**, 355–365 (1987)
26. Nievinksi, I.C.L., Souza, M., Goldfeld, P., Augusto, D.A., Rogrigues, J.R.P., Carvalho, L.M.: Parallel implementation of a two-level algebraic ILU(k)-based domain decomposition preconditioner. Tendencias em Matematics Aplicada e Computatcional **19**, 59–77 (2018)
27. de la Riva, A., Rodrigo, C., Gaspar, F.: An efficient multigrid solver for isogeometric analysis (2018). arXiv:1806.05848v1
28. Saad, Y.: ILUT: a dual threshold incomplete LU factorization. Numer. Linear Algebra Appl. **1**, 387–402 (1994)
29. Saad, Y.: SPARSKIT: a basic tool kit for sparse matrix computations (1994)
30. Sampath, R.S., Biros, G.: A parallel geometric multigrid method for finite elements on octree meshes. SIAM J. Sci. Comput. **32**, 1361–1392 (2010)
31. van Slingerland, P., Vuik, C.: Fast linear solver for diffusion problems with applications to pressure computation in layered domains. Comput. Geosci. **18**, 343–356 (2014)
32. Sogn, J., Takacs, S.: Robust multigrid solvers for the biharmonic problem in isogeometric analysis. Comput. Methods Appl. Mech. Eng. **77**, 105–124 (2019)
33. Tang, J.M., Saad, Y.: Domain-decomposition-type methods for computing the diagonal of a matrix inverse. SIAM J. Sci. Comput. **33**, 2823–2847 (2011)
34. Sangalli, G., Tani, M.: Isogeometric preconditioners based on fast solvers for the Sylvester equation. SIAM J. Sci. Comput. **38**, 3644–3671 (2016)
35. Tielen, R., Möller, M., Vuik, C.: Efficient multigrid based solvers for Isogeometric Analysis. In: Proceedings of the 6th European Conference on Computational Mechanics and the 7th European Conference on Computational Fluid Dynamics, Glasgow (2018)
36. Tielen, R., Möller, M., Vuik, C.: Efficient p-multigrid solvers for multipatch geometries in Isogeometric Analysis. In: Proceedings of the 3rd Conference on Isogeometric Analysis and Applications, Delft, The Netherlands (2018)
37. Tielen, R., Möller, M., Göddeke, D., Vuik, C.: p-multigrid methods and their comparison to h-multigrid methods within isogeometric analysis. Comput. Methods Appl. Mech. Eng. **372**, 113347 (2020)
38. Tielen, R., Möller, M., Vuik, C.: A direct projection to low-order level for p-multigrid methods in isogeometric analysis. In: Proceedings of the 13th Conference on European Numerical Mathematics and Advanced Applications (ENUMATH). Egmond aan Zee, The Netherlands (2019)
39. Trottenberg, U., Oosterlee, C., Schüller, A.: Multigrid. Academic, New York (2001)

Completeness Characterization of Type-I Box Splines

Nelly Villamizar, Angelos Mantzaflaris, and Bert Jüttler

Abstract We present a completeness characterization of box splines on three-directional triangulations, also called *type-I* box spline spaces, based on edge-contact smoothness properties. For any given type-I box spline, of specific maximum degree and order of global smoothness, our results allow to identify the local linear subspace of polynomials spanned by the box spline translates. We use the global super-smoothness properties of box splines as well as the additional super-smoothness conditions at edges to characterize the spline space spanned by the box spline translates. Subsequently, we prove the completeness of this space with respect to the local polynomial space induced by the box spline translates. The completeness property allows the construction of hierarchical spaces spanned by the translates of box splines for any polynomial degree on multilevel type-I grids. We provide a basis for these hierarchical box spline spaces under explicit geometric conditions of the domain.

1 Introduction

Box splines are locally supported piecewise polynomial functions defined on uniform grids. They were first introduced by de Boor and DeVore in [10], and are considered a generalization of the univariate B-spline functions to the multivariate setting. From a geometric point of view, box splines can be seen as density functions of the shadows of higher dimensional boxes and half-boxes [35]. We remark that they can also be studied as a special case of the so-called simplex splines [7]. Box splines possess a number of useful properties that make them well-suited for applications [13, 34].

N. Villamizar (✉)
Swansea University, Swansea, United Kingdom
e-mail: n.y.villamizar@swansea.ac.uk

A. Mantzaflaris
Inria Sophia Antipolis–Méditerranée, Université Côte d'Azur, Nice, France
e-mail: angelos.mantzaflaris@inria.fr

B. Jüttler
Johannes Kepler University Linz & RICAM, Linz, Austria
e-mail: Bert.Juettler@jku.at

C. Manni and H. Speleers (eds.), *Geometric Challenges in Isogeometric Analysis*,
Springer INdAM Series 49, https://doi.org/10.1007/978-3-030-92313-6_12

For instance, it has been shown that box splines have small support (a few cells of the underlying grid), they are non-negative, form a partition of unity, and are refinable, i.e., the box spline spaces on refined grids are nested [3, 4].

Box splines can be defined from an arbitrary set of directional vectors in $\mathbb{R}^n$, but of particular interest for Geometric Design are box splines surfaces which are defined on uniform triangulations of the plane. In this article, we focus on *type-I box splines*. They are splines defined on three-directional meshes that are commonly known as *type-I triangulations* of $\mathbb{R}^2$.

From the rich literature on box splines, we mention a few monographs and survey articles which include [4–6, 12, 35], and a few representative publications on two specific topics. Firstly, a substantial number of results on the *approximation power* of box splines is described in the literature, e.g., [11, 30, 36, 38]. Secondly, several publications discuss techniques for the *efficient manipulation* of box spline bases. Sablonnière pioneered local Bernstein-Bézier representations for a variety of splines, including type-I box splines, see [37] and references therein. A general stable evaluation algorithm is devised in [27]. In [26] the problem of efficient evaluation of box splines is addressed by making use of the local Bernstein representation of basis functions on each triangle. Recent applications of box splines include surface fitting [24], and solving linear elasticity problems in isogeometric analysis [20]. In other areas of mathematics, the theory of box splines has been proved useful to compute the volume of polytopes, and to deal with the integration of continuous functions over polytopes [44].

In this article, we are interested in the linear spaces of spline functions generated by the translates of any fixed type-I box spline. These spline spaces share good approximation properties. For example, the set of translates of any type-I box spline form a partition of unity on $\mathbb{R}^2$, they are globally, and also locally, linearly independent. These properties of type-I box splines were studied by Dahmen and Micchelli in [14] and Jia in [21]. In particular, they investigated the linear independence of translates of a box spline in [9, 22].

The linear independence property implies that the set of translates of any type-I box spline constitutes a basis for the spline space they span. In general this property is not satisfied for box spline functions associated to other uniform partitions, and that makes type-I box splines particularly relevant for applications. For instance, the set of translates of box splines defined from a set of four directional vectors, the so-called type-II box splines, are linearly dependent. For a concise treatment of type-II box splines, further references, and alternative proofs of box spline properties, see [29, Chap. 12] and [5, Chap. 2].

A second interesting feature of type-I box splines is that, although for any fixed type-I box spline of degree d and order of smoothness r, the box spline translates form a basis, in general these translates do not generate all possible piecewise polynomial functions of degree $\leq d$ and global smoothness r over the three-directional mesh. More precisely, the box spline translates span a proper subspace of the space of C^r-continuous spline functions $\mathscr{S}_d^r(G)$ of degree at most d on a three-directional mesh G (cf. Fig. 1), for any $d > 1$. Thus, if the domain is taken as the entire mesh G, or as an infinite collection of triangles in G, then both spaces are infinite dimensional.

However, if the domain is restricted to a finite collection of triangles Ω, as it is the usual setting in practice, then their finite dimension differ. Explicit dimension formulas in terms of the combinatorics of the domain Ω are well known for both spaces, dim $\mathcal{S}_d^r(\Omega)$ can be computed using homological methods [33], or Bernstein-Bézier methods as in [5, Chap. 2].

The understanding of the space spanned by (a finite set of) translates of type-I box splines in terms of continuity and polynomial or spline function inclusion is paramount for inducing the approximation power of the considered finite dimensional space. We refer to [11] for early results in this direction based on truncated power bases and [38] for results based on the construction of quasi-interpolants. In our work we deduce hierarchical space and basis constructions for adaptively refined type-I grids, thus extending the aforementioned approximation results to the hierarchical setting.

In the present paper we provide a characterization for the space spanned by box splines translates based on supersmooth conditions across the edges of the underlying partition. We prove that for any fixed degree d and order of global smoothness r, the space of splines satisfying these extra local smoothness conditions is precisely the space spanned by the translates of the corresponding type-I box spline. From this we deduce that the type-I box spline spaces are complete with respect to the local polynomial space induced by the box spline translates. The proof of this result uses the Fourier transform of box splines, as well as the algebraic properties of the Bernstein-Bézier representations of type-I box splines. We generalize the classic definition of spline space in Definition 1 to link the spline polynomial pieces to specific polynomial subspaces of $\mathbb{R}[x, y]$ and present the main result in the paper in Sect. 4.

Furthermore, in Sect. 5, we apply the completeness characterization of box splines into the construction of hierarchical spline spaces based on local refinements of a type-I triangulation. Hierarchical splines constitute a well-established approach to adaptive refinement in geometric modeling [15] and numerical analysis [39, 43]. Hierarchical tensor-product spline spaces were introduced by Kraft in [28] using a selection mechanism for B-splines. The method has been refined leading to spline basis with better approximation properties, such as the partition of unity property, strong stability and full approximation power [18, 19, 41, 45]. It has also been adapted to Powell-Sabin splines [40], Zwart-Powell elements and B-spline-type basis functions for cubic splines on regular grids [46]. In an earlier article, we constructed a hierarchical basis for quartic C^2-continuous box splines [42]. Quartic hierarchical box splines spaces have also been studied and used for surface fitting applications by Kang, Chen and Deng in [24, 25]. Truncated hierarchical type-I box splines were considered in [20, 23] in connection to isogeometric analysis applications. Other subdivision schemes has been explored in [16, 32]. A C^1-continuous scheme based on cubic half-box splines was presented in [1].

The results we present in this article generalize our previous work [42] on quartic box splines. Our results apply to type-I box splines of any polynomial degree with no restriction on the symmetry of their support.

The remainder of this paper is organized as follows. In Sect. 2 we introduce the relevant notation for type-I triangulations, spline functions and the directional derivatives. Section 3 concerns the definition and properties of type-I box spline spaces. We define the space of translates and recall existing results on local and global smoothness of these functions. In Sect. 4 we prove Lemma 2 which is the main result in the paper, and corresponds to the edge-contact characterization for type-I box splines. In Sect. 5 we construct the hierarchical type-I meshes and the corresponding hierarchical box spline spaces. This construction follows the approach presented in [31, 42]. We conclude the paper with some final remarks in Sect. 6.

2 Preliminaries

Throughout this article we assume that G is the uniform type-I triangulation of the real plane $\mathbb{R}^2$, see Fig. 1. This triangulation is obtained by drawing in the north-east diagonals in the bi-infinity grid with grid lines at the integers. This triangulation of the plane is associated to three directional vectors, namely $\boldsymbol{e}_1 = (1, 0)$, $\boldsymbol{e}_2 = (0, 1)$ and $\boldsymbol{e}_3 = (1, 1)$, and therefore is also called a three directional mesh. Each line of G is parallel to one of these vectors and go through the points of the integer grid $\mathbb{Z}^2$.

We consider G as the union of T, E and V, where T denotes the collection of triangles in G, which are considered as open sets in $\mathbb{R}^2$, E is the collection of all edges, and $V = \mathbb{Z}^2$ is the collection of all vertices. The set of edges is the disjoint union $E = E_1 \sqcup E_2 \sqcup E_3$, where E_i is the set of edges that are parallel to the vector $\boldsymbol{e}_i$. The edges in E_i are called edges of *type i*.

The combinatorial closure of a triangle $\triangle \in T$, denoted by $\hat{\triangle}$, is the set consisting of the vertices and edges of $\triangle$, and $\triangle$ itself. Analogously, $\hat{\varepsilon}$ of an edge $\varepsilon \in E$ is the set consisting of the edge itself and its two vertices.

A multicell domain M is the triangulation in $\mathbb{R}^2$ induced by a finite set of triangles $\{\triangle_1, \ldots, \triangle_m\} \subset T$, i.e.,

$$M = \bigcup_{i=1}^{m} \hat{\triangle}_i .$$

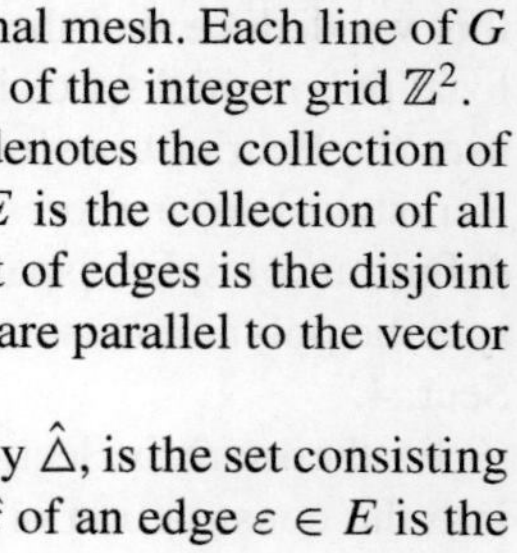

Fig. 1 Uniform type-I triangulation (or three-directional mesh) of $\mathbb{R}^2$ associated to the directional vectors $\boldsymbol{e}_1 = (1, 0)$, $\boldsymbol{e}_2 = (0, 1)$ and $\boldsymbol{e}_3 = (1, 1)$. We denote this grid as G

This means, that for every triangle $\triangle_i \in M$, all the vertices and edges of $\triangle_i$ are considered as elements of M. The subspace of $\mathbb{R}^2$ defined by the (topological) closure $\overline{\triangle}_i$ of the triangles $\triangle_i$ defining the multicell domain M will be denoted by M^*, namely

$$M^* = \bigcup_{\triangle \in M} \overline{\triangle}. \tag{1}$$

Given a multicell domain M, the *diamond of an edge* ε is defined as the union of all (at most two) triangles of M which have ε as an edge, that is

$$\Diamond(\varepsilon) = \bigcup_{\triangle \in M,\, \varepsilon \in \hat{\triangle}} \hat{\triangle}.$$

Similarly, the *diamond of a vertex* ν is defined by

$$\Diamond(\nu) = \bigcup_{\triangle \in M,\, \nu \in \hat{\triangle}} \hat{\triangle},$$

which is the union of the (at most six) triangles $\triangle$ in M such that ν is a vertex of $\hat{\triangle}$.

Notice that the diamond $\diamond(\cdot)$, of an edge or a vertex, depends on the multicell domain M. From the context, it will be clear the particular domain we are considering in each case.

We denote by $\mathbb{R}[x, y]$ the space of bivariate polynomials over the real numbers, and for $d \geq 0$, $\mathscr{P}_d \subseteq \mathbb{R}[x, y]$ is the set of all bivariate polynomials in x and y of total degree $\leq d$. In our presentation, the polynomial pieces that define the splines are taken from a finite vector subspace $\mathscr{V}$ of $\mathbb{R}[x, y]$. This subspace $\mathscr{V}$ is not necessarily the same as $\mathscr{P}_d$ for any polynomial degree d, it may be a proper linear subspace. In this setting, we define the space of continuous splines $\mathbb{P}(M, \mathscr{V})$ on a multicell domain M as follows.

Definition 1 Given a multicell domain M, and a vector subspace $\mathscr{V} \subseteq \mathbb{R}[x, y]$, we define $\mathbb{P}(M, \mathscr{V})$ as the *set of piecewise polynomials* functions on M, i.e.,

$$\mathbb{P}(M, \mathscr{V}) = \left\{ f \in C^0(M^*) \colon f|_{\triangle} \in \mathscr{V}|_{\triangle} \text{ for each triangle } \triangle \in M \right\},$$

where M^* is as defined in Eq. (1), and $f|_{\triangle}$ denotes the restriction of the function f to the triangle $\triangle$, and $\mathscr{V}|_{\triangle}$ is the restriction to $\triangle$ of the polynomials in $\mathscr{V}$ (seen as functions on $\mathbb{R}^2$).

In particular, when $\mathscr{V} = \mathscr{P}_d$, the space $\mathbb{P}(M, \mathscr{V})$ coincides with the usual space of C^0-continuous splines (or piecewise polynomial functions) on M of degree at most d.

For any index $\mathbf{s} = (s_1, s_2, s_3) \in \mathbb{Z}^3_{\geq 0}$, that we also called *regularity vector*, we consider the *mixed directional derivative operator*

$$\begin{aligned} D_{\boldsymbol{s}}\colon \mathbb{R}[x,y] &\to \mathbb{R}[x,y] \\ p &\mapsto (\nabla\cdot\boldsymbol{e}_1)^{s_1}(\nabla\cdot\boldsymbol{e}_2)^{s_2}(\nabla\cdot\boldsymbol{e}_3)^{s_3}(p)\,. \end{aligned}$$

For a given multicell domain M and a polynomial vector space $\mathscr{V}$, we extend the operator $D_{\boldsymbol{s}}$ to elements $f \in \mathbb{P}(M,\mathscr{V})$ by applying $D_{\boldsymbol{s}}$ to the restrictions $f|_{\triangle}$, namely

$$(D_{\boldsymbol{s}} f)|_{\triangle} = D_{\boldsymbol{s}}(f|_{\triangle})\,, \text{ for each triangle } \triangle \in M.$$

If $\boldsymbol{x}_0$ is a point on the boundary of triangle $\triangle$, i.e., if $\boldsymbol{x}_0$ is a vertex or a point on an edge of $\triangle$, we define $(D_{\boldsymbol{s}} f|_{\triangle})|_{\boldsymbol{x}_0}$ as $\lim_{\boldsymbol{x}\to\boldsymbol{x}_0}(D_{\boldsymbol{s}} f|_{\triangle})|_{\boldsymbol{x}}$. By an abuse of notation, we say that for $D_{\boldsymbol{s}} f \in C^0(M^*)$ if $(D_{\boldsymbol{s}} f|_{\triangle})|_{\boldsymbol{x}_0} = (D_{\boldsymbol{s}} f|_{\triangle'})|_{\boldsymbol{x}_0}$ for any pair of triangles $\triangle, \triangle' \in M$ such that $\boldsymbol{x}_0 \in \overline{\triangle} \cap \overline{\triangle}'$.

Definition 2 For a given index set $I \subset \mathbb{Z}^3_{\geq 0}$, and a vector space of functions $\mathscr{V} \subseteq \mathbb{R}[x,y]$, we define the space of functions $\mathbb{D}_I(M,\mathscr{V})$ on a multicell domain M by

$$\mathbb{D}_I(M,\mathscr{V}) = \big\{ f \in \mathbb{P}(M,\mathscr{V})\colon D_{\boldsymbol{s}} f \in C^0(M^*) \text{ for all } \boldsymbol{s} \in I \big\},$$

where M^* is as defined in Eq. (1).

Remark 1 Using the notation in Definition 2 we have

$$\mathbb{D}_{\emptyset}(M,\mathscr{V}) = \mathbb{P}(M,\mathscr{V})\,.$$

If $\mathscr{S}^r_d(M)$ denotes the space of globally C^r-continuous spline functions on M of degree at most d, then $\mathscr{S}^r_d(M)$ can be written as $\mathbb{D}_I(M,\mathscr{P}_d)$, where $I = \big\{\boldsymbol{s} \in \mathbb{Z}^3_{\geq 0}\colon s_1+s_2+s_3 \leq r\big\}$.

In this paper, for a given multicell domain M, we shall consider piecewise polynomial functions, or splines, on M with a specific order of smoothness associated to each of the three directions associated to the grid G. Namely, for a regularity vector $\boldsymbol{d} = (d_1,d_2,d_3) \in \mathbb{Z}^3_{\geq 0}$ we define the index sets:

$$\begin{aligned} I^{\boldsymbol{d}}_1 &= \big\{\boldsymbol{s} \in \mathbb{Z}^3_{\geq 0}\colon s_2+s_3 \leq d_1\big\}, \\ I^{\boldsymbol{d}}_2 &= \big\{\boldsymbol{s} \in \mathbb{Z}^3_{\geq 0}\colon s_1+s_3 \leq d_2\big\}, \\ I^{\boldsymbol{d}}_3 &= \big\{\boldsymbol{s} \in \mathbb{Z}^3_{\geq 0}\colon s_1+s_2 \leq d_3\big\}, \end{aligned}$$

and consider the spline space $\mathbb{S}^{\boldsymbol{d}}(M,\mathscr{V})$ defined as follows.

Definition 3 For a multicell domain M of the three directional grid G, a vector space $\mathscr{V} \subseteq \mathbb{R}[x,y]$, and a vector $\boldsymbol{d} \in \mathbb{Z}^3$, the *spline space with edge smoothness* $\boldsymbol{d}$ *on* M denoted $\mathbb{S}^{\boldsymbol{d}}(M,\mathscr{V})$ is defined as the set of piecewise polynomial functions on M such that the derivatives of order $\boldsymbol{s} \in I^{\boldsymbol{d}}_i$ are continuous across the edges of type i for $i = 1, 2, 3$. More precisely,

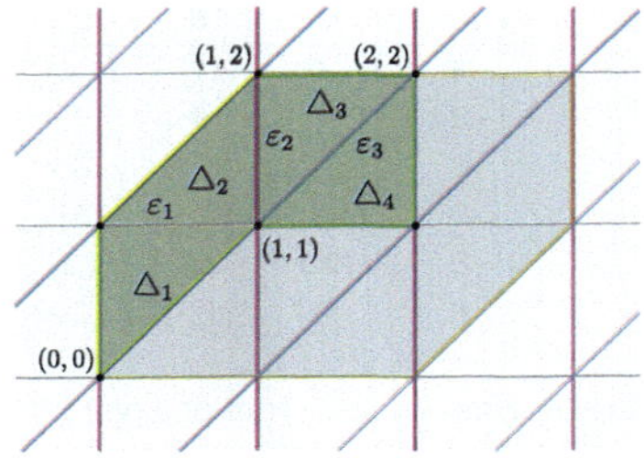

Fig. 2 Multicell domain $M = \cup_{i=1}^{4} \hat{\Delta}_i$, with edges $\varepsilon_i = \overline{\Delta}_i \cap \overline{\Delta}_{i+1}$ for $i = 1, 2, 3$

$$\begin{aligned}\mathbb{S}^{d}(M, \mathscr{V}) &= \left\{ f \in \mathbb{P}(M, \mathscr{V})\colon f|_{\Diamond(\varepsilon)^*} \in C^{d_i}\left(\Diamond(\varepsilon)^*\right) \text{ for every } \varepsilon \in E_i \cap M \right.\\ &\qquad\qquad \left. \text{and } i \in \{1, 2, 3\} \right\},\\ &= \left\{ f \in \mathbb{P}(M, \mathscr{V})\colon f|_{\Diamond(\varepsilon)^*} \in \mathbb{D}_{I_i^d}\left(\Diamond(\varepsilon)^*, \mathscr{V}\right) \text{ for every } \varepsilon \in E_i \cap M \right.\\ &\qquad\qquad \left. \text{and } i \in \{1, 2, 3\} \right\},\end{aligned}$$

where $\Diamond(\varepsilon)^* = \bigcup_{\Delta \in \diamond(\varepsilon) M} \overline{\Delta}$ as defined in Eq. (1).

Later in this paper (see Definition 6 below), we shall introduce a spline space but with smoothness conditions at the vertices of the domain M, the notation in Definition 3 will be particularly convenient for that purpose.

In the following example we illustrate Definition 3 for a specific multicell domain in the grid G and a regularity vector $\boldsymbol{d} \in \mathbb{Z}^3_{\geq 0}$.

Example 1 Let M be the multicell domain in Fig. 2. It is composed of four triangles denoted $\Delta_1, \ldots, \Delta_4$, we take $\mathscr{V} = \mathscr{P}_2$ (the polynomials in $\mathbb{R}[x, y]$ of degree at most 2), and the regularity vector $\boldsymbol{d} = (0, 1, 0)$. Then

$$I_1^{\boldsymbol{d}} = \{(i, 0, 0)\colon i \in \mathbb{Z}_{\geq 0}\},\ I_2^{\boldsymbol{d}} = \{(0, j, 0), (1, j, 0), (0, j, 1)\colon j \in \mathbb{Z}_{\geq 0}\},$$
$$\text{and } I_3^{\boldsymbol{d}} = \{(0, 0, k)\colon k \in \mathbb{Z}_{\geq 0}\}.$$

If we define $f \in \mathbb{P}(M, \mathscr{P}_2)$ by $f|_{\hat{\Delta}_i} = f_i$, where

$$\begin{aligned} f_1 &= x^2; & f_3 &= (y - 2x)(y - 2);\\ f_2 &= (x - y + 1)^2; & f_4 &= 2(x - 1)(y - x) + 2x - y^2. \end{aligned} \tag{2}$$

Then f is an element in $\mathbb{S}^{d}(M, \mathscr{P}_2)$. In fact, if we put $g_{1,2} = (f_1, f_2) = f|_{\diamond(\varepsilon_1)}$, since $f_1 - f_2 = (y - 1)(2x - y + 1)$ then $D_{\boldsymbol{s}}\, g_{1,2}$ is a continuous function on $\diamond(\varepsilon_1)$ for $\boldsymbol{s} \in I_1^{d}$. Similarly, if we put $g_{i,i+1} = (f_i, f_{i+1})$ it is easy to check that $D_{\boldsymbol{s}}\, g_{2,3}$ and $D_{\boldsymbol{t}}\, g_{3,4}$ are continuous functions for every $\boldsymbol{s} \in I_2^{\boldsymbol{d}}$ and $\boldsymbol{t} \in I_3^{\boldsymbol{d}}$ on $\diamond(\varepsilon_2)$ and $\diamond(\varepsilon_3)$, respectively. ◇

In an analogous way as we defined a spline space associated to smoothness along the edges (Definition 3) of a multicell domain, we will introduce a space of splines

ST(△,△′)	A	B	C	D	E	F
A	∅	{1}	{1,2}	{1,2,3}	{2,3}	{3}
B	{1}	∅	{2}	{2,3}	{1,2,3}	{1,3}
C	{1,2}	{2}	∅	{3}	{1,3}	{1,2,3}
D	{1,2,3}	{2,3}	{3}	∅	{1}	{1,2}
E	{2,3}	{1,2,3}	{1,3}	{1}	∅	{2}
F	{3}	{1,3}	{1,2,3}	{1,2}	{2}	∅

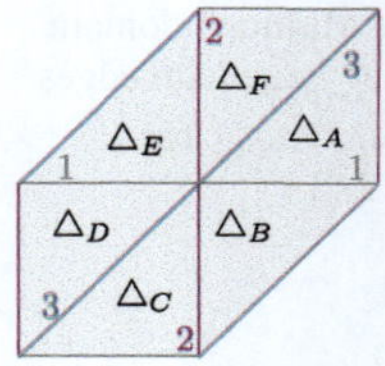

Fig. 3 Smoothness types of a pair of triangles $\triangle, \triangle'$ such that $\overline{\triangle} \cap \overline{\triangle}' \neq \emptyset$. We can identify $\triangle, \triangle'$ with two triangles in the picture on the right. The type $\mathrm{ST}(\triangle, \triangle')$ is the corresponding index set shown in the table on the left side, which is constructed according to the shortest edge-connected chain between them (see Definition 4)

with additional smoothness at the vertices of a given multivariate domain M. We prepare this definition by listing the possible vertex-vertex contact configurations $\overline{\triangle} \cap \overline{\triangle}' = \{\nu\}$ between any pair of triangles $\triangle, \triangle' \in T$. First we need the following definitions.

Definition 4 Two triangles $\triangle$ and $\triangle'$ in the grid G, are said to be *edge-connected* if there is a collection of triangles $\triangle_0, \triangle_1, \ldots, \triangle_m \in T$ such that $\triangle = \triangle_0$, $\triangle' = \triangle_m$ and $\overline{\triangle}_{i-1} \cap \overline{\triangle}_i \in E$ for every $i = 1, \ldots m$. Such a collection of triangles $\triangle_0, \triangle_1, \ldots, \triangle_m$ is called an *edge-connected chain* between $\triangle$ and $\triangle'$.

Definition 5 If $\triangle, \triangle' \in T$ are triangles such that $\overline{\triangle} \cap \overline{\triangle}' \neq \emptyset$, we define the *smoothness type* $\mathrm{ST}(\triangle, \triangle') \subseteq \{1, 2, 3\}$ as the set of edge-types that are in the shortest edge-connected chain in G between $\triangle$ and $\triangle'$. If $\triangle = \triangle'$ we define $\mathrm{ST}(\triangle, \triangle') = \emptyset$.

For any given pair of triangles $\triangle, \triangle' \in T$ with a non-empty intersection, we can identify them with a pair of triangles from A to F in Fig. 3, and their smoothness type $\mathrm{ST}(\triangle, \triangle')$ becomes one of the subsets listed in the table on the left of Fig. 3.

We now use Definitions 4 and 5 to introduce the space of *strongly regular splines* associated to a multicell domain M in the three directional grid G.

Definition 6 For a multicell domain M in G, a vector space $\mathscr{V} \subseteq \mathbb{R}[x, y]$, and a regularity vector $\boldsymbol{d} = (d_1, d_2, d_3)$, we define the set $\hat{\mathbb{S}}^{\boldsymbol{d}}(M, \mathscr{V})$ of *strongly regular splines on* M as follows,

$$\hat{\mathbb{S}}^{\boldsymbol{d}}(M, \mathscr{V}) = \Big\{ f \in \mathbb{P}(M, \mathscr{V}) : \; f|_U \in \mathbb{D}_I(U, \mathscr{V}) \text{ for } \triangle, \triangle' \in M, \overline{\triangle} \cap \overline{\triangle}' \neq \emptyset,$$

$$U = \overline{\triangle} \cup \overline{\triangle}', \text{ and } I = \bigcap_{i \in \mathrm{ST}(\triangle, \triangle')} I_i^{\boldsymbol{d}} \Big\}.$$

Namely, the elements in $\hat{\mathbb{S}}^{\boldsymbol{d}}(M, \mathscr{V})$ are the splines $f \in \mathbb{P}(M, \mathscr{V})$ such that for any pair of triangles $\triangle, \triangle' \in M$ such that $\overline{\triangle} \cap \overline{\triangle}' \neq \emptyset$, the derivatives $D_{\boldsymbol{s}}$ of f are C^0-smooth for every $\boldsymbol{s} \in \bigcap_{i \in \mathrm{ST}(\triangle, \triangle')} I_i^{\boldsymbol{d}}$. A spline $f \in \mathbb{P}(M, \mathscr{V})$ that satisfies this property is called *strongly regular*. (Notice that an edge-connected chain is composed of triangles in the grid G and are not necessarily in M.)

The set $\hat{\mathbb{S}}^{\boldsymbol{d}}(M, \mathscr{V})$ is the linear space of splines *with edge and vertex smoothness* $\boldsymbol{d}$ *on the multicell domain* M.

Example 2 Let M be the multicell domain in Fig. 2, $\mathscr{V} = \mathscr{P}_2$ and $\boldsymbol{d} = (0, 1, 0)$ as in Example 1. It is easy to check that the piecewise function f defined in Eq. (2) is in $\hat{\mathbb{S}}^{\boldsymbol{d}}(M, \mathscr{V})$. For instance, if we take the triangles Δ_1 and Δ_4, the smoothness type $\mathrm{ST}(\Delta_1, \Delta_4) = \{1, 2, 3\}$. Then $I = \bigcap_{i=1}^{3} I_i^{\boldsymbol{d}} = \{(0, 0, 0)\}$, and in fact $f_1(1, 1) = f_4(1, 1)$. Similarly, taking the triangles Δ_2 and Δ_4, we get $\mathrm{ST}(\Delta_2, \Delta_4) = \{2, 3\}$ and $I = \bigcap_{i=2}^{3} I_i^{\boldsymbol{d}} = \{(0, 0, k) : k = 0, 1\}$. The polynomials f_2 and f_4 and also their derivatives $\partial(f_i)/\partial(x - y)$, for $i = 2$ and 3, have the same value at $(1, 1)$.

In contrast, if g is the function on M defined by $g|_{\Delta_i} = g_i$ with $g_1 = 0, g_2 = y - 1$, $g_3 = x^2 - 2x + y$ and $g_4 = x^2 - y$, then g is also in $\mathbb{S}^{\boldsymbol{d}}(M, \mathscr{V})$, but it is not in $\hat{\mathbb{S}}^{\boldsymbol{d}}(M, \mathscr{V})$. In fact, for the triangles Δ_2 and Δ_4, the derivatives $\partial g_2/\partial(x - y) = -1$ and $\partial g_4/\partial(x - y) = 2x + 1$. Then $g|_U \notin \mathbb{D}_I(U, \mathscr{V})$ for $U = \overline{\Delta}_2 \cup \overline{\Delta}_4$. ◊

Remark 2 From Definitions 3 and 6, it is clear that both $\hat{\mathbb{S}}^{\boldsymbol{d}}(M, \mathscr{V})$ and $\mathbb{S}^{\boldsymbol{d}}(M, \mathscr{V})$ are contained in the space of splines that are globally C^r-continuous on M, where $r = \min\{d_1, d_2, d_3\}$. Moreover, they are both contained in $\mathbb{D}_I(M, \mathscr{V})$ for $I = I_1^{\boldsymbol{d}} \cap I_2^{\boldsymbol{d}} \cap I_3^{\boldsymbol{d}}$ (see Definition 2), and $\hat{\mathbb{S}}^{\boldsymbol{d}}(M, \mathscr{V}) \subseteq \mathbb{S}^{\boldsymbol{d}}(M, \mathscr{V})$. By Example 2 we also know that the set of strongly regular splines $\hat{\mathbb{S}}^{\boldsymbol{d}}(M, \mathscr{V})$ may be properly contained in the spline space $\mathbb{S}^{\boldsymbol{d}}(M, \mathscr{V})$.

In the following we give a sufficient condition for the equality $\mathbb{S}^{\boldsymbol{d}}(M, \mathscr{V}) = \hat{\mathbb{S}}^{\boldsymbol{d}}(M, \mathscr{V})$. For this we introduce the concept of over concave vertices and kissing triangles.

Definition 7 For a multicell domain M in the three-directional grid G we say that a vertex $\nu \in V$ on the boundary of M is *over-concave* if star $(\nu) \setminus M$ consists of a single triangle, where

$$\mathrm{star}\,(\nu) = \{\Delta \in T : \nu \in \hat{\Delta}\} \cup \{\varepsilon \in E : \nu \in \hat{\varepsilon}\} \cup \{\nu\}.$$

Since we work exclusively in the three directional grid G then star (ν) consists of 6 triangles, 6 edges and ν itself. An example of an over-concave vertex is illustrated in Fig. 4.

Definition 8 Two non edge-connected triangles Δ and Δ' in M are called *kissing triangles* if $\overline{\Delta} \cap \overline{\Delta}' = \{\nu\}$ for a vertex $\nu \in V$.

In Fig. 3, for instance, in a multicell domain composed by only two triangles, the triangles $\{A, E\}$, $\{A, D\}$, and $\{A, C\}$ are kissing triangles.

Proposition 1 *If M is a multicell domain in the three directional grid G, such that it does not have kissing triangles nor over-concave boundary vertices, then $\mathbb{S}^{\boldsymbol{d}}(M, \mathscr{V}) = \hat{\mathbb{S}}^{\boldsymbol{d}}(M, \mathscr{V})$, i.e., all splines with edge smoothness $\boldsymbol{d}$ on M are strongly regular.*

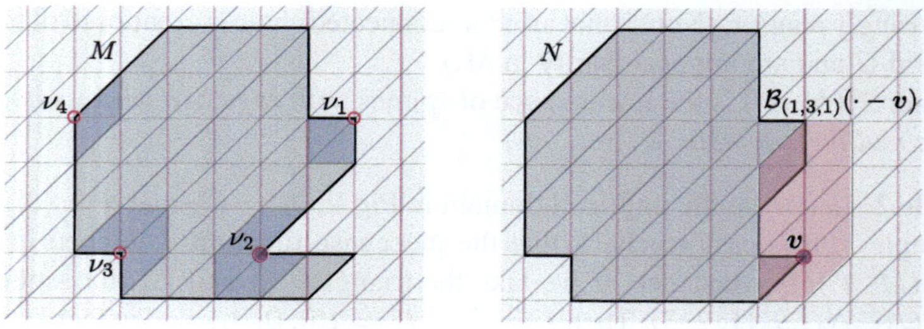

Fig. 4 On the left, the shadowed regions around the vertices $\nu_i \in M$ correspond to the triangles in star $(\nu_i) \cap M$, respectively; notice that in this multicell domain M, by Definition 7, the only *over-concave vertex* is ν_2. On the right, the multicell domain N does not have over-concave vertices; the figure also illustrates Definition 13 from Sect. 4, the shadowed region corresponds to the support of the box spline $\mathscr{B}_{(1,3,1)}(\cdot - \boldsymbol{v})$ considered in Example 5

Proof Let Δ and Δ' be two triangles in the multicell domain M, and denote $U = \overline{\Delta} \cup \overline{\Delta}'$. If $\overline{\Delta} \cap \overline{\Delta}' = \varepsilon \in E_i$ then $U = \diamond(\varepsilon)^*$. By Definition 5, the smoothness type of this pair of triangles is $\mathrm{ST}(\Delta, \Delta') = \{i\}$, and the index set associated to U in Definition 6 is simply $I = I_i^d$. Then, by Definition 3 and Definition 6, we have $\hat{\mathbb{S}}^d(M, \mathscr{V}) \subseteq \mathbb{S}^d(M, \mathscr{V})$. Conversely, let us take $f \in \mathbb{S}^d(M, \mathscr{V})$. By hypothesis, M does not have kissing triangles and since we have checked that the continuity conditions restricted to two triangles are equivalent if the triangles have a common edge, then let us assume that $\overline{\Delta} \cap \overline{\Delta}' = \{\nu\}$ for a vertex $\nu \in V$, where Δ and Δ' are edge-connected. We want to check that $f|_U \in \mathbb{D}_I(U, \mathscr{V})$. By Definition 8, we know that here exist an edge-connected chain of triangles $\Delta_0, \Delta_1, \ldots, \Delta_m$ in $M \cap$ star (ν) such that $\Delta = \Delta_0$ and $\Delta' = \Delta_m$. The smoothness type $\mathrm{ST}(\Delta, \Delta')$ is contained in every set of edge-types in any edge-connected chain in G between Δ and Δ'. In particular, $\mathrm{ST}(\Delta, \Delta')$ is a subset of the edge-types in the chain $\Delta_0, \Delta_1, \ldots, \Delta_m$. Since $f \in \mathbb{S}^d(M, \mathscr{V})$ then $D_{\boldsymbol{s}} f$ is continuous at ν for every $\boldsymbol{s} \in I_j^d$ and every j in the edge-types of the chain $\Delta_0, \Delta_1, \ldots, \Delta_m$. In particular, $D_{\boldsymbol{s}} f|_U$ is a continuous function for every $\boldsymbol{s} \in I_j^d$ and every $j \in \mathrm{ST}(\Delta, \Delta')$. It follows that $f \in \hat{\mathbb{S}}^d(M, \mathscr{V})$, and hence $\mathbb{S}^d(M, \mathscr{V}) \subseteq \hat{\mathbb{S}}^d(M, \mathscr{V})$ as required. □

3 Box Splines on Type-I Triangulations

In this section we define box splines on the uniform type-I triangulation G defined in Sect. 2, Fig. 1. This triangulation has vertices at all lattice points $(i, j) \in \mathbb{Z}^2$.

Definition 9 If β is a real-valued function on $\mathbb{R}^2$, we denote by supp (β) the support β, and it is defined as the set of points $\boldsymbol{x} \in \mathbb{R}^2$ such that $\beta(\boldsymbol{x}) \neq 0$.

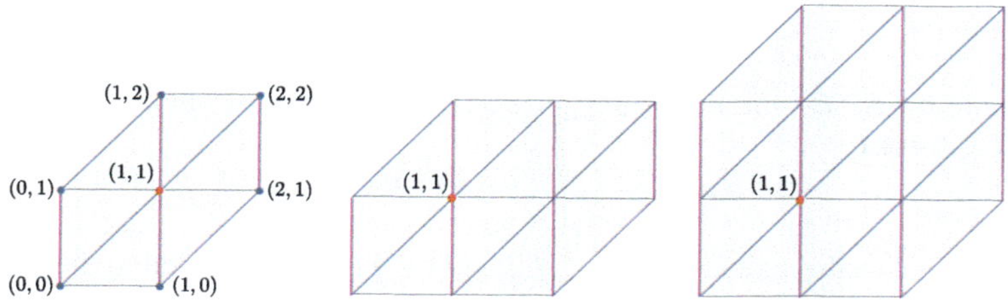

Fig. 5 Support of the Courant hat function $\mathscr{B}_1$, it corresponds to star (ν) where ν is the vertex $(1, 1)$ (left), support of the box splines $\mathscr{B}_{(2,1,1)}$ (center) and $\mathscr{B}_{(2,2,1)}$ (right)

Recall from Definition 7, that the *star of a vertex* $\nu \in V$, denoted star (ν), is composed of ν and all the triangles $\Delta \in T$ and all edges $\varepsilon \in E$ which have ν as one of their vertices.

Definition 10 If $\boldsymbol{n} = (n_1, n_2, n_3) \in \mathbb{Z}^3$ is a triple of integers such that $n_i \geq 1$, the type-I box spline $\mathscr{B}_{\boldsymbol{n}}$ associated to $\boldsymbol{n}$ is defined recursively by

$$\mathscr{B}_{\boldsymbol{n}}(\boldsymbol{x}) = \int_0^1 \mathscr{B}_{\boldsymbol{n}-\boldsymbol{e}_i}(\boldsymbol{x} - t\boldsymbol{e}_i)dt\,,$$

for $\boldsymbol{x} \in \mathbb{R}^2$ and $i \in \{1, 2, 3\}$ such that $\boldsymbol{n} - \boldsymbol{e}_i \geq \mathbf{1} = (1, 1, 1)$; the function $\mathscr{B}_{\mathbf{1}}$ is the classical *Courant hat function* with support on the star of the vertex $(1, 1)$ given in Fig. 5. More precisely, $\mathscr{B}_{\mathbf{1}}$ is the piecewise linear function on $\mathbb{R}^2$ satisfying $\mathscr{B}_{\mathbf{1}}(1, 1) = 1$ and $\mathscr{B}_{\mathbf{1}}(i, j) = 0$ for every $(i, j) \in \mathbb{Z}^2 \setminus \{(1, 1)\}$.

The coordinates n_i of $\boldsymbol{n}$ denote the number of convolutions of $\mathscr{B}_{\mathbf{1}}$ along the directions $\boldsymbol{e}_i$. The support of the box spline $\mathscr{B}_{\boldsymbol{n}}$ is the zonotope in $\mathbb{R}^2$ formed by the Minkowski sum of the direction vectors $\boldsymbol{e}_i$ taken n_i times, for $i = 1, 2, 3$, respectively (see Fig. 5 for an example). Moreover, for every $\boldsymbol{n}$, the box spline $\mathscr{B}_{\boldsymbol{n}}$ is strictly positive for all $\boldsymbol{x}$ in the interior of its support, and zero otherwise [29, Theorem 12.2].

From the general theory of type-I box splines, it follows that the box spline $\mathscr{B}_{\boldsymbol{n}}$ is independent of the order in which the vectors $\boldsymbol{e}_i$ appear in the recursive construction of $\mathscr{B}_{\boldsymbol{n}}$ in Definition 10. This result follows immediately from the formula for the Fourier transform of a type-I box spline [29, Theorem 12.6].

For any $\boldsymbol{n}$, the box spline $\mathscr{B}_{\boldsymbol{n}}$ is a piecewise polynomial function on three directional triangulation G, and each polynomial is of total degree $n = |\boldsymbol{n}| - 2$, where $|\boldsymbol{n}| = n_1 + n_2 + n_3$. It is also well-known that $\mathscr{B}_{\boldsymbol{n}}$ is a C^r-continuous function on $\mathbb{R}^2$, where $r = \min_{i=1,2,3}\{|\boldsymbol{n}| - n_i - 2\}$ [29, Theorem 2.4].

There is a rich literature on type-I box splines, a detailed construction and the proof of structural and smoothness properties can be found for instance in [29, Chap. 12] or [12]. In particular, it is known that each convolution along a direction $\boldsymbol{e}_i$ increases the continuity of the box spline with respect to differentiation in that direction by one [29, Theorem 12.3]. Thus, following the notation introduced in Sect. 2, if $\boldsymbol{d} = (n_2 + n_3 - 2,\ n_1 + n_3 - 2,\ n_1 + n_2 - 2)$ then

$$\mathscr{B}_{\boldsymbol{n}} \in \mathbb{S}^d(G, \mathscr{P}_n)\,. \tag{3}$$

We denote by $B_{\boldsymbol{n}}(G)$ the set of integer *translates of the box spline* $\mathscr{B}_{\boldsymbol{n}}$, which is defined as the set

$$B_{\boldsymbol{n}}(G) = \left\{\mathscr{B}_{\boldsymbol{n}}(\,\cdot - \boldsymbol{v}) : \boldsymbol{v} \in \mathbb{Z}^2\right\}. \tag{4}$$

The set $B_{\boldsymbol{n}}(G)$ is also called the set of *shifted box splines* associated to the direction vector $\boldsymbol{n}$.

Remark 3 Notice that the translates in $B_{\boldsymbol{n}}(G)$ have distinct support, it is the zonotope which is the support of $\mathscr{B}_{\boldsymbol{n}}$ shifted by $\boldsymbol{v}$. In fact, the set $B_{\boldsymbol{n}}(G)$ is (globally) linearly independent [29, Theorem 12.19], i.e., if

$$\sum_{\boldsymbol{v}\in\mathbb{Z}^2} a_{\boldsymbol{v}} \mathscr{B}_{\boldsymbol{n}}(\boldsymbol{x} - \boldsymbol{v}) = 0, \text{ for all } \boldsymbol{x} \in \mathbb{R}^2$$

then $a_{\boldsymbol{v}} = 0$ for all $\boldsymbol{v} \in \mathbb{Z}^2$. Furthermore, it has been shown that the translates in $B_{\boldsymbol{n}}(G)$ are also locally linearly independent, i.e., if A is an open set, then the shifted box splines

$$\left\{\mathscr{B}_{\boldsymbol{n}}(\cdot - \boldsymbol{v})\colon \operatorname{supp}\left(\mathscr{B}_{\boldsymbol{n}}(\cdot - \boldsymbol{v})\right) \cap A \neq \emptyset\right\}$$

are linearly independent [8, 22]. Here, supp (β) denotes the support of the function β (Definition 9).

We now introduce the definition of the set of active box splines on a given multicell domain M in the type-I triangulation G.

Definition 11 If $M \subseteq G$ is a multicell domain, we define

$$\operatorname{Supp}_{\boldsymbol{n}}(M) = \left\{\boldsymbol{v} \in \mathbb{Z}^2\colon \operatorname{supp}\left(\mathscr{B}_{\boldsymbol{n}}(\cdot - \boldsymbol{v})\right) \cap M^* \neq \emptyset\right\},$$

and the set of *active box splines* $\mathscr{B}_{\boldsymbol{n}}$ on M by

$$B_{\boldsymbol{n}}(M) = \left\{\mathscr{B}_{\boldsymbol{n}}(\cdot - \boldsymbol{v})\big|_M \colon \boldsymbol{v} \in \operatorname{Supp}_{\boldsymbol{n}}(M)\right\},$$

where M^* is the closure of M in $\mathbb{R}^2$ as defined in Eq. (1).

In particular, if the multicell domain $M = \hat{\Delta}$, for a triangle $\Delta \in T$, then $B_{\boldsymbol{n}}(\hat{\Delta})$ is the set of translates $\mathscr{B}_{\boldsymbol{n}}(\,\cdot - \boldsymbol{v}) \in B_{\boldsymbol{n}}(G)$ whose support contains Δ. We will denote the number of elements in $B_{\boldsymbol{n}}(\hat{\Delta})$ by $\phi(\boldsymbol{n})$.

Notice that this number $\phi(\boldsymbol{n})$ can be computed by counting the number of different triangles in the support of $\mathscr{B}_{\boldsymbol{n}}$ which are translations (and not a reflection) of Δ. This number is exactly half of the total number of different triangles in the support of $\mathscr{B}_{\boldsymbol{n}}$. Thus,

$$\phi(\boldsymbol{n}) = n_1 n_2 + n_1 n_3 + n_2 n_3\,, \tag{5}$$

for $\boldsymbol{n} = (n_1, n_2, n_3) \in \mathbb{Z}^3_{\geq 1}$.

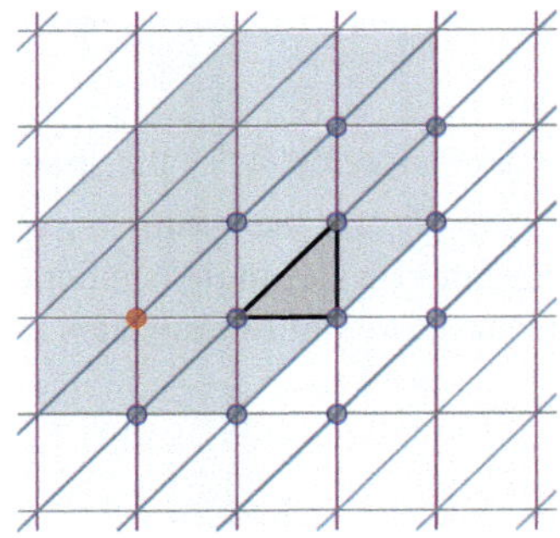

Fig. 6 The 12 lattice points correspond to the 1-ring of a triangle $\triangle \in T$ associated to the box spline $\mathscr{B}_{\mathbf{2}} = \mathscr{B}_{(2,2,2)}$. Each of these points is the reference point $(1, 1) - \boldsymbol{v}$ for the translate $\mathscr{B}_{\mathbf{2}}(\cdot - \boldsymbol{v})$ such that $\triangle \subseteq \operatorname{supp}(\mathscr{B}_{\mathbf{2}}(\cdot - \boldsymbol{v}))$

Definition 12 If $\mathscr{B}_{\boldsymbol{n}}(\cdot - \boldsymbol{v})$ is the translate of the type-I box spline $\mathscr{B}_{\boldsymbol{n}}$ by $\boldsymbol{v} \in \mathbb{Z}^2$, then we take $(1, 1) - \boldsymbol{v}$ as the point of reference of $\operatorname{supp}\big(\mathscr{B}_{\boldsymbol{n}}(\cdot - \boldsymbol{v})\big)$ in the lattice. For a triangle $\triangle \in T$, we define the *1-ring neighborhood* of $\triangle$ as the set of *reference lattice points* $(1, 1) - \boldsymbol{v}$ such that $\operatorname{supp}\big(\mathscr{B}_{\boldsymbol{n}}(\cdot - \boldsymbol{v})\big) \cap \triangle \neq \emptyset$.

For instance, if $\mathbf{2} = (2, 2, 2)$ then the elements in $B_{\mathbf{2}}(\hat{\triangle})$ are the translates $\mathscr{B}_{\mathbf{2}}(\cdot - \boldsymbol{v})$ associated to the $\phi(\boldsymbol{n}) = 12$ lattice points in a 1-ring neighborhood of $\triangle$ in the grid G, which are shown in Fig. 6.

Notice that to any type-I box spline $\mathscr{B}_{\boldsymbol{n}}$ and a triangle $\triangle \in T$ we can associate a linear space of polynomials. Namely, extending by linearity, we can take $\mathscr{V}_{\boldsymbol{n}}|_{\triangle}$ as the space generated by the restrictions of the active box splines $\mathscr{B}_{\boldsymbol{n}}$ to the triangle $\triangle$, i.e.,

$$\mathscr{V}_{\boldsymbol{n}}|_{\triangle} = \operatorname{span} B_{\boldsymbol{n}}(\hat{\triangle})\,. \tag{6}$$

By Equations (6) and (3), we see that $\mathscr{V}_{\boldsymbol{n}}|_{\triangle}$ is a linear subspace of $\mathscr{P}_n \subseteq \mathbb{R}[x, y]$, for any $\triangle \in T$ and $n = n_1 + n_2 + n_3 - 2$.

We now prove that for any $\boldsymbol{n} \in \mathbb{Z}^3$, the subspace $\mathscr{V}_{\boldsymbol{n}}|_{\triangle} \subseteq \mathscr{P}_n$ is independent of the choice of the triangle $\triangle \in T$. The proof of this result is a generalization of [42, Proposition 27], where we consider the case $\mathbf{2} = (2, 2, 2)$ and the box spline $\mathscr{B}_{\mathbf{2}}$.

Proposition 2 *Let* $\boldsymbol{n} = (n_1, n_2, n_3) \in \mathbb{Z}^3_{\geq 1}$ *with* $n = n_1 + n_2 + n_3 - 2$. *Then,* $\mathscr{V}_{\boldsymbol{n}}|_{\triangle} = \mathscr{V}_{\boldsymbol{n}}|_{\triangle'}$ *for any pair of triangles* $\triangle$ *and* $\triangle'$ *in* T.

Proof Let $\triangle$ and $\triangle'$ be two triangles in the three-directional grid G. Denote by G^0 the triangulation of $\mathbb{R}^2$ obtained by the lines parallel to the vectors $\boldsymbol{e}_1' = (a, 0)$, $\boldsymbol{e}_2' = (0, a)$ and $\boldsymbol{e}_1' = (a, a)$, for a fixed number $a \in \mathbb{Z}$. Notice that for any $a \in \mathbb{Z}$, we can see the grid G as a refinement of a grid G^0. Denote by $\triangle_a$ the triangle in G^0 with vertices at $(0, 0)$, $(0, a)$ and (a, a). In particular, let us take $\ell \in \mathbb{Z}_+$ and $a = 2^\ell$ in such a way that the translate $\tilde{\triangle} = \triangle_a - \boldsymbol{v}$ of $\triangle_a$ by a vector $\boldsymbol{v} \in \mathbb{Z}^2$ contains both $\triangle$ and $\triangle'$. Then, $B_{\boldsymbol{n}}(1/2^\ell\, \cdot)$ is the correspondent box spline associated to $\boldsymbol{n}$ in the grid G^0. By the refinement equation for box splines [29, Theorem 12.9], there exists a finite sequence $\{c_{\boldsymbol{v}}\}_{\boldsymbol{v}\in\mathbb{Z}^2}$ such that

$$B_{\boldsymbol{n}}(1/2^{\ell}\,\cdot) = \sum_{\boldsymbol{v}\in\mathbb{Z}^2} c_{\boldsymbol{v}} B_{\boldsymbol{n}}(\cdot - \boldsymbol{v})\,. \tag{7}$$

Let us denote by $\mathscr{V}_{\boldsymbol{n}}^0|_{\tilde{\triangle}}$ the span of the box spline translates $\mathscr{B}_{\boldsymbol{n}}(1/2^{\ell}\cdot - \boldsymbol{v})$ restricted to $\tilde{\triangle}$ in G^0. By the symmetry of the box splines supports, the number $\phi(\boldsymbol{n})$ of active translates on a triangle is independent of the grid and of the given triangle in the grid. Furthermore, these translates are linearly independent (see Remark 3). Thus,

$$\dim \mathscr{V}_{\boldsymbol{n}}^0|_{\tilde{\triangle}} = \dim \mathscr{V}_{\boldsymbol{n}}|_{\triangle} = \dim \mathscr{V}_{\boldsymbol{n}}|_{\triangle'}. \tag{8}$$

Taking the restrictions to $\triangle$ and $\triangle'$, Eq. (7) implies $\big(\mathscr{V}_{\boldsymbol{n}}^0|_{\tilde{\triangle}}\big)|_{\triangle} \subseteq \mathscr{V}_{\boldsymbol{n}}|_{\triangle}$ and $\big(\mathscr{V}_{\boldsymbol{n}}^0|_{\tilde{\triangle}}\big)|_{\triangle'} \subseteq \mathscr{V}_{\boldsymbol{n}}|_{\triangle'}$.

Since $\mathscr{V}_{\boldsymbol{n}}^0|_{\tilde{\triangle}}$ is a polynomial subspace of $\mathscr{P}_n$, we have $\mathscr{V}_{\boldsymbol{n}}^0|_{\tilde{\triangle}} = (\mathscr{V}_{\boldsymbol{n}}^0|_{\tilde{\triangle}})|_{\triangle} = (\mathscr{V}_{\boldsymbol{n}}^0|_{\tilde{\triangle}})|_{\triangle'}$. In particular, both $(\mathscr{V}_{\boldsymbol{n}}^0|_{\tilde{\triangle}})|_{\triangle}$ and $(\mathscr{V}_{\boldsymbol{n}}^0|_{\tilde{\triangle}})|_{\triangle'}$ have the same dimension as $\mathscr{V}_{\boldsymbol{n}}^0|_{\tilde{\triangle}}$. The statement follows by applying Eq. (8). □

Proposition 2 implies that for a given $\boldsymbol{n} \in \mathbb{Z}^3_{\geq 1}$ the restrictions of the translates of the box spline $\mathscr{B}_{\boldsymbol{n}}$ to a triangle $\triangle \in T$ define a polynomial space which is independent of $\triangle$. From now on, we will denote such polynomial space as $\mathscr{V}_{\boldsymbol{n}}$.

Remark 4 For any $\triangle \in T$ and $\boldsymbol{n} \in \mathbb{Z}^3_{\geq 1}$, Remark 3 implies that $B_{\boldsymbol{n}}(\hat{\triangle}) \subseteq \mathscr{P}_n$ is a linearly independent set. Here $\mathscr{P}_n$ is as before, the set of bivariate polynomials of degree at most $n = n_1 + n_2 + n_3 - 2$, and $\boldsymbol{n} = (n_1, n_2, n_3)$. The number of elements in $B_{\boldsymbol{n}}(\hat{\triangle}) = \phi(\boldsymbol{n})$ (Eq. (5)) is the dimension of the space of polynomials $\dim \mathscr{V}_{\boldsymbol{n}}$ associated to the type-I box $\mathscr{B}_{\boldsymbol{n}}$. Thus, $\dim(\mathscr{V}_{\boldsymbol{n}}) = n_1n_2 + n_1n_3 + n_2n_3 \leq \binom{n_1+n_2+n_3}{2} = \dim \mathscr{P}_n$.

For $n_i \geq 1$, equality holds only for $\boldsymbol{n} = \mathbf{1} = (1, 1, 1)$, which corresponds to polynomial space $\mathscr{V}_{\mathbf{1}}$ associated to the Courant hat function $\mathscr{B}_{\mathbf{1}}$ in Definition 10. For any other $\boldsymbol{n} \in \mathbb{Z}^3$ and the corresponding box spline $\mathscr{B}_{\boldsymbol{n}}$, the polynomial space $\mathscr{V}_{\boldsymbol{n}}$ is a proper subspace of $\mathscr{P}_{\boldsymbol{n}}$.

4 Characterization of Box Spline Spaces

Let $\triangle \in T$, $\boldsymbol{n} \in \mathbb{Z}^3_{\geq 1}$, and consider $\mathscr{V}_{\boldsymbol{n}} = \operatorname{span} B_{\boldsymbol{n}}(\hat{\triangle})$. As observed in Remark 4, the set $B_{\boldsymbol{n}}(\hat{\triangle})$ of the box spline translates with support on $\triangle$ is linearly independent, and hence the restriction of a polynomial $f \in \mathscr{V}_{\boldsymbol{n}}$ to $\triangle$ has a unique representation

$$f|_{\triangle}(\boldsymbol{x}) = \sum_{\beta\in B_n(\hat{\triangle})} \lambda_{\triangle}^{\beta}(f|_{\triangle})\beta(\boldsymbol{x}), \quad \boldsymbol{x} \in \overline{\triangle}, \tag{9}$$

for coefficients $\lambda_{\triangle}^{\beta}(f|_{\triangle}) \in \mathbb{R}$.

Lemma 1 *Let $\boldsymbol{n} = (n_1, n_2, n_3) \in \mathbb{Z}^3_{\geq 1}$, f and f' in $\mathcal{V}_{\boldsymbol{n}}$, and $M = \hat{\triangle} \cup \hat{\triangle}'$, such that $\triangle, \triangle' \in T$ and $\overline{\triangle} \cap \overline{\triangle}' \in E_i$ for some $i \in \{1, 2, 3\}$. If $(f|_\triangle, f'|_{\triangle'}) \in \mathbb{S}^{\boldsymbol{d}}(M, \mathcal{V}_{\boldsymbol{n}})$, then there exist polynomials $g, h \in \mathcal{V}_{\boldsymbol{n}}$ such that*

$$(f|_\triangle, f'|_{\triangle'}) = (g|_\triangle, f'|_{\triangle'}) + (h|_\triangle, 0|_{\triangle'}), \tag{10}$$

with $\lambda^\beta_\triangle(g|_\triangle) = \lambda^\beta_{\triangle'}(f'|_{\triangle'})$ for every $\beta \in B_{\boldsymbol{n}}(\hat{\triangle}) \cap B_{\boldsymbol{n}}(\hat{\triangle}')$.

Proof Define

$$g|_\triangle = \sum_{\beta \in B(\hat{\triangle}) \cap B(\hat{\triangle}')} \lambda^\beta_{\triangle'}(f'|_{\triangle'})\beta|_\triangle + \sum_{\beta \in B(\hat{\triangle}) \setminus B(\hat{\triangle}')} 0 \cdot \beta|_\triangle.$$

Then $g|_\triangle \in \mathcal{V}_{\boldsymbol{n}}|_\triangle$, and extending by linearity we can see $g|_\triangle$ as the restriction to $\triangle$ of a polynomial g in $\mathcal{V}_{\boldsymbol{n}} \subseteq \mathbb{R}[x, y]$. By construction, the pair $(g|_\triangle, f'|_{\triangle'})$ satisfies the required condition, and taking $h = f - g$ we obtain Eq. (10). □

Lemma 2 *Let $\boldsymbol{n} = (n_1, n_2, n_3) \in \mathbb{Z}^3_{\geq 1}$, f and f' in $\mathcal{V}_{\boldsymbol{n}}$, and $\triangle, \triangle' \in T$, such that $\overline{\triangle} \cap \overline{\triangle}' \in E_i$ for some $i \in \{1, 2, 3\}$. Then, for $M = \hat{\triangle} \cup \hat{\triangle}'$ the following two statements are equivalent:*

(i) The pair $(f|_\triangle, f'|_{\triangle'}) \in \mathbb{S}^{\boldsymbol{d}}(M, \mathcal{V}_{\boldsymbol{n}})$, with $\boldsymbol{d} = (n_2 + n_3 - 2, n_1 + n_3 - 2, n_1 + n_2 - 2)$.
(ii) For every $\beta \in B_{\boldsymbol{n}}(\hat{\triangle}) \cap B_{\boldsymbol{n}}(\hat{\triangle}')$, it holds $\lambda^\beta_\triangle(f|_\triangle) = \lambda^\beta_{\triangle'}(f'|_{\triangle'})$.

Proof (ii) ⇒ (i) Let $(f, f') \in \mathcal{V}_{\boldsymbol{n}}$ be a pair of polynomials satisfying (ii). Then

$$(f|_\triangle, f'|_{\triangle'}) = \sum_{\beta \in B_{\boldsymbol{n}}(\hat{\triangle}) \cap B_{\boldsymbol{n}}(\hat{\triangle}')} \lambda^\beta_\triangle(f|_\triangle)\beta|_{\triangle \cup \triangle'} + (h|_\triangle, 0|_\triangle) + (0|_{\triangle'}, h'|_{\triangle'}), \tag{11}$$

where

$$h|_\triangle = \sum_{\beta \in B_{\boldsymbol{n}}(\hat{\triangle}) \setminus B_{\boldsymbol{n}}(\hat{\triangle}')} \lambda^\beta_\triangle(f|_\triangle)\beta|_\triangle, \text{ and } h'|_{\triangle'} = \sum_{\beta \in B_{\boldsymbol{n}}(\hat{\triangle}') \setminus B_{\boldsymbol{n}}(\hat{\triangle})} \lambda^\beta_\triangle(f'|_{\triangle'})\beta|_\triangle.$$

Since $\beta \in \mathbb{S}^{\boldsymbol{d}}(M, \mathcal{V}_{\boldsymbol{n}})$ for every $\beta \in B_{\boldsymbol{n}}(M)$, then the first term in Eq. (11) is in $\mathbb{S}^{\boldsymbol{d}}(M, \mathcal{V}_{\boldsymbol{n}})$. Now, if $\beta \in B(\hat{\triangle}) \setminus B(\hat{\triangle}')$ then $(\beta|_\triangle, 0|_{\triangle'}) = \beta|_M$, and similarly for $\beta \in B(\hat{\triangle}') \setminus B(\hat{\triangle})$. Hence, the last two terms in Eq. (11) are also in $\mathbb{S}^{\boldsymbol{d}}(M, \mathcal{V}_{\boldsymbol{n}})$, and (i) follows.

(i) ⇒ (ii) By Lemma 1 it is enough to consider $(f|_\triangle, 0|_{\triangle'}) \in \mathbb{S}^{\boldsymbol{d}}(M, \mathcal{V}_{\boldsymbol{n}})$. We need to show that $\lambda^\beta_\triangle(f|_\triangle) = 0$ for every $\beta \in B(\hat{\triangle}) \cap B(\hat{\triangle}')$. To simplify, by an abuse of notation we will denote $\mathrm{Supp}_{\boldsymbol{n}}(\hat{\triangle}) \cap \mathrm{Supp}_{\boldsymbol{n}}(\hat{\triangle}')$ simply by $\mathrm{Supp}_{\boldsymbol{n}}(\hat{\varepsilon}_3)$.

Let $\overline{\triangle} \cap \overline{\triangle}' = \varepsilon_i \in E_i$, and denote by b_i the barycentric coordinate relative to $\triangle$ which vanishes at the edge ε_i. Let $n = n_1 + n_2 + n_3$. Then $(f|_\triangle, 0|_{\triangle'}) \in \mathbb{S}^{\boldsymbol{d}}(M, \mathcal{V}_{\boldsymbol{n}})$ if and only if the polynomial $b_i^{n-n_i-1}$ divides f, see [2, Lemma 2.2]. Notice that the latter condition is satisfied if and only if the Bernstein-Bézier coefficients $c_{jk\ell}$ of $f|_\triangle$ are zero for every $0 \le j \le n - n_1 - 2$.

Without loss of generality we can assume that $i = 3$ and $\triangle$ is the triangle with vertices at $(0, 0)$, $(1, 0)$, and $(1, 1)$. Thus, $b_3 = x - y$. We will prove that if $f \in \mathscr{V}_{\boldsymbol{n}}$ and $b_3^{n-n_3-1}$ divides f then $\lambda_\triangle^\beta(f|_\triangle) = 0$ for every $\beta \in B(\hat{\triangle}) \cap B(\hat{\triangle}')$. We proceed by induction on n, with $n_i \ge 1$.

The induction base is $n = 3$, with $\boldsymbol{n} = (1, 1, 1)$, $n_3 = 1$ and $n - n_3 - 1 = 1$. We have

$$B_{\boldsymbol{n}}(\hat{\triangle} \cup \hat{\triangle}') = \{\beta_j\}_{j=0}^3,$$

with $\beta_0 = \mathscr{B}_{\boldsymbol{n}}$, and $\beta_i = \mathscr{B}_{\boldsymbol{n}}(\cdot + \boldsymbol{e}_i)$ for $i = 1, 2, 3$. Thus, the only translates with support on $\triangle$ are β_0, β_2 and β_3. Since $\beta_0 = b_1$, $\beta_2 = b_3$, and $\beta_3 = b_2$, then

$$f|_\triangle = \lambda_\triangle^{\beta_0}(f|_\triangle)b_1|_\triangle + \lambda_\triangle^{\beta_2}(f|_\triangle)b_3|_\triangle + \lambda_\triangle^{\beta_3}(f|_\triangle)b_2|_\triangle.$$

Since the polynomials b_1, b_2, b_3 are linearly independent, and by hypothesis $b_3|f$, it follows $\lambda_\triangle^{\beta_i}(f|_\triangle) = 0$ for $i = 0, 3$. This proves the statement for the case $n = 3$.

Let us assume that the result is true for every $m \le n$ for some $n \ge 3$. Let $\boldsymbol{n} = (n_1, n_2, n_3)$ with $n + 1 = n_1 + n_2 + n_3$. We consider three cases, in the first two cases only one of the indices n_i is ≥ 2 and in the last case at least two of the indexes are ≥ 2.

Case 1. Suppose $n_1 \ge 2$, and $n_2 = n_3 = 1$.

We have $(x - y)^{n_1}|f$ and $D_{\boldsymbol{e}_1} f \equiv 0 \mod (x - y)^{n_1 - 1}$. Notice that Eq. (9) may be rewritten as

$$f|_\triangle = \sum_{\boldsymbol{v}\in\mathbb{Z}^2} \lambda_{\boldsymbol{v}}(f)\mathscr{B}_{\boldsymbol{n}}(\cdot - \boldsymbol{v})\big|_\triangle, \tag{12}$$

where $\lambda_{\boldsymbol{v}} = 0$ for every $\boldsymbol{v} \notin \mathrm{Supp}_{\boldsymbol{n}}(\triangle)$, and $\lambda_{\boldsymbol{v}}(f) = \lambda_\triangle^\beta(f|_\triangle)$ for $\beta \in B(\hat{\triangle})$ and $\beta = \mathscr{B}_{\boldsymbol{n}}(\cdot - \boldsymbol{v})$.

Moreover, we know that for any linear combination of box splines

$$\begin{aligned} D_{\boldsymbol{e}_i} \sum_{\boldsymbol{v}\in\mathbb{Z}^2} a_{\boldsymbol{v}}\mathscr{B}_{\boldsymbol{n}}(\cdot - \boldsymbol{v}) &= \sum_{\boldsymbol{v}\in\mathbb{Z}^2} a_{\boldsymbol{v}}\big(\mathscr{B}_{\boldsymbol{m}}(\cdot - \boldsymbol{v}) - \mathscr{B}_{\boldsymbol{m}}(\cdot - \boldsymbol{v} - \boldsymbol{e}_1)\big) \\ &= \sum_{\boldsymbol{v}\in\mathbb{Z}^2} (a_{\boldsymbol{v}+\boldsymbol{e}_i} - a_{\boldsymbol{v}})\mathscr{B}_{\boldsymbol{m}}(\cdot - \boldsymbol{v} + \boldsymbol{e}_i), \end{aligned} \tag{13}$$

for $a_{\boldsymbol{v}} \in \mathbb{R}$ for every $\boldsymbol{v} \in \mathbb{Z}^2$, and $\boldsymbol{m} = (n_1 - 1, n_2, n_3)$, see [29, Lemma 12.3]. Thus,

$$\begin{aligned} D_{\boldsymbol{e}_1} \sum_{\boldsymbol{v}\in\mathbb{Z}^2} \lambda_{\boldsymbol{v}}(f)\mathscr{B}_{\boldsymbol{n}}(\cdot - \boldsymbol{v})\big|_\triangle &= \sum_{\boldsymbol{v}\in\mathbb{Z}^2} \big(\lambda_{\boldsymbol{v}+\boldsymbol{e}_1}(f) - \lambda_{\boldsymbol{v}}(f)\big)\mathscr{B}_{\boldsymbol{m}}(\cdot - \boldsymbol{v} + \boldsymbol{e}_1)\big|_\triangle \\ &\equiv 0 \mod (x - y)^{n_1 - 1}, \end{aligned}$$

where $\boldsymbol{m} = (n_1 - 1, 1, 1)$. By induction hypothesis $\lambda_{\boldsymbol{v}+\boldsymbol{e}_1}(f) = \lambda_{\boldsymbol{v}}(f)$ for every $\boldsymbol{v} \in \mathrm{Supp}_{\boldsymbol{m}}(\hat{\varepsilon}_3)$.

Notice that, since $\boldsymbol{n} = (n_1, 1, 1)$ then the elements $\boldsymbol{v} \in \mathrm{Supp}_{\boldsymbol{v}}(\hat{\triangle})$ are either a multiple $t\boldsymbol{e}_1$ of $\boldsymbol{e}_1$ or of the form $\boldsymbol{e}_2 + t\boldsymbol{e}_1$, for $t \in \mathbb{Z}$. Thus, we have

$$f|_{\triangle} = \lambda_{\boldsymbol{e}_1}(f) \sum_{t\in\mathbb{Z}} \mathscr{B}_{\boldsymbol{n}}(\cdot - t\boldsymbol{e}_1)|_{\triangle} + \lambda_{\boldsymbol{e}_2}(f) \sum_{t\in\mathbb{Z}} \mathscr{B}_{\boldsymbol{n}}(\cdot - \boldsymbol{e}_2 + t\boldsymbol{e}_1)|_{\triangle},$$

for constants $\lambda_{\boldsymbol{e}_i}(f) \in \mathbb{R}$, for $i = 1, 2$. Moreover, $\mathscr{B}_{\boldsymbol{n}}(\cdot - t\boldsymbol{e}_1)|_{\overline{\varepsilon}_1} = 0$ for every t. In particular, they are zero at the vertex $(0, 0)$. But $f(0, 0) = 0$, and

$$1 = \sum_{\boldsymbol{v}\in\mathbb{Z}^2} \mathscr{B}_{\boldsymbol{n}}(\cdot - \boldsymbol{v}) = \sum_{t\in\mathbb{Z}} \mathscr{B}_{\boldsymbol{n}}(\cdot - t\boldsymbol{e}_1) + \mathscr{B}_{\boldsymbol{n}}(\cdot - \boldsymbol{e}_2 + t\boldsymbol{e}_1).$$

Then $f|_{(0,0)} = \lambda_{\boldsymbol{e}_2}(f) \sum_{t\in\mathbb{Z}} \mathscr{B}_{\boldsymbol{n}}(\cdot - \boldsymbol{e}_2 + t\boldsymbol{e}_1)|_{(0,0)} = \lambda_{\boldsymbol{e}_2}(f) = 0$. We obtain $\lambda_{\boldsymbol{e}_1}(f) = 0$ by considering the restriction at the vertex $(1, 1)$.

Case 2. Suppose $n_3 \geq 2$ and $n_1 = n_2 = 1$.

By hypothesis $(x - y)|f$, and so $D_{\boldsymbol{e}_3} f \equiv 0 \mod (x - y)$. Following the same argument as above, $\lambda_{\boldsymbol{v}+\boldsymbol{e}_3}(f) = \lambda_{\boldsymbol{v}}(f)$ for every $\boldsymbol{v} \in \mathrm{Supp}_{\boldsymbol{n}}(\hat{\varepsilon}_3)$. Thus, there is a constant $\lambda_{\boldsymbol{e}_3}$ which is equal to $\lambda_{\boldsymbol{v}}(f)$ for every $\boldsymbol{v} \in \mathrm{Supp}_{\boldsymbol{n}}(\hat{\varepsilon}_3)$, and

$$f|_{\varepsilon_3} = \lambda_{\boldsymbol{e}_3}(f) \sum_{\boldsymbol{v}\in\mathbb{Z}^2} \mathscr{B}_{\boldsymbol{n}}(\cdot - \boldsymbol{v})\Big|_{\varepsilon_3}.$$

Since $f|_{\varepsilon_3} = 0$, and $\sum_{\boldsymbol{v}\in\mathbb{Z}^2} \mathscr{B}_{\boldsymbol{n}}(\cdot - \boldsymbol{v})\Big|_{\varepsilon_3} = 1$, it follows $\lambda_{\boldsymbol{e}_3}(f) = 0$.

Case 3. At least two of the indices n_i are ≥ 2, say $n_1, n_2 \geq 2$. By hypothesis $(x - y)^{n_1+n_2-1}|f$, and so $D_{\boldsymbol{e}_i} f \equiv 0 \mod (x - y)^{n_1+n_2-2}$ for $i = 1, 2$. Similarly as before, rewrite $f|_{\triangle}$ as in Equation (12), and consider Eq. (13) with $i = 1$. Thus,

$$\begin{aligned} D_{\boldsymbol{e}_1} \sum_{\boldsymbol{v}\in\mathbb{Z}^2} \lambda_{\boldsymbol{v}}(f)\mathscr{B}_{\boldsymbol{n}}(\cdot - \boldsymbol{v})\Big|_{\triangle} &= \sum_{\boldsymbol{v}\in\mathbb{Z}^2} \big(\lambda_{\boldsymbol{v}+\boldsymbol{e}_1}(f) - \lambda_{\boldsymbol{v}}(f)\big)\mathscr{B}_{\boldsymbol{m}}(\cdot - \boldsymbol{v} + \boldsymbol{e}_1)\Big|_{\triangle} \\ &\equiv 0 \mod (x - y)^{n_1+n_2-2}. \end{aligned}$$

By induction hypothesis $\lambda_{\boldsymbol{v}+\boldsymbol{e}_1}(f) = \lambda_{\boldsymbol{v}}(f)$ for every $\boldsymbol{v} \in \mathrm{Supp}_{\boldsymbol{m}}(\hat{\varepsilon}_3)$. Similarly, by considering $D_{\boldsymbol{e}_2} f$ we get $\lambda_{\boldsymbol{v}+\boldsymbol{e}_2}(f) = \lambda_{\boldsymbol{v}}(f)$ for every $\boldsymbol{v} \in \mathrm{Supp}_{\boldsymbol{m}'}(\hat{\varepsilon}_3)$, where $\boldsymbol{m}' = (n_1, n_2 - 1, n_3)$. This implies that the coefficients $\lambda_{\boldsymbol{v}}(f)$ are equal to a constant λ, for every $\boldsymbol{v} \in \mathrm{Supp}_{\boldsymbol{n}}(\hat{\varepsilon}_3)$.

Therefore, $f|_{\varepsilon_3} = \lambda \sum_{\boldsymbol{v}\in\mathbb{Z}^2} \mathscr{B}_{\boldsymbol{n}}(\cdot - \boldsymbol{v})|_{\varepsilon_3}$. But $f|_{\varepsilon_3} = 0$, and $\sum_{\boldsymbol{v}\in\mathbb{Z}^2} \mathscr{B}_{\boldsymbol{n}}(\cdot - \boldsymbol{v}) = 1$, in particular when taking the restriction to ε_3. Hence, $0 = \lambda = \lambda_{\triangle}^{\beta}(f)$, for every $\beta \in B(\hat{\triangle}) \cap B(\hat{\triangle}')$.

An analogous proof applies when the edge of intersection between $\overline{\triangle}$ and $\overline{\triangle'}$ is parallel to one of the other two vectors $\boldsymbol{e}_1$ or $\boldsymbol{e}_2$. □

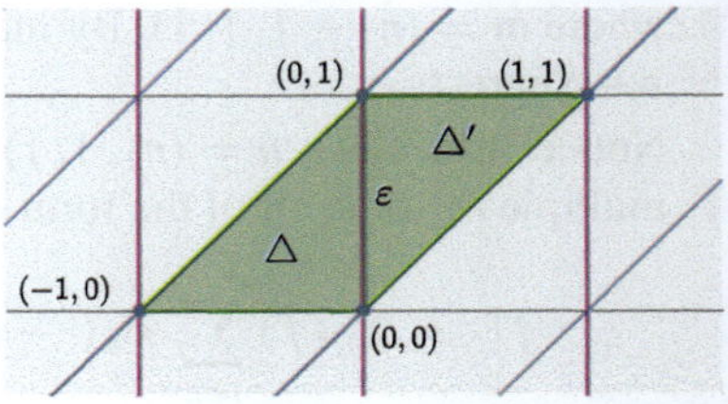

Fig. 7 The triangles Δ and Δ' in G define the multicell domain M in Example 3, the triangles share the edge $\varepsilon \in E_2$ that is parallel to the directional vector $\boldsymbol{e}_2$

The edge-contact property plays a fundamental role for the construction of hierarchical spline spaces, we present that construction for any type-I box spline in Sect. 5. The following examples illustrate the special nature of type-I box splines, as depicted in Lemma 2; simply requiring continuity of a spline space will not yield such intrinsic properties.

Example 3 Let $M = \hat{\Delta} \cup \hat{\Delta}'$ be the multicell domain in Fig. 7. Take the set $B = \{f_1, \ldots, f_4\}$ of spline functions f_i on M defined as

$$f_1 = \left(y^2 + 3x^2 + 4x,\ y^2 + x\right), \quad f_2 = \left(4y^2,\ 4y^2\right),$$

$$f_3 = \left(x^2,\ 0\right), \ \text{and}\ f_4 = \left(0,\ x^2\right).$$

Each pair of polynomials f_i is defined on Δ and Δ', respectively. Then, B is contained in the space of splines $\mathbb{S}^{\mathbf{0}}(M, \mathscr{Q})$, where $\mathbf{0} = (0, 0, 0)$, and $\mathscr{Q} \subseteq \mathscr{P}_2$ is the linear subspace of polynomial spanned by x, x^2, and y^2.

We have

$$\begin{aligned} \mathscr{Q}|_{\Delta} &= B|_{\Delta} = \left\{y^2 + 3x^2 + 4x,\ 4y^2,\ x^2\right\}\big|_{\Delta}, \text{ and} \\ \mathscr{Q}|_{\Delta'} &= B|_{\Delta'} = \left\{y^2 + x,\ 4y^2,\ x^2\right\}\big|_{\Delta'}. \end{aligned}$$

Then, $g = \left(5 \cdot (y^2 + 3x^2 + 4x) + 4y^2,\ y^2 + x + 2 \cdot (4y^2)\right) \in \mathbb{S}^{\mathbf{0}}(M, \mathscr{Q})$, but g is not an element in span (B). $\Diamond$

The following example illustrates the importance of the spline space $\mathbb{S}^d(M, \mathscr{V}_{\boldsymbol{n}})$ (Definition 3) in the proof of Lemma 2.

Example 4 Let $M = \hat{\Delta} \cup \hat{\Delta}'$ be the multicell domain defined from the triangles Δ and Δ' such that $\overline{\Delta} \cap \overline{\Delta}' = \varepsilon \in E_3$ in Fig. 8. Take $\boldsymbol{n} = (2, 1, 1)$, and consider the linear space $\mathscr{V}_{\boldsymbol{n}} = \text{span}\, B_{\boldsymbol{n}}(\hat{\Delta})$. The support and Bernstein coefficients of the box spline $\mathscr{B}_{(2,1,1)}$ are displayed in Fig. 8. The generators of $\mathscr{V}_{(2,1,1)}$ can easily be described by restricting the translates $\mathscr{B}_{(2,1,1)}(\cdot - \boldsymbol{v}) \in B_{\boldsymbol{n}}(M)$ to Δ' and using the Bernstein coefficients of $\mathscr{B}_{(2,1,1)}$. Namely, $\mathscr{V}_{(2,1,1)}$ is generated by

$$f_1 = 2(x - y)y + y^2\,; \quad f_2 = 2(1 - x)y + y^2\,;$$

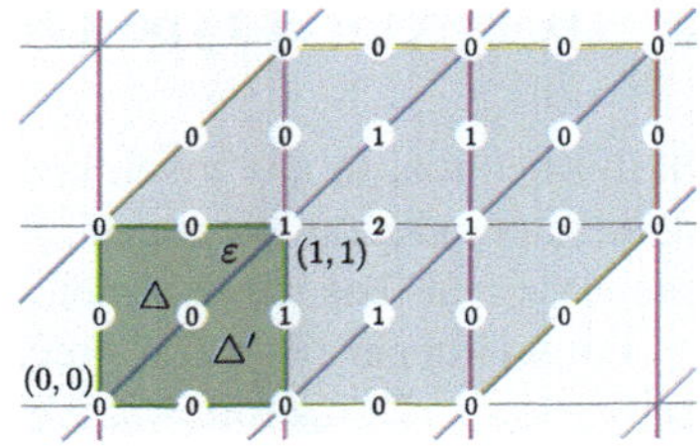

Fig. 8 Bernstein coefficients of the box spline $2\mathscr{B}_{(2,1,1)}$

$$f_3 = (x - y)^2 ;$$

$$f_4 = (1 - x)^2 + 4(1 - x)(x - y) + (x - y)^2 + 2(1 - x)y + 2(x - y)y ;$$

$$\text{and} \quad f_5 = (1 - x)^2 .$$

In particular,

$$f = 4(x - y)y + 4(1 - x)(x - y) + (x - y)^2 = f_1 - f_2 + f_4 - f_5 \in \mathscr{V}_{(2,1,1)} .$$

Let us notice that $g = \left(0|_{\Delta}, f|_{\Delta'}\right) \in \mathbb{S}^{\mathbf{0}}(M, \mathscr{V}_{\boldsymbol{n}})$, but $g \notin \operatorname{span} \mathscr{B}_{\boldsymbol{n}}(M)$. In fact, for $\beta(\cdot) = \mathscr{B}_{\boldsymbol{n}}(\cdot)$ we have $\beta|_{\Delta'} = f_1$, and $\lambda^{\beta}_{\Delta'}(f|_{\Delta'}) = 1$, but $\lambda^{\beta}_{\Delta}(0|_{\Delta}) = 0$.

On the other hand, $\varepsilon \in E_3$, i.e., ε is an edge parallel to the directional vector $\boldsymbol{e}_3$, and g is not a C^1-continuous spline on M. Thus, $g \notin \mathbb{S}^{\boldsymbol{d}}(M, \mathscr{V}_{\boldsymbol{n}})$ for $\boldsymbol{d} = (0, 1, 1)$.$\diamond$

Definition 13 A multicell domain M is *admissible* with respect to a type-I box spline $\mathscr{B}_{\boldsymbol{n}}$ if the support of any $\beta \in B_{\boldsymbol{n}}(M)$ is a connected set, and there exist no over-concave vertices (Definition 7), and no kissing triangles in M (Definition 8).

Example 5 The multicell domain on the left of Fig. 4 is not admissible for any type-I box spline $\mathscr{B}_{\boldsymbol{n}}$ because it has an over-concave vertex ν_2. If the over-concave vertex is eliminated by adding one more triangle, we obtain the domain N shown on the right of Fig. 4. The multicell domain N is admissible with respect to the box splines $\mathscr{B}_1$ and $\mathscr{B}_{(1,2,1)}$ for example, but N it is not an admissible domain with respect to the box spline $\mathscr{B}_{(1,3,1)}$ because $\operatorname{supp} \mathscr{B}_{(1,3,1)}(\cdot - \boldsymbol{v})|_N$ is not connected.

In view of Lemma 2 we arrive to the following completeness result.

Corollary 1 (Completeness of Type-I Box Splines). *If $M \subseteq G$ is an admissible domain, and $\boldsymbol{n} = (n_1, n_2, n_3) \in \mathbb{Z}^3_{\geq 1}$, then the generating set $B_{\boldsymbol{n}}(M)$ is complete for $\mathbb{S}^{\boldsymbol{d}(\boldsymbol{n})}(M, \mathscr{V}_{\boldsymbol{n}})$, where $\boldsymbol{d}(\boldsymbol{n}) = (n_2 + n_3 - 2, n_1 + n_3 - 2, n_1 + n_2 - 2)$.*

5 Hierarchical Type-I Box Splines

In this section, to any given box spline $\mathscr{B}_n$ we associate a space of *special splines* defined on a hierarchical grid. We construct a hierarchical basis for such space and prove that this basis is complete under certain assumptions on the domain hierarchy.

For an integer $N \geq 0$, we recursively define a sequence of three-directional grids $G^1, G^2, \ldots, G^N$ as follows. We take $G^1 = G$ as the three-directional grid with vertices $\mathbb{Z}^2$ introduced in Sect. 2, Fig. 1. The hierarchical grids

$$G^\ell \quad \text{for } \ell = 2, \ldots, N$$

are defined recursively, in such a way that $G^{\ell+1}$ is obtained from G^ℓ by one global, uniform dyadic refinement step. More precisely, the grid $G^{\ell+1} = \frac{1}{2}G^\ell$ is the triangulation of $\mathbb{R}^2$ with vertices at points $1/2^\ell(k, k)$, obtained by drawing in the lines $x = k/2^\ell$, $y = k/2^\ell$, and $x - y = k/2^\ell$, for all $k \in \mathbb{Z}$. Thus, to construct $G^{\ell+1}$, every triangle in the grid G^ℓ is split into four smaller ones, as illustrated in Fig. 9. The index ℓ will be called the *level* of the grid, and the number N specifies the number thereof. Each grid G^ℓ is a uniform three-directional grid, similarly as for the grid $G = G^1$ before, we consider the triangles in this grid as open sets in $\mathbb{R}^2$.

Let Ω be a domain of $\mathbb{R}^2$ whose boundary $\partial\Omega$ is the union of edges from the grid G^N. We define a hierarchical multicell domain H associated to the domain Ω as follows.

A *nested sequence of subdomains* of Ω is defined as a collection of domains $\mathscr{M}^\ell$ such that

$$\emptyset = \mathscr{M}^0 \subseteq \mathscr{M}^1 \subseteq \cdots \subseteq \mathscr{M}^N = \Omega,$$

where $\mathscr{M}^\ell = \bigcup_{\triangle \in M^\ell} \overline{\triangle}$, and $M^\ell \subseteq G^\ell$ is a multicell domain in the grid G^ℓ, for for each $\ell = 1, \ldots, N$.

Thus, for each level ℓ, the boundary $\partial\mathscr{M}^\ell$ is a union of edges of the grid G^ℓ. The difference between two successive subdomains, denoted $\mathscr{D}^\ell$, is defined as the closure

$$\mathscr{D}^\ell = \overline{\mathscr{M}^\ell \setminus \mathscr{M}^{\ell-1}}.$$

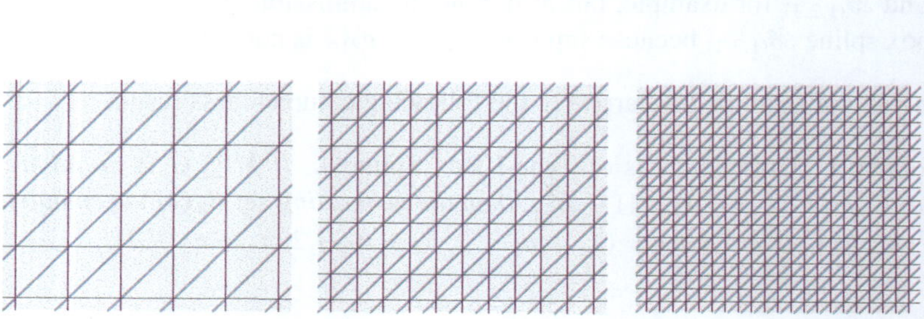

Fig. 9 Three levels of three-directional hierarchical grids

The associated refined domain of level $D^\ell \subseteq G^\ell$ is defined as $D^\ell = T^\ell(\mathscr{D}^\ell)$, where $T^\ell(\cdot)$ is the triangulation operator which restrict the grid G^ℓ to a given subset of the plane $\mathbb{R}^2$. More precisely,

$$T^\ell(\mathscr{Q}) = \{\hat{\triangle} \in G^\ell : \triangle \subset \mathscr{Q}\}.$$

The *hierarchical multicell domain* H associated to Ω is then the collection of triangles form all levels of the refinement area

$$H = \bigcup_{\ell=1}^{N} D^\ell.$$

Using this notation, the domain Ω can be written as the union $\Omega = \bigcup_{\triangle \in H} \overline{\triangle}$.

Definition 14 Let H be a three-directional hierarchical multicell domain associated to a domain $\Omega \subseteq \mathbb{R}^2$, and let $\mathscr{B}_{\boldsymbol{n}}$ be a box spline, for some triple $\boldsymbol{n} = (n_1, n_2, n_3) \in \mathbb{Z}^3_{\geq 1}$. We define $\mathbb{P}(H, \mathscr{V}_{\boldsymbol{n}})$ as the set of piecewise polynomial functions on H associated to $\mathscr{B}_{\boldsymbol{n}}$, i.e.,

$$\mathbb{P}(H, \mathscr{V}_{\boldsymbol{n}}) = \left\{ f \in C^0(\Omega) \colon f|_\triangle \in \mathscr{V}_{\boldsymbol{n}}|_\triangle \text{ for each triangle } \triangle \in H \right\}.$$

If $\boldsymbol{d} = (n_2 + n_3 - 2, n_1 + n_3 - 2, n_1 + n_2 - 2)$, the *hierarchical box spline space with edge smoothness* $\boldsymbol{d}$ *on* H is defined as the set

$$\mathbb{S}^{\boldsymbol{d}}(H, \mathscr{V}_{\boldsymbol{n}}) = \Big\{ f \in \mathbb{P}(H, \mathscr{V}_{\boldsymbol{n}}) \colon f|_{\Diamond(\varepsilon)} \in C^{d_i}\left(\Diamond(\varepsilon)^*\right) \text{ for every } \varepsilon \in E_i \cap H \text{ and } i \in \{1, 2, 3\} \Big\}. \quad (14)$$

For $\boldsymbol{d}$ as above, we define the linear space of *hierarchical box splines with edge and vertex smoothness* $\boldsymbol{d}$ *on the hierarchical multicell domain* H as the set

$$\hat{\mathbb{S}}^{\boldsymbol{d}}(H, \mathscr{V}_{\boldsymbol{n}}) = \Big\{ f \in \mathbb{P}(H, \mathscr{V}_{\boldsymbol{n}}) : f|_U \in \mathbb{D}_I(U, \mathscr{V}_{\boldsymbol{n}}) \text{ for } \triangle, \triangle' \in M^\ell, \text{ for some } \ell,\ \overline{\triangle} \cap \overline{\triangle}' \neq \emptyset,\ U = \overline{\triangle} \cup \overline{\triangle}',\ \text{and } I = \bigcap_{i \in \mathrm{ST}(\triangle, \triangle')} I_i^{\boldsymbol{d}} \Big\}. \quad (15)$$

In Eq. (14), the diamond of and edge $\varepsilon \in H$ is taken over the multicell domain M^ℓ such that ε is an edge of the a triangle $\triangle \in M^\ell$. Namely,

$$\Diamond(\varepsilon) = \bigcup_{\triangle \in M^\ell, \varepsilon \in \hat{\triangle}} \hat{\triangle}.$$

Let us fix $\boldsymbol{n} = (n_1, n_2, n_3) \in \mathbb{Z}^3_{\geq 1}$ and the corresponding box spline $\mathscr{B}_{\boldsymbol{n}}$.

For each level $\ell = 1, \ldots, N$ we denote by $B_{\boldsymbol{n}}^{\ell}$ the set of translates of $\mathscr{B}_{\boldsymbol{n}}$ with respect to the grid G^{ℓ}. In particular, we have $B_{\boldsymbol{n}}^{1} = B_{\boldsymbol{n}}(G)$ as defined in Eq. (4), and recursively

$$B_{\boldsymbol{n}}^{\ell+1} = \left\{\beta(2\cdot\,) \colon \beta \in B_{\boldsymbol{n}}^{\ell}\right\}.$$

In an analogous way as we introduce $B_{\boldsymbol{n}}(M)$ in Definition 11 for a multicell domain M in the three-directional grid G, we now define the set $B_{\boldsymbol{n}}^{\ell}(M^{\ell})$ of active box splines on the multicell domain M^{ℓ} in G^{ℓ} as follows,

$$B_{\boldsymbol{n}}^{\ell}(M^{\ell}) = \left\{\beta|_{M^{\ell}} \colon \beta \in B_{\boldsymbol{n}}^{\ell}\,, \text{ and } \operatorname{supp}\beta \cap \mathscr{M}^{\ell} \neq \emptyset\right\}.$$

A set of linearly independent box splines on a hierarchical multicell domain H can be constructed by a selection procedure analogous to that proposed by Kraft in [28] in the context of tensor-product B-splines.

For all levels ℓ, we select box splines translates, and define the sets K^{ℓ} as follows

$$K^{\ell} = \left\{\beta^{\ell} \in B_{\boldsymbol{n}}^{\ell}(M^{\ell}) \colon \operatorname{supp}\beta^{\ell} \cap \mathscr{M}^{\ell-1} = \emptyset\right\}.$$

The collection of these box splines translates in the levels $\ell = 1, \ldots, N$ forms a *hierarchical box splines basis* given by

$$K = \bigcup_{\ell=1}^{N} K^{\ell}. \tag{16}$$

The linear independence of the functions in K is implied by the local linear independence of the box splines at each level, see [28].

Then the question of completeness of hierarchical type-I box spline spaces can be stated as follows. For a given hierarchical multicell domain H, does the basis K in Eq. (16) span the hierarchical box spline space $\mathbb{S}^{d}(H, \mathscr{V}_{\boldsymbol{n}})$ defined in Eq. (14)? Does K span $\hat{\mathbb{S}}^{d}(H, \mathscr{V}_{\boldsymbol{n}})$ defined in Eq. (15)?

In the following theorem we address the first question, and provide a sufficient condition for the completeness of the hierarchical spline basis.

Theorem 1 *Let H be a three-directional hierarchical multicell domain, and let $\mathscr{B}_{\boldsymbol{n}}$ be a box spline, for some triple $\boldsymbol{n} = (n_1, n_2, n_3) \in \mathbb{Z}_{\geq 1}^{3}$. The basis K in Eq. (16) spans the hierarchical box spline space $\mathbb{S}^{d}(H, \mathscr{V}_{\boldsymbol{n}})$ if each multicell domain M^{ℓ} of H is admissible (Definition 13) with respect to the grid level ℓ.*

Proof The proof follows standard arguments already presented in [17, 31] for the case of hierarchical tensor B-spline bases. Let $\Omega \subset \mathbb{R}^2$ be the domain associated to H. We prove by induction on the levels ℓ that every spline function $s \in \mathbb{S}^{d}(H, \mathscr{V}_{\boldsymbol{n}})$ admits a representation

$$s = (h^1 + \cdots + h^N)|_{\Omega}, \tag{17}$$

where $h^{\ell} \in \operatorname{span} B_{\boldsymbol{n}}^{\ell}(M^{\ell})$, and

$$h^{\ell}|_{\mathscr{M}^{\ell}} = s|_{\mathscr{M}^{\ell}} - (h^{1} + \ldots + h^{\ell-1})|_{\mathscr{M}^{\ell}}. \tag{18}$$

for $\ell = 1, \ldots, N$.

For any given level ℓ, all functions $h^{k}|_{\mathscr{M}^{\ell}}$ of lower levels $k < \ell$ are contained in $\mathbb{S}^{d}(M^{\ell}, \mathscr{V}_{\boldsymbol{n}})$. This follows from the relation

$$\operatorname{span} B_{\boldsymbol{n}}^{k}(M^{k})|_{M^{\ell}} \subseteq \mathbb{S}^{\boldsymbol{d}}(M^{\ell}, \mathscr{V}_{\boldsymbol{n}}).$$

It follows that $s|_{\mathscr{M}^{\ell}} \in \mathbb{S}^{\boldsymbol{d}}(M^{\ell}, \mathscr{V}_{\boldsymbol{n}})$. Consequently, the right-hand side of Equation (18) is contained in $\mathbb{S}^{d}(M^{\ell}, \mathscr{V}_{\boldsymbol{n}})$. Since the multicell domain M^{ℓ} is admissible, we conclude that $h^{\ell} \in \operatorname{span} B_{\boldsymbol{n}}^{\ell}(M^{\ell})$ according to Corollary 1. In particular, choosing $\ell = N$ in Eq. (18) we get Eq. (17).

Moreover, the construction of the functions h^{ℓ} ensures that

$$h^{\ell}|_{\mathscr{M}^{\ell-1}} = 0|_{\mathscr{M}^{\ell-1}}.$$

Since the box splines possess the property of local linear independence we can conclude that $h^{\ell} \in \operatorname{span} K^{\ell}$. This completes the proof. □

Theorem 1 is a consequence of Corollary 1, and therefore of Lemma 2. This result generalizes the completeness property of the space of translates of the quartic box spline $\mathscr{B}_{\boldsymbol{2}}$ proved in [42, Theorem 26].

6 Concluding Remarks

Remark 5 The contact characterization property proved in Lemma 2 implies the completeness of the space spanned by the translates of type-I box splines $\mathscr{B}_{\boldsymbol{n}}$ on a multicell domain M with respect to the spline space $\mathbb{S}^{\boldsymbol{d}(\boldsymbol{n})}(M, \mathscr{V}_{\boldsymbol{n}})$, where $\boldsymbol{d}(\boldsymbol{n}) = (n_2 + n_3 - 2, n_1 + n_3 - 2, n_1 + n_2 - 2)$, and $\boldsymbol{n} = (n_1, n_2, n_3)$. This result holds whenever M is an admissible domain, i.e., whenever M does not have over-concave vertices nor kissing triangles, and supp (β) is connected for all translates β of $\mathscr{B}_{\boldsymbol{n}}$ which have support on M. This is a sufficient condition, but it will also be interesting to prove necessary conditions to achieve this completeness property of the type-I box spline. In this direction, a complete characterization of the vertex-vertex contact plays a crucial role. This calls for exploring the algebraic formulation of super-smoothness at vertices in order to proving an analogous to Lemma 2 for vertex-vertex contact of type-I box splines.

Remark 6 We partially undertake the vertex-vertex contact question raised in the previous remark, for low degrees. In particular, we have verified algorithmically the vertex-vertex contact (that is, Lemma 2 for the case $\overline{\triangle} \cap \overline{\triangle}' \in V$ and the space $\hat{\mathbb{S}}^{d}(M, \mathscr{V}_{\boldsymbol{n}})$) for all type-I box splines up to $\boldsymbol{n} = (4, 4, 4)$.

The approach is based on the fact that both statements (i) and (ii) of Lemma 2 yield linear relations on the coefficients $\lambda_{\triangle}^{\beta}(f|_{\triangle})$ and $\lambda_{\triangle'}^{\beta}(f'|_{\triangle'})$. We can express each of the two statements as matrices $A^{(i)}$ and $A^{(ii)}$. The latter matrix has the form

$$A^{(ii)} = \begin{bmatrix} I_p & \mathbf{0}_q & -PI_p & \mathbf{0}_q \end{bmatrix},$$

where P is a permutation matrix, $p = \#B_{\boldsymbol{n}}(\hat{\Delta}) \cap B_{\boldsymbol{n}}(\hat{\Delta}')$, and $q = \phi(\boldsymbol{n}) - p$.

The matrix[1] $A^{(i)}$ can be computed as a product $A^{(i)} = C_v L_M$. The factor C_v describes the Bézier continuity conditions at the common vertex v, corresponding to the smoothness type $\mathrm{ST}(\Delta, \Delta')$ (Definition 5) and to the regularity vector $\boldsymbol{d} = (n_2 + n_3 - 2, n_1 + n_3 - 2, n_1 + n_2 - 2)$. The right factor L_M is the Bernstein–Bézier representation of the translates of $\mathscr{B}_{\boldsymbol{n}}$, considered independently on each of the two triangles of M:

$$L_M = \begin{bmatrix} L_\Delta & 0 \\ 0 & L_{\Delta'} \end{bmatrix}$$

or, equivalently, a change of basis matrix of the space $C^{-1}(M^*)$.

Showing that Lemma 2 holds in this setting (and for a fixed $\boldsymbol{n}$) is done by showing that $A^{(i)}$ and $A^{(ii)}$ are equivalent matrices. Indeed, in all our computations we obtain $A^{(ii)}$ as the reduced row echelon form of $A^{(i)}$ for type-I box splines up to total degree 12. This verifies that the matrices are equivalent and so the vertex-vertex contact lemma holds in these cases.

Remark 7 In the discussion of completeness of hierarchical type-I box splines spaces there are two main differences to the original approach, which was formulated for tensor-product splines.

First, the translates of a box spline do not span the whole space of bivariate polynomials of a given total degree. For this reason, this special polynomial subspace had to be identified. In some sense this situation generalizes the tensor-product case, where the B-splines span a polynomial space of a given (coordinate-wise) bi-degree, instead of the the space of bivariate polynomials of a given total degree.

Second, the constraints on the domains are entirely different, due to the differences in the characterization of contacts between polynomial pieces. For bivariate tensor-product splines, both edge-edge and vertex-vertex contacts could be characterized easily by the equality of spline coefficients. In the case of type-I box splines, we proved the characterization solely for edge-edge contacts. Consequently, the completeness of hierarchical splines requires more severe restrictions to the hierarchical grid.

We have alleviated these extra restrictions by proving algorithmically the vertex-vertex contact for small polynomial degrees (total degree up to 12). By proving a characterization of the vertex-vertex contact of box splines for arbitrary degree, as we indicated in Remark 5, we could relax these restrictions on the hierarchical grids and construct more general hierarchical type-I spline space. We leave this as a future research direction.

[1] Observe that $\ker A^{(i)} = \hat{\mathbb{S}}^{\boldsymbol{d}}(M, \mathscr{V}_{\boldsymbol{n}})$.

Acknowledgements We would like to thank the anonymous referees for carefully reading our initial submission providing many helpful comments which led to an improved presentation of the results in this paper. N. Villamizar and A. Mantzaflaris wish to thank the support by The Alliance Hubert Curien Programme 2021, Project number: 515492678. N. Villamizar was supported by the UK Engineering and Physical Sciences Research Council (EPSRC) New Investigator Award EP/V012835/1.

References

1. Barendrecht, P., Sabin, M., Kosinka, J.: A bivariate C^1 subdivision scheme based on cubic half-box splines. Comput. Aided Geometr. Design **71**, 77–89 (2019)
2. Billera, L.: Homology of smooth splines: generic triangulations and a conjecture of Strang. Trans. Am. Math. Soc. **310**, 325–340 (1988)
3. Böhm, W.: Subdividing multivariate splines. Comput. Aided Des. **15**, 345–352 (1983)
4. Boor, C.D., Ron, A.: Box splines revisited: convergence and acceleration methods for the subdivision and the cascade algorithms. J. Approx. Theory **150**, 1–23 (2008)
5. Chui, C.: Multivariate Splines. Society for Industrial and Applied Mathematics (SIAM), Philadelphia (1988)
6. Dæhlen, M., Lyche, T.: Box splines and applications. In: Hagen, H., Roller, D. (eds.), Geometric Modeling, pp. 35–93. Springer, Berlin (1991)
7. Dahmen, W., Micchelli, C.: Recent progress in multivariate splines. In: Chui, C.K., Schumaker, L.L., Ward, J. (eds.), Approximation Theory IV, pp. 27–121. Academic, New York (1983)
8. Dahmen, W., Micchelli, C.: Line average algorithm: A method for the computer generation of smooth surfaces. Comput. Aided Geometr. Design **2**, 77–85 (1985)
9. Dahmen, W., Micchelli, C.: On the local linear independence of translates of a box spline. Stud. Math. **82**, 243–263 (1985)
10. de Boor, C., DeVore, R.: Approximation by smooth multivariate splines. Trans. Am. Math. Soc. **276**, 775–788 (1983)
11. de Boor, C., Höllig, K.: Bivariate box splines and smooth pp functions on a three direction mesh. J. Comput. Appl. Math. **9**, 13–28 (1983)
12. de Boor, C., Höllig, K., Riemenschneider, S.: Box Splines. Applied Mathematical Sciences, vol. 98. Springer, New York (1993)
13. Entezari, A., Nilchian, M., Unser, M.: A box spline calculus for the discretization of computed tomography reconstruction problems. IEEE Trans. Med. Imaging **31**, 1532–1541 (2012)
14. Dahmen, W., Micchelli, C.: On the solution of certain systems of partial difference equations and linear dependence of translates of box splines. Trans. Am. Math. Soc. **292**, 305–320 (1985)
15. Forsey, D., Bartels, R.: Hierarchical B-spline refinement. Comput. Graph. **22**, 205–212 (1988)
16. Gérot, C.: Elementary factorisation of box spline subdivision. Adv. Comput. Math. **45**, 153–171 (2019)
17. Giannelli, C., Jüttler, B.: Bases and dimensions of bivariate hierarchical tensor-product splines. J. Comput. Appl. Math. **239**, 162–178 (2013)
18. Giannelli, C., Jüttler, B., Speleers, H.: THB-splines: the truncated basis for hierarchical splines. Comput. Aided Geometr. Design **29**, 485–498 (2012)
19. Giannelli, C., Jüttler, B., Speleers, H.: Strongly stable bases for adaptively refined multilevel spline spaces. Adv. Comput. Math. **40**, 459–490 (2014)
20. Giannelli, C., Kanduč, T., Pelosi, F., Speleers, H.: An immersed-isogeometric model: application to linear elasticity and implementation with THBox-splines. J. Comput. Appl. Math. **349**, 410–423 (2019)
21. Jia, R.: Linear independence of translates of a box spline. J. Approx. Theory **40**, 158–160 (1984)

22. Jia, R.: Local linear independence of the translates of a box spline. Constr. Approx. **1**, 175–182 (1985)
23. Kanduč, T., Giannelli, C., Pelosi, F., Speleers, H.: Adaptive isogeometric analysis with hierarchical box splines. Comput. Methods Appl. Mech. Eng. **316**, 817–838 (2017)
24. Kang, H., Chen, F., Deng, J.: Hierarchical B-splines on regular triangular partitions. Graph. Models **76**, 289–300 (2014)
25. Kang, H., Chen, F., Deng, J.: Hierarchical box splines. In: 14th International Conference on Computer-Aided Design and Computer Graphics, pp. 73–80 (2015)
26. Kim, M., Peters, J.: Fast and stable evaluation of box-splines via the BB-form. Numer. Algorithms **50**, 381–399 (2009)
27. Kobbelt, L.: Stable evaluation of box-splines. Numer. Algorithms **14**, 377–382 (1997)
28. Kraft, R.: Adaptive und linear unabhängige Multilevel B-Splines und ihre Anwendungen. Ph.D. thesis, Universität Stuttgart (1998)
29. Lai, M.J., Schumaker, L.L.: Spline Functions on Triangulations. Encyclopedia of Mathematics and its Applications, vol. 110. Cambridge University Press, Cambridge (2007)
30. Lyche, T., Manni, C., Sablonnière, P.: Quasi-interpolation projectors for box splines. J. Comput. Appl. Math. **221**, 416–429 (2008)
31. Mokriš, D., Jüttler, B., Giannelli, C.: On the completeness of hierarchical tensor-product B-splines. J. Comput. Appl. Math. **271**, 53–70 (2014)
32. Mourrain, B.: On the dimension of spline spaces on planar T-meshes. Math. Comput. **83**, 847–871 (2014)
33. Mourrain, B., Villamizar, N.: Homological techniques for the analysis of the dimension of triangular spline spaces. J. Symb. Comput. **50**, 564–577 (2013)
34. Pelosi, F., Giannelli, C., Manni, C., Sampoli, M.L., Speleers, H.: Splines over regular triangulations in numerical simulation. Comput. Aided Des. **82**, 100–111 (2017)
35. Prautzsch, H., Boehm, W.: Box splines. In: Farin, G., Hoschek, J., Kim, M.S., (eds.), Handbook of Computer Aided Geometric Design, pp. 255–282. Elsevier (2002)
36. Ron, A., Sivakumar, N.: The approximation order of box spline spaces. Proc. Am. Math. Soc. **117**, 473–482 (1993)
37. Sablonnière, P.: Bernstein-Bézier methods for the construction of bivariate spline approximants. Comput. Aided Geometr. Design **2**, 29–36 (1985)
38. Sablonnière, P.: Quasi-interpolants associated with H-splines on a three-direction mesh. J. Comput. Appl. Math. **66**, 433–442 (1996)
39. Schillinger, D., Dedè, L., Scott, M.A., Evans, J.A., Borden, M.J., Rank, E., Hughes, T.J.R.: An isogeometric design-through-analysis methodology based on adaptive hierarchical refinement of NURBS, immersed boundary methods, and T-spline CAD surfaces. Comput. Methods Appl. Mech. Eng. **249–252**, 116–150 (2012)
40. Speleers, H., Dierckx, P., Vandewalle, S.: Quasi-hierarchical Powell-Sabin B-splines. Comput. Aided Geometr. Design **26**, 174–191 (2009)
41. Speleers, H., Manni, C.: Effortless quasi-interpolation in hierarchical spaces. Numer. Math. **132**, 155–184 (2016)
42. Villamizar, N., Mantzaflaris, A., Jüttler, B.: Characterization of bivariate hierarchical quartic box splines on a three-directional grid. Comput. Aided Geometr. Design **41**, 47–61 (2016)
43. Vuong, A., Giannelli, C., Jüttler, B., Simeon, B.: A hierarchical approach to adaptive local refinement in isogeometric analysis. Comput. Methods Appl. Mech. Eng. **200**, 3554–3567 (2011)
44. Xu, Z.: Multivariate splines and polytopes. J. Approx. Theory **163**, 377–387 (2011)
45. Zeng, C., Deng, F., Li, X., Deng, J.: Dimensions of biquadratic and bicubic spline spaces over hierarchical T-meshes. J. Comput. Appl. Math. **287**, 162–178 (2015)
46. Zore, U., Jüttler, B.: Adaptively refined multilevel spline spaces from generating systems. Comput. Aided Geometr. Design **31**, 545–566 (2014)

THU-Splines: Highly Localized Refinement on Smooth Unstructured Splines

Xiaodong Wei

Abstract We present a novel method named *truncated hierarchical unstructured splines* (THU-splines) that supports both local h-refinement and unstructured quadrilateral meshes. In a THU-spline construction, an unstructured quadrilateral mesh is taken as the input control mesh, where the degenerated-patch method [20] is adopted in irregular regions to define C^1-continuous bicubic splines, whereas regular regions only involve C^2 B-splines. Irregular regions are then smoothly joined with regular regions through the truncation mechanism [31], leading to a globally smooth spline construction. Subsequently, local refinement is performed following the truncated hierarchical B-spline construction [11] to achieve a flexible refinement without propagating to unanticipated regions. Challenges lie in refining transition regions where a mixed types of splines play a role. THU-spline basis functions are globally C^1-continuous and are non-negative everywhere except near extraordinary vertices, where slight negativity is inevitable to retain refinability of the spline functions defined using the degenerated-patch method. Such functions also have a finite representation that can be easily integrated with existing finite element or isogeometric codes through Bézier extraction.

1 Introduction

Isogeometric analysis (IGA) was proposed to tightly integrate computer-aided design (CAD) with engineering simulation via employing the same spline-based basis in both sides [9, 12]. Local refinement and extraordinary vertices[1] (EVs) have been two of the most intensively studied topics in IGA to boost computational efficiency and support complex real-world geometries. Indeed, various methods have been proposed in each of the two topics. In the area of local refinement, T-splines [2,

[1] An extraordinary vertex is an interior vertex shared by other than four quadrilateral faces.

X. Wei (✉)
Institute of Mathematics, École Polytechnique Fédérale de Lausanne, Lausanne, Switzerland
e-mail: xiaodong.wei@epfl.ch

C. Manni and H. Speleers (eds.), *Geometric Challenges in Isogeometric Analysis*, Springer INdAM Series 49, https://doi.org/10.1007/978-3-030-92313-6_13

24, 25] and (truncated) hierarchical B-splines [10, 11, 14, 27] are two of the mostly studied methods. On the other hand, in the area of supporting EVs, recent efforts have been devoted to recovering optimal approximation properties, such as geometrically continuous construction [13], manifold splines [4, 18], the degenerated patch (D-patch) method [19, 20, 26], blended construction in 3D [31] and subdivision methods [16, 28].

Moreover, several methods have also been proposed to support both local refinement and EVs at the same time, such as isogeometric spline forests [23], truncated hierarchical Catmull-Clark subdivision [29, 30], and analysis-suitable unstructured T-splines (ASUT-splines) with applications to Kirchhoff-Love shells [8]. Particularly, ASUT-splines combine analysis-suitable T-splines [3, 15, 21] with the local D-patch construction [26], where the D-patch method is adopted to define C^1-continuous splines around EVs while using regular C^2-continuous B-splines away from EVs. While ASUT-splines show superior performance in Kirchhoff-Love shell problems [8], the need to resolve T-mesh constraints generally leads to refinement propagation beyond the region of interest. This becomes even more cumbersome when EVs are involved.

Therefore, we propose an alternative to ASUT-splines in this paper, namely *truncated hierarchical unstructured splines* (THU-splines), by combining truncated hierarchical B-splines (THB-splines) with the local D-patch construction. Unlike T-splines, hierarchical B-splines (HB-splines) have a definite termination condition to drive local refinement. Moreover, there is no need for HB-splines to resolve mesh constraints to guarantee desired properties for the underlying spline functions. These advantages make HB-splines be able to locally refine regions of interest flexibly without propagation. However, applying THB-like refinement to the local D-patch construction is challenging because various types of spline functions are defined in a local D-patch construction. These different splines need to be well organized such that all hierarchical splines defined on each element can be evaluated in a unified manner. Also due to this challenge, proving properties of the underlying hierarchical basis becomes technically much more involving. While we postpone most of the theoretical study in a forthcoming paper, we study how THU-splines retain an invariant geometric mapping in a constructive manner, in order to provide insights of how the truncation in THU-splines works. Moreover, we will use examples to show the numerical evidence of other claimed properties, such as refinability and partition of unity.

The paper is organized as follows. We first briefly review two closely related methods, that is, THB-splines in Sect. 2 and local D-patch construction in Sect. 3. In Sect. 4, we introduce our proposed method THU-splines. Section 5 presents several numerical examples to show flexibility, effectiveness and efficiency of THU-splines. In the end, we conclude the paper and comment on the future work in Sect. 6.

2 THB-Splines

We briefly review THB-splines in this section. THB-splines [11] were proposed as an enhancement of hierarchical B-splines (HB-splines) [14] to reduce support overlapping of B-splines from different hierarchical levels, which in turn leads to a sparse stiffness matrix with reduced band width. Moreover, THB-spline basis functions are piecewise polynomials that form a non-negative partition of unity.

In what follows, we explain the key steps of THB-spline construction in the unvariate setting, including initialization, selection, and truncation. While these key ideas are the same in other dimension cases, interested readers are referred to [11, 17] for details.

2.1 *Initialization*

To start with, a set of n^0 univariate B-splines $\mathscr{B}^0 = \{B_i^0\}_{i=1}^{n^0}$ of degree p ($n^0,\ p \in \mathbb{N}^+$), defined on the knot vector $\Xi^0 = \{\xi_i^0\}_{i=1}^{n^0+p+1}$, is given and initialized as the initial set of THB-spline basis functions, that is,

$$\mathscr{H}^0 = \mathscr{B}^0. \tag{1}$$

The initial domain serves as the entire domain Ω of THB-splines, i.e., $\Omega \equiv \Omega^0 = (\xi_1^0, \xi_{n+p+1}^0)$.

To facilitate explanation, we introduce a series of nested spaces spanned by the B-splines,

$$\mathscr{B}^0 \subset \mathscr{B}^1 \subset \cdots \subset \mathscr{B}^{\ell_{\max}}, \tag{2}$$

where $\ell_{\max}$ is the predefined maximum level, and the set $\mathscr{B}^\ell = \{B_i^\ell\}_{i=1}^{n^\ell}$ ($1 \le \ell \le \ell_{\max}$) is defined on the knot vector $\Xi^\ell = \{\xi_i^\ell\}_{i=1}^{n^\ell+p+1}$. Note that the degree p is fixed at all levels. As will become clear shortly, only a small subset of $\mathscr{B}^\ell$ is used and the full set $\mathscr{B}^\ell$ is never generated in practice. Ξ^ℓ is obtained by bisecting all the nonzero knot spans of $\Xi^{\ell-1}$, so knot vectors are nested as well,

$$\Xi^0 \subset \Xi^1 \subset \cdots \subset \Xi^{\ell_{\max}}. \tag{3}$$

A hierarchical construction involves a nested sequence of hierarchical domains,

$$\Omega \equiv \Omega^0 \supseteq \Omega^1 \supseteq \cdots \supseteq \Omega^{\ell_{\max}}. \tag{4}$$

This nested sequence implies a series of local refinements and it drives where to perform local refinement. For instance, once a set of THB-splines is initialized, we use $\mathscr{H}^0$ to solve the problem of interest. According to a certain a posteriori error estimator, a subdomain of Ω^0, which is in fact the domain at Level 1 (i.e., Ω^1), is marked to perform local refinement. Once Ω^1 is determined, THB-splines $\mathcal{H}^1$ can be constructed accordingly.

THB-splines are constructed level by level, so we only need to study a generic two-level construction for selection and truncation: given the THB-splines $\mathcal{H}^\ell$ that has been constructed up to Level ℓ ($0 \leq \ell < \ell_{\max}$) and $\Omega^{\ell+1}$ (i.e., the subdomain of Ω^ℓ to be refined), we discuss how to build $\mathcal{H}^{\ell+1}$.

2.2 Selection

The selection mechanism, first proposed in [14] and later generalized in [27], is aimed to ensure linear independence of resulting hierarchical basis functions. It is based on the relation between the support of a candidate B-spline and *active* subdomains of Ω^ℓ and $\Omega^{\ell+1}$. The selection at Level ℓ states that a B-spline B_i^ℓ is selected only if its support has a nonzero intersection with $\Omega_a^\ell := \Omega^\ell \backslash \Omega^{\ell+1}$, the *active* subdomain at Level ℓ. All the selected Level-ℓ B-splines are collected in $\mathcal{B}_a^\ell$,

$$\mathcal{B}_a^\ell = \{B_i^\ell \in \mathcal{B}^\ell : \operatorname{supp} B_i^\ell \cap \Omega_a^\ell \neq \varnothing\}. \tag{5}$$

where $\operatorname{supp} B_i^\ell := (\xi_i^\ell, \xi_{i+p+1}^\ell)$ is the support of B_i^ℓ. On the other hand, at Level $\ell + 1$, a B-spline is selected only if its support is fully contained in $\Omega^{\ell+1}$. All the selected Level-($\ell + 1$) B-splines form the set $\mathcal{B}_{a,0}^{\ell+1}$,

$$\mathcal{B}_{a,0}^{\ell+1} = \{B_i^{\ell+1} \in \mathcal{B}^{\ell+1} : \operatorname{supp} B_i^{\ell+1} \subseteq \Omega^{\ell+1}\}. \tag{6}$$

Selected B-splines are called *active*, and the complementary set of B-splines (i.e., $\mathcal{B}^\ell \backslash \mathcal{B}_a^\ell$ and $\mathcal{B}^{\ell+1} \backslash \mathcal{B}_{a,0}^{\ell+1}$) are *passive*. A prefix "H" will be added (i.e., H-active or H-passive) to emphasize the hierarchical refinement when the context is not evident; see Sect. 4. The subscript "0" in $\mathcal{B}_{a,0}^{\ell+1}$ indicates the initial selection at Level $\ell + 1$ with the entire $\Omega^{\ell+1}$ being active. Certain functions in $\mathcal{B}_{a,0}^{\ell+1}$ may latter become passive when Level $\ell + 2$ is constructed from Level $\ell + 1$ and the active Level-($\ell + 1$) subdomain changes from $\Omega^{\ell+1}$ to $\Omega_a^{\ell+1} \equiv \Omega^{\ell+1} \backslash \Omega^{\ell+2}$. $\mathcal{I}^\ell$, $\mathcal{A}^\ell$, $\mathcal{I}^{\ell+1}$ and $\mathcal{A}_0^{\ell+1}$ are introduced to denote the index sets of $\mathcal{B}^\ell$, $\mathcal{B}_a^\ell$, $\mathcal{B}^{\ell+1}$ and $\mathcal{B}_{a,0}^{\ell+1}$, respectively.

2.3 Truncation

The truncation mechanism, which enables reduced overlapping and partition of unity, plays the key role in THB-splines. We need the *refinability relationship* to explain the idea. Refinability states that a Level-ℓ B-spline B_i^ℓ can be expressed as a linear combination of certain Level-($\ell + 1$) B-splines,

$$B_i^\ell = \sum_{j \in \mathscr{C}_i^{\ell+1}} h_{ij}^{\ell+1} B_j^{\ell+1}, \quad \text{with} \quad \mathscr{C}_i^{\ell+1} := \{j \in \mathscr{I}^{\ell+1} : \operatorname{supp} B_j^{\ell+1} \subset \operatorname{supp} B_i^\ell\}, \tag{7}$$

where $h_{ij}^{\ell+1}$ are coefficients obtained from the knot insertion algorithm [5], and $B_j^{\ell+1}$ ($j \in \mathscr{C}_i^{\ell+1}$) are called the *children* (or H-children, children in the hierarchical refinement) of B_i^ℓ. In fact, refinability is the reason for Eq. (2) to hold.

The truncation mechanism applies to B_i^ℓ when some of its children are active, and it discards the active children from Eq. (7). As a result, the truncated function of B_i^ℓ is defined as

$$\operatorname{trun}_H B_i^\ell = \sum_{j \in \mathscr{C}_i^{\ell+1} \setminus \mathscr{A}_0^{\ell+1}} h_{ij}^{\ell+1} B_j^{\ell+1}, \tag{8}$$

which is also referred to as the *inter-level truncation* in this paper. Equivalently speaking, the coefficient $h_{ij}^{\ell+1}$ in Eq. (7) is set to be 0 by truncation if $B_j^{\ell+1}$ is active. Note that when all the children are passive (i.e., $\mathscr{C}_i^{\ell+1} \cap \mathscr{A}_0^{\ell+1} = \varnothing$), we have $\mathscr{C}_i^{\ell+1} \setminus \mathscr{A}_0^{\ell+1} \equiv \mathscr{C}_i^{\ell+1}$ and thus $\operatorname{trun}_H B_i^\ell \equiv B_i^\ell$. In other words, Eq. (8) can represent a generic definition including both truncated and non-truncated B-splines.

In summary, after selection of $\mathscr{B}_a^\ell$ and $\mathscr{B}_{a,0}^{\ell+1}$, truncation is applied to certain B-splines in $\mathscr{B}_a^\ell$, leading to a set of truncated (active) B-splines at Level ℓ,

$$\mathscr{T}_a^\ell = \{\operatorname{trun}_H B_i^\ell : B_i^\ell \in \mathscr{B}_a^\ell\}. \tag{9}$$

The THB-spline basis functions up to Level $\ell + 1$ are now obtained as

$$\mathscr{H}^{\ell+1} = \mathscr{H}^{\ell-1} \cup \mathscr{T}_a^\ell \cup \mathscr{B}_{a,0}^{\ell+1}, \tag{10}$$

where note that $\mathscr{H}^{\ell-1} := \varnothing$ when $\ell = 0$.

In Eq. (10), the local refinement is realized by enriching basis functions and it solely depends on $\mathscr{B}_{a,0}^{\ell+1}$. $\mathscr{H}^{\ell-1}$ and $\mathscr{T}_a^\ell$ are merely adjustments of existing basis functions to accommodate $\mathscr{B}_{a,0}^{\ell+1}$. Therefore, local refinement is effective only when $\mathscr{B}_a^{\ell+1} \neq \varnothing$; otherwise we end up with $\mathscr{H}^{\ell+1} \equiv \mathscr{H}^\ell$. A non-empty $\mathscr{B}_{a,0}^{\ell+1}$ implies that $\Omega^{\ell+1}$ must be able to cover the support of at least one Level-($\ell + 1$) B-spline; see Eq. (6). This is indeed the only requirement for local refinement of THB-splines. It has a definite termination condition and it indicates a highly localized refinement with no propagation beyond the region of interest.

3 Local D-Patch Construction

THB-splines are built upon B-splines that can only take structured meshes as input. However, our goal is to perform truncated hierarchical refinement on *unstructured* quad meshes that involve EVs. As refinability is the key requirement for building

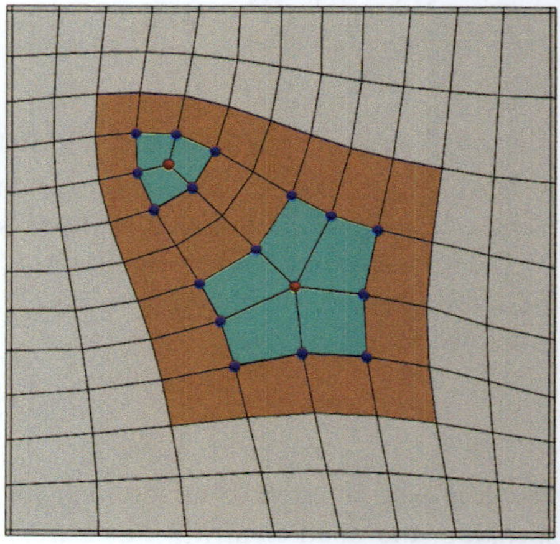

Fig. 1 An unstructured quad mesh taken as the input control mesh. EVs and interface vertices are marked as red and blue dots, respectively. Irregular elements are shaded blue and they are also transition elements in this mesh. Transition regular elements are shaded orange, whereas all the other elements are non-transition regular elements

hierarchical splines, we choose the D-patch method that retains this property while producing C^1-continuous splines in irregular regions, which we call *D-patch splines* in the following. D-patch splines have shown optimal convergence in solving the 2nd order PDEs (partial differential equations), near-optimal convergence in the 4th order problems [26], as well as superior performance in Kirchhoff-Love shells [8].

In contrast to the *global* D-patch construction [19, 20] that leads to reduced continuity everywhere and proliferation of degrees of freedom, a *local* D-patch construction defines D-patch splines only in irregular regions while keeping B-splines elsewhere. Such a local construction has been explored with global refinement [26] as well as local refinement via T-splines [8].

Here we review the local D-patch construction following [8], including Bézier extraction on unstructured meshes, D-patch construction around EVs, smooth transition to join D-patches with regular B-splines, and global refinement. Interested readers are referred to related literature [8, 20, 26] for further details.

3.1 Bézier Extraction

We first introduce several terminologies of unstructured meshes with the reference to Fig. 1. An element is called an *irregular element* if any of its vertices[2] is an EV; otherwise it is regular. An element is a *boundary element* if any of its vertices lies on the boundary; otherwise it is an interior element. The *one-ring neighborhood* of a vertex includes all the elements sharing this vertex; recursively, the n-ring ($n \geq 2$) neighborhood is the $(n-1)$-ring neighborhood together with those sharing vertices

[2] We use "vertex", "edge" and "face" to emphasize mesh connectivity (or topology), whereas using "point" to carry the position or geometry information. "Element" and "face" are used interchangeably.

with the $(n-1)$-ring neighborhood. The neighborhood of an element can be defined similarly. We adopt the following assumptions regarding the input mesh to simplify discussion.

- An irregular element can only have one EV;
- A boundary vertex can only be shared by either one or two elements; and
- An irregular element must not lie in the three-ring neighborhood of any boundary vertex.

Each vertex corresponds to a control point, and we call such points *V-points* (i.e., vertex points). Each edge is assigned with a non-negative real number called the knot span or knot interval, which is used to define element (parametric) domains as well as spline functions. For simplicity, we adopt a semi-uniform configuration of knot intervals where all edges have a unit knot interval except for those perpendicular to the boundary, which have a zero knot interval. As a result, edges around EVs all have the same knot intervals, which indeed is a necessary assumption for the D-patch construction [20].

Note that given the third assumption regarding the input mesh, Bézier extraction is never needed for elements near the boundary. Without loss of generality, we next explain how to extract a Bézier representation for an irregular element Ω_e [22]. Figure 2a shows as a generic irregular element Ω_e, where $P_{e,i}$ $(i = 1, \ldots, 4)$ are its 4 vertices, and $P_{e,1}$ is an EV. In the bicubic case, Bézier extraction is to find the 4×4 Bézier control points $Q_{e,j}$ $(j = 1, \ldots, 16)$ corresponding to Ω_e. They can be divided into face, edge and vertex Bézier points, and they are called *FB-points*, *EB-points*, and *VB-points*, respectively; see Fig. 2b. The 4 FB-points are computed as convex combinations of the 4 element corners, that is,

$$\begin{bmatrix} Q_{e,6} \\ Q_{e,7} \\ Q_{e,11} \\ Q_{e,10} \end{bmatrix} = \begin{bmatrix} a & b & c & b \\ b & a & b & c \\ c & b & a & b \\ b & c & b & a \end{bmatrix} \begin{bmatrix} P_{e,1} \\ P_{e,2} \\ P_{e,3} \\ P_{e,4} \end{bmatrix}, \quad \text{with} \quad a = \frac{4}{9},\ b = \frac{2}{9},\ c = \frac{1}{9}. \tag{11}$$

As shown in Fig. 2c, d, EB- and VB-points are simply an average of their neighboring FB-points, for examples,

$$\begin{aligned} Q_{e,9} &= Q_{e+1,3} = \frac{1}{2} Q_{e,10} + \frac{1}{2} Q_{e+1,7}, \\ Q_{e,1} &= \frac{1}{N} \sum_{i=1}^{N} Q_{i,6}, \end{aligned} \tag{12}$$

where $N \in \mathbb{N}^+$ is the number of faces sharing $P_{e,1}$, also called the valence of $P_{e,1}$.

Now let $\boldsymbol{Q}_e := [Q_{e,1}, \ldots, Q_{e,16}]^T$ denote the vector of Bézier points of Ω_e, and $\boldsymbol{P}_v$ and $\boldsymbol{P}_f$ contain the V-points and FB-points of the one-ring neighborhood of Ω_e, respectively. From Eqs. (11), (12), we observe that $\boldsymbol{Q}_e$ can be obtained through either

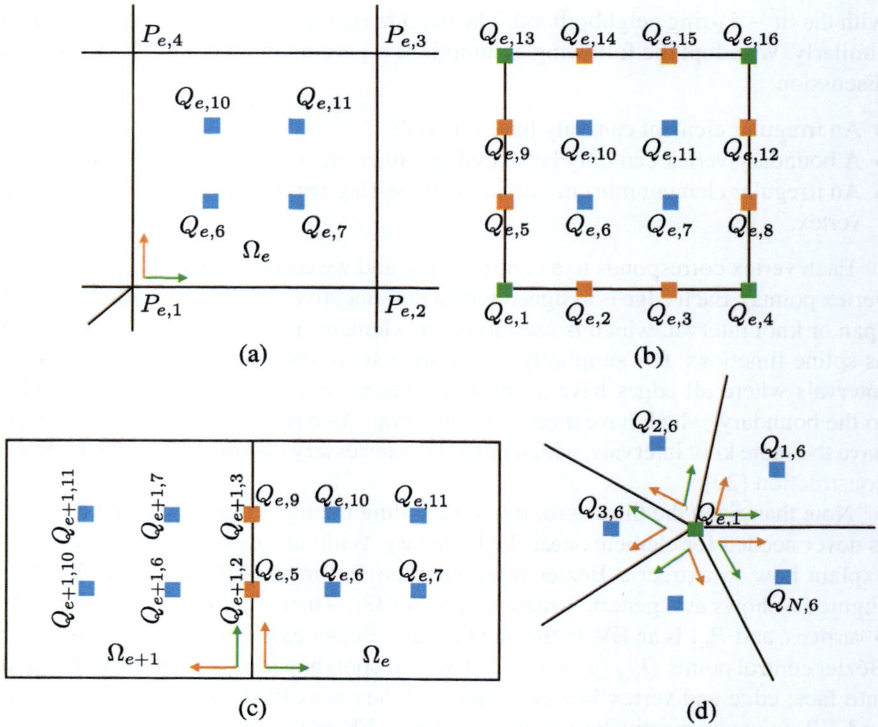

Fig. 2 Computing 4×4 Bézier control points of Ω_e. **a** The irregular element Ω_e and its four vertices $P_{e,i}$ $(i = 1, \ldots, 4)$, where $P_{e,1}$ is an EV; **b** the 16 Bézier control points $Q_{e,j}$ $(j = 1, \ldots, 16)$ of Ω_e, where blue, orange and green squares are face, edge and vertex Bézier points, respectively; **c** computing an EB-point (edge Bézier point) as an average of neighboring FB-points (face Bézier points); and **d** computing a VB-point (vertex Bézier point) as an average of neighboring FB-points

$$\boldsymbol{Q}_e = \boldsymbol{A}_e \boldsymbol{P}_f, \tag{13}$$

or

$$\boldsymbol{Q}_e = \boldsymbol{E}_e \boldsymbol{P}_v, \tag{14}$$

where the matrix $\boldsymbol{A}_e$ contains the averaging coefficients in Eq. (12), and $\boldsymbol{E}_e$ is the extraction matrix with coefficients from both Eqs. (11), (12). When $N = 4$, Eqs. (13), (14) indicate Bézier extraction for a regular bicubic B-spline element.

The Bézier representation of Ω_e is written as

$$\boldsymbol{S}(u, v)|_{\Omega_e} = \boldsymbol{Q}_e^T \boldsymbol{b}(u, v), \quad (u, v) \in [0, 1]^2, \tag{15}$$

where $\boldsymbol{b}(u, v) := [b_1(u, v), \ldots, b_{16}(u, v)]^T$ is the vector of bicubic Bernstein polynomials. $b_j(u, v)$ is expressed as,

$$b_j(u, v) = N_{(j-1)\%4}(u)N_{(j-1)/4}(v), \quad (u, v) \in [0, 1]^2, \tag{16}$$

where "%" and "/" denote remainder and quotient operations, respectively; and

$$N_k(t) = \binom{3}{k} t^k (1-t)^{3-k}, \quad k = 0, \ldots, 3. \tag{17}$$

Alternatively, the surface patch $\boldsymbol{S}(u, v)|_{\Omega_e}$ can also be expressed in terms of $\boldsymbol{P}_v$ and the associated spline functions $\boldsymbol{B}_v(u, v)$,

$$\boldsymbol{S}(u, v)|_{\Omega_e} = \boldsymbol{P}_v^T \boldsymbol{B}_v(u, v). \tag{18}$$

On the other hand, Eqs. (15), (18) must be equivalent, leading to

$$\boldsymbol{P}_v^T \boldsymbol{B}_v(u, v) \equiv \boldsymbol{Q}_e^T \boldsymbol{b}(u, v) = \boldsymbol{P}_v^T \boldsymbol{E}_e^T \boldsymbol{b}(u, v), \tag{19}$$

where we have used Eq. (14) to arrive at the last term. According to the arbitrariness of $\boldsymbol{P}_v$, we conclude that

$$\boldsymbol{B}_v(u, v) = \boldsymbol{E}_e^T \boldsymbol{b}(u, v). \tag{20}$$

Similarly, the surface patch can also be written with FB-points $\boldsymbol{P}_f$,

$$\boldsymbol{S}(u, v)|_{\Omega_e} = \boldsymbol{P}_f^T \boldsymbol{B}_f(u, v), \quad \text{with} \quad \boldsymbol{B}_f(u, v) = \boldsymbol{A}_e^T \boldsymbol{b}(u, v). \tag{21}$$

The central idea of Bézier extraction is to express each spline function as a linear combination of Bernstein polynomials, which enables general spline functions to be easily integrated with existing finite element and isogeometric codes. While in regular elements, $\boldsymbol{B}_v$ and $\boldsymbol{B}_f$ are actually bicubic C^2 and C^1 B-splines, respectively, they are only C^0-continuous in irregular elements. The D-patch construction essentially adjusts $\boldsymbol{E}_e^T$ (or $\boldsymbol{A}_e^T$) to yield smooth spline functions, which we discuss in the following.

3.2 D-Patch Method

In a local D-patch construction, the D-patch method is only adopted in regions around EVs, where we need to add necessary face-based control points and define their associated splines. Face-based control points are referred to as *F-points* in the following. Splines associated with V-points and F-points are referred to as *V-splines* and *F-splines*, respectively.

Enrichment via Adding F-Points. We introduce *enriched elements* to guide where to add F-points. Initially, all irregular elements are defined as enriched elements in the input mesh, whereas all the other elements are non-enriched elements. Enriched

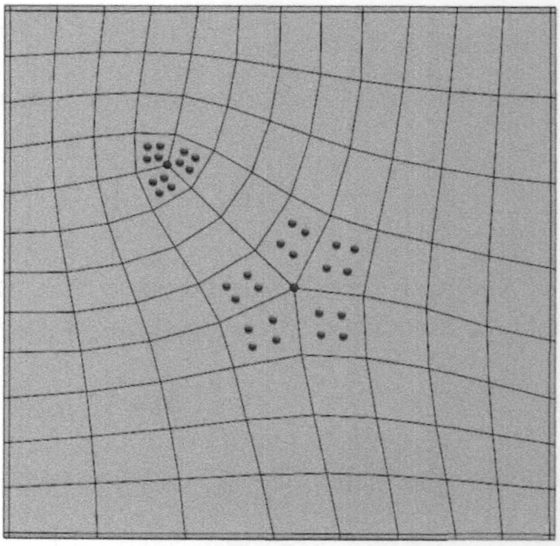

Fig. 3 The control mesh of a local D-patch representation corresponding to the input in Fig. 1, where blue dots are F-points, whereas red dots are EVs and they are passive

elements in refined meshes follow an "inheritance" rule, which will be discussed in Sect. 3.4.

Four F-points are then added to each enriched element. In fact, F-points are equivalent to FB-points, and they are obtained according to Eq. (11) in the input mesh. As a result, the control mesh of a local D-patch representation consists of both V-points and F-points; see Fig. 3. However, while all the F-points of enriched elements are active in the sense that they are used in the local D-patch representation, not all V-points are active. A V-point is *passive* if all its one-ring neighboring elements are enriched elements; otherwise it is *active*. A prefix "D" will be added to emphasize the context of local D-patch representation when necessary to distinguish those in hierarchical refinement. Clearly, all EVs are passive according to this definition. Essentially, passive V-points are replaced by their neighboring F-points. Note that such a control mesh corresponds to the analysis space discussed in [26].

Definition of D-Patch Splines. We next briefly discuss the smooth D-patch splines around an EV that is shared by N elements Ω_e, $e = 1, 2, \dots, N$. Interested readers are referred to [8, 20, 26] for details. D-patch splines are developed based on the C^0 version in Eq. (21). This essentially involves smoothing the underlying surface patches around the EV, so we first proceed from the perspective of Bézier control points. Starting from Eq. (13), we refine the Bézier patch via a 2×2 split,

$$\bar{\boldsymbol{Q}}_e^{ij} = \boldsymbol{D}_{ij} \boldsymbol{Q}_e, \tag{22}$$

where $\boldsymbol{D}_{ij}$ $(i, j = 1, 2)$ is the split matrix obtained from the well-known de Casteljau's algorithm [6], and $\bar{\boldsymbol{Q}}_e^{ij}$ represents control points of the refined Bézier patch corresponding to the subelement Ω_e^{ij}; see Fig. 4. In particular, $\bar{Q}_{1,1}^{11} = \bar{Q}_{2,1}^{11} = \cdots = \bar{Q}_{N,1}^{11} =: \bar{Q}_{EV}$.

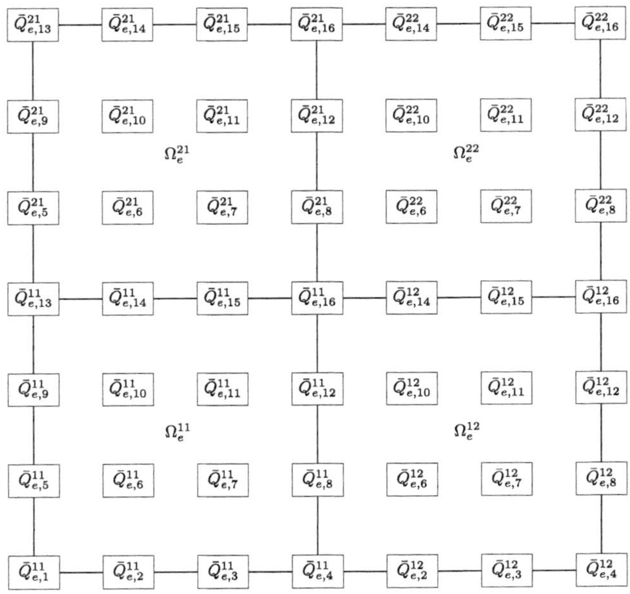

Fig. 4 2×2 split of the Bézier element in Fig. 2b

In the D-patch method, a smooth representation is achieved by adjusting all $\bar{Q}^{11}_{e,k}$, $e \in \{1, \ldots, N\}, k \in \{1, 2, \ldots, 16\}\backslash\{11, 12, 15, 16\}$, which are collected into a vector $\bar{\boldsymbol{Q}}^{11}_s$. This is done through a linear projection matrix $\boldsymbol{\Pi}$,

$$\tilde{\boldsymbol{Q}}^{11}_s = \boldsymbol{\Pi}\,\bar{\boldsymbol{Q}}^{11}_s. \tag{23}$$

As a result, for every $e \in \{1, \ldots, N\}$, $\tilde{Q}^{11}_{e,1} = \tilde{Q}^{11}_{e,2} = \tilde{Q}^{11}_{e,5} = \tilde{Q}^{11}_{e,6} \equiv \bar{Q}_{EV}$, and the five points $\tilde{Q}^{11}_{e,3}$, $\tilde{Q}^{11}_{e,7}$, $\tilde{Q}^{11}_{e,9}$, $\tilde{Q}^{11}_{e,10}$ and $\bar{Q}_{EV}$ are coplanar; see indices in Fig. 4. The reminder of $\bar{\boldsymbol{Q}}^{ij}_e$ stays the same, and the resulting Bézier surface patches are smoothly joined together around the EV. The smoothing matrix $\boldsymbol{\Pi}$ corresponds to the idempotent version of the analysis space [8, 26]. Interested readers are referred there for further details. Note that although $\boldsymbol{\Pi}$ contains negative coefficients and thus leads to slightly negative spline functions, it guarantees refinability, which is the fundamental property required in a hierarchical spline construction.

Now we discuss the F-splines defined on an irregular element Ω_e, which are associated with the F-points of the one-ring neighborhood of Ω_e. They are piecewise-defined on the four subelements (Ω^{ij}_e, $i, j = 1, 2$) as

$$\tilde{\boldsymbol{B}}_f(u, v)|_{\Omega^{ij}_e} = \boldsymbol{A}^T_e \boldsymbol{D}^T_{ij} \boldsymbol{\Pi}^T_{e,ij} \boldsymbol{b}(u, v). \tag{24}$$

where the entries of $\boldsymbol{\Pi}_{e,ij}$ come from $\boldsymbol{\Pi}$ with an proper arrangement for Ω_e^{ij}, and in particular, $\boldsymbol{\Pi}_{e,22}$ is an identity matrix. Comparing Eqs. (21), (24), we observe that by introducing matrices $\boldsymbol{D}_{ij}$ and $\boldsymbol{\Pi}_{e,ij}$, C^1 continuity is built into $\tilde{\boldsymbol{B}}_f$. Moreover, we have $\tilde{\boldsymbol{B}}_f \equiv \boldsymbol{B}_f$ outside of the irregular element Ω_e thanks to the 2×2 split, or equivalently speaking, the presence of $\boldsymbol{D}_{ij}$. Mostly importantly, it has also been proved that $\tilde{\boldsymbol{B}}_f$ are refinable in the sense of Eq. (7); see [19]. In the reminder of the paper, we omit "$\sim$" over $\tilde{\boldsymbol{B}}_f$ to keep notations simple.

3.3 Smooth Transition

Next, we introduce how to smoothly join enriched and non-enriched elements in a local D-patch representation. The following terms are needed to facilitate the discussion. An interior edge is called an *interface edge* if it is shared by an enriched element and a non-enriched element. Vertices of interface edges are interface vertices. An element is then called a *transition element* if it contains an interface vertex. Note that transition elements can be regular or irregular; see Fig. 1. Now let $\mathscr{V}$ and $\mathscr{F}$ denote the index sets of V-points and F-points, respectively, and $\mathscr{V}_a$ be the active subset of $\mathscr{V}$.

We now discuss the splines associated with the active V-points ($\mathscr{V}_a$), which include interface points ($\mathscr{V}_t$) and the active regular points ($\mathscr{V}_r$) that do not lie on the interface, that is, $\mathscr{V}_a = \mathscr{V}_t \cup \mathscr{V}_r$. Splines associated with $\mathscr{V}_r$ are simply bicubic C^2 B-splines, whereas those associated with $\mathscr{V}_t$ need *truncation* to enable a smooth transition between enriched and non-enriched elements. It is related but not the same as that introduced in the context of THB-splines. It was first proposed in blended B-splines [31] to recover the optimal convergence property when using unstructured hexahedral meshes. To distinguish, we refer to the truncation in a local D-patch representation as the *same-level truncation*. Recall that it is called the inter-level truncation in the context of THB-splines.

For the V-spline $B_{v,i}$ associated with an interface point P_i ($i \in \mathscr{V}_t$), we can express it in terms of its one-ring neighboring F-splines $B_{f,j}$,

$$B_{v,i} = \sum_{j \in \mathscr{F}_i} d_{ij} B_{f,j}, \tag{25}$$

where the coefficients d_{ij} are obtained from Eq. (11) and they are illustrated in Fig. 5a, $\mathscr{F}_i \subseteq \mathscr{F}$ is the index set of the F-points on the one-ring neighborhood of P_i, and $B_{f,j}$ are the children (or D-children, children in the local D-patch representation) of $B_{v,i}$. $B_{f,j}$ is defined by Eq. (24) if its F-point lies in an irregular element; otherwise it is defined by Eq. (21). Note that not all $B_{f,j}$ are associated with the F-points of enriched elements. As P_i is an interface point, there must exist at least one non-enriched element in its one-ring neighborhood. F-points on non-enriched elements

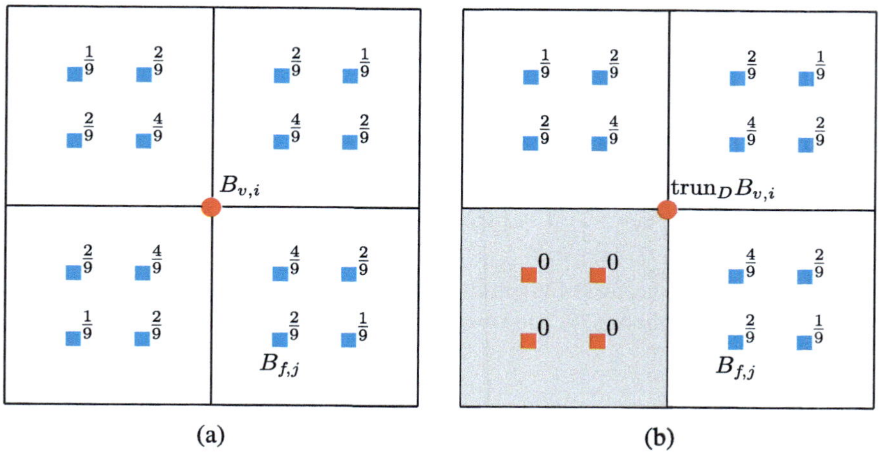

Fig. 5 The same-level truncation of $B_{v,i}$ that is associated with an interface point. **a** The V-spline $B_{v,i}$ is expressed in terms of one-ring neighboring F-splines $B_{f,j}$, where the numbers are the coefficients d_{ij}; and **b** truncation of $B_{v,i}$ by setting certain $d_{ij} = 0$, where the gray element is enriched, and the red and blue squares are active and passive F-points, respectively

are not part of the control mesh in a local D-patch representation, and thus they are passive. In contrast, F-points of enriched elements are active.

The same-level truncation for $B_{v,i}$ states that d_{ij} is set to be 0 if $B_{f,j}$ is active. In other words, active children of $B_{v,i}$ are discarded,

$$\text{trun}_D B_{v,i} = \sum_{j \in \mathscr{F}_i \setminus \mathscr{F}_a} d_{ij} B_{f,j}, \tag{26}$$

where $\mathscr{F}_a$ is the index set of active F-points. As a result of truncation, $\text{trun}_D B_{v,i}$ is always a linear combination of regular C^1 B-splines. Figure 5b shows an example of $\text{trun}_D B_{v,i}$, where the red squares are active F-points as the gray element is enriched whereas the blue squares are passive and only serve as an auxiliary purpose to define $\text{trun}_D B_{v,i}$. Therefore, the splines associated with interface points are truncated B-splines defined according to Eq. (26). From an element-wise perspective, truncation has no influence outside of transition elements in the sense that $\text{trun}_D B_{v,i} \equiv B_{v,i}$.

On the other hand, given an active non-interface point P_k ($k \in \mathscr{V}_r$), its associated spline $B_{v,k}$ is a C^2 B-spline and can be also expressed in the form of Eq. (26), where, however, truncation is trivial as $\mathscr{F}_k \setminus \mathscr{F}_a = \mathscr{F}_k$, because we have $\mathscr{F}_k \cap \mathscr{F}_a = \varnothing$ by definition. In other words, $\text{trun}_D B_{v,k} \equiv B_{v,k}$ holds for $k \in \mathscr{V}_r$.

To this end, all spline functions of a local D-patch construction have been discussed, which form a partition of unity and are linearly independent [8]. The local D-patch representation is written as

$$S = \sum_{i\in\mathscr{F}_a} Q_i\, B_{f,i} + \sum_{i\in\mathscr{V}_t} P_i\, \mathrm{trun}_D B_{v,i} + \sum_{i\in\mathscr{V}_r} P_i\, B_{v,i} = \sum_{i\in\mathscr{F}_a} Q_i\, B_{f,i} + \sum_{i\in\mathscr{V}_a} P_i\, \mathrm{trun}_D B_{v,i}, \tag{27}$$

where we recall that $\mathscr{V}_a = \mathscr{V}_r \cup \mathscr{V}_t$, and Q_i and P_i are (active) F-points and V-points, respectively. The corresponding matrix form is given as,

$$\boldsymbol{S} = \begin{bmatrix} \boldsymbol{Q}^T & \boldsymbol{P}^T \end{bmatrix} \begin{bmatrix} \boldsymbol{B}_f \\ \boldsymbol{B}_v \end{bmatrix}, \tag{28}$$

where $\boldsymbol{B}_v$ and $\boldsymbol{B}_f$ are vectors of V-splines and F-splines, respectively, and $\boldsymbol{B}_v$ includes both regular C^2 B-splines ($\mathscr{V}_r$) and truncated B-splines ($\mathscr{V}_t$).

3.4 Global Refinement

Finally, we discuss how to perform global h-refinement for a local D-patch representation. In the absence of EVs, h-refinement of a B-spline patch is straightforward through the knot insertion algorithm, where the underlying geometric mapping stays the same during refinement. We call such a refinement the *consistent refinement*. However, consistent refinement is usually not available when EVs are involved. Currently, only two methods[3] have this property: the D-patch method and subdivision. In fact, subdivision implies a large family of methods that feature an infinite piecewise definition of splines in irregular regions. Interested readers are referred to [16, 28] for some of the latest developments that aim to recover optimal convergence rates.

Consistent global refinement of a local D-patch representation consists of a topological step and a geometric step. The topological step deals with mesh connectivity as well as inheritance of enriched elements, whereas the geometric step computes control point coordinates of the refined mesh. Elementwise definitions of basis functions are discussed as well after the topological step.

Topological Step. In the topological step, every element with a nonzero (parametric) measure is equally subdivided into 4 child elements. If a zero-measure element has a pair of opposite edges with nonzero knot intervals, it is subdivided into 2 child elements at the midpoints of such edges. Recall that irregular elements in the input mesh are marked as enriched elements. Subsequently in a refined mesh, elements in the $(2^k + 1)$-ring neighborhood of each EV are marked as enriched elements, where $k \geq 1$ is number of global refinements. We observe the following: (1) regular elements in a refined mesh can be enriched elements; (2) child elements of a enriched element are always enriched elements; and (3) for a transition element that is not enriched, some of its child elements are also enriched, which are in fact the $(2^k + 1)^{\text{th}}$-ring neighborhood of an EV. Such an inheritance pattern of enriched elements ensures that the irregular region in the initial local D-patch representation is maintained throughout refinements, which is the key to guaranteeing refinability [26].

[3] We only focus on methods based on quadrilaterals.

Elementwise Definitions of Basis Functions. Once topological step is finished, basis functions of a refined mesh can be defined accordingly. Their definitions vary among different element types: regular elements that are not transition or enriched (type-R), non-transition enriched elements (type-E), and transition elements (type-T). A type-R element only involves 16 C^2 B-splines. C^1-continuous splines are defined on a type-E element, which are D-patch splines if the element is irregular and 16 C^1 B-splines otherwise. On the other hand, a type-T element (no matter if the element is enriched or not) may involve a mix of all kinds of splines, such as V-splines (including C^2 B-splines and truncated B-splines) and F-splines (including C^1 B-splines and C^1 D-patch splines). As V-splines ($\boldsymbol{B}_v$) can be expressed as linear combinations of F-splines ($\boldsymbol{B}_f$), we adopt the following matrix form to compactly organize all these splines on a transition element,

$$\begin{bmatrix} \boldsymbol{B}_v \\ \boldsymbol{B}_{f,a} \end{bmatrix} = \begin{bmatrix} \boldsymbol{0} & \boldsymbol{T} \\ \boldsymbol{I} & \boldsymbol{0} \end{bmatrix} \begin{bmatrix} \boldsymbol{B}_{f,a} \\ \boldsymbol{B}_{f,p} \end{bmatrix}, \quad \text{with} \quad \boldsymbol{B}_f = \begin{bmatrix} \boldsymbol{B}_{f,a} \\ \boldsymbol{B}_{f,p} \end{bmatrix}, \tag{29}$$

where coefficients of the matrix $\boldsymbol{T}$ come from Eq. (26), $\boldsymbol{I}$ is an identity matrix, and subscripts "a" and "p" again denote active and passive splines, respectively.

Geometric Step. Positions of control points in a refined mesh are determined in the geometric step. Overall, all the coordinates are determined through the knot insertion algorithm. As has been discussed in [31], this simple and unified computation is enabled by the (same-level) truncation.

The computation of point coordinates has two cases: B-spline refinement for non-enriched elements, and Bézier refinement for enriched elements. First, for a surface patch restricted to a non-enriched element (no matter if the element is transition or not), it is simply a B-spline patch when we only focus on its V-points. All the V-points of refined patches are obtained through the knot insertion algorithm applied to such a B-spline patch. Moreover, when the non-enriched element is a transition element, recall that some of its child elements are enriched elements, whose F-points are added according to Eq. (11).

Second, for a surface patch restricted to an enriched element, the corresponding local D-patch representation is first converted to its Bézier representation. Refinement is then applied to this Bézier patch using de Casteljau's algorithm. In each of the resulting child Bézier patches, the 4 interior Bézier control points are added to the corresponding element (which must be an enriched element by definition) as the F-points.

To this end, both basis functions and control points are defined for a refined mesh. Its local D-patch representation has the same form as Eq. (28).

4 THU-Splines

We introduce the proposed method in this section, namely *truncated hierarchical unstructured splines* (THU-splines). Following the construction of THB-splines, we discuss the three steps in constructing THU-splines: initialization, selection, and truncation. Among them, truncation becomes particularly complex due to the presence of two kinds of truncation: the same-level truncation and the inter-level truncation.

4.1 Initialization

THU-splines are initialized with the local D-patch construction on an input unstructured quad mesh $\mathscr{M}^0$. For the ease of discussion, we introduce a series of consecutively refined meshes, $\{\mathscr{M}^\ell\}_{\ell=0}^{\ell_{\max}}$, where $\ell_{\max} \in \mathbb{N}^+$ is the predefined maximum number of refinements, and $\mathscr{M}^\ell$ is a global refinement of $\mathscr{M}^{\ell-1}$ $(1 \le \ell \le \ell_{\max})$ according to Sect. 3.4. Note that global refinement is never actually performed and these globally refined meshes are introduced only to simplify explanation. Correspondingly, a local D-patch basis $\mathscr{B}^\ell$, which only consists of D-active splines, is constructed on each $\mathscr{M}^\ell$. The initial THU-spline basis $\mathscr{H}^0$ is given by $\mathscr{B}^0$. Moreover, with a closely related proof in [26, Proposition 4.4], we conjecture that refinability holds for $\{\mathscr{B}^\ell\}$, for which we will show numerical evidence in Sect. 5.

Let $\mathscr{F}^\ell$ and $\mathscr{V}_a^\ell$ denote the index sets of F-points and active V-points of $\mathscr{M}^\ell$, respectively. Note that $\mathscr{F}^\ell = \mathscr{F}_a^\ell \cup \mathscr{F}_p^\ell$, where passive F-points ($\mathscr{F}_p^\ell$) are introduced as an auxiliary means to define truncated B-splines associated with interface points, and splines associated with $\mathscr{F}_p^\ell$ do not belong to $\mathscr{B}^\ell$.

4.2 Selection

The same as in Sect. 2, we focus on constructing $\mathscr{H}^{\ell+1}$ from $\mathscr{H}^\ell$. Following Eq. (4), we assume that a series of nested hierarchical domains $\{\Omega^\ell\}_{\ell=0}^{\ell_{\max}}$ is given to guide local refinement. However, we need to define the "$\supset$" operation in Eq. (4) for unstructured meshes. Let $\Omega^0 := \{\Omega_e^0\}_{e=1}^{n^0}$ be a collection of all n^0 quad elements in $\mathscr{M}^0$. Ω^ℓ $(1 \le \ell \le \ell_{\max})$ is defined similarly but does not necessarily contain all elements of $\mathscr{M}^\ell$. $\Omega^\ell \supset \Omega^{\ell+1}$ means that every element in $\Omega^{\ell+1}$ is a child element of a certain element in Ω^ℓ. In other words, $\Omega^{\ell+1}$ is obtained by locally refining certain elements in Ω^ℓ; after refinement, the parent elements of those in $\Omega^{\ell+1}$ are no longer used and thus they are marked as passive. The active subdomain of Ω_a^ℓ consists of all active elements at Level ℓ.

It is left to define the support of every basis function to apply the selection mechanism. Given a generic basis function B_i^ℓ (whose control point is denoted as P_i^ℓ), its support, $\operatorname{supp} B_i^\ell$, is defined as a collection of certain elements in $\mathscr{M}^\ell$. There are

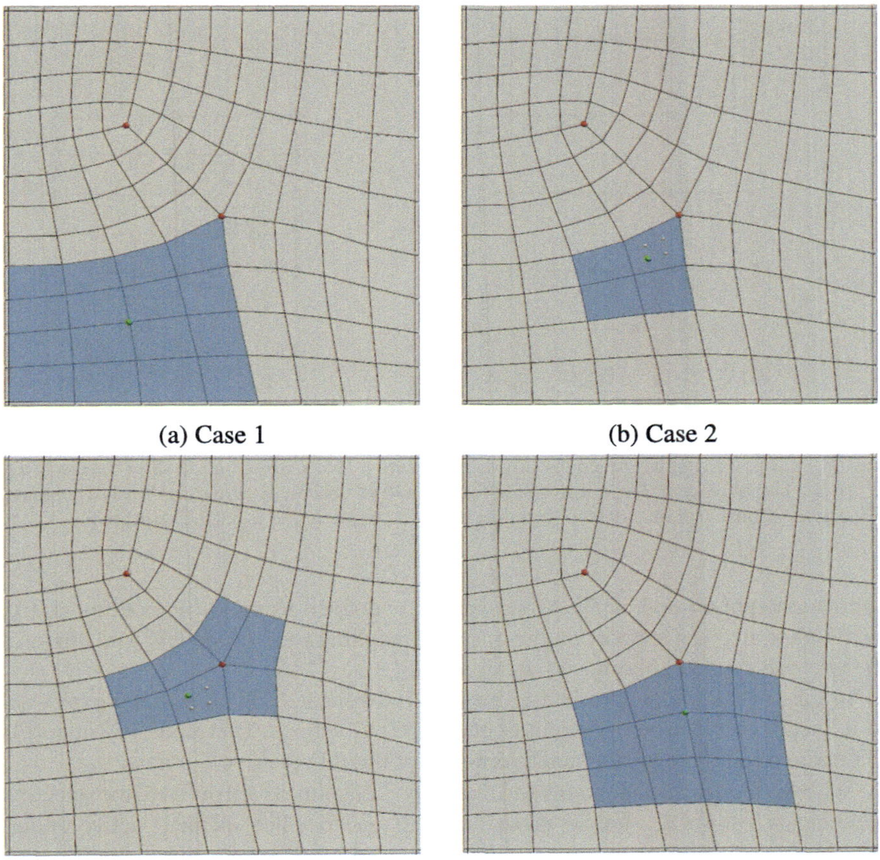

Fig. 6 Support of different spline functions, where red dots are extraordinary points, and blue shades are the supports of the green dots

four cases depending on the type of P_i^ℓ: (1) a V-point that is not on the interface, (2) an F-point on a regular element, (3) an F-point on an irregular element, and (4) an interface point; see Fig. 6.

In Case (1), B_i^ℓ is a bicubic C^2 B-spline, so suppB_i^ℓ is the two-ring neighborhood of P_i^ℓ. In Case (2), B_i^ℓ is a bicubic C^1 B-spline with doubly repeated knots, and suppB_i^ℓ is the one-ring neighborhood of P_i^ℓ. When P_i^ℓ is an F-point, its one-ring neighborhood means the one-ring neighborhood of its closest V-point. As a special case, the F-point $Q_{e,11}$ in Fig. 2b, although being on an irregular element, also falls into Case (2), whereas the other three ($Q_{e,6}$, $Q_{e,7}$ and $Q_{e,10}$) belong to Case (3).

Case (3) must involve a neighboring EV, and B_i^ℓ is defined as a D-patch spline. suppB_i^ℓ consists of two parts. First, the same as Case (2), B_i^ℓ has support on the

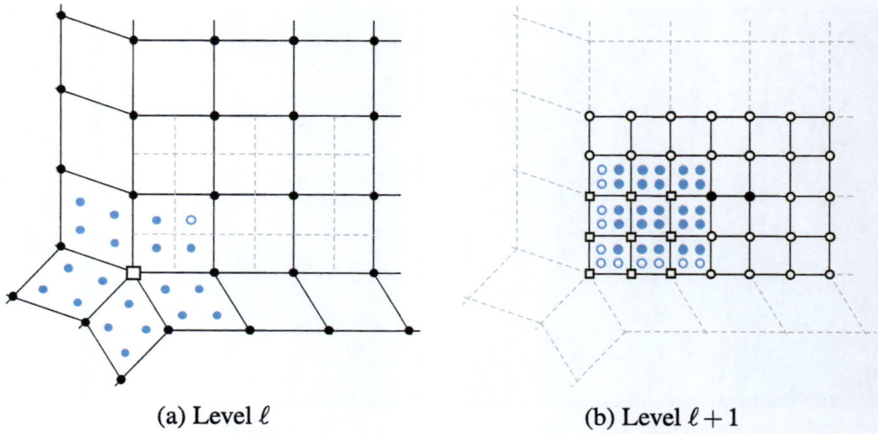

Fig. 7 Selection of spline functions at two consecutive levels. Dots and squares represent D-active and D-passive points, respectively. Filled and hollow dots are H-active and H-passive points, respectively. Black and blue dots (or squares) stand for V-points and F-points, respectively

one-ring neighborhood of P_i^ℓ. Moreover, B_i^ℓ is defined using the smoothing matrix such that it also has support on the one-ring neighborhood of the EV. The union of the two one-ring neighborhoods yields $\mathrm{supp} B_i^\ell$.

In Case (4), B_i^ℓ (more precisely, $\mathrm{trun}_D B_i^\ell$) is a truncated B-spline that is expressed as a linear combination of certain D-passive C^1 B-splines. Therefore, its support is the union of all the one-ring neighborhoods of these D-passive F-points.

With the hierarchical domains and the support of splines well defined, the selection mechanism follows the same as those in Eqs. (5), (6) to collect all the H-active splines as $\mathscr{B}_a^\ell$ and $\mathscr{B}_{a,0}^{\ell+1}$; see an example in Fig. 7. Note that during selection, only D-active splines at each level can be selected to be H-active or H-passive, whereas D-passive ones only serve as an auxiliary purpose to help define truncated B-splines and they never enter the selection procedure. In other words, D-passive functions are neither H-active nor H-passive.

Finally, recall that $\Omega^{\ell+1}$ needs to be large enough such that at least one Level-$(\ell+1)$ spline function can be added to $\mathscr{H}^{\ell+1}$. In the THU-spline construction, this condition can be easily guaranteed if the parent elements of those in $\Omega^{\ell+1}$ are a collection of one-ring neighborhoods of certain Level-ℓ points. In this manner, $\Omega^{\ell+1}$ at least contains the two-ring neighborhood of a Level-$(\ell+1)$ point. As the largest support in the above four cases is a two-ring neighborhood, a Level-$(\ell+1)$ spline can always be added.

4.3 Truncation

The inter-level truncation follows the same as in Eq. (8). Certain H-active Level-ℓ functions are truncated. As a result, $\mathscr{B}_a^\ell$ is changed to $\mathscr{T}_a^\ell$, and $\mathscr{H}^{\ell+1}$ is constructed according to Eq. (10). Although the expressions stay the same as those in THB-splines, evaluation of THU-splines becomes much more involving because there are various types of splines rather than a single type as in THB-splines.

Without loss of generality, we study the THU-splines defined on a transition element $\Omega_e^{\ell+1}$, where we consider two levels (Levels ℓ and $\ell+1$) to simplify explanation. Following [7], we evaluate all levels of H-active functions ($\boldsymbol{H}_a^\ell$ and $\boldsymbol{H}_a^{\ell+1}$) with the functions (both H-active and H-passive) at the highest level, and we write it in the matrix form,

$$\boldsymbol{H}_a := \begin{bmatrix} \boldsymbol{H}_a^\ell \\ \boldsymbol{H}_a^{\ell+1} \end{bmatrix} = \begin{bmatrix} \boldsymbol{0} & \boldsymbol{C} \\ \boldsymbol{I} & \boldsymbol{0} \end{bmatrix} \begin{bmatrix} \boldsymbol{H}_a^{\ell+1} \\ \boldsymbol{H}_p^{\ell+1} \end{bmatrix}, \tag{30}$$

where the coefficients of $\boldsymbol{C}$ come from Eq. (8), and spline functions in $\boldsymbol{H}_a^\ell$ and $\boldsymbol{H}_*^{\ell+1}$ ($* \in \{a, p\}$) are from $\mathscr{B}_a^\ell$ and $\mathscr{B}^{\ell+1}$, respectively. Note that the inter-level truncation has been incorporated for $\boldsymbol{H}_a^\ell$ through the zero submatrix corresponding to $\boldsymbol{H}_a^{\ell+1}$.

We further split $\boldsymbol{H}_a^\ell$ and $\boldsymbol{H}_*^{\ell+1}$ as V-splines ($\boldsymbol{H}_{v,a}^\ell$, $\boldsymbol{H}_{v,*}^{\ell+1}$) and F-splines ($\boldsymbol{H}_{f,a}^\ell$, $\boldsymbol{H}_{f,*}^{\ell+1}$). As a result, Eq. (30) becomes

$$\begin{bmatrix} \boldsymbol{H}_{v,a}^\ell \\ \boldsymbol{H}_{f,a}^\ell \\ \boldsymbol{H}_{v,a}^{\ell+1} \\ \boldsymbol{H}_{f,a}^{\ell+1} \end{bmatrix} = \begin{bmatrix} \boldsymbol{0} & \boldsymbol{0} & \boldsymbol{C}_{11} & \boldsymbol{C}_{12} \\ \boldsymbol{0} & \boldsymbol{0} & \boldsymbol{C}_{21} & \boldsymbol{C}_{22} \\ \boldsymbol{I} & \boldsymbol{0} & \boldsymbol{0} & \boldsymbol{0} \\ \boldsymbol{0} & \boldsymbol{I} & \boldsymbol{0} & \boldsymbol{0} \end{bmatrix} \begin{bmatrix} \boldsymbol{H}_{v,a}^{\ell+1} \\ \boldsymbol{H}_{f,a}^{\ell+1} \\ \boldsymbol{H}_{v,p}^{\ell+1} \\ \boldsymbol{H}_{f,p}^{\ell+1} \end{bmatrix}, \tag{31}$$

where $\boldsymbol{C}_{21} = \boldsymbol{0}$ because an F-spline can only have F-splines as its children. Moreover, $\boldsymbol{H}_{v,*}^{\ell+1}$ can be evaluated according to Eq. (29), that is, $\boldsymbol{H}_{v,*}^{\ell+1} = \boldsymbol{T}_*^{\ell+1} \boldsymbol{B}_{f,p}^{\ell+1}$, where $\boldsymbol{B}_{f,p}^{\ell+1}$ is a vector of D-passive F-splines that do not belong to $\mathscr{B}^{\ell+1}$. Substituting this into Eq. (31), we have

$$\boldsymbol{H}_a = \begin{bmatrix} \boldsymbol{H}_{v,a}^\ell \\ \boldsymbol{H}_{f,a}^\ell \\ \boldsymbol{H}_{v,a}^{\ell+1} \\ \boldsymbol{H}_{f,a}^{\ell+1} \end{bmatrix} = \begin{bmatrix} \boldsymbol{0} & \boldsymbol{0} & \boldsymbol{C}_{11} & \boldsymbol{C}_{12} \\ \boldsymbol{0} & \boldsymbol{0} & \boldsymbol{0} & \boldsymbol{C}_{22} \\ \boldsymbol{I} & \boldsymbol{0} & \boldsymbol{0} & \boldsymbol{0} \\ \boldsymbol{0} & \boldsymbol{I} & \boldsymbol{0} & \boldsymbol{0} \end{bmatrix} \begin{bmatrix} \boldsymbol{0} & \boldsymbol{0} & \boldsymbol{T}_a^{\ell+1} \\ \boldsymbol{I} & \boldsymbol{0} & \boldsymbol{0} \\ \boldsymbol{0} & \boldsymbol{0} & \boldsymbol{T}_p^{\ell+1} \\ \boldsymbol{0} & \boldsymbol{I} & \boldsymbol{0} \end{bmatrix} \begin{bmatrix} \boldsymbol{H}_{f,a}^{\ell+1} \\ \boldsymbol{H}_{f,p}^{\ell+1} \\ \boldsymbol{B}_{f,p}^{\ell+1} \end{bmatrix} =: \boldsymbol{M}_f \boldsymbol{B}_f^{\ell+1}, \tag{32}$$

where $\boldsymbol{B}_f^{\ell+1}$ contains the full set of F-splines (including both D-active ones $\boldsymbol{B}_{f,a}^{\ell+1} := [\boldsymbol{H}_{f,a}^{\ell+1,T}, \boldsymbol{H}_{f,p}^{\ell+1,T}]^T$, and D-passive ones $\boldsymbol{B}_{f,p}^{\ell+1}$) defined on the transition element at Level $\ell+1$. Equation (32) implies that THU-splines $\boldsymbol{H}_a$ defined on a transition element can be evaluated in this unified manner as linear combinations of F-splines. The key is to obtain elementwise $\boldsymbol{M}_f$. With the two kinds of truncation, it can be shown that each column sum of $\boldsymbol{M}_f$ equals to one. Therefore, THU-splines $\boldsymbol{H}_a$ form a partition of unity. While we postpone the related proofs to a forthcoming paper, we will show numerical evidence in Sect. 5.

When $\Omega_e^{\ell+1}$ is a non-transition enriched element (i.e., type-E), only F-splines $\boldsymbol{H}_{f,*}^{\ell+1}$ are involved. In other words, we have $\boldsymbol{H}_{v,*}^{\ell+1} = \boldsymbol{0}$ in Eq. (31). In contrast, when $\Omega_e^{\ell+1}$ is a type-R element, only C^2 B-splines $\boldsymbol{H}_{v,*}^{\ell+1}$ are involved and they not truncated. Moreover, $\boldsymbol{H}_{f,*}^{\ell+1} = \boldsymbol{0}$ as no F-splines are involved. In fact, this latter case coincides with standard THB-splines.

4.4 Consistent Refinement with THU-Splines

We discuss how THU-splines can achieve a consistent refinement in this section. We approach it in a constructive manner to show how the two kinds of truncation work together to retain this property.

We assume that only two levels (i.e., ℓ and $\ell + 1$) are involved to provide insights of construction while keeping notations simple. Globally refined meshes ($\mathscr{M}^\ell$ and $\mathscr{M}^{\ell+1}$) and and their bases ($\mathscr{B}^\ell$ and $\mathscr{B}^{\ell+1}$) are in place to help discussion. Also recall the following notations for index sets at Level $L \in \{\ell, \ell+1\}$:

- $\mathscr{F}_a^L$ and $\mathscr{F}_p^L$: D-active and D-passive F-points/splines, respectively;
- $\mathscr{V}_a^L$: D-active V-points/splines; and
- $\mathscr{A}^L$: H-active points/splines (which can be F- or V-points/splines).

In what follows, we start with Level $\ell + 1$ alone and then add back Level ℓ. Without loss of generality, we study a transition element at Level $\ell + 1$, $\Omega_e^{\ell+1}$, which can be enriched or non-enriched. Its surface patch can be written as follows with a full set of Level-($\ell + 1$) F-splines,

$$S_e^{\ell+1} = \sum_{i \in \mathscr{F}_a^{\ell+1}} B_{f,i}^{\ell+1} Q_i^{\ell+1} + \sum_{i \in \mathscr{F}_p^{\ell+1}} B_{f,i}^{\ell+1} Q_i^{\ell+1}, \tag{33}$$

which is essentially the same as Eq. (21). Next, we express $S_e^{\ell+1}$ in terms of D-active points/splines only. A D-passive F-point is obtained through its neighboring D-active V-points at the same level, so for $i \in \mathscr{F}_p^{\ell+1}$, we have

$$Q_i^{\ell+1} = \sum_{j \in \mathscr{V}_a^{\ell+1}} d_{ji}^{\ell+1} P_j^{\ell+1}, \quad \sum_{j \in \mathscr{V}_a^{\ell+1}} d_{ji}^{\ell+1} = 1, \tag{34}$$

which is in fact the same as Eq. (13), with $d_{ji}^{\ell+1} \in \{4/9, 2/9, 1/9, 0\}$. In fact, there are only four of $d_{ji}^{\ell+1}$ that are nonzero. Substituting Eq. (34)–(33), we have

$$
\begin{aligned}
S_e^{\ell+1} &= \sum_{i\in\mathscr{F}_a^{\ell+1}} B_{f,i}^{\ell+1} Q_i^{\ell+1} + \sum_{i\in\mathscr{F}_p^{\ell+1}} B_{f,i}^{\ell+1} \sum_{j\in\mathscr{V}_a^{\ell+1}} d_{ji}^{\ell+1} P_j^{\ell+1} \\
&= \sum_{i\in\mathscr{F}_a^{\ell+1}} B_{f,i}^{\ell+1} Q_i^{\ell+1} + \sum_{j\in\mathscr{V}_a^{\ell+1}} P_j^{\ell+1} \sum_{i\in\mathscr{F}_p^{\ell+1}} d_{ji}^{\ell+1} B_{f,i}^{\ell+1} \\
&= \sum_{i\in\mathscr{F}_a^{\ell+1}} B_{f,i}^{\ell+1} Q_i^{\ell+1} + \sum_{j\in\mathscr{V}_a^{\ell+1}} P_j^{\ell+1} \text{trun}_D B_{v,j}^{\ell+1},
\end{aligned}
\tag{35}
$$

where note that we have used the definition of the same level in the last equation; see Eq. (26). Equation (35) is the local D-patch representation and it is indeed equivalent to Eq. (27).

Now we add to Eq. (35) the refinement relation between Levels $\ell+1$ and ℓ. A Level-($\ell+1$) D-active V-point $P_j^{\ell+1}$ ($j \in \mathscr{V}_a^{\ell+1}$) is computed with the Level-ℓ D-active V-points,

$$
P_j^{\ell+1} = \sum_{j\in\mathscr{V}_a^{\ell}} h_{ji}^{\ell+1} P_j^{\ell}, \quad \sum_{j\in\mathscr{V}_a^{\ell}} h_{ji}^{\ell+1} = 1, \tag{36}
$$

where $h_{ji}^{\ell+1}$ are the same coefficients as Eq. (7) and only a small number of them are nonzero. On the other hand, a Level-($\ell+1$) D-active F-point is computed from Level-ℓ F-points (both D-active and D-passive), so for $i \in \mathscr{F}_a^{\ell+1}$, we have

$$
Q_i^{\ell+1} = \sum_{j\in\mathscr{F}_a^{\ell}} h_{ji}^{\ell+1} Q_j^{\ell} + \sum_{j\in\mathscr{F}_p^{\ell}} h_{ji}^{\ell+1} Q_j^{\ell}, \quad \sum_{j\in\mathscr{F}_a^{\ell}\cup\mathscr{F}_p^{\ell}} h_{ji}^{\ell+1} = 1, \tag{37}
$$

which is merely applying the B-spline knot insertion to doubly repeated knot vectors. When $j \in \mathscr{F}_p^{\ell}$, Q_j^{ℓ} is computed according to Eq. (34) (replacing $\ell+1$ with ℓ).

We further divide $\mathscr{F}_a^L$ and $\mathscr{V}_a^L$ ($L \in \{\ell, \ell+1\}$) into H-active and H-passive subsets, leading to $\mathscr{A}_f^L := \mathscr{F}_a^L \cap \mathscr{A}^L$, $\mathscr{A}_v^L := \mathscr{V}_a^L \cap \mathscr{A}^L$, $\mathscr{N}_f^L := \mathscr{F}_a^L \backslash \mathscr{A}^L$, and $\mathscr{N}_v^L := \mathscr{V}_a^L \backslash \mathscr{A}^L$. Next, starting from Eq. (35), we have

$$
\begin{aligned}
S_e^{\ell+1} &= \sum_{i\in\mathscr{F}_a^{\ell+1}} B_{f,i}^{\ell+1} Q_i^{\ell+1} + \sum_{j\in\mathscr{V}_a^{\ell+1}} P_j^{\ell+1} \text{trun}_D B_{v,j}^{\ell+1} \\
&= \sum_{i\in\mathscr{A}_f^{\ell+1}} B_{f,i}^{\ell+1} Q_i^{\ell+1} + \sum_{i\in\mathscr{N}_f^{\ell+1}} B_{f,i}^{\ell+1} Q_i^{\ell+1} \\
&\quad + \sum_{j\in\mathscr{A}_v^{\ell+1}} P_j^{\ell+1} \text{trun}_D B_{v,j}^{\ell+1} + \sum_{j\in\mathscr{N}_v^{\ell+1}} P_j^{\ell+1} \text{trun}_D B_{v,j}^{\ell+1}.
\end{aligned}
\tag{38}
$$

We keep the terms corresponding to H-active subsets unchanged but further expand H-passive ones using Eqs. (36), (37),

$$
\begin{aligned}
S_{e,\mathcal{N}}^{\ell+1} &:= \sum_{i\in\mathcal{N}_f^{\ell+1}} B_{f,i}^{\ell+1} Q_i^{\ell+1} + \sum_{j\in\mathcal{N}_v^{\ell+1}} P_j^{\ell+1} \mathrm{trun}_D B_{v,j}^{\ell+1} \\
&= \sum_{i\in\mathcal{N}_f^{\ell+1}} B_{f,i}^{\ell+1} \left(\sum_{j\in\mathcal{F}_a^\ell} h_{ji}^{\ell+1} Q_j^\ell + \sum_{j\in\mathcal{F}_p^\ell} h_{ji}^{\ell+1} Q_j^\ell \right) \\
&\quad + \sum_{i\in\mathcal{N}_v^{\ell+1}} \left(\sum_{j\in\mathcal{V}_a^\ell} h_{ji}^{\ell+1} P_j^\ell \right) \mathrm{trun}_D B_{v,i}^{\ell+1} \\
&= \sum_{i\in\mathcal{N}_f^{\ell+1}} B_{f,i}^{\ell+1} \left(\sum_{j\in\mathcal{F}_a^\ell} h_{ji}^{\ell+1} Q_j^\ell + \sum_{j\in\mathcal{F}_p^\ell} h_{ji}^{\ell+1} \sum_{k\in\mathcal{V}_a^\ell} d_{kj}^{\ell+1} P_k^\ell \right) \\
&\quad + \sum_{i\in\mathcal{N}_v^{\ell+1}} \left(\sum_{j\in\mathcal{V}_a^\ell} h_{ji}^{\ell+1} P_j^\ell \right) \mathrm{trun}_D B_{v,i}^{\ell+1}.
\end{aligned}
\tag{39}
$$

Notice that according to the selection mechanism of hierarchical splines, a Level-ℓ spline (F- or V-spline) is H-passive if and only if all its children are H-active; in other words, if any of its children is passive, then the spline is active. Therefore in the last equation in Eq. (39), we observe that when $i \in \mathcal{N}_f^{\ell+1}$ and $h_{ji}^{\ell+1} \neq 0$, Q_j^ℓ must be H-active, i.e., $j \in \mathcal{A}_f^\ell$. Similarly, when $i \in \mathcal{N}_f^{\ell+1}$, $h_{ji}^{\ell+1} \neq 0$ and $d_{kj}^{\ell+1} \neq 0$, we have $k \in \mathcal{A}_v^\ell$ for P_k^ℓ; when $i \in \mathcal{N}_v^{\ell+1}$ and $h_{ji}^{\ell+1} \neq 0$, we must have $j \in \mathcal{A}_v^\ell$ for P_j^ℓ. Then after proper rearrangement, Eq. (39) becomes

$$
\begin{aligned}
S_{e,\mathcal{N}}^{\ell+1} &= \sum_{j\in\mathcal{A}_f^\ell} Q_j^\ell \sum_{i\in\mathcal{N}_f^{\ell+1}} h_{ji}^{\ell+1} B_{f,i}^{\ell+1} + \sum_{k\in\mathcal{A}_v^\ell} P_k^\ell \sum_{i\in\mathcal{N}_f^{\ell+1}} B_{f,i}^{\ell+1} \sum_{j\in\mathcal{F}_p^\ell} h_{ji}^{\ell+1} d_{kj}^{\ell+1} \\
&\quad + \sum_{j\in\mathcal{A}_v^\ell} P_j^\ell \sum_{i\in\mathcal{N}_v^{\ell+1}} h_{ji}^{\ell+1} \mathrm{trun}_D B_{v,i}^{\ell+1} \\
&= \sum_{j\in\mathcal{A}_f^\ell} Q_j^\ell \sum_{i\in\mathcal{N}_f^{\ell+1}} h_{ji}^{\ell+1} B_{f,i}^{\ell+1} + \sum_{k\in\mathcal{A}_v^\ell} P_k^\ell \sum_{i\in\mathcal{N}_f^{\ell+1}} B_{f,i}^{\ell+1} \sum_{j\in\mathcal{F}_p^\ell} h_{ji}^{\ell+1} d_{kj}^{\ell+1} \\
&\quad + \sum_{j\in\mathcal{A}_v^\ell} P_j^\ell \sum_{i\in\mathcal{N}_v^{\ell+1}} h_{ji}^{\ell+1} \sum_{k\in\mathcal{F}_p^{\ell+1}} d_{ik}^{\ell+1} B_{f,k}^{\ell+1},
\end{aligned}
\tag{40}
$$

where we have used the definition of $\mathrm{trun}_D B_{v,i}^{\ell+1}$ in the last term. Let $s_{ki}^{\ell+1} := \sum_{j\in\mathcal{F}_p^\ell} h_{ji}^{\ell+1} d_{kj}^{\ell+1}$ and $t_{jk}^{\ell+1} := \sum_{i\in\mathcal{N}_v^{\ell+1}} h_{ji}^{\ell+1} d_{ik}^{\ell+1}$. We have

$$
\begin{aligned}
S_{e,\mathcal{N}}^{\ell+1} &= \sum_{j\in\mathscr{A}_f^\ell} Q_j^\ell \sum_{i\in\mathcal{N}_f^{\ell+1}} h_{ji}^{\ell+1} B_{f,i}^{\ell+1} + \sum_{k\in\mathscr{A}_v^\ell} P_k^\ell \sum_{i\in\mathcal{N}_f^{\ell+1}} s_{ki}^{\ell+1} B_{f,i}^{\ell+1} + \sum_{j\in\mathscr{A}_v^\ell} P_j^\ell \sum_{k\in\mathscr{F}_p^{\ell+1}} t_{jk}^{\ell+1} B_{f,k}^{\ell+1} \\
&= \sum_{j\in\mathscr{A}_f^\ell} Q_j^\ell \sum_{i\in\mathcal{N}_f^{\ell+1}} h_{ji}^{\ell+1} B_{f,i}^{\ell+1} + \sum_{j\in\mathscr{A}_v^\ell} P_j^\ell \left(\sum_{i\in\mathcal{N}_f^{\ell+1}} s_{ji}^{\ell+1} B_{f,i}^{\ell+1} + \sum_{i\in\mathscr{F}_p^{\ell+1}} t_{ji}^{\ell+1} B_{f,i}^{\ell+1} \right) \\
&= \sum_{j\in\mathscr{A}_f^\ell} Q_j^\ell \mathrm{trun}_H B_{f,j}^\ell + \sum_{j\in\mathscr{A}_v^\ell} P_j^\ell \mathrm{trun}_{HD} B_{v,j}^\ell,
\end{aligned}
\tag{41}
$$

with

$$
\mathrm{trun}_H B_{f,j}^\ell = \sum_{i\in\mathcal{N}_f^{\ell+1}} h_{ji}^{\ell+1} B_{f,i}^{\ell+1}, \tag{42}
$$

and

$$
\mathrm{trun}_{HD} B_{v,j}^\ell := \sum_{i\in\mathcal{N}_f^{\ell+1}} s_{ji}^{\ell+1} B_{f,i}^{\ell+1} + \sum_{i\in\mathscr{F}_p^{\ell+1}} t_{ji}^{\ell+1} B_{f,i}^{\ell+1}. \tag{43}
$$

Equation (42) is merely a THB-spline that has a knot vector with doubly repeated knots, whereas Eq. (43) implies a mix of the same-level truncation and the inter-level truncation encoded in the coefficients $s_{ji}^{\ell+1}$ and $t_{ji}^{\ell+1}$. Moreover, Eqs. (42), (43) correspond to $\boldsymbol{H}_{f,a}^\ell$ and $\boldsymbol{H}_{v,a}^\ell$ in Eq. (32), respectively. Although truncated splines in Eqs. (42), (43) have different forms, the key idea is essentially the same: A truncated spline is always expressed as a linear combination of certain passive splines.

Finally, with Eqs. (38), (41), the surface patch $S_e^{\ell+1}$ of the element $\Omega_e^{\ell+1}$ can be written in terms of H-active splines only,

$$
\begin{aligned}
S_e^{\ell+1} = &\sum_{i\in\mathscr{A}_f^{\ell+1}} B_{f,i}^{\ell+1} Q_i^{\ell+1} + \sum_{i\in\mathscr{A}_v^{\ell+1}} P_i^{\ell+1} \mathrm{trun}_D B_{v,i}^{\ell+1} \\
&+ \sum_{i\in\mathscr{A}_f^\ell} Q_i^\ell \mathrm{trun}_H B_{f,i}^\ell + \sum_{i\in\mathscr{A}_v^\ell} P_i^\ell \mathrm{trun}_{HD} B_{v,i}^\ell,
\end{aligned}
\tag{44}
$$

which is also the general form of a THU-spline representation (with two levels). When $\Omega_e^{\ell+1}$ is a Type-R (non-transition and non-enriched) element, only THB-splines are involved, i.e., $\mathscr{A}_f^L = \varnothing$, $\mathrm{trun}_D B_{v,i}^{\ell+1} = B_{v,i}^{\ell+1}$ and $\mathrm{trun}_{HD} B_{v,i}^\ell = B_{v,i}^\ell$. When $\Omega_e^{\ell+1}$ is a Type-E (non-transition and enriched) element, only F-splines are involved, i.e., $\mathscr{A}_v^L = \varnothing$.

5 Numerical Examples

In this section, we present several examples to show its effectiveness and efficiency. We start with a patch test on a domain of unit square, where linear elasticity is solved using THU-splines. The unstructured quad mesh in Fig. 1 is taken as the input.

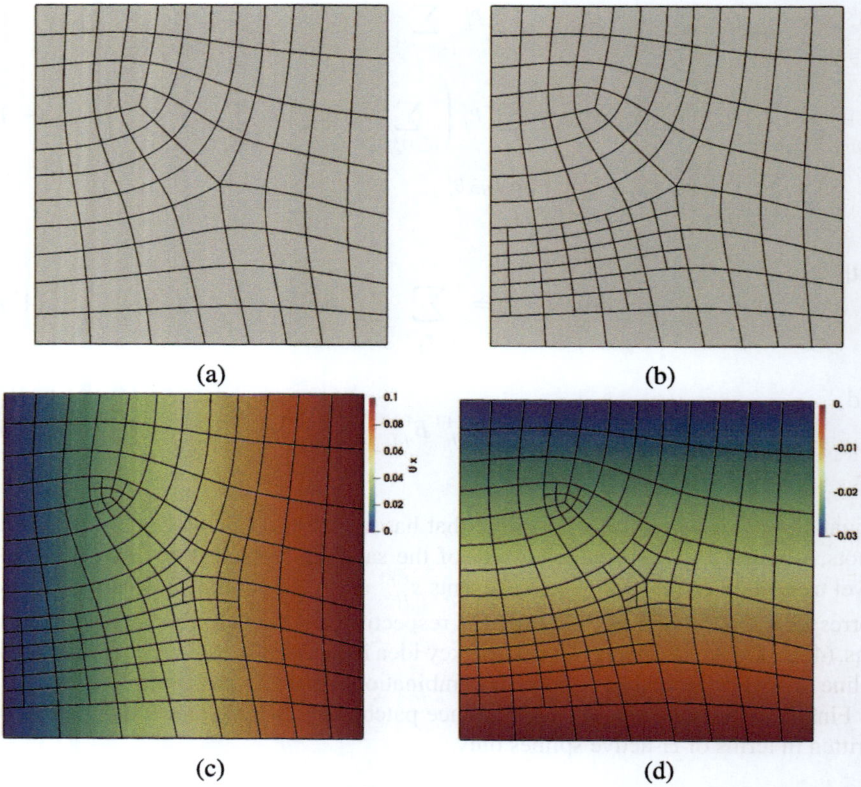

Fig. 8 The patch test using THU-splines on a unit square, where linear elasticity is solved. **a** The spline mesh of the initial local D-patch representation, whose input is given in Fig. 1; **b** the THU-spline mesh after several elements in **a** are refined; and **c**, **d** displacement (u_x and u_y) results on the Bézier mesh

Accordingly, an initial local D-patch representation is constructed, with its high-order spline mesh shown in Fig. 8a. Next, we locally refine certain elements using THU-splines, which is intended to be general to include all types of elements: irregular, transition, and regular elements; see the resulting THU-spline mesh in Fig. 8b. We use it to perform the patch test. Dirichlet boundary conditions $u_x = 0$, $u_x = 0.1$, and $u_y = 0$ are imposed on the left, right, and lower boundaries, respectively, whereas free traction is applied to all the other boundaries. Isotropic and homogeneous material is assumed, where the Young's modulus E and the Poisson's ratio are given as $E = 1$ and $\nu = 0.3$, respectively. The displacement results are shown on the Bézier mesh in Fig. 8c, d, where recall that each irregular element in Fig. 8b needs to be split into 2×2 Bézier elements for the smooth D-patch construction. Moreover, we find that the L^2-norm error of strains (ϵ_{xx}, ϵ_{yy} and ϵ_{xy}) is in the order of 10^{-16}. In other words, the patch is passed with machine precision.

Successfully passing the patch test shows the applicability of THU-splines in engineering analysis. The numerical results also show supportive evidence regarding important properties of THU-splines. First, the involved THU-spline functions sum up to one and thus form a partition of unity. This confirms that the two kinds of truncation work as expected. Second, the involved THU-splines are linearly independent and thus constitute a basis. However, all these properties need to be rigorously studied, which is also the primary task of the future work.

Next, we use THU-splines to adaptively solve the biharmonic equation with the homogeneous Dirichlet and Neumann boundary conditions,

$$\begin{cases} \Delta^2 u = f \text{ in } \Omega, \\ u = 0 \text{ on } \partial\Omega, \\ \nabla u \cdot \boldsymbol{n} = 0 \text{ on } \partial\Omega, \end{cases} \tag{45}$$

where $\boldsymbol{n}$ is the outwards normal of $\partial\Omega$. This is a 4th order PDE that requires C^1 continuity of the underlying basis functions. We again solve the problem on a unit square domain and start with the mesh given in Fig. 8a. For the sake of convergence study, we use the following manufactured solution that has a large gradient across the line $x - y = 0$,

$$u(x, y) = ax^2(1-x)^2y^2(1-y)^2 \tanh(b(y-x)), \quad (x, y) \in [0, 1]^2, \tag{46}$$

where a and b are two constants that control the behavior of the solution, and particularly, b controls the "sharpness" of the gradient; see Fig. 9a, where we use $a = 10^3$ and $b = 10\sqrt{2}$. H^1 semi-norm error is used to drive local refinement to efficiently capture the gradient feature. Global refinement is taken as a reference. The convergence plot is summarized in Fig. 9b with respect to the square root of degrees of freedom (DOF). The approximate solution u^h and the element-wise H^1 semi-norm error on the final locally refined mesh are shown in Fig. 9c and d, respectively. We observe that compared to global refinement, THU-splines can achieve the same accuracy with much fewer DOF.

6 Conclusion and Future Work

In this paper, we have presented the construction method of THU-splines, a novel method that features a global smooth construction with the presence of EVs and supports flexible local refinement through the idea of THB-splines. Numerical evidence shows that THU-spline functions are refinable and form a partition of unity. Moreover, through solving the biharmonic equation with a manufactured solution, we show the efficiency of THU-splines compared to global refinement.

However, much remains to be studied to fully unleash the power of THU-splines. First of all, we will rigorously prove properties of THU-splines, including refinability,

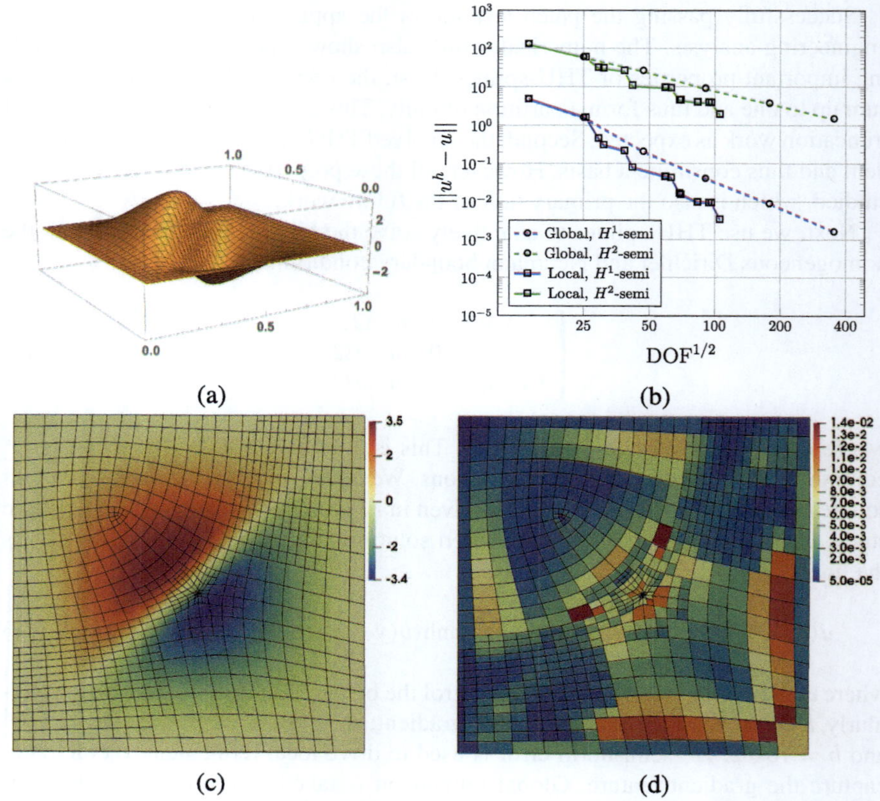

Fig. 9 Adaptively solving the biharmonic equation using THU-splines. **a** The graph of the manufactured solution, **b** the convergence plot with respect to $\mathrm{DOF}^{\frac{1}{2}}$, **c** the approximate solution obtained on a locally refined mesh, and **d** the element-wise H^1 semi-norm error

partition of unity, and linear independence. Once these foundations are ready, we will apply THU-splines to Kirchhoff-Love shell problems, where a posteriori error estimate is needed to automatically guide the adaptive analysis of THU-splines. The recent work reported in [1] may well fit this purpose with the adaptation in irregular regions.

Acknowledgements X. Wei is partially supported by the ERC AdG project CHANGE n. 694515, as well as the Swiss National Science Foundation project HOGAEMS n.200021_188589.

References

1. Antolin, P., Buffa, A., Coradello, L.: A hierarchical approach to the a posteriori error estimation of isogeometric Kirchhoff plates and Kirchhoff-Love shells. Comput. Methods Appl. Mech. Eng. **363**, 112,919 (2020)
2. Bazilevs, Y., Calo, V.M., Cottrell, J.A., Evans, J., Hughes, T.J.R., Lipton, S., Scott, M.A., Sederberg, T.W.: Isogeometric analysis using T-splines. Comput. Methods Appl. Mech. Eng. **199**, 229–263 (2010)
3. Beirão da Veiga, L., Buffa, A., Cho, D., Sangalli, G.: Analysis-suitable T-splines are dual-compatible. Comput. Methods Appl. Mech. Eng. **249–252**, 42–51 (2012)
4. Bercovier, M., Matskewich, T.: Smooth Bézier Surfaces over Unstructured Quadrilateral Meshes. Springer International Publishing (2017)
5. Boehm, W.: Inserting new knots into B-spline curves. Comput. Aided Des. **12**, 199–201 (1980)
6. Boehm, W., Müller, A.: On de Casteljau's algorithm. Comput. Aided Geomet. Design **16**, 587–605 (1999)
7. Bornermann, P.B., Cirak, F.: A subdivision-based implementation of the hierarchical B-spline finite element method. Comput. Methods Appl. Mech. Eng. **253**, 584–598 (2013)
8. Casquero, H., Wei, X., Toshniwal, D., Li, A., Hughes, T.J.R., Kiendl, J., Zhang, Y.: Seamless integration of design and Kirchhoff-Love shell analysis using analysis-suitable unstructured T-splines. Comput. Methods Appl. Mech. Eng. **360**, 112,765 (2020)
9. Cottrell, J.A., Hughes, T.J.R., Bazilevs, Y.: Isogemetric Analysis: Toward Integration of CAD and FEA. Wiley, New York (2009)
10. Forsey, D.R., Bartels, R.H.: Hierarchical B-spline refinement. Comput. Graph. **22**, 205–212 (1988)
11. Giannelli, C., Jüttler, B., Speleers, H.: THB-splines: the truncated basis for hierarchical splines. Comput. Aided Geomet. Design **29**, 485–498 (2012)
12. Hughes, T.J.R., Cottrell, J.A., Bazilevs, Y.: Isogeometric analysis: CAD, finite elements, NURBS, exact geometry and mesh refinement. Comput. Methods Appl. Mech. Eng. **194**, 4135–4195 (2005)
13. Kapl, M., Vitrih, V., Jüttler, B., Birner, K.: Isogeometric analysis with geometrically continuous functions on two-patch geometries. Comput. Math. Appl. **70**, 1518–1538 (2015)
14. Kraft, R.: Adaptive and linearly independent multilevel B-splines. In: Méhauté, A.L., Rabut, C., Schumaker, L.L., (eds.) Surface Fitting and Multiresolution Methods, pp. 209–218. Vanderbilt University Press (1997)
15. Li, X., Scott, M.A.: Analysis-suitable T-splines: characterization, refineability, and approximation. Math. Models Methods Appl. Sci. **24**, 1141–1164 (2014)
16. Li, X., Wei, X., Zhang, Y.: Hybrid non-uniform recursive subdivision with improved convergence rates. Comput. Methods Appl. Mech. Eng. **352**, 606–624 (2019)
17. Lyche, T., Manni, C., Speleers, H.: Foundations of spline theory: B-splines, spline approximation, and hierarchical refinement. In: Lyche, T., Manni, C., Speleers, H. (eds.), Splines and PDEs: From Approximation Theory to Numerical Linear Algebra. Lecture Notes in Mathematics, vol. 2219, pp. 1–76. Springer International Publishing (2018)
18. Majeed, M., Cirak, F.: Isogeometric analysis using manifold-based smooth basis functions. Comput. Methods Appl. Mech. Eng. **316**, 547–567 (2017)
19. Nguyen, T., Peters, J.: Refinable C^1 spline elements for irregular quad layout. Comput. Aided Geometr. Design **43**, 123–130 (2016)
20. Reif, U.: A refineable space of smooth spline surfaces of arbitrary topological genus. J. Approx. Theory **90**, 174–199 (1997)
21. Scott, M.A., Li, X., Sederberg, T.W., Hughes, T.J.R.: Local refinement of analysis-suitable T-splines. Comput. Methods Appl. Mech. Eng. **213–216**, 206–222 (2012)
22. Scott, M.A., Simpson, R.N., Evans, J.A., Lipton, S., Bordas, S.P.A., Hughes, T.J.R., Sederberg, T.W.: Isogeometric boundary element analysis using unstructured T-splines. Comput. Methods Appl. Mech. Eng. **254**, 197–221 (2013)

23. Scott, M.A., Thomas, D.C., Evans, E.J.: Isogeometric spline forests. Comput. Methods Appl. Mech. Eng. **269**, 222–264 (2014)
24. Sederberg, T.W., Cardon, D.L., Finnigan, G.T., North, N.S., Zheng, J., Lyche, T.: T-spline simplification and local refinement. ACM Trans. Graph. **23**, 276–283 (2004)
25. Sederberg, T.W., Zheng, J., Bakenov, A., Nasri, A.: T-splines and T-NURCCs. ACM Trans. Graph. **22**, 477–484 (2003)
26. Toshniwal, D., Speleers, H., Hughes, T.J.R.: Smooth cubic spline spaces on unstructured quadrilateral meshes with particular emphasis on extraordinary points: geometric design and isogeometric analysis considerations. Comput. Methods Appl. Mech. Eng. **327**, 411–458 (2017)
27. Vuong, A.V., Giannelli, C., Jüttler, B., Simeon, B.: A hierarchical approach to adaptive local refinement in isogeometric analysis. Comput. Methods Appl. Mech. Eng. **200**, 3554–3567 (2011)
28. Wei, X., Li, X., Zhang, Y., Hughes, T.J.R.: Tuned hybrid nonuniform subdivision surfaces with optimal convergence rates. Int. J. Numer. Meth. Eng. **122**, 2117–2144 (2021)
29. Wei, X., Zhang, Y., Hughes, T.J.R., Scott, M.A.: Truncated hierarchical Catmull-Clark subdivision with local refinement. Comput. Methods Appl. Mech. Eng. **291**, 1–20 (2015)
30. Wei, X., Zhang, Y., Hughes, T.J.R., Scott, M.A.: Extended truncated hierarchical Catmull-Clark subdivision. Comput. Methods Appl. Mech. Eng. **299**, 316–336 (2016)
31. Wei, X., Zhang, Y., Toshniwal, D., Speleers, H., Li, X., Manni, C., Evans, J.A., Hughes, T.J.R.: Blended B-spline construction on unstructured quadrilateral and hexahedral meshes with optimal convergence rates in isogeometric analysis. Comput. Methods Appl. Mech. Eng. **341**, 609–639 (2018)

HexGen and Hex2Spline: Polycube-Based Hexahedral Mesh Generation and Spline Modeling for Isogeometric Analysis Applications in LS-DYNA

Yuxuan Yu, Xiaodong Wei, Angran Li, Jialei Ginny Liu, Jeffrey He, and Yongjie Jessica Zhang

Abstract In this paper, we present two software packages, HexGen and Hex2Spline, that seamlessly integrate geometry design with isogeometric analysis (IGA) in LS-DYNA. Given a boundary representation of a solid model, HexGen creates a hexahedral mesh by utilizing a semi-automatic polycube-based mesh generation method. Hex2Spline takes the output hexahedral mesh from HexGen as the input control mesh and constructs volumetric truncated hierarchical splines. Through Bézier extraction, Hex2Spline transfers spline information to LS-DYNA and performs IGA therein. We explain the underlying algorithms in each software package and use a rod model to explain how to run the software. We also apply our software to several other complex models to test its robustness. Our goal is to provide a robust volumetric modeling tool and thus expand the boundary of IGA to volume-based industrial applications.

Y. Yu · A. Li · J. G. Liu · Y. J. Zhang (✉)
Department of Mechanical Engineering, Carnegie Mellon University, Pittsburgh, PA, USA
e-mail: jessicaz@andrew.cmu.edu

Y. Yu
e-mail: yuxuany1@andrew.cmu.edu

A. Li
e-mail: angranl@andrew.cmu.edu

J. G. Liu
e-mail: jialeil@andrew.cmu.edu

X. Wei
Institute of Mathematics, École Polytechnique Fédérale de Lausanne, Lausanne, Switzerland
e-mail: xiaodong.wei@epfl.ch

J. He
Department of Mechanical Engineering, Northwestern University, Evanston, IL, USA
e-mail: jeffreyhe2022@u.northwestern.edu

C. Manni and H. Speleers (eds.), *Geometric Challenges in Isogeometric Analysis*, Springer INdAM Series 49, https://doi.org/10.1007/978-3-030-92313-6_14

1 Introduction

Isogeometric analysis (IGA) [19] is a computational technique that integrates computer aided design (CAD) with simulation methods such as finite element analysis (FEA). It adopts the idea of design-through-analysis and enables direct analysis of the designed geometry. IGA has many advantages over traditional FEA such as exact, smooth geometric representation and superior numerical performance. Many software packages have been developed for IGA and there are mainly two directions. The first direction is to incorporate IGA with commercial finite element software. For example, the user subroutine UEL in Abaqus is used to define IGA elements and perform IGA in Abaqus [21, 22]. The second direction is to develop open software packages. GeoPDEs [7] and igatools [29] work on NURBS (Non-Uniform Rational B-Spline) patches and provide a general framework to implement IGA methods. PetIGA [6] is another framework for IGA based on PETSc [3]. Tensor-product B-spline bases of arbitrary dimension and polynomial degree. GISmo [27] supports a variety of spline functions including hierarchical and truncated hierarchical B-splines (THB-splines) of arbitrary polynomial degree. These software packages help boost the use of IGA in engineering applications. However, these packages are analysis-oriented. Currently, there is no available toolkit from the geometric modeling side, especially for volumetric parameterization. Therefore, the motivation of our work is to develop a geometric modeling tool to bridge the gap between geometric design and IGA analysis.

There are two major challenges in volumetric parameterization: control mesh generation and volumetric spline construction. A control mesh is generally an unstructured hexahedral (hex) mesh. Various strategies have been proposed in the literature [44] for unstructured hex mesh generation, such as grid-based or octree-based [34, 35], medial surface [30, 31], plastering [4, 38], whisker weaving [10] and vector field-based methods [28]. These methods have created hex meshes for certain geometries, but are not robust and reliable for arbitrary geometries. The polycube-based method [14, 39] is another appealing approach for all-hex meshing. A smooth harmonic field [41] was used to generate polycubes for arbitrary genus geometries. Boolean operations [24] were introduced to cope with arbitrary genus geometries. In [25], polycube structure was generated based on the skeleton branches of a geometric model. For these methods, the hex mesh quality is directly affected by the polycube structure and mapping distortion. Computing the polycube structure with a low-distortion mapping remains an open problem for arbitrary geometries. It is essential to improve the mesh quality for analysis by using methods such as pillowing, smoothing and optimization [32, 33, 45, 46]. Pillowing is a sheet insertion technique that eliminates the situations where two neighboring hex elements share more than one face. Smoothing and optimization are used to further improve mesh quality by relocating vertices. In our software, we implement all the above mentioned methods for quality improvement.

The second ingredient in volumetric parameterization is volumetric spline construction. Several algorithms have been developed. The initial development of IGA was based on NURBS. Since it adopts a global tensor-product structure, it does not support local refinement. T-splines were initially developed to support local refinement for surfaces [36, 37]. For solid models, the rational T-spline basis functions were used to convert unstructured hex meshes to solid T-splines [42]. Boolean operations [24] and skeletons [25] are used to create hex meshes, which are later converted to T-meshes. However, local refinement using T-splines may require extensive mesh manipulation to satisfy desired properties such as linear independence. Hierarchical splines are the alternative to T-spline to avoid this issue. Several techniques were then developed based on hierarchical B-splines (HB-splines) [11, 40], such as THB-splines [12, 13].

In this paper, we integrate our semi-automatic polycube-based mesh generation with the volumetric truncated hierarchical spline construction (TH-spline3D) [43] to perform IGA on volumetric models in LS-DYNA. The developed software packages (https://github.com/CMU-CBML/HexGen_Hex2Spline) feature: (1) semi-automatic polycube-based all-hex mesh generation from a CAD model; (2) TH-spline3D construction on hex meshes; and (3) Bézier extraction for LS-DYNA. We first overview the entire pipeline and explain the algorithm behind each module of the pipeline. We then provide various examples to explain how to run the software package. The main objective of the software package is to make our pipeline accessible to industrial and academic communities who are interested in real-world engineering applications. Our software favors versatility over efficiency. We will use a concrete example to go through all the steps in running the software. In particular, when user intervention is needed, we will explain details of the involved manual work.

The paper is outlined as follows. In Sect. 2, we overview the pipeline. In Sect. 3, we present the HexGen software package that conducts semi-automatic polycube-based all-hex mesh generation from a CAD file. In Sect. 4, we talk about Hex2Spline that constructs TH-spline3D on hex meshes and performs Bézier extraction for IGA in LS-DYNA. Finally, in Sect. 5, we demonstrate several complex models using our software package.

2 Pipeline Design

Our pipeline incorporates two software packages to bridge the gap between the input CAD model with IGA in LS-DYNA, as shown in Fig. 1. We first use the HexGen software package to build an all-hex mesh for the CAD model. With a high quality all-hex mesh generated, we then use the Hex2Spline software package to construct TH-spline3D and extract Bézier information for LS-DYNA.

As shown in Fig. 1, we first generate a triangle mesh from the CAD model by using the free software LS-PrePost, which is a pre and post-processor for LS-DYNA. Then we use centroidal Voronoi tessellation (CVT) segmentation [17] to create a polycube structure [39], which is used to generate all-hex meshes via parametric mapping [9] and octree subdivision [46]. The quality of the all-hex mesh is evaluated to ensure that the resulting volumetric spline model can be used in IGA. In case that

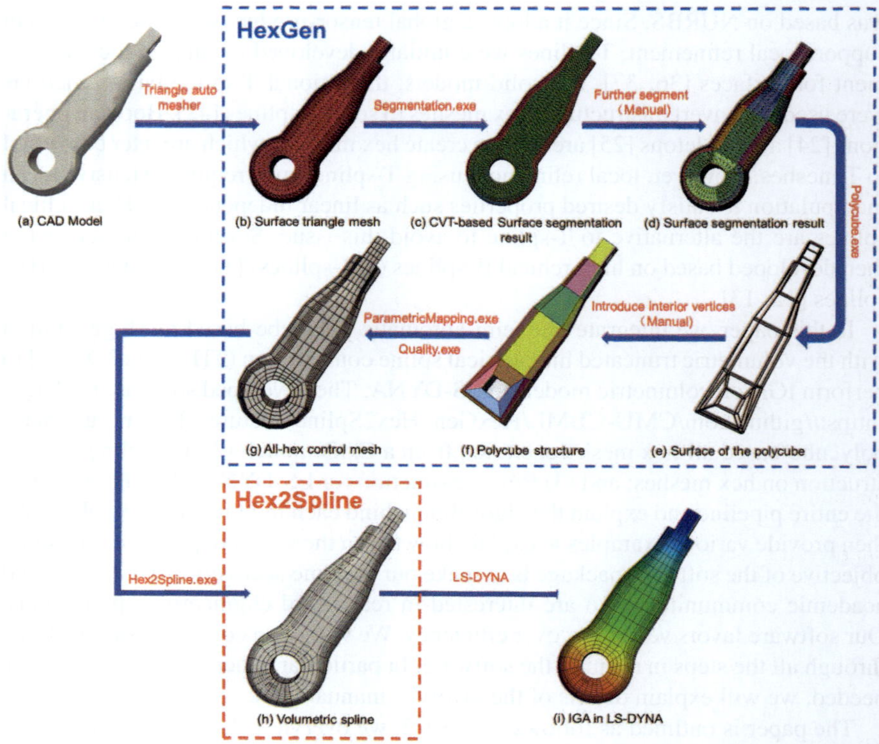

Fig. 1 The HexGen software package and the Hex2Spline software package. The red text on each arrow shows the operation between two steps. The manual work is involved in **further segmentation** and **introducing interior vertices**

a poor quality hex mesh is generated, the program has several quality improvement functions, including pillowing [45], smoothing, and optimization [32]. Each quality improvement function can be run independently and one can use these functions to improve the mesh quality.

Once a good quality hex mesh is obtained, one can run the Hex2Spline program to build volumetric splines. In particular, TH-spline3D is built on the unstructured hex mesh and it also supports local refinement. The Hex2Spline can output the Bézier extraction information of TH-spline3D in a format that can be imported into LS-DYNA to perform IGA. Currently, our software only has a command-line interface (CLI). Users need to specify necessary options via the command line to run the software. In Sects. 3 and 4, we will explain the algorithms implemented in each software as well as how to run the software in detail.

3 HexGen: Polycube-Based Hex Mesh Generation

Surface segmentation, polycube construction, parametric mapping, and octree subdivision are used together in the HexGen software package to construct an all-hex mesh from the boundary representation given by the input CAD model. Given an triangle mesh generated from the CAD model, we first use surface segmentation to divide the mesh into several surface patches that satisfy the polycube structure constraints, which will be discussed in Sect. 3.1. Then, the corner vertices, edges and face information of each surface patch are extracted from the surface segmentation result to construct a polycube structure. Each component of the polycube structure is topologically equivalent to a cube. Finally, we generate the all-hex mesh through parametric mapping and octree subdivision. Quality improvement techniques can be used to further improve the mesh quality.

In this section, we introduce the main algorithm for each module of the HexGen software package, namely surface segmentation, polycube construction, parametric mapping and octree subdivision, and quality improvement. We use a rod model (see Fig. 1) to explain how to run CLI for each module. We also discuss the user intervention that is involved in the semi-automatic polycube-based hex mesh generation.

3.1 Surface Segmentation

The surface segmentation in the pipeline framework is implemented based on CVT segmentation [17]. CVT segmentation is used to classify vertices into different groups by minimizing an energy function. Each group is called a Voronoi region $\{V_j\}$ and it has a corresponding center called a generator $\{g_j\}$. The Voronoi region and its corresponding generator are updated iteratively in the minimization process. In [17], each element of the surface triangle mesh is assigned to one of the six Voronoi regions $\{V_j\}_{j=1}^{6}$ based on the normal vector $\varkappa_{\mathscr{T}(i)}$ of the surface, where $\mathscr{T}(i)$ is the ith element of the surface triangle mesh $\mathscr{T}$. The initial generators of the Voronoi regions are the three principal normal vectors and their opposite normals vectors ($\pm X$, $\pm Y$, $\pm Z$). Two energy functions and their corresponding distance functions are used together in [17]. The classical energy function and its corresponding distance function provide initial Voronoi regions and generators. Then the harmonic boundary-enhanced (HBE) energy function and its corresponding distance function are applied to eliminate non-monotone boundaries [17], where the direction changes. As mentioned in [17], the non-monotone boundaries may lead to extreme distortions when mapped to corresponding polycube edges. The detailed definitions of energy functions and their corresponding distance functions are described in [17] with formulae (2, 4, 6). Here, we summarize the surface segmentation process in **Surface Segmentation Algorithm**.

In the pseudocode of **Surface Segmentation Algorithm**, we describe two energy minimization processes, which are combined together to yield a monotone segmen-

Surface Segmentation Algorithm

Input: Manifold triangular surface mesh $\mathscr{T}$, weighting factor ω
Output: Manifold triangular surface mesh including segmentation information
1: Calculate the unit normals for each triangle of the triangle mesh
2: Use six principal axes ($\pm X$, $\pm Y$, $\pm Z$) as the initial generators $\{g_j\}_{j=1}^{6}$
CVT step:
3: **while** classical energy not converged **do**
4: Associate each triangle with $\{g_j\}_{j=1}^{6}$ by using classical distance functions
5: Update each generator g_j based on the information of each triangle in the group $\{V_j\}$ by using the classical energy function
6: **end while**
HBE CVT step:
7: Use CVT results as the input for HBE CVT
8: **while** HBE energy not converged **do**
9: Associate each triangle with $\{g_j\}_{j=1}^{6}$ by using the HBE distance functions controlled by weighting factor ω
10: Update each generator g_j based on the information of each triangle in the group $\{V_j\}$ by using the HBE energy function
11: **end while**

tation. When we use the HBE distance function to define Voronoi regions, we use a weighting factor ω to control the balance between the classical distance and the boundary-enhanced term (see Eq. 4 in [17]).

Based on **Surface Segmentation Algorithm**, we implement and organize the code into a CLI program (Segmentation.exe), which can segment a given triangle mesh into 6 Voronoi regions. Users can give options through the command line to run Segmentation.exe. Taking the rod model as an example, we first generate a triangle mesh from its CAD model by using LS-PrePost. Then we segment the triangle mesh by running the following command:

```
Segmentation.exe -i rod_tri.k -o rod_initial_write.k -
  m rod_manual.txt -l 0.1
```

There are four options used in the command:

- **–i**: Surface triangle mesh of the input geometry (rod_tri.k);
- **–o**: Output segmentation result (rod_initial_write.k);
- **–m**: Input file with user intervention (rod_manual.txt); and
- **–l**: Weighting factor ω used in **HBE distance function**.

The input and output are .k files, which can be read by LS-PrePost. We refer the reader to [26], which explains the file format. We use **–l** to control the balance between the distance and the boundary-enhanced term. The weighting factor ω can be assigned any arbitrary positive value; however, to obtain the best segmentation behavior, ω must take small value. We find that when $\omega = 0.1$, the segmentation result of the rod model has fewer zig-zags and outliers. Users need to do a trial and error to obtain a good weighting factor. Note that zig-zags and outliers may still exist regardless of the choice of ω. To fix this issue, user intervention is needed to prepare a file that stores the correct segmentation result for such elements. Segmentation.exe can read this file through option **–m** to improve the segmentation result. The snippets of the input text file for the rod model are given in Appendix A1.

Once we get the initial segmentation result, we need to further segment each Voronoi region into several patches to satisfy the topological constraints for polycube construction (see Fig. 1d).

The following three conditions should be satisfied during the further segmentation, see [17, 18]: (1) two patches with opposite orientations (e.g., +X and –X) cannot share a boundary; (2) each corner vertex must be shared by more than two patches; and (3) each patch must have four boundaries. Note that we define the corner vertex as a vertex locating at the corner of the cubic region in the model.

The further segmentation is done manually by using the patch ID reassigning function in LS-PrePost. The detailed operation is shown in Appendix A2.

In addition to the issue with zig-zags and outliers, the algorithm has several limitations. For example, **Surface Segmentation Algorithm** cannot guarantee a good quality polycube structure, which will affect the quality of hex mesh. Elements with small or negative scaled Jacobian in a hex mesh may appear. Some adjustments on the polycube structure and quality improvement are needed as a follow-up step.

3.2 Polycube Construction

In this section, we discuss the detailed algorithm of polycube construction using the segmented triangle mesh. A polycube consisting of multiple cubes is topologically equivalent to the original geometry. Several automatic polycube construction algorithms have been proposed in the literature [16, 17, 23], but it is challenging to generalize these methods to general CAD models. To achieve versatility for real industrial applications, we develop a semi-automatic polycube construction software based on the segmented surface. However, for some complex geometries, it may slow down the process because of the potentially heavy user intervention.

The key information we need for a polycube is its corners and the connectivity relationship among them. For the surface of polycube, we can automatically get the corners and build their connectivity based on the segmentation result by using **Polycube Boundary Surface Construction Algorithm**. However, it is usually difficult to obtain inner vertices and their connectivity as we only have a surface input without any information about the interior volume. Indeed, this is also the place that involves user intervention, where we use LS-PrePost to manually build the interior connectivity. The detailed operation is shown in Appendix A3. The process of obtaining inner vertices and their connectivity is done cube by cube. The inner vertices are first obtained and simultaneously used to build the interior connectivity. In the future, we can separate this process and make the step of building the interior connectivity automatic. As the auxiliary information for this user intervention, **Polycube Boundary Surface Construction Algorithm** will output corners and connectivity of the segmented surface patches into .k file. Finally, the generated polycube structure is the cubic regions splitting the volumetric domain of the geometry.

Polycube Boundary Surface Construction Algorithm

Input: Manifold triangular surface mesh including segmentation information
Output: The boundary surface of the polycube structure
Computing the corners step:

```
for each vertex v_i do
    Get the number of patches n which the vertex v_i surrounding
    if n ≥ 3 then
        Mark the vertex as a corner v_i^c
    end if
end for
Output file including corner coordinates
```

Computing the index array step:

```
for each patches {S_j} do
    Find its four corners v_i^c which define a quad 𝒬_j
end for
Extract edge information from 𝒬_j
Output files of connectivity relationship including edges and faces
```

We implement and organize the code into a CLI program (PolyCube.exe) based on **Polycube Boundary Surface Construction Algorithm**. For the rod model, we run the following command to extract the corners and their connectivity for the boundary surface of its polycube:

```
PolyCube.exe -i rod_initial_read.k -o
    rod_polycube_structure.k -c 1
```

There are four options used in the command:

- **–i**: Surface triangle mesh with the segmentation information (rod_initial_read.k);
- **–o**: Polycube surface connectivity (rod_polycube_structure.k) for polycube construction in LS-PrePost; and
- **–c**: Control indicator if some additional file needs to be output
  - -c 0: No output; and
  - -c 1: Output corner points, edges, faces of polycube structure.

The output .k file contains the corners and their connectivity for the boundary surface of the polycube (see Fig. 2a). Users need to import it into LS-PrePost and manually create interior corners and corresponding connectivity (see Fig. 2b) to build a polycube structure (see Fig. 1f). We also provide option **–c** to output the corners, edges, and faces of the polycube structure if users intend to use other software to build a polycube structure. Users can find their file format in Appendix 4.

3.3 *Parametric Mapping and Octree Subdivision*

After the polycube is constructed, we need to build a bijective mapping between the input triangle mesh and the boundary surface of the polycube structure. In our

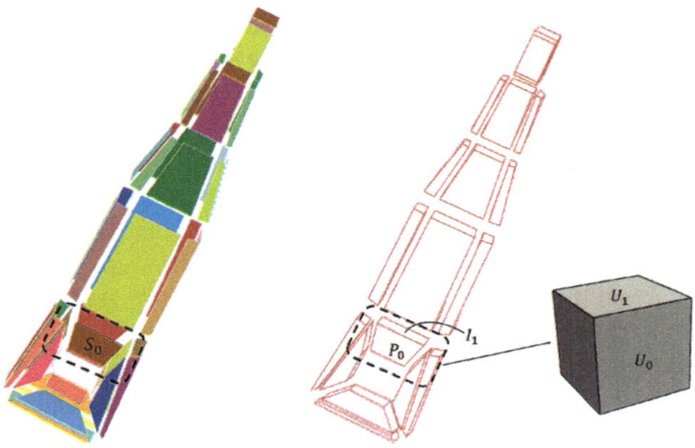

Fig. 2 The polycube construction and the parametric mapping process. **a** The boundary surface of the polycube generated by **Polycube Boundary Surface Construction Algorithm**; **b** The interior corners and connectivity manually created in LS-PrePost to generate polycube structure. We split the polycube into multiple cubes and use each individual cube as the parametric domain. S_0, P_0 and U_0 are used for parametric mapping. I_1 and U_1 are used for linear interpolation to create interior vertices of the mesh

software, we implement the same idea as in [25] to use a unit cube as the parametric domain for polycube structure. As a result, we can construct a generalized polycube structure (see Fig. 2b) that can align with the given geometry better and generate a high quality hex mesh.

In the pseudocode of **Parametric Mapping Algorithm**, we describe how the segmented surface mesh, the polycube structure and the unit cube are combined to create a (volume) parametric mapping and octree subdivision. Let $\{S_i\}_{i=1}^N$ be the segmented surface patches coming from the segmentation result (see Fig. 2a). Each segmented surface patch corresponds to one boundary surface of the polycube P_i $(1 \leq i \leq N)$ (see Fig. 2b), where N is the number of the boundary surface. There are also interior surfaces, denoted by I_j $(1 \leq j \leq M)$, where M is the number of the interior surface. The union of $\{P_i\}_{i=1}^N$ and $\{I_j\}_{j=1}^M$ is the set of surfaces of the polycube structure. For the parametric domain, let $\{U_k\}_{k=1}^6$ denote the six surface patches of the unit cube (see Fig. 2b).

Each cubic region in the polycube structure represents one volumetric region of the geometry and has a unit cube as its parametric domain. Figure 2(b) shows the example of one cubic region and its corresponding volume domain of the geometry marked in the dashed rectangle. Therefore, for each cube in the polycube structure, we can find its boundary surface P_i and map the segmented surface patch S_i to its corresponding parametric surface U_k of the unit cube. To map S_i to U_k, we first map its corresponding boundary edges of S_i to the boundary edges of U_k. Then we get the parameterization of S_i by using the cotangent Laplace operator to compute the harmonic function [8, 46]. Note that for an interior surface I_j of the polycube structure, we skip the parametric mapping step.

An all-hex mesh can be obtained from this surface parameterization combined with the octree subdivision. We generate the hex element for each cubic region in the following process. To obtain vertex coordinates on the segmented patch S_i, we first subdivide the unit cube (see Fig. 2b) recursively to get their parametric coordinates. The physical coordinates can be obtained by using the parametric mapping, which has a one-to-one correspondence between the parametric domain U_k and the physical domain S_i. To obtain the vertices on the interior surface of the cubic region, we skip the parametric mapping step and directly use the linear interpolation to calculate the physical coordinates. Figure 2 shows the example of the rod model. A composition of mappings among S_0, P_0 and U_0 is done to build parametric mapping and obtain vertex coordinates on the surface S_0. I_1 and U_1 are combined for linear interpolation to obtain the vertices on the interior surface of the cubic region. Finally, the vertices inside the cubic region are calculated by linear interpolation. The entire all-hex mesh is built by going through all the cubic regions.

Based on **Parametric Mapping Algorithm**, we implement and organize the code into a CLI program (ParametricMapping.exe) that can generate an all-hex mesh by combining parametric mapping with the octree subdivision. Here, we run the following command to generate an all-hex mesh for the rod model:

```
ParametricMapping.exe -i rod_indexPatch_read.k -p
rod_polycube_structure.k -o rod_hex.vtk -s 2
```

There are three options used in the command:

- **–i**: Surface triangle mesh of the input geometry with segmentation information (rod_indexPatch_read.k);
- **–o**: Unstructured hex mesh (rod_hex.vtk);
- **–p**: Polycube structure (rod_polycube_structure.k); and
- **–s**: Octree level.

We use **–i** to set the segmentation file generated in Sect. 3.1 and use **–p** to set the polycube structure created in Sect. 3.2. Option **–s** is used to set the level of recursive subdivision to be applied. There is no subdivision if we set **–s** to be 0. In the rod model, we set **–s** to be 2 to create a level-2 all-hex mesh. The output all-hex mesh is stored in the VTK format (see Fig. 1(g)) and it can be visualized in Paraview [1].

3.4 Quality Improvement

If the quality of the hex mesh is not satisfactory, quality improvement needs to be applied to the hex mesh. We integrate three quality improvement techniques in the software package, namely pillowing, smoothing and optimization. Users can improve mesh quality through the command line options before building volumetric splines.

Parametric Mapping Algorithm

Input: Segmented triangle mesh $\mathscr{T} = \{S_i\}_{i=1}^{N}$, polycube structure
Output: All-hex mesh

Find boundary surfaces $\{P_i\}_{i=1}^{N}$ and interior surfaces $\{I_j\}_{j=1}^{M}$ in the polycube structure
Surface parameterization step:
for each cube in the polycube structure **do**
  Create a unit cube $\{U_k\}_{k=1}^{6}$ as the parametric domain
  for each surface in the cube **do**
    if it is a boundary surface P_i **then**
      Get the surface parameterization $f : S_i \rightarrow U_k \subset \mathbb{R}^2$
    end if
  end for
end for
Parametric mapping and octree subdivision step:
for each cube in the polycube structure **do**
  Subdivide the unit cube $\{U_k\}_{k=1}^{6}$ recursively to get parametric coordinates v_{para}
  for each surface in the cube **do**
    if it is a boundary surface P_i **then**
      Obtain physical coordinates using $f^{-1}(v_{para})$
    else if it is an interior surface I_j **then**
      Obtain physical coordinates using linear interpolation
    end if
  end for
  Obtain interior vertices in the cubic region using linear interpolation
end for
Combine hex elements from each cubic region

We first use pillowing to insert one layer around the boundary [46]. By using the pillowing technique, we ensure that each element has at most one face on the boundary, which can help improve the mesh quality around the boundary. After pillowing, smoothing and optimization [46] are used to further improve mesh quality. For smoothing, different relocation methods are applied to three types of vertices: vertices on sharp edges on the boundary, vertices on the boundary surface, and interior vertices. Note that we perform smoothing on the surface before smoothing the volume. For each sharp-edge vertex, we first detect its two neighboring vertices on the curve, and then calculate their middle point. For each vertex on the boundary surface, we calculate the area center of its neighboring boundary quadrilaterals (quads). For each interior vertex, we calculate the weighted volume center of its neighboring hex elements as the new position. We relocate a vertex in an iterative way. Each time the vertex moves only a small step towards the new position and this movement is done only if the new location results in an improved local Jacobian.

If there are still poor quality elements after smoothing, we run the optimization whose objective function is the Jacobian. Each vertex is then moved toward an optimal position that maximizes the worst Jacobian. We present **Quality Improvement Algorithm** for quality improvement. Here, we show how to improve mesh quality for the rod model. We first run the following command to perform pillowing on the rod model:

```
Quality.exe -i rod_hex.vtk -Q -m 1 -n 1 -o
    rod_hex_pillow.vtk
```

There are four options used in the command:

- **–i**: Unstructured hex mesh (rod_hex.vtk);
- **–o**: The hex mesh after quality improvement (rod_hex_pillow.vtk);
- **–m**: Improvement method. Pillowing when **–m 1**; and
- **–n**: Number of pillowing layer.

Option **–n** allows users to specify the number of layers to be inserted. With **–n 1**, we insert one layer around the boundary, which is enough to ensure each element have at most one face on the boundary. The result is shown in Fig. 3a.

After pillowing, we can use the following command to smooth the mesh:

```
Quality.exe -i rod_hex_pillow.vtk -Q -m 2 -p 0.001 -n
    50 -s 2 -o rod_hex_pillow_lap.vtk
```

There are seven options used in the command:

- **–i**: The input unstructured hex mesh in the vtk format (rod_hex_pillow.vtk);
- **–o**: The output hex mesh after quality improvement (rod_hex_pillow_lap.vtk);
- **–m**: Improvement method. Smoothing when **–m 2**; and
- **–s**: Sharp feature preservation
  - **–s 0**: No sharp features are preserved;
  - **–s 1**: Detect sharp features automatically, and set tolerance **–t**; and
  - **–s 2**: Manually select sharp feature points and store the indices in the "sharp.txt" file.
- **–t**: Tolerance for automatically detecting sharp features;
- **–p**: Step size for smoothing; and
- **–n**: Number of steps for smoothing.

By using the above command, we relocate a vertex only if the new location will improve the local scaled Jacobian. Option **–s** is used to preserve sharp features. Here, sharp feature detection is only based on the mesh normal information. Therefore, it is not robust for complex geometries and manual work is needed to adjust sharp features. When we use automatic detection with option **–s 1**, we need to set a tolerance **–t**. There is no typical number for this option. We need to do a trial and error to get the optimal value. For the rod model, we set it to be 0.8. However, some sharp features may not be detected regardless of the tolerance. User intervention is needed with the option **–s 2** if automatic detection is not satisfactory. Through this command line option, Quality.exe will read an input file that includes the user-defined sharp features. The snippets of the related file is shown in Appendix 5.

Then we run the optimization step by using the command:

```
Quality.exe -i rod_hex_pillow_smooth.vtk -Q -m 3 -p
    0.001 -n 15 -o rod_hex_pillow_smooth_opt.vtk
```

There are five options used in the command:

(a) (b)

Fig. 3 Mesh quality improvement. **a** The original mesh before quality improvement; **b** The mesh after pillowing, smoothing and optimization. Part of the mesh is removed to show the interior. The elements on the original cross section are depicted in green color while the inserted boundary layer is depicted in red color

- **–i**: Unstructured hex mesh (rod_hex_pillow_smooth.vtk);
- **–o**: The hex mesh after quality improvement (rod_hex_pillow_smooth_opt.vtk);
- **–t**: Tolerance related to sharp feature preservation;
- **–m**: Improvement method. Optimization when **–m 3**;
- **–p**: Step size for optimization; and
- **–n**: Number of steps for optimization.

The quality improvement result for the rod model is shown in Fig. 3 with a boundary layer created using pillowing, followed up by smoothing and optimization.

4 Hex2Spline: Unstructured Spline Construction

With the generated hex mesh as the input control mesh, now we present the Hex2Spline software package to build TH-spline3D. TH-spline3D can define spline functions on arbitrarily unstructured hex meshes. It further supports local refinement for adaptive IGA. Hex2Spline can output the Bézier information of constructed volumetric splines, which can be easily used in LS-DYNA or any other existing IGA frameworks. In the following, we introduce the main algorithm for each component of the Hex2Spline software package, including blending functions on an unstructured hex mesh, TH-spline3D with local refinement, and Bézier extraction.

4.1 Blending Functions on an Unstructured Hex Mesh

In this section, we describe how to build blending functions on an all-hex mesh. Hex2Spline supports arbitrarily unstructured all-hex mesh. In the following, we denote Ω_e as an hex element indexed by e. There are three types of elements in the hex mesh: boundary elements, interior regular elements and interior irregular elements. Note that we don't distinguish regular and irregular boundary elements as

Quality Improvement Algorithm

```
Input: Hex mesh, step number n
Output: Hex mesh with improved quality
Pillowing:
  Insert an outer layer to the input mesh
Smoothing:
  while iteration ¡ n do
    for each vertex v_i do
      if v_i is on a sharp edge then
        if improving the local Jacobian then
          Relocate v_i a small step with step size p towards the middle of its neighboring vertices
        end if
      else if v_i is on the boundary surface then
        if improving the local Jacobian then
          Relocate v_i a small step with step size p towards the area center of its neighboring boundary quads
        end if
      else if v_i is an interior vertex then
        if improving the local Jacobian then
          Relocate v_i a small step towards the weighted volume center of its neighboring elements
        end if
      end if
    end for
  end while
Optimization:
  while iteration ¡ n do
    for each negative Jacobian element do
      if improving the local Jacobian then
        Relocate v_i to maximize the Jacobian
      end if
    end for
    Relocate vertices where Jacobian is minimum to maximize the worst Jacobian
  end while
```

we treat them all the same. The element is defined as a boundary element if it contains a boundary vertex; otherwise, it is an interior element. For an interior element, if it contains an extraordinary edge,[1] we call it an irregular element; otherwise, it is a regular element. In the following, we discuss the main algorithm for building blending functions on boundary elements and interior irregular elements. A regular interior element is a special case of an irregular interior element whose blending functions are merely tricubic B-splines. Note that the following construction only applies to tricubic splines with uniform knot intervals (i.e., the same knot interval for every edge). The resulting volume is C^0-continuous around extraordinary points and edges and C^2-continuous elsewhere. **Blending Functions Algorithm (Interior)** shows the pseudocode to obtain the blending functions defined on an interior irregular element. They are obtained through the Bézier extraction matrix **M**. **M** can be

[1] An extraordinary edge is an interior edge shared by other than four hexahedral elements.

Blending Functions Algorithm (Interior)

```
Input: An interior element Ω_e and its local control mesh 𝒩
Output: The blending functions B_e
Obtain the Bézier transformation matrix M_{64×N}:
for each Bézier point Q_{e,i} of Ω_e, where i = 0, ..., 63 do
    if Q_{e,i} is body point then
        Compute its coordinates based on P
    else if Q_{e,i} is face, edge, corner point then
        Compute its coordinates by averaging the nearest body points
    end if
end for
Assemble matrix M such that Q_e = MP
Obtain the blending functions B_e:
B_e = M^T b
```

obtained by computing the 64 Bézier control points $\mathbf{Q}_e$ from a local spline control mesh $\mathcal{N}$ that consists of Ω_e and its one-ring neighborhood. Each of these Bézier points is obtained by a convex combination of the vertices in the local control mesh, and we have $\mathbf{Q}_e = \mathbf{MP}$, where $\mathbf{P}$ is the vector of vertices in the local control mesh. Then the blending functions on Ω_e are defined by using the transpose of $\mathbf{M}$, that is, $\mathbf{B}_e = \mathbf{M}^T\mathbf{b}$, where

$$\begin{aligned}\mathbf{b} =&[N_0(u)N_0(v)N_0(w), \ldots, N_0(u)N_1(v)N_0(w), N_1(u)N_1(v)N_0(w),\\ &\ldots, N_0(u)N_0(v)N_1(w), N_1(u)N_0(v)N_1(w), \ldots, N_3(u)N_3(v)N_3(w)]^T,\end{aligned}$$

is the vector of 64 tricubic Bernstein polynomials. Each univariate cubic Bernstein polynomial is given as $N_k(t) = \binom{3}{k}(1-t)^{3-k}t^k$ $(k = 0, \ldots, 3)$. We refer the reader to [43] for the coefficients of $\mathbf{M}$. Note that we only need $\mathbf{M}$ to define $\mathbf{B}_e$, and we do not actually compute Bézier points.

Blending Functions Algorithm (Boundary) shows the pseudocode to define blending functions on a boundary element. The body Bézier points and the Bézier points on the interior corners, edges or faces are obtained the same way as in **Blending Functions Algorithm (Interior)**, while the Bézier points on the boundary is defined using only the boundary quadrilateral mesh. The 16 Bézier points on a boundary face can be obtained by convex combinations of the vertices on the local quad control mesh. The detailed computation method is explained in [43]. We finally get all the Bézier points as $\mathbf{Q}_e = \mathbf{MP}$. The blending functions are then defined by $\mathbf{B}_e = \mathbf{M}^T\mathbf{b}$. Note that **Blending Functions Algorithm (Boundary)** can also be used to preserve sharp features in the mesh by adjusting the Bézier extraction matrix $\mathbf{M}$. We refer the reader to [43] for details.

Blending Functions Algorithm (Boundary)

Input: A boundary element Ω_e and its local control mesh $\mathcal{N}$; sharp feature information
Output: The blending functions $\mathbf{B}_e$
Obtain the Bézier transformation matrix $\mathbf{M}_{64\times N}$:
for each Bézier point $Q_{e,i}$ of Ω_e, where $i = 0, \ldots, 63$ **do**
  if $Q_{e,i}$ corresponds to a sharp corner P_k **then**
    $Q_{e,i} = P_k$
  else if $Q_{e,i}$ is on a sharp edge **then**
    Compute its coordinates as a convex combination of the two end points of the sharp edge
  else if $Q_{e,i}$ is on the boundary **then**
    if $Q_{e,i}$ is a face point **then**
      Compute its coordinates based on P_k
    else if $Q_{e,i}$ is an edge or a corner point **then**
      Compute its coordinates by averaging the nearest boundary face points
    end if
  else if $Q_{e,i}$ is a body point or on an interior surface **then**
    Calculate $Q_{e,i}$ the same as **Blending Functions Algorithm (Interior)**
  end if
end for
Assemble matrix $\mathbf{M}$ such that $\mathbf{Q}_e = \mathbf{MP}$
Obtain the blending functions $\mathbf{B}_e$:
$\mathbf{B}_e = \mathbf{M}^T\mathbf{b}$

4.2 TH-Spline3D for Local Refinement

The next step is to introduce local refinement to achieve computational efficiency and accuracy. TH-spline3D employs a hierarchical structure and uses the truncation mechanism to perform local refinement. Global refinement is also supported since it is a special case of local refinement, and it is done by Catmull-Clark subdivision for solids [2, 5].

TH-spline3D Algorithm shows the pseudocode to construct TH-spline3D based on the blending functions developed in Sect. 4.1. Locally refined meshes as well as spline functions between different levels are related through Catmull-Clark subdivision for solids.

The program allows users to specify a list of target elements to be locally refined. This is enabled by reading a series of user-defined files to the program, each of which contains indices of target elements at a certain level. The input mesh is treated as the level-0 mesh by default, one needs to provide a file named "lev_rfid.txt" to refine certain level-0 elements. As a result, a level-1 mesh is generated that can be used to define multi-level local refinement. One can check the hierarchical control meshes and add more elements in the "lev_rfid.txt" file to further refine the mesh.

The remaining procedure to construct TH-spline3D is automatic and can be divided into three steps: (i) refine target elements by Catmull-Clark subdivision for solids, (ii) select certain blending functions from hierarchical meshes, and (iii) truncate some blending functions on the coarse mesh. Then with the help of **Blending Functions Algorithm (Interior)** and **Blending Functions Algorithm (Boundary)**,

TH-spline3D Algorithm

Input: Initial control points $\mathbf{P}^0$ and their associated blending functions $\mathscr{B}^0$, element Ω_e and its local control mesh $\mathscr{N}$
Output: The truncated blending functions $\mathscr{B}_{\text{TH-spline3D}}$
Refined meshes via local refinement:
1: **if** local refinement is needed **then**
2: Generate mesh $\mathscr{M}^{\ell+1}$ from mesh $\mathscr{M}^{\ell}$ based on Catmull-Clark subdivision [2, 5]
3: **end if**
4: Assemble matrix $\mathbf{C}$ so that $\mathbf{P}^{\ell+1} = \mathbf{C}\mathbf{P}^{\ell}$
Construct the hierarchical blending functions:
5: Compute the Bézier control points on hierarchical structure $\mathbf{Q}^{\ell+1} = \mathbf{M}\mathbf{P}^{\ell+1}$
6: Construct the hierarchical B-splines on the hierarchical control meshes by using **Blending Functions Algorithm (Interior)** or **Blending Functions Algorithm (Boundary)**
TH-spline3D Construction:
7: Select the blending function to be active
8: Truncate chosen blending functions
9: Collect all the blending functions ($\mathscr{B}^{\ell+1}_{\text{TH-spline3D}}$) up to Level $\ell + 1$

we can construct TH-spline3D on hierarchical control meshes. We refer the reader to [43] for details.

4.3 *Bézier Extraction for LS-DYNA*

After blending functions are defined based on Bernstein polynomials, the Bézier information of the constructed volumetric splines can be written in the BEXT file for LS-DYNA. The program will also output files for the visualization of Bézier mesh in Paraview [1]. The BEXT file contains all the control points and the Bézier extraction matrix $\mathbf{M}^T$ for each Bézier element. We reduce the file size by using both sparse and dense formats to write $\mathbf{M}^T$. The matrix is output row by row. In a sparse format, only non-zeros of a row are output, where an index is paired with each non-zero coefficient to indicate its column location in the matrix. On the other hand, an entire row is output in the dense format without additional column indices. The choice of the two formats depend on the number of non-zeros in a row. The sparse format is favored when the row only has a few non-zeros. The snippets of the BEXT format file is shown in Appendix 6.

4.4 *Applying Hex2Spline to the Rod Model*

Based on the above algorithms, we implement and organize the code into a CLI program (Hex2Spline.exe) that can construct TH-spline3D on an unstructured hex mesh and extract Bézier information for analysis. During the spline construction, users can specify if refinement is needed. In the end, Hex2Spline generates the BEXT file for LS-DYNA. The program also supports sharp feature preservation. All the available options for this program are explained as follows:

- **–i**: Unstructured hex mesh (rod_hex.vtk);
- **–o**: The BEXT file for LS-DYNA (rod_hex_BEXT.txt);
- **–S**: Spline construction mode;
- **–s**: Sharp feature preservation;
  - **–s 0**: No sharp features need to be preserved;
  - **–s 1**: Detect sharp feature automatically, and set tolerance **–t**; and
  - **–s 2**: Manually select sharp feature points and store them in "sharp.txt".
- **–g**: Set the level of global refinement;
- **–l**: Enable local refinement; and
- **–t**: Tolerance related to sharp feature preservation.

Here, we apply local refinement to create a hierarchical mesh of the rod model (see Fig. 1(g)) by using the command:

```
Hex2Spline.exe -i rod_hex.vtk -S -s 2 -l -o rod_hex.
   BEXT
```

In the command, we use the option **–l** to switch on the local refinement mode and construct TH-spline3D with local refinement. Unlike global refinement, users need to prepare a "lev_rfid.txt" file to specify indices of target elements. The snippets of the "lev_rfid.txt" file are shown in Appendix 7. Figure 4b shows the spline construction with one level local refinement. Users can perform further refinements level by level. For example, users can edit the "lev_rfid.txt" file to include more elements and use

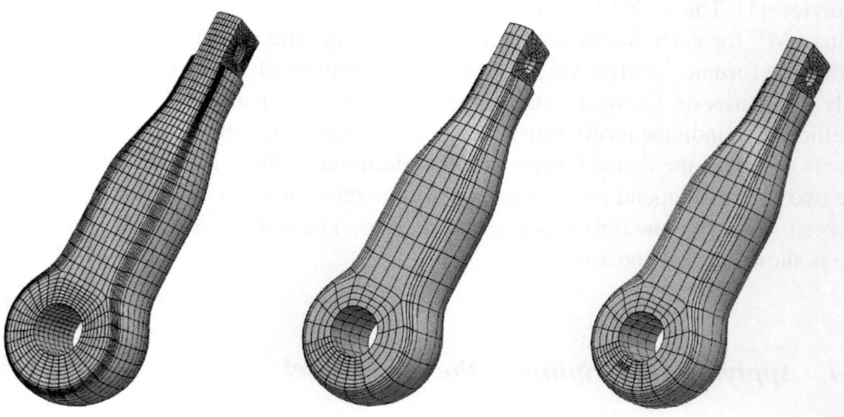

(a) (b) (c)

Fig. 4 The visualization of the output Bézier mesh with: **a** global refinement; **b** one level local refinement; and **c** two levels of local refinement

the same command to perform two levels of local refinement and the result is shown in Fig. 4c:

```
Hex2Spline.exe -i rod_hex.vtk -S -s 2 -l 2
```

Users can also use the following command to perform spline construction with one level global refinement and the result is shown in Fig. 4a:

```
Hex2Spline.exe -i rod_hex.vtk -o rod_hex -S -s 2 -g 1
```

Here we use the option **–g** to switch on global refinement mode and set the argument to 1 to construct spline with one level global refinement.

5 Applications Using HexGen and Hex2Spline

The algorithms discussed in Sects. 3 and 4 are implemented in C++. The Eigen library [15] and Intel MKL [20] are used for matrix and vector operations and numerical linear algebra. We also take advantage of openMP to support multi-threading computation. We use a compiler-independent building system (CMake) and a version-control system (Git) to support software development. We have compiled the source code into two software packages,

- **Hex2Gen software package**:
  - **Segmentation module (Segmentation.exe);**
  - **Polycube construction module (Polycube.exe);**
  - **All-hex mesh generation module (ParametricMapping.exe); and**
  - **Quality improvement module (Quality.exe).**
- **Hex2Spline software package**:
  - **Volumetric spline construction module (Hex2Spline.exe).**

The software is open-source and can be found in the following Github link (https://github.com/CMU-CBML/HexGen_Hex2Spline).

We have applied the software packages to several models and generated all-hex meshes with good quality. For each model, we show the HBECVT based segmentation result, further segmentation result, corresponding polycube structure, and the all-hex mesh. These models include: two types of mount and hepta models (Fig. 5); engine mount and lower arm from Honda Co. along with rockerarm (Fig. 6); ant, bust, and fertility models (Fig. 7); and the joint model from Honda Co. (Fig. 8). Table 1 shows the statistics of all testing models. We use the scaled Jacobian to evaluate the quality of all-hex meshes. From Table 1, we can observe that the obtained all-hex meshes have good quality (minimal Jacobian > 0.1). Figures 5, 6, 7 and 8a show HBECVT segmentation results of testing models, we can observe that the initial segmentation results generated by the HBECVT do not satisfy the topological constraints for polycube construction. We need to further segment each Voronoi region into several patches. The generated polycubes (Figs. 5, 6, 7 and 8b) align with the

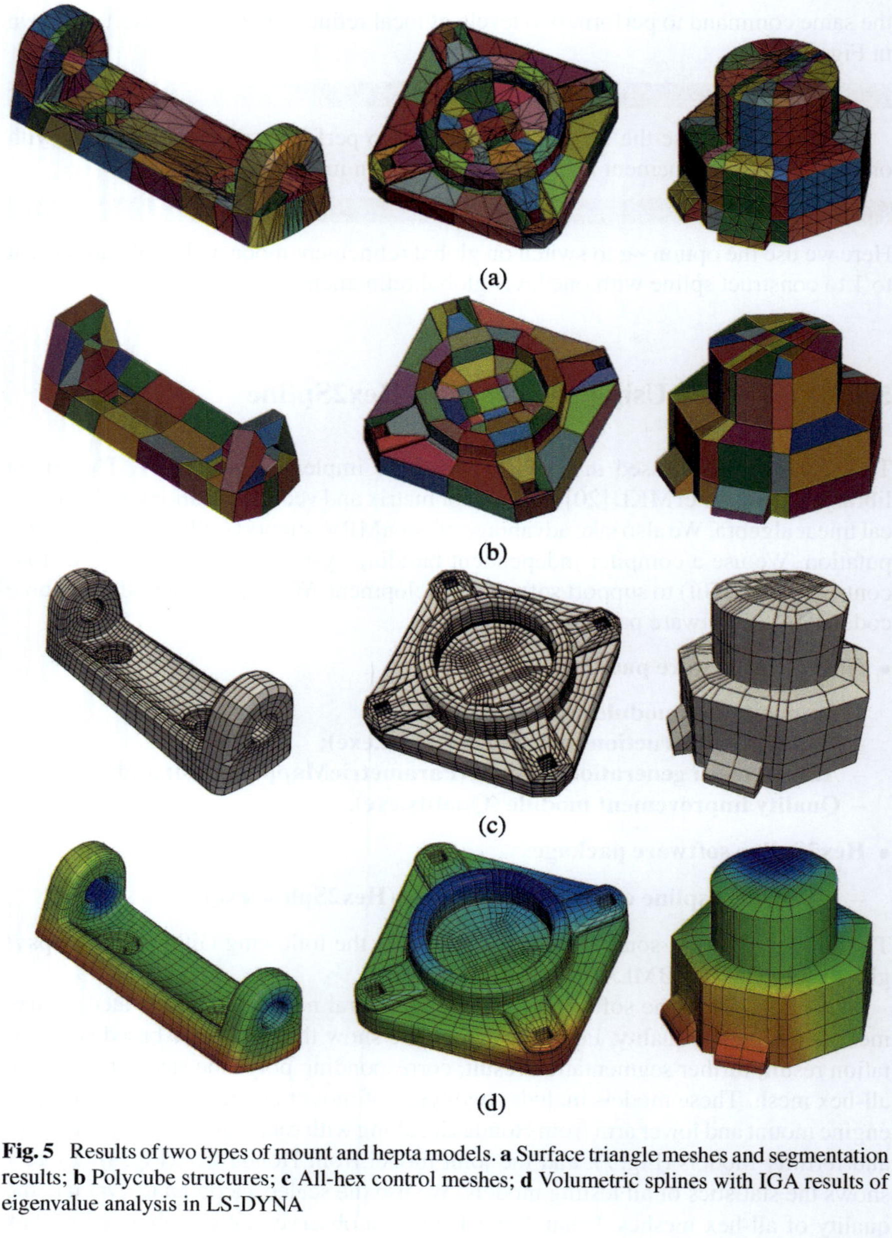

Fig. 5 Results of two types of mount and hepta models. **a** Surface triangle meshes and segmentation results; **b** Polycube structures; **c** All-hex control meshes; **d** Volumetric splines with IGA results of eigenvalue analysis in LS-DYNA

given geometry better, which in turn induces less mesh distortion and yields a mesh of better quality.

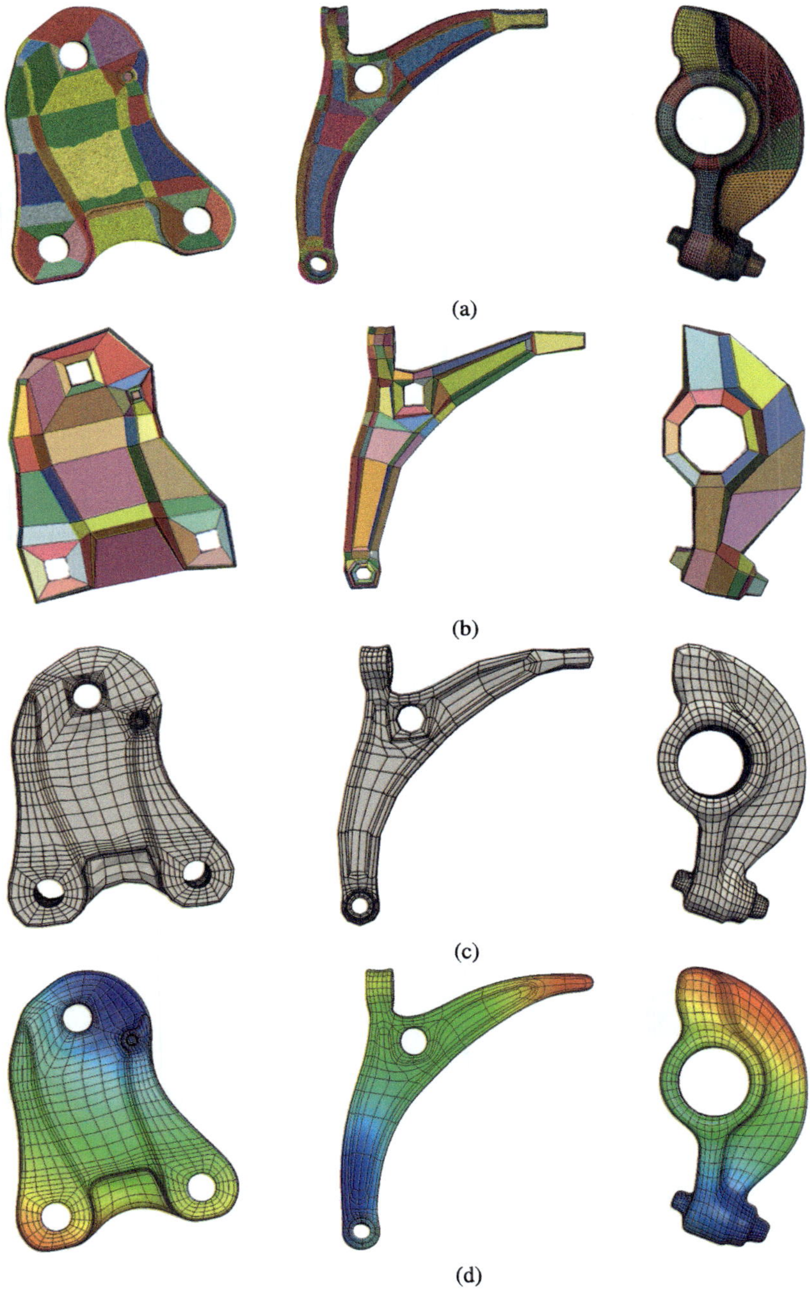

Fig. 6 Results of engine mount, lower arm, and rockerarm model. **a** Surface triangle meshes and segmentation results; **b** Polycube structures; **c** All-hex control meshes; **d** Volumetric splines with IGA results of eigenvalue analysis in LS-DYNA

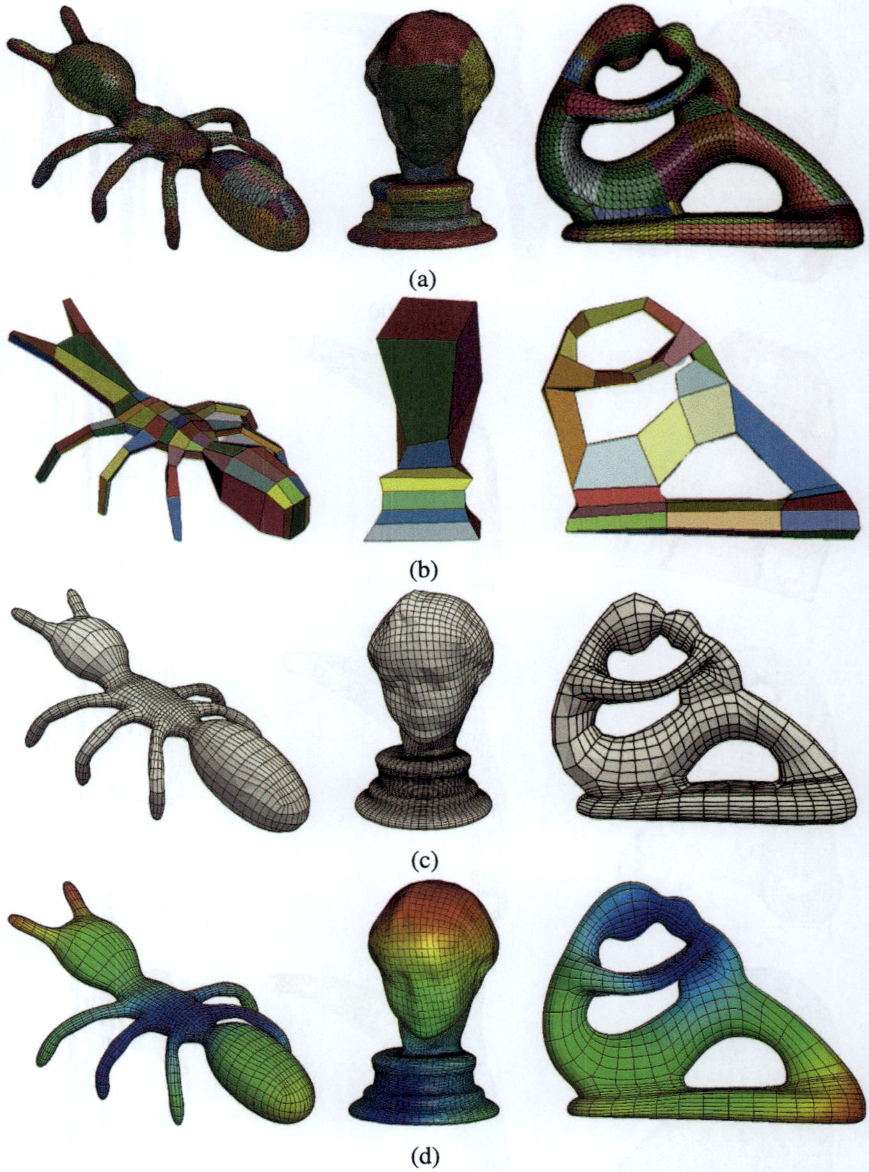

Fig. 7 Results of ant, bust, and fertility model. **a** Surface triangle meshes and segmentation results; **b** Polycube structures; **c** All-hex control meshes; **d** Volumetric splines with IGA results of eigenvalue analysis in LS-DYNA

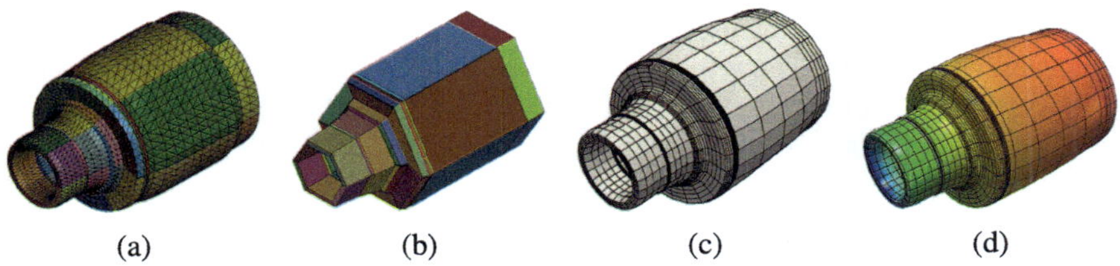

Fig. 8 Results of joint model. **a** Surface triangle meshes and segmentation results; **b** Polycube structures; **c** All-hex control meshes; **d** Volumetric splines with IGA results of solving Poisson equation in LS-DYNA

Table 1 Statistics of all the tested models

Model	Input triangle mesh (vertices, elements)	Octree levels	Output hex mesh (vertices, elements)	Worst Jacobian	Hausdorff distance
Rod (Fig. 1)	(2,238, 4,480)	2	(1,815, 1,280)	0.32	1.45e-2
Mount1 (Fig. 5)	(886, 1,782)	2	(5,849, 4,480)	0.16	1.82e-2
Mount2 (Fig. 5)	(895, 1,802)	2	(7,945, 6,208)	0.14	2.65e-2
Hepta (Fig. 5)	(676, 1,348)	1	(1,259, 944)	0.24	1.98e-2
Engine mount (Fig. 6)	(57,487, 114,982)	2	(7,338, 5,888)	0.11	2.23e-2
Lower arm (Fig. 6)	(165,201, 330,410)	1	(1,996, 1,328)	0.10	3.21e-2
Rockerarm (Fig. 6)	(11,655, 23,310	2	(6,880, 5,696)	0.10	2.26e-2
Ant (Fig. 7)	(7,216, 14,428)	2	(7,711, 6,176)	0.17	1.62e-2
Bust (Fig. 7)	(12,596, 25,188)	4	(103,299, 98,816)	0.11	3.03e-2
Fertility (Fig. 7)	(6,622, 13,256)	2	(4,983, 4,000)	0.20	2.59e-2
Joint (Fig. 8)	(3,806, 7,612)	2	(7,512, 6,144)	0.10	2.66e-2

After generating all-hex meshes (Figs. 5, 6, 7 and 8c), we tested all the models for IGA by using TH-spline3D. The result of the spline construction is C^0-continuous around extraordinary points and edges and C^2-continuous elsewhere. To show the surface accuracy, we calculated the Hausdorff distance between our generated surface and the input boundary while all the models were scaled to make the maximum edge length of the bounding box equal to 1.0 [25]. From Table 1, we can observe that our resulting parameterization has small Hausdorff distance which can be further reduced via local refinement. Next, Bézier elements are extracted for the IGA analysis (Figs. 5,

6, 7 and 8d). For each testing model in Figs. 5, 6 and 7, we use LS-DYNA to perform eigenvalue analysis and show the first mode result. For the testing model in Fig. 8, we show the result of solving a Poisson problem in LS-DYNA. From the results we can observe that our algorithm yields valid TH-spline3D for IGA applications in LS-DYNA.

6 Conclusion and Future Work

In this paper, we present two open source software packages (HexGen and Hex2Spline) for IGA applications in LS-DYNA. The main goal of HexGen and Hex2Spline is to make our pipeline accessible to industrial and academic communities who are interested in real-world engineering applications. The all-hex mesh generation program (HexGen) can generate all-hex meshes. It consists of four executable files, namely segmentation module (Segmentation.exe), polycube construction module (Polycube.exe), all-hex mesh generation module (ParametricMapping.exe) and quality improvement module (Quality.exe). The volumetric spline construction program (Hex2Spline.exe) is developed based on the spline construction method in [43]. Users can generate a volumetric spline model given any unstructured hex mesh and output a BEXT file to perform IGA in LS-DYNA. Both programs are compiled in executable files and can be easily run in the Command Prompt (cmd). The rod model is used to explain how to use these two programs in detail. We also tested our software package using several other models.

In conclusion, we integrate our hex mesh generation and volumetric spline construction techniques and develop a software platform to create IGA models for LS-DYNA. Our software also has limitations that we will address in our future work. First, the hex mesh generation module is semi-automatic and needs user intervention to create polycube structure. We will improve the underneath algorithm and make polycube construction more automatic. In addition, our software cannot guarantee to generate good quality hex mesh for complex geometry. Therefore, in the future we will expand our software package to use hex-dominant meshing methods to create hybrid meshes for IGA applications. To further improve the surface accuracy, we will address the following two challenges in the future: (i) including a refitting function to better fit the resulting spline models to the input CAD model, and (ii) integrating two separate software packages more closely.

Acknowledgements Y. Yu, A. Li, J. Liu and Y. Zhang were supported in part by Honda funds. X. Wei is partially supported by the ERC AdG project CHANGE n. 694515, as well as the Swiss National Science Foundation project HOGAEMS n.200021_188589. We also acknowledge the open source scientific library Eigen and its developers. The authors would like to thank Kenji Takada for providing the CAD geometries of engine mount, lower arm and joint. The authors would also like to thank Attila P. Nagy and David J. Benson for various fruitful discussions about the commercial software LS-DYNA.

Appendix

A1 Input Text File to Correct Segmentation Result from Segmentation.exe

In this section, we describe the data format of the input text file used in Segmentation.exe to correct the segmentation result. One can prepare this file to move elements on the wrong patch to the desired patch. In this text file, each row has two values to define this modification of one element (see List 1). The first value indicates the element index in the triangular mesh and the second value is the desired patch index. Segmentation.exe can read this file through option **–m** to improve the segmentation result.

```
355 1
356 1
361 1
362 1
365 1
366 1
369 1
370 1
495 6
496 6
499 6
500 6
```

List 1 Snippets of the input text file. The first column is the element index and the second is the desired patch index.

A2 Further Segmentation in LS-PrePost

In Sect. 3.1 of the paper, we present CVT-based surface segmentation results. However, the segmentation results do not satisfy the topological constraints for polycube construction. To resolve this problem, we perform further segmentation manually by using the patch ID reassigning function in LS-PrePost. There are four steps to achieve this (see Fig. 9): (1) click **move/copy** tab; (2) select elements; (3) reassign the patch ID; and (4) click the **Apply** button to finish. By repeating the same operation of reassigning elements to different patches, we generate the final segmentation result which satisfies the polycube structure constraints.

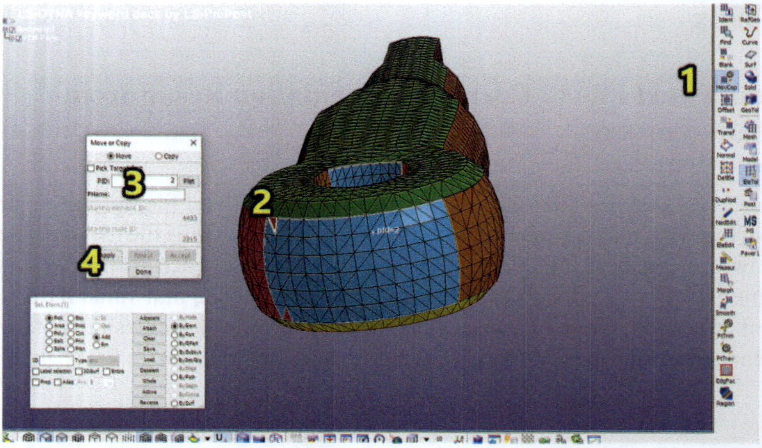

Fig. 9 The detailed operations for further segmentation

A3 Building the Interior Connectivity of Polycube Structure in LS-PrePost

In this section, we use the rod model to introduce how to build the interior connectivity of polycube structure in LS-PrePost. There are four steps to create one cubic region (see Fig. 10): (1) click **EleEdt** tab; (2) select **Elem Type** as Hexa; (3) select eight nodes to define cubic regions, you can also check the selection on the float box; and (4) click the **Accept** button to finish. By repeating the same operation, we generate the polycube structure with multiple cubic regions to split the volumetric domain of the geometry.

A4 Output Text File from Polycube.exe

In CLI program (PolyCube.exe), we output a .k file which contains the corners and their connectivity for the boundary surface of the polycube. It can be directly opened by LS-PrePost. If one intends to use other software to build a polycube structure, we also provide option to output the corners, edges, and faces of the polycube in three separate text files (see Lists 2–4). In the corner file, each row depicts the associated vertex (v_i) information. The first value indicates the index of the vertex in the triangles mesh, the last three values are its x, y, z coordinates (x_i, y_i, z_i). In the edge file, each row uses the indices of two corners to define the edge between them. These indices should agree with the corner file. The face file stores the information of boundary faces on polycube structure. Each row contains four vertex indices in counter-clockwise order to define the connectivity of one face.

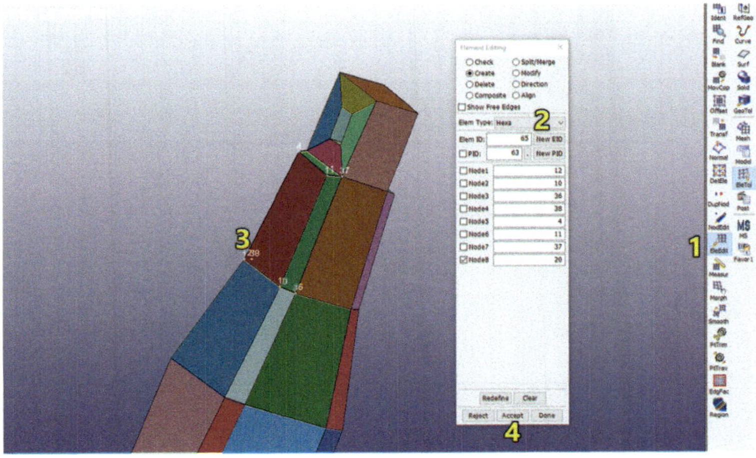

Fig. 10 The detailed operations to build the interior connectivity of polycube structure

```
21 4.06622 0.0052 2.31336
143 16.4588 0.7317 2.29912
153 4.0543 3.3803 2.31406
371 16.4604 2.67402 4.12666
391 13.5206 0.65268 2.07416
```

List 2 Snippets of the corner file. The first column is the vertex index and last three columns are their corresponding coordinates.

```
71,446
436,446
436,442
371,442
1464,1639
1601,1639
```

List 3 Snippets of the edge file. Each row is the indices of edges.

```
71,446,436,442
1464,1639,1601,1503
1439,1664,1639,1464
1744,1917,1877,1784
1246,1784,1877,153
```

List 4 Snippets of the face file. Each row is the indices of four vertices defining the connectivity of one face.

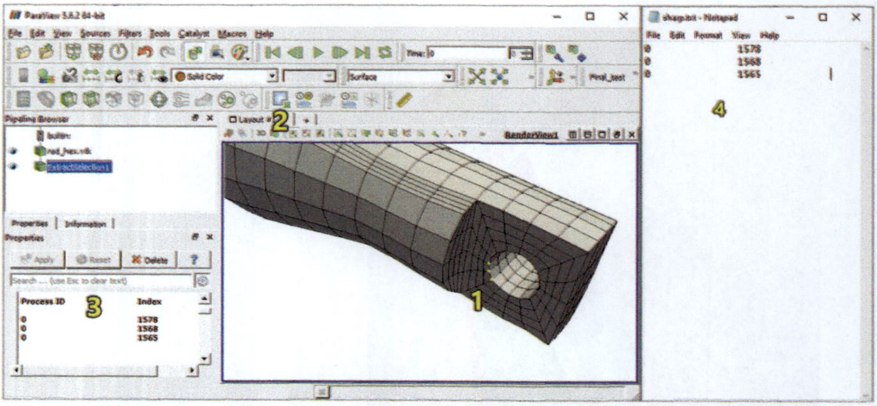

Fig. 11 The detailed operations to create user-defined sharp features

A5 Sharp Feature File for Quality.exe and Hex2Spline.exe

This section describes how to manually define sharp features for Quality.exe and Hex2Spline.exe. One can prepare an input file including the sharp feature information with the help of Paraview. There are four steps (see Fig. 11): (1) select the points along the sharp feature; (2) use **Extract selection** function to extract the indices of these nodes; (3) check the selected points information under "Properties"; and (4) copy the index to "sharp.txt" file and use it as the input sharp feature file for Quality.exe and Hex2Spline.exe.

A6 Input BEXT File for LS-DYNA

In this section, we describe the data format of BEXT file for IGA in LS-DYNA. The BEXT file consists of two parts to store the spline information: (1) control point (Fig. 12A); and (2) Bézier element including the indices of control points supported by this Bézier element (Fig. 12B) and the Bézier extraction matrix (Fig. 12C). The Bézier extraction matrix is output row by row and the format of each row depends on the number of non-zero values in this row. If this row has less than 20 non-zeros, a sparse format is used to store their column indices and values; otherwise, a dense format is used to store all values in this row. To distinguish between two formats, a sparse row begins with **s** while a dense row begins with **d**.

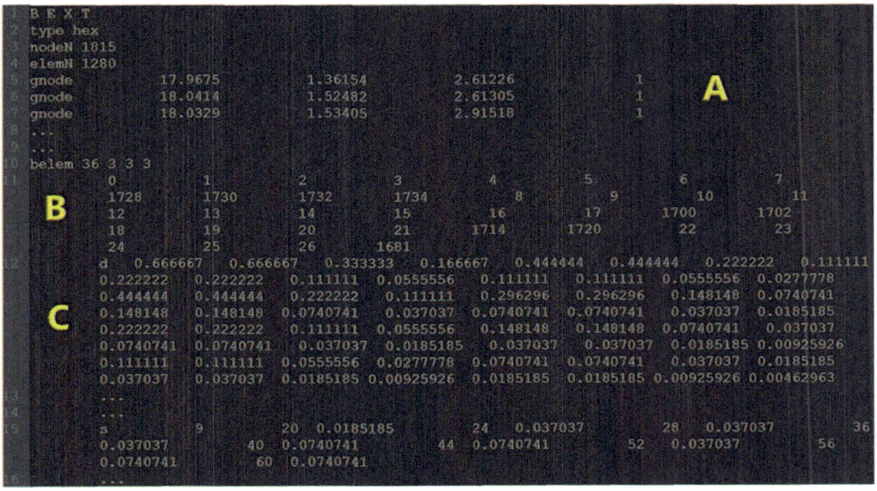

```
B E X T
type hex
nodeN 1815
elemN 1280
gnode     17.9675     1.36154     2.61226     1
gnode     18.0414     1.52482     2.61305     1
gnode     18.0329     1.53405     2.91518     1
...
...
belem 36 3 3 3
      0         1         2         3         4         5         6         7
      1728      1730      1732      1734      8         9         10        11
      12        13        14        15        16        17        1700      1702
      18        19        20        21        1714      1720      22        23
      24        25        26        1681
    d  0.666667  0.666667  0.333333  0.166667  0.444444  0.444444  0.222222  0.111111
    0.222222  0.222222  0.111111  0.0555556  0.111111  0.111111  0.0555556  0.0277778
    0.444444  0.444444  0.222222  0.111111  0.296296  0.296296  0.148148  0.0740741
    0.148148  0.148148  0.0740741  0.037037  0.0740741  0.0740741  0.037037  0.0185185
    0.222222  0.222222  0.111111  0.0555556  0.148148  0.148148  0.0740741  0.037037
    0.0740741  0.0740741  0.037037  0.0185185  0.037037  0.037037  0.0185185  0.00925926
    0.111111  0.111111  0.0555556  0.0277778  0.0740741  0.0740741  0.037037  0.0185185
    0.037037  0.037037  0.0185185  0.00925926  0.0185185  0.0185185  0.00925926  0.00462963
    ...
    ...
    s         9     20  0.0185185     24  0.037037     28  0.037037     36
    0.037037        40  0.0740741     44  0.0740741    52  0.037037     56
    0.0740741       60  0.0740741
    ...
```


Fig. 12 Snippets of the BEXT file

A7 Local Refinement File for Hex2Spline.exe

In this section, we describe the data format of the local refinement file "lev_rfid.txt" used in Hex2Spline.exe to specify indices of target elements. In this text file, each row has two values to define this modification of one element (see List 5). The first value indicates the local refinement level and the second value is the element index in the triangular mesh. Hex2Spline.exe can read this file when we use option **–l**.

```
0    950
0    948
0    934
0    933
0    932
0    935
0    949
0    951
0    958
0    959
0    956
0    957
0    942
0    943
0    941
0    940
1    85
1    102
1    103
1    84
```

List 5 Snippets of the input text file

References

1. Ahrens, J., Geveci, B., Law, C.: Paraview: An end-user tool for large data visualization. In: The Visualization Handbook, pp. 717–731. Butterworth-Heinemann, Burlington (2005)
2. Bajaj, C., Schaefer, S., Warren, J., Xu, G.: A subdivision scheme for hexahedral meshes. Vis. Comput. **18**, 343–356 (2002)
3. Balay, S., Gropp, W.D., McInnes, L.C., Smith, B.F.: Efficient management of parallelism in object-oriented numerical software libraries. In: Arge, E., Bruaset, A.M., Langtangen, H.P. (eds.), Modern Software Tools for Scientific Computing, pp. 163–202. Springer (1997)
4. Blacker, T.D., Stephenson, M.B.: Paving: A new approach to automated quadrilateral mesh generation. Int. J. Numer. Meth. Eng. **32**, 811–847 (1991)
5. Burkhart, D., Hamann, B., Umlauf, G.: Iso-geometric finite element analysis based on Catmull-Clark subdivision solids. Comput. Graph. Forum **29**, 1575–1584 (2010)
6. Dalcin, L., Collier, N., Vignal, P., Côrtes, A.M.A., Calo, V.M.: PetIGA: A framework for high-performance isogeometric analysis. Comput. Methods Appl. Mech. Eng. **308**, 151–181 (2016)
7. De Falco, C., Reali, A., Vázquez, R.: GeoPDEs: a research tool for isogeometric analysis of PDEs. Adv. Eng. Softw. **42**, 1020–1034 (2011)
8. Eck, M., DeRose, T., Duchamp, T., Hoppe, H., Lounsbery, M., Stuetzle, W.: Multiresolution analysis of arbitrary meshes. In: Proceedings of the 22nd Annual Conference on Computer Graphics and Interactive Techniques, pp. 173–182. ACM (1995)
9. Floater, M.S.: Parametrization and smooth approximation of surface triangulations. Comput. Aided Geometr. Design **14**, 231–250 (1997)
10. Folwell, N., Mitchell, S.: Reliable whisker weaving via curve contraction. Eng. Comput. **15**, 292–302 (1999)
11. Forsey, D.R., Bartels, R.H.: Hierarchical B-spline refinement. Comput. Graph. **22**, 205–212 (1988)
12. Giannelli, C., Jüttler, B., Speleers, H.: THB-splines: the truncated basis for hierarchical splines. Comput. Aided Geometr. Design **29**, 485–498 (2012)
13. Giannelli, C., Jüttler, B., Speleers, H.: Strongly stable bases for adaptively refined multilevel spline spaces. Adv. Comput. Math. **40**, 459–490 (2014)
14. Gregson, J., Sheffer, A., Zhang, E.: All-Hex mesh generation via volumetric polycube deformation. Comput. Graph. Forum **30**, 1407–1416 (2011)
15. Guennebaud, G., Jacob, B.: Eigen v3 (2010). http://eigen.tuxfamily.org
16. He, Y., Wang, H., Fu, C., Qin, H.: A divide-and-conquer approach for automatic polycube map construction. Comput. Graph. **33**, 369–380 (2009)
17. Hu, K., Zhang, Y.J.: Centroidal Voronoi tessellation based polycube construction for adaptive all-hexahedral mesh generation. Comput. Methods Appl. Mech. Eng. **305**, 405–421 (2016)
18. Hu, K., Zhang, Y.J., Liao, T.: Surface segmentation for polycube construction based on generalized centroidal Voronoi tessellation. Comput. Methods Appl. Mech. Eng. **316**, 280–296 (2017)
19. Hughes, T.J.R., Cottrell, J.A., Bazilevs, Y.: Isogeometric analysis: CAD, finite elements, NURBS, exact geometry, and mesh refinement. Comput. Methods Appl. Mech. Eng. **194**, 4135–4195 (2005)
20. Intel: Math kernel library. https://software.intel.com/en-us/intel-mkl
21. Lai, Y., Liu, L., Zhang, Y.J., Chen, J., Fang, E., Lua, J.: Rhino 3D to Abaqus: A T-spline based isogeometric analysis software framework. In: Bazilevs, Y., Takizawa, K., (eds.), Advances in Computational Fluid-Structure Interaction and Flow Simulation, Modeling and Simulation in Science, Engineering and Technology, pp. 271–281. Birkhäuser (2016)
22. Lai, Y., Zhang, Y.J., Liu, L., Wei, X., Fang, E., Lua, J.: Integrating CAD with Abaqus: a practical isogeometric analysis software platform for industrial applications. Comput. Math. Appl. **74**, 1648–1660 (2017)

23. Lin, J., Jin, X., Fan, Z., Wang, C.: Automatic polycube-maps. In: Advances in Geometric Modeling and Processing. Lecture Notes in Computer Science, vol. 4975, pp. 3–16. Springer, Berlin (2008)
24. Liu, L., Zhang, Y., Hughes, T.J.R., Scott, M.A., Sederberg, T.W.: Volumetric T-spline construction using boolean operations. Eng. Comput. **30**, 425–439 (2014)
25. Liu, L., Zhang, Y., Liu, Y., Wang, W.: Feature-preserving T-mesh construction using skeleton-based polycubes. Comput. Aided Design **58**, 162–172 (2015)
26. Livermore Software Technology Corporation: LS-DYNA Keyword User's Manual (2007)
27. Mantzaflaris, A.: An overview of geometry plus simulation modules. In: Slamanig, D., Tsigaridas, E., Zafeirakopoulos, Z., (eds.) Mathematical Aspects of Computer and Information Sciences. Lecture Notes in Computer Science, vol. 11989, pp. 453–456. Springer International Publishing (2020)
28. Nieser, M., Reitebuch, U., Polthier, K.: CubeCover - parameterization of 3D volumes. Comput. Graph. Forum **30**, 1397–1406 (2011)
29. Pauletti, M.S., Martinelli, M., Cavallini, N., Antolin, P.: Igatools: an isogeometric analysis library. SIAM J. Sci. Comput. **37**, C465–C496 (2015)
30. Price, M.A., Armstrong, C.G.: Hexahedral mesh generation by medial surface subdivision: Part II. Solids with flat and concave edges. Int. J. Numer. Methods Eng. **40**, 111–136 (1997)
31. Price, M.A., Armstrong, C.G., Sabin, M.A.: Hexahedral mesh generation by medial surface subdivision: Part I. Solids with convex edges. Int. J. Numer. Methods Eng. **38**, 3335–3359 (1995)
32. Qian, J., Zhang, Y.: Automatic unstructured all-hexahedral mesh generation from B-Reps for non-manifold CAD assemblies. Eng. Comput. **28**, 345–359 (2012)
33. Qian, J., Zhang, Y., Wang, W., Lewis, A.C., Qidwai, M.A.S., Geltmacher, A.B.: Quality improvement of non-manifold hexahedral meshes for critical feature determination of microstructure materials. Int. J. Numer. Meth. Eng. **82**, 1406–1423 (2010)
34. Schneiders, R.: A grid-based algorithm for the generation of hexahedral element meshes. Eng. Comput. **12**, 168–177 (1996)
35. Schneiders, R.: An algorithm for the generation of hexahedral element meshes based on an octree technique. In: 6th International Meshing Roundtable, pp. 195–196 (1997)
36. Sederberg, T.W., Cardon, D.L., Finnigan, G.T., North, N.S., Zheng, J., Lyche, T.: T-spline simplification and local refinement. ACM Trans. Graph. **23**, 276–283 (2004)
37. Sederberg, T.W., Zheng, J., Bakenov, A., Nasri, A.: T-splines and T-NURCCs. ACM Trans. Graph. **22**, 477–484 (2003)
38. Staten, M., Kerr, R., Owen, S., Blacker, T.: Unconstrained paving and plastering: Progress update. In: Proceedings of the 15th International Meshing Roundtable, pp. 469–486 (2006)
39. Tarini, M., Hormann, K., Cignoni, P., Montani, C.: Polycube-maps. ACM Trans. Graph. **23**, 853–860 (2004)
40. Vuong, A.V., Giannelli, C., Jüttler, B., Simeon, B.: A hierarchical approach to adaptive local refinement in isogeometric analysis. Comput. Methods Appl. Mech. Eng. **200**, 3554–3567 (2011)
41. Wang, W., Zhang, Y., Liu, L., Hughes, T.J.R.: Trivariate solid T-spline construction from boundary triangulations with arbitrary genus topology. Comput. Aided Design **45**, 351–360 (2013)
42. Wang, W., Zhang, Y., Xu, G., Hughes, T.J.R.: Converting an unstructured quadrilateral/hexahedral mesh to a rational T-spline. Comput. Mech. **50**, 65–84 (2012)
43. Wei, X., Zhang, Y., Hughes, T.J.R.: Truncated hierarchical tricubic C^0 spline construction on unstructured hexahedral meshes for isogeometric analysis applications. Comput. Math. Appl. **74**, 2203–2220 (2017)
44. Zhang, Y.: Geometric Modeling and Mesh Generation from Scanned Images. Chapman and Hall/CRC (2016)
45. Zhang, Y., Bajaj, C.L., Xu, G.: Surface smoothing and quality improvement of quadrilateral/hexahedral meshes with geometric flow. Commun. Numer. Methods Eng. **25**, 1–18 (2009)
46. Zhang, Y., Wang, W., Hughes, T.J.R.: Solid T-spline construction from boundary representations for genus-zero geometry. Comput. Methods Appl. Mech. Eng. **249**, 185–197 (2012)

C^1 Triangular Isogeometric Analysis of the von Karman Equations

Mehrdad Zareh and Xiaoping Qian

Abstract In this paper, we report the use of rational Triangular Bézier Splines (rTBS) to numerically solve the von Karman equations, a system of fourth order PDEs. C^1 smoothness of the mesh, generated by triangular Bézier elements, enables us to directly solve the von Karman systems of equations without using a mixed formulation. Numerical results of benchmark problems show high accuracy and expected convergence rate in L^1, H^1 and H^2 norm for quadratic and cubic triangular Bézier elements. Results of this study show that triangular isogeometric analysis can efficiently and accurately solve systems of high order PDEs.

1 Introduction

Linear and nonlinear structural analysis of plates have been addressed by various formulations. In linear analysis, the most appealing theories are Kirchhoff-Love for thin plates in which transverse shear deformation is neglected, and Reissner-Mindlin theory for moderately thick plates.

Von Karman, in 1910, introduced a system of fourth order elliptic equations to mathematically model the nonlinear behavior of plates in large deflections. Such fourth order partial differential equations (PDEs) have also been applied in other nonlinear problems such as multiphysics modeling, ionic polymer metal composites [19] and the growth of biological tissues [22]. Fourth order PDEs and nonlinearity cause complexity in numerically solving von Karman equations [4]. The finite element method (FEM) is a widely used numerical simulation method in solid and fluid mechanics mostly because of its well-established mathematics, ability to approximate model geometries and its generality in numerical resolution, i.e. the solution and its derivatives can be estimated at any location; therefore, FEM has been an

M. Zareh · X. Qian (✉)
University of Wisconsin-Madison, Madison, USA
e-mail: qian@engr.wisc.edu

M. Zareh
e-mail: zarehbannadk@engr.wisc.edu

C. Manni and H. Speleers (eds.), *Geometric Challenges in Isogeometric Analysis*, Springer INdAM Series 49, https://doi.org/10.1007/978-3-030-92313-6_15

attractive numerical technique to approximate the solution of von Karman equations [4, 5, 17, 20, 23].

For such fourth order PDEs, C^1 smoothness over the mesh is essential, which can be achieved by using relatively sophisticated finite elements (FE) such as Bogner-Fox-Schmit and Argyris FE. However, these elements are complex to implement and computationally expensive [4, 17]. Hence, alternative methods such as smooth splines in the context of isogeometric analysis (IGA) have been explored.

Several techniques in the framework of IGA have been developed and used to solve high order PDEs, particularly fourth order PDEs. In [3, 12, 13], nonuniform rational B-splines (NURBS) were implemented to represent shell geometries and to discretize high order PDEs of Kirchhoff-Love theory. Despite its strengths, the tensor product nature of NURBS makes them unattractive for representing complex geometries and for local mesh refinement. This problem motivated research on using multiple patches instead of a single patch. Although the technique can be effective [1, 8, 30], enforcing C^1 over the mesh constructed by multiple patches is challenging. This issue has been addressed by developing novel formulations to create C^1 splines on unstructured quadrilateral meshes [10, 11, 18, 26]. Besides unstructured quadrilateral elements, triangular elements have been explored in the context of IGA. [16, 18, 24] employed Powell-Sabin B-splines to solve the equations of fourth order PDEs and Kirchhoff-Love plate theory. Flexibility in meshing makes Powell-Sabin B-splines desirable compared with NURBS. In the same context, rational Bézier triangles were used in [15] for solving the Kirchhoff plate problem. They impose smoothness constraints by the method of Lagrange multipliers. In addition to the discussed methods, a few other strategies provide alternative tools to cope with the C^1 requirement.

Nonconforming FE with C^0 penalty [4] and discontinuous Galerkin FE [5] have been employed as alternatives to eliminate the C^1 requirement. Mixed finite elements have also been used to solve high order PDEs [6, 19, 20, 23]; in this method additional variables are introduced into the original problem to lower the order of PDEs and, subsequently, relax the need for C^1. However, mixed FEM significantly increases the DOFs. This can cause difficulties in convergence for nonlinear problems. Moreover, introducing a new variable can change the original problem, leading to physically irrelevant solutions in some cases, e.g., geometries with re-entrant corners. In such geometries numerical solutions from mixed FEM and C^1-FEM (direct solution), for example, for a biharmonic PDE are different. In a mathematical explanation, the direct solution from C^1-FEM is in H^2 space; however, the solution from mixed FEM is not necessarily in H^2 space. This inconsistency implies that the mixed FEM solution cannot consistently converge to the correct solution that is otherwise achievable with the C^1-smooth modeling of the original problem [7, 21].

In this study, we employ C^1 rational Triangular Bézier Splines (rTBS) in the framework of triangular isogeometric analysis (TIGA) to solve von Karman equations without introducing new variables. TIGA was developed in [9, 25, 28, 29] and has been implemented to solve various low and high order PDEs [25, 27, 33] such as Kirchhof-Love plate and shell equations [31, 32]. The availability of C^1 smoothness in TIGA enables us to efficiently and accurately solve high order PDEs. This paper presents numerical examples to show the validity and optimal convergence rate in the numerical solution of von Karman equations using TIGA.

2 Rational Triangular Bézier Splines

In this section, we describe some background technologies on how to achieve C^1 smooth Bézier elements. More details can be found in [29].

We use triangular Bézier elements to discretize both geometry and solution field. A Bézier curve is defined by Bernstein basis functions; the d-degree Bernstein polynomials are given by

$$\psi_{I,d}(\zeta) = \frac{d!}{i!j!}\zeta^i(1-\zeta)^j, \quad |I| = i+j = d. \tag{1}$$

Here I refers to a double index i, j. In this study we implement Bézier triangles based on bivariate Bernstein polynomials. The bivariate extension of equation (1) describes the bivariate Bernstein polynomials:

$$\psi_{I,d}(\zeta) = \frac{d!}{i!j!k!}\zeta_1^i\zeta_2^j\zeta_3^k, \quad |I| = i+j+k = d.$$

Here I refers to a triple index i, j, k. Moreover $\zeta_1, \zeta_2, \zeta_3$ are the barycentric coordinates of the point $\zeta \in \mathbb{R}^2$: every point (s, t) in a fixed triangle τ defined by vertices v_1, v_2, v_3 (see Fig. 1) can be uniquely defined by

$$(s, t) = \zeta_1 v_1 + \zeta_2 v_2 + \zeta_3 v_3, \quad \zeta_1 + \zeta_2 + \zeta_3 = 1.$$

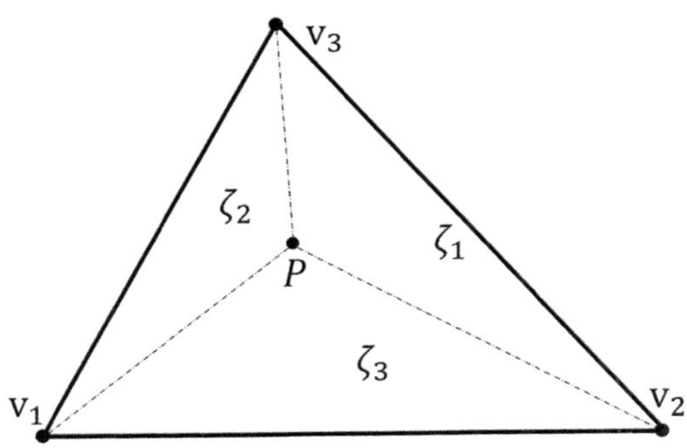

Fig. 1 Barycentric coordinates of a point, $P = (s, t)$, in a triangle, τ. $\zeta_1 = \frac{area(V_3V_2P)}{area(V_1V_2V_3)}$, $\zeta_2 = \frac{area(V_3V_1P)}{area(V_1V_2V_3)}$, $\zeta_3 = \frac{area(V_1V_2P)}{area(V_1V_2V_3)}$

Now, a triangular Bézier patch can be defined by

$$x(\zeta) = \sum_{|I|=d} p_I \psi_{I,d}(\zeta),$$

where the p_I represent the control points. By introducing weights into the above formula, a rational Bézier triangle is defined by

$$x(\zeta) = \sum_{|I|=d} p_I \Psi_{I,d}(\zeta),$$

where

$$\Psi_{I,d} = \frac{w_I \psi_{I,d}}{\sum_{|I|=d} w_I \psi_{I,d}}. \tag{2}$$

Here w_I represents the weight of the control point p_I. Following the isoparametric concept, the same bivariate Bernstein basis on a triangle τ with vertices v_1, v_2, v_3 is used for defining a d-degree polynomial function f over τ as

$$f(\zeta) = \sum_{|I|=d} b_I \Psi_{I,d}(\zeta).$$

The b_I (or b_{ijk}) refer to the Bézier ordinates of f; their corresponding array of domain points are given by

$$q_{ijk} = \frac{iv_1 + jv_2 + kv_3}{d}, \quad i + j + k = d.$$

The control polygon of the function f is defined by the points (q_{ijk}, b_{ijk}). Figure 2 presents an example of a triangular Bézier patch and the corresponding domain points of the Bézier ordinates.

Due to the presence of second order PDEs in weak form of the von Karman equations, smoothness within and between elements is the key requirement in applying the finite element method for these equations. One way to relax this requirement is the introduction of an additional variable into the equation, which can cause expensive computation and inaccuracy in some cases. In this study, we use the C^1 triangular isogeometric method; therefore, we can achieve the required smoothness. We describe how high continuity is attained between triangular Bézier patches. Two degree-d polynomials f and $\tilde{f}$ join r times differentiably across the interface of two triangles $\tau = \{v_1, v_2, v_3\}$ and $\tilde{\tau} = \{v_4, v_2, v_3\}$ if and only if [14] for $(j + k + \rho = d,\ \rho = 0, \ldots, r,)$:

$$\tilde{b}_{\rho,j,k} - \sum \frac{\rho!}{\mu!\nu!\kappa!} b_{\mu,k+\nu,j+\kappa} \zeta_1^{\mu}, \zeta_2^{\nu}, \zeta_3^{\kappa} = 0, \tag{3}$$

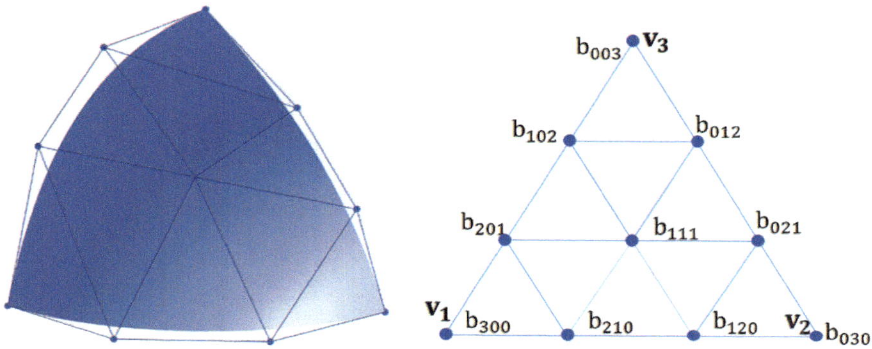

(a) Triangular Bézier patch.

(b) Associated domain points of the Bézier ordinates b_{ijk} in $\{v_1, v_2, v_3\}$.

Fig. 2 Triangular Bézier patch and domain points

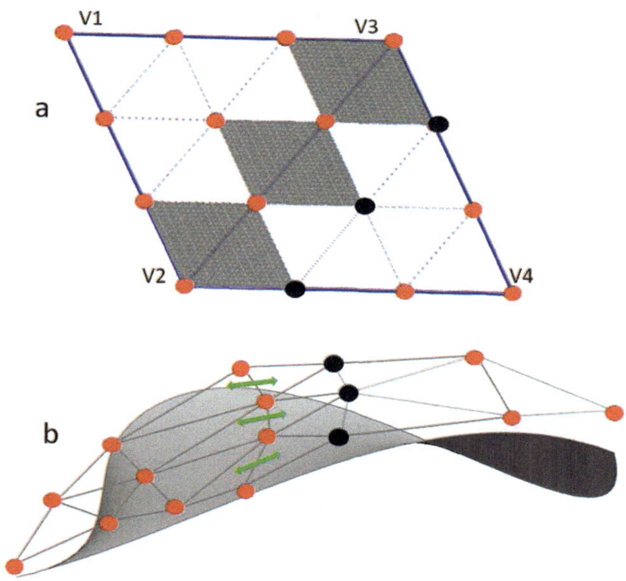

Fig. 3 Dependent (black) and independent nodes (red). **a** Two domain triangles, where smoothness constraints are imposed on the shaded triangles. **b** For two Bézier patches with C^1 smoothness, the control points in the triangle pairs (indicated by $\longleftrightarrow$) are coplanar. For clarity in visualization, the control net is shifted up

where ζ_1, ζ_2, ζ_3 represent the barycentric coordinates of vertex v_4 in relation to triangle τ. Figure 3 shows two triangular Bézier patches with C^1 smoothness across the boundary of patches. The red solids are free nodes whose values are independently computed; the values of black solids (dependent nodes) are determined by

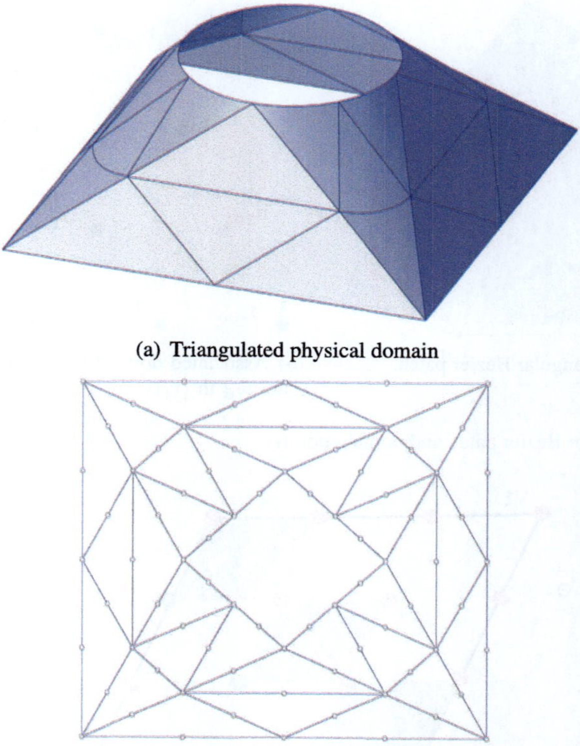

(a) Triangulated physical domain

(b) Triangulated parametric domain

Fig. 4 Illustration of a triangulated physical and parametric domain

applying the smoothness constraints between red and blacked nodes. The smoothness constraints are applied over the gray area; moreover, this figure demonstrates the coplanarity of the control points in these triangle pairs.

Having a parametric domain $\hat{\Omega}$ and its triangulation $\hat{T}$ (Fig. 4), the spline spaces of piecewise d-degree polynomials $\hat{T}$ are defined by [14];

$$S_d^r(\hat{T}) = \left\{ f \in C^r(\hat{\Omega}) : f|_\tau \in P,\ \forall \tau \in \hat{T} \right\},$$

where τ is an arbitrary triangle in $\hat{T}$, r is the continuity order of the spline space over $\hat{\Omega}$ and P is the space of polynomials of degree d. When a spline has higher smoothness across some edges or at some vertices, it is called superspline; the associated space is given by [14];

$$S_d^{r,\rho}(\hat{T}) = \left\{ f \in S_d^r(\hat{T}) : f \in C^{r_v}(V)\ \forall v \in V \ \&\ f \in C^{r_e}(e)\ \forall e \in E \right\}.$$

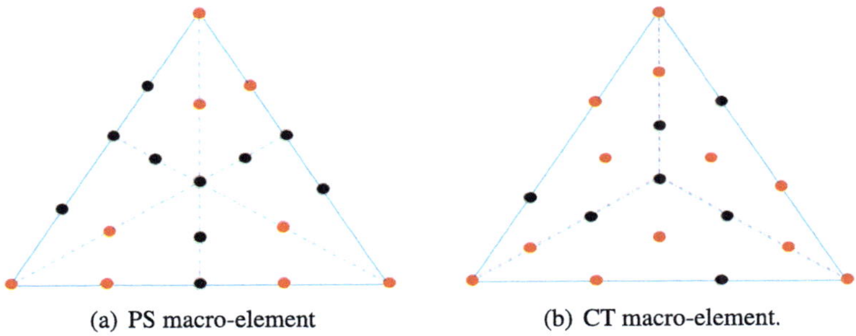

Fig. 5 Splitting methods used in this study. Red and black nodes represent independent and dependent nodes, respectively

All vertices and edges in $\hat{T}$ are represented by V and E, and $\rho := \{\rho_v\}_{v \in V} \cup \{\rho_e\}_{e \in E}$ with $r \leq \rho_v, \rho_e \leq d$ for each $v \in V$ and $e \in E$.

Before applying smoothness constraints, a minimal determining set (MDS) is framed, such set contains all free domain points. In this study, we use the direct construction (DC) method to construct MDS, in which a set of free domain points are directly chosen based on the connectivity of the triangle elements. Reference [29] details the DC method and the alternative method of Gaussian elimination to build MDS.

Imposing condition (3) directly on the triangles is a conventional technique to create C^r spline spaces on a triangulated domain $\hat{\Omega}(\hat{T})$. Despite being straightforward, this direct method requires the degree of the polynomial to be much higher than r, i.e. $d \geq 3r + 2$ [29]. In this work, triangles in $\hat{T}$ are split into multiple microtriangles before imposing the smoothness constraints on the microtriangles. Clough-Tocher (CT) and Powell-Sabin (PS) methods are implemented for the splitting; these methods do not need such high-degree polynomials to provide continuity, e.g., CT cubic elements and PS quadratic elements can create C^1 smoothness. In the CT splitting method, each vertex of a triangle is connected to its centroid point, forming three micro-triangles. The PS method splits each macro-triangle into six micro-triangles with centroid point as the interior split point. Edges are then bisected (see Fig. 5).

In order to evaluate the convergence rate, we use the smooth-refine-smooth procedure [28]. The key point of this method is the relation between the sequence of triangulations in refinements, such that any refined triangulation (mesh) is a subset of the pre-refined triangulation. In this technique, smoothness constraints are applied before refining the mesh in order that a smooth pre-refinement map is constructed; then, further refinement followed by C^r smoothness constraints are applied. The smooth-refine-smooth procedure ensures that control points do not relocate as they have already satisfied the continuity conditions. The obtained mesh is C^r smooth, and the geometric map remains the same; therefore, inconsistency is avoided for all subsequent refinements.

3 Formulations and Methods

This section presents the von Karman equations, the corresponding mixed formulation, the weak formulations and finite element discretizations. As discussed in the introduction, C^1 smoothness is required for directly solving the original von Karman equations due to the high order PDEs. On the other hand, such high order derivatives do not appear in the weak form of mixed FEM; therefore, C^0 smoothness, available in conventional FEM, is sufficient for solving the mixed formulation.

3.1 Von Karman Equations

The canonical form of the von Karman equations with simply supported boundary condition is described by

$$\begin{cases} \Delta^2 u = [u, v] + f, \\ \Delta^2 v = -\frac{1}{2}[u, u], \end{cases} \tag{4}$$

where the bi-harmoinc operator and the von Karman bracket are defined by

$$\Delta^2 u \equiv \frac{\partial^4 u}{\partial x^4} + 2\frac{\partial^4 u}{\partial x^2 \partial y^2} + \frac{\partial^4 u}{\partial y^4},$$

$$[u, v] \equiv \frac{\partial^2 u}{\partial x^2}\frac{\partial^2 v}{\partial y^2} + \frac{\partial^2 u}{\partial y^2}\frac{\partial^2 v}{\partial x^2} - 2\frac{\partial^2 u}{\partial x \partial y}\frac{\partial^2 v}{\partial x \partial y}.$$

The above equations have fourth order PDEs. Because of their complications in a classical finite element discretisation, the mixed form of the von Karman equations has been extensively used, which does not have high order PDEs. For completeness, the next section presents the mixed form; however, we note that the focus of this paper is on using C^1 smoothness for the von Karman equations with fourth order PDEs.

3.2 Mixed Form of the von Karman Equations

By introducing an additional variable, $\bar{u}$, into the Eq. (4), the mixed formulation is defined by

$$\begin{cases} \Delta u = \bar{u}, \\ \Delta \bar{u} = [u, v] + f, \\ \Delta v = \bar{v}, \\ \Delta \bar{v} = -\frac{1}{2}[u, u], \end{cases} \tag{5}$$

where Δ is the Laplacian operator, and the von Karman bracket was defined in the previous section.

As we can observe, the above formulation does not have high order PDEs, i.e., fourth order PDEs do not appear in the equations. The next section provides the finite element discretisations of the above equations.

3.3 *Finite Element Discretisations*

The strong forms of the von Karman equations are multiplied by test functions to obtain the weak formulations.

3.3.1 Weak Form of the Mixed Formulation

The following Eq. represents the weak form of (5) for $f \in L^2(\Omega)$ and $u, v, \bar{u}, \bar{v} \in S := H_0^1(\Omega)$:

$$\begin{cases} a(u, w) + m(\bar{u}, w) = 0, \quad \forall w \in S, \\ a(\bar{u}, \mu) + b(u, v, \mu) + b(v, u, \mu) = l(w), \quad \forall \mu \in S, \\ a(v, \tau) + m(\bar{v}, \tau) = 0, \quad \forall \tau \in S, \\ a(\bar{v}, \phi) - b(u, u, \phi) = 0, \quad \forall \phi \in S, \end{cases} \tag{6}$$

where the bilinear a, trilinear b and l are given by ($\forall \beta, \gamma, \varphi \in S$, and $\operatorname{cof}\left(D^2\beta\right)$ is the cofactor matrix of the Hessian of β):

$$\begin{aligned} a(\beta, \gamma) &:= -\int_\Omega \nabla\beta . \nabla\gamma d\Omega, \\ b(\beta, \gamma, \varphi) &:= \frac{1}{2}\int_\Omega \operatorname{cof}\left(D^2\beta\right) D\gamma \cdot D\varphi d\Omega, \\ l(w) &:= \int_\Omega f \cdot w d\Omega. \end{aligned}$$

3.3.2 Weak Form of the Original Formulation

The weak formulation of (4) is given by (for $f \in L^2(\Omega)$ and $u, v \in S := H_0^2(\Omega)$):

$$\begin{cases} \hat{a}(u, w) + b(u, v, w) + b(v, u, w) = l(w), \quad \forall w \in S, \\ \hat{a}(v, \phi) - b(u, u, \phi) = 0, \quad \forall \phi \in S. \end{cases} \tag{7}$$

The bilinear $\hat{a}$ is defined by

$$\hat{a}(\beta, \gamma) := \int_{\Omega} D^2\beta : D^2\gamma d\Omega,$$

As it can be observed, the weak form (7) requires the second derivative of the solution. Therefore, C^1 smoothness between and within elements is needed in order to obtain a compatible finite element method to meet computability. We proceed with (7) to describe the next steps in solving the von Karman equations; the same procedure can be applied to (6).

After discretization, the following element matrices are defined:

$$K_e := \int_{\Omega_e} H_e^T H_e d\Omega, \tag{8}$$

$$F_e := \int_{\Omega_e} f\Phi_i d\Omega, \tag{9}$$

$$B_e^m := \left[b_{i,j}^m\right]_{1\le i,j\le n}, 1 \le m \le n, \tag{10}$$

where (n is the number of nodes in the element)

$$H_e = \begin{bmatrix} \Phi_{1,xx} & \Phi_{2,xx} & \cdots & \Phi_{n,xx} \\ \Phi_{1,yy} & \Phi_{2,yy} & \cdots & \Phi_{n,yy} \end{bmatrix},$$

and

$$b_{i,j}^k = \frac{1}{2}\int_{\Omega}\left(\left(\frac{\partial^2\Phi_i}{\partial y^2}\frac{\partial\Phi_j}{\partial x} - \frac{\partial^2\Phi_i}{\partial x\partial y}\frac{\partial\Phi_j}{\partial y}\right)\frac{\partial\Phi_k}{\partial x} + \left(\frac{\partial^2\Phi_i}{\partial x^2}\frac{\partial\Phi_j}{\partial y} - \frac{\partial^2\Phi_i}{\partial x\partial y}\frac{\partial\Phi_j}{\partial x}\right)\frac{\partial\Phi_k}{\partial y}\right)d\Omega.$$

In this study, the displacement, u, and airy stress, v, of the structure are piecewise polynomial functions. Following the Galerkin method and FE discretization, the weighted basis functions introduced in Eq. (2) are plugged in equations (8)–(10).

The element matrices form the global matrices, e.g., F is assembled from F_e. Finally, the discrete form is given by

$$\begin{cases} Ku + \left(u^T B^m v\right)_{1\le m\le N} + \left(v^T B^m u\right)_{1\le m\le N} = F, \\ Kv - \left(u^T B^m u\right)_{1\le m\le N} = 0, \end{cases}$$

where

$$\left(u^T B^m v\right)_{1\le m\le N} := \left[u^T B^1 v \;\; u^T B^2 v \;\; \cdots \;\; u^T B^N v\right]^T,$$
$$u = (u_1, \ldots, u_N)^T, \quad v = (v_1, \ldots, v_N)^T.$$

Due to the nonlinearity, we use Newton's iterative methods based on the following formulation:

$$R(u,v) = \left(Ku + \left(u^T B^m v\right)_{1\le m\le N} + \left(v^T B^m u\right)_{1\le m\le N} - F,\, Kv - \left(u^T B^m u\right)_{1\le m\le N}\right).$$

The solution can be compacted into $Z^n = (u^n, v^n)$. The Jacobian of R is needed; it is defined by (Bv and Bu are computed by using $\left(u^T B^m v\right)_{1\le m\le N}$):

$$J(Z) = \begin{bmatrix} K + Bv & Bu \\ -Bu & K \end{bmatrix}.$$

Then, in each iteration:

$$Z^{n+1} = Z^n + \delta^n,$$

where δ^n is obtained by solving the following system:

$$J(Z^n)\delta^n = -R(Z^n).$$

4 Numerical Results

In this section numerical examples are presented. First, the presented method is verified against an analytical solution. Also, the convergence rate is investigated. Finally, we compare the results obtained from both mixed FEM and isogeometric analysis.

4.1 Convergence Study

The considered benchmark problem is a unit square with simply supported boundary condition. We use

$$\begin{aligned} u &= x^3(1-x)^3 y^3(1-y)^3, \\ v &= \sin(\pi x)\sin(\pi y). \end{aligned}$$

as exact solution in Eq. (4) to obtain f and g (g is added to the right hand side of the bottom of equations (5) and (4)). Therefore, in our numerical method, f and g are implemented as inputs, and u and v are approximated. Figure 6 shows the mesh and numerical solution. This mesh is built from 432 elements (microtriangles) created by the PS method; the red solid nodes are independent nodes (free) whose values

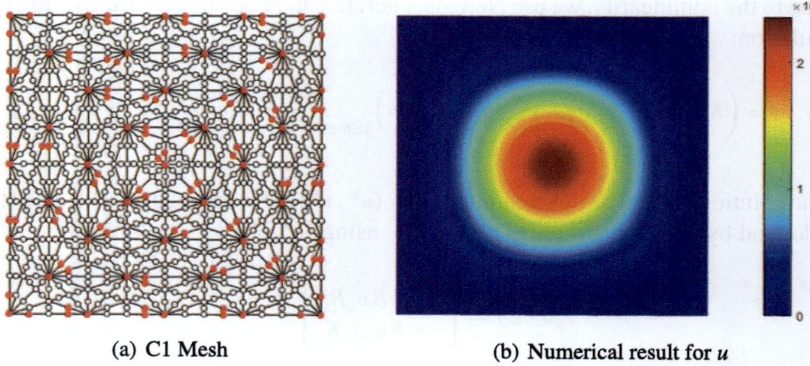

Fig. 6 Benchmark Example: **a** mesh with 432 elements created by the PS method; the red solid nodes are independent nodes and white nodes are dependent nodes, **b** numerical solution for u using C^1 constraints

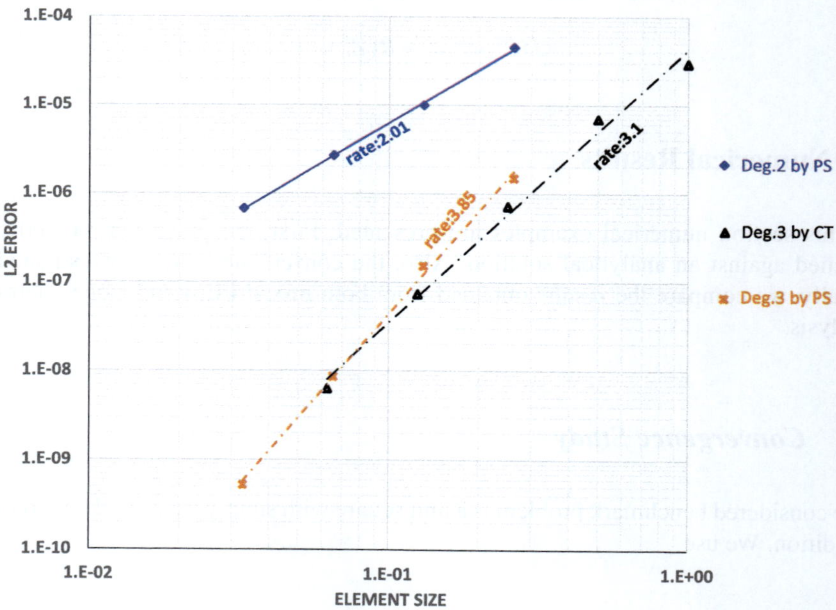

Fig. 7 L^2 error analysis. The best convergence rate (quadratic) is obtained for TIGA using quadratic triangular elements. Using cubic elements, the convergence rate is 3 for TIGA-CT and 4 for TIGA-PS (optimal rate is expected to be 4)

determine the values of the white nodes (dependent nodes) through the smoothness constraints, as explained in Sect. 2.

We consider three error norms in our convergence analysis: the L^2 norm, and the H^1 and H^2 seminorm. First, we define the following error function:

$$e = u^h - u^a,$$

where u^h and u^a represent the numerical and exact solutions. We use e in computing the error norms.

The convergence plot for the L^2 error is illustrated in Fig. 7; we compute this norm by

$$\|e\|_{L^2(\Omega)} = \sqrt{\int_\Omega e \cdot e d\Omega}.$$

The best convergence rate is expected to be as follows: a rate of 2 for quadratic elements and a rate of 4 for cubic elements. We can observe in Fig. 7 that this best convergence rate is obtained for both quadratic (rate of 2) and cubic elements (rate of 4) created by the PS method. We note that in the PS method, an optimal rate of 3 $(p + 1 = 3)$ is not attainable for quadratic elements [2, 24].

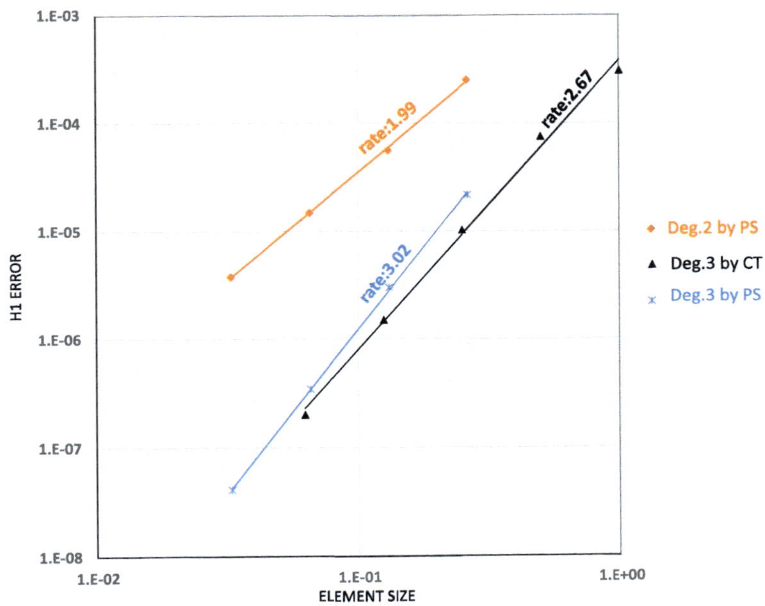

Fig. 8 H^1 error analysis. The best convergence rates are 2 and 3 for quadratic and cubic elements, respectively. We can observe that these optimal convergence rates are obtained for quadratic elements (rate of 2) and for cubic PS elements (rate of 3). For cubic CT elements a sub-optimal rate, 2.67, is obtained

Figure 8 presents the convergence plot for the H^1 error analysis. For this analysis, we use the seminorm given by

$$| e |_{H^1(\Omega)} = \sqrt{\int_{\Omega} (\nabla e \cdot \nabla e) d\Omega}.$$

The best convergence rate is expected to be 2 and 3 for quadratic and cubic elements, respectively. We can observe in Fig. 8 that the optimal convergence rate is obtained for quadratic elements (rate of 2). For cubic elements, a sub-optimal rate (rate of 2.67) is obtained when triangular elements are created by the CT method, but for cubic elements created by the PS method the optimal rate (rate of 3) is obtained.

The convergence plot for the H^2 error analysis is illustrated in Fig. 9. For this analysis, we use the seminorm defined by

$$| e |_{H^2(\Omega)} = \sqrt{\int_{\Omega} (d^2 e \cdot d^2 e) d\Omega},$$

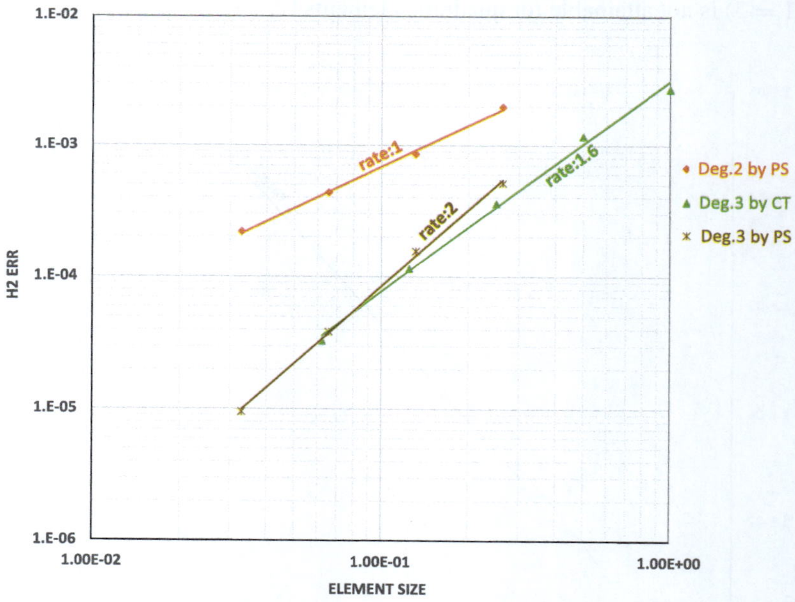

Fig. 9 H^2 error analysis. The best convergence rates are 1 and 2 for quadratic and cubic elements, respectively. We can observe that these optimal convergence rates are obtained for quadratic elements (rate of 1) and cubic PS elements (rate of 2). For cubic CT elements a sub-optimal rate, 1.6, is obtained

where d^2e is given by $d^2e = \begin{bmatrix} \frac{\partial^2 e}{\partial x^2} \\ \frac{\partial^2 e}{\partial x^2} \end{bmatrix}$. The best convergence rate is expected to be 1 and 2 for quadratic and cubic elements, respectively. We can observe in Fig. 9 that the optimal convergence rate is obtained for quadratic elements (rate of 1). For cubic elements, a sub-optimal rate (rate of 1.6) is obtained when triangular elements are created by the CT method, but the optimal rate (rate of 2) is obtained for cubic elements created by the PS method.

4.2 *Efficiency*

Figure 10 compares the efficiency of C^1 TIGA with mixed FEM. One can observe that C^1 TIGA is much more efficient than mixed FEM. For instance, assuming the H^1 error to be 10^{-6}, for quadratic elements, mixed FEM needs (7688 DOFs) 48% more DOFs than TIGA does (5204 DOFs). For cubic elements, mixed FEM needs (2600 DOFs) 150% and 490% more DOFs than TIGA-CT (1040 DOFs) and TIGA-PS (440 DOFs) does, respectively.

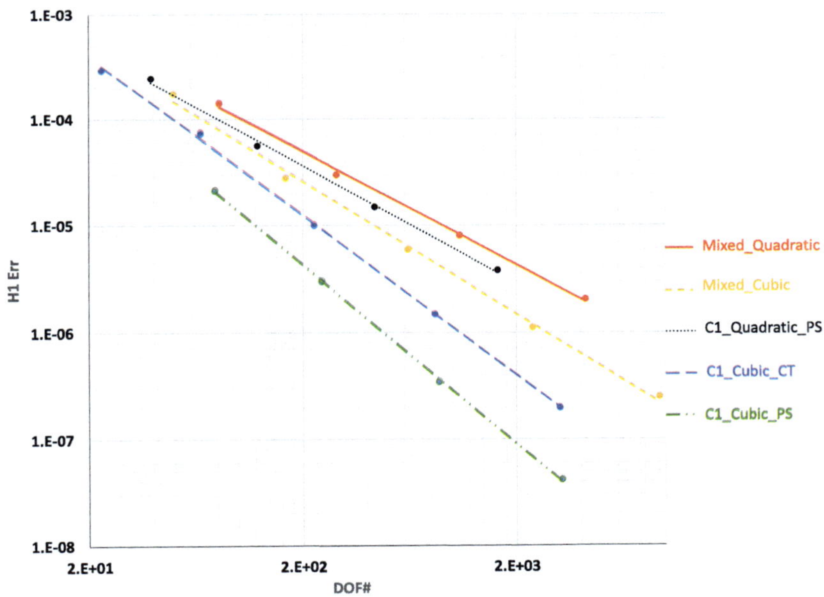

Fig. 10 H^1 error analysis with respect to number of DOF. Assuming the H^1 error to be 10^{-6}, for quadratic elements, mixed FEM needs (7688 DOFs) 48% more DOFs than TIGA does (5204 DOFs). For cubic elements, mixed FEM needs (2600 DOFs) 150% and 490% more DOFs than TIGA-CT (1040 DOFs) and TIGA-PS (440 DOFs) does, respectively

4.3 Accuracy and Reliability in Complex Geometries

Figures 11, 12 and 13 compare the reliability of the mentioned techniques. Each figure shows a geometry with a re-entrant corner, a C^0 mesh, a C^1 mesh created by the PS method, along with the results obtained by applying simply supported boundary condition and unity as the right hand side function, f. In mixed FEM, all nodes are free, while in C^1 TIGA, there are free nodes (red solid nodes) and dependent nodes (white nodes) in the mesh.

As it can be observed, on non-convex domains with re-entrant corners, the numerical results from mixed FEM (left) and TIGA (right) converge to different solutions; it is worth noting that we compare results after conducting mesh independence studies for each case. In Fig. 11, the maximum of the approximated u is 0.016 and 0.008 for mixed FEM (1601 nodes) and C^1 TIGA (1251 free nodes), respectively, i.e., the

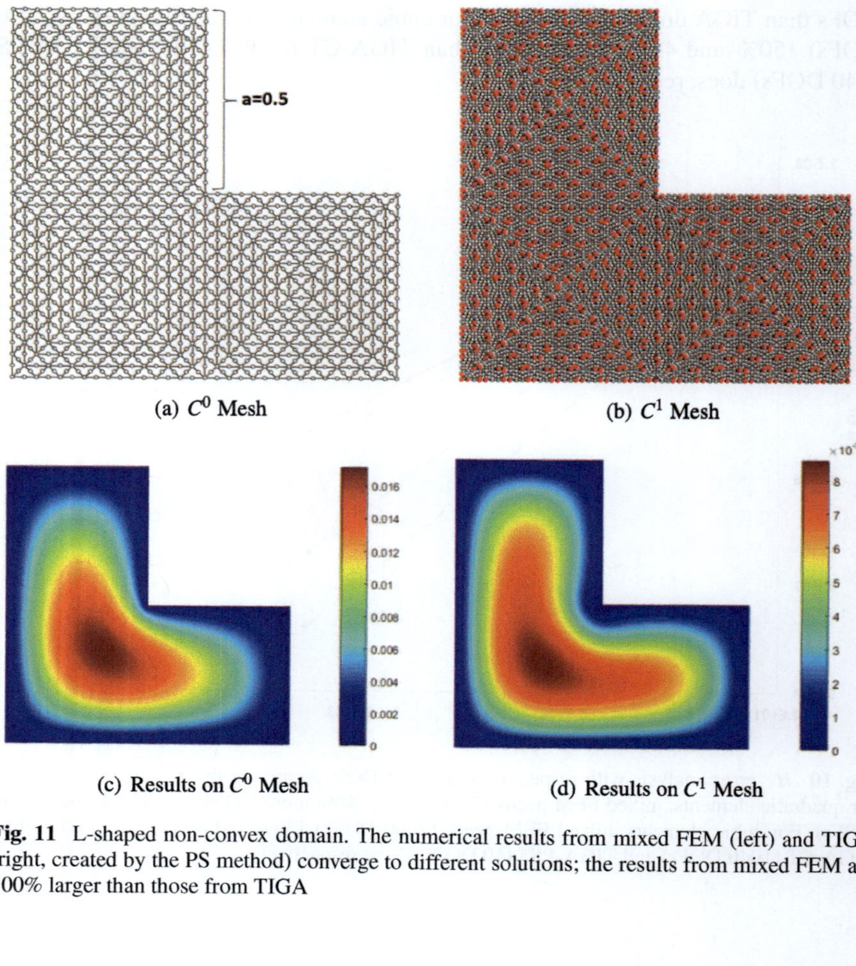

Fig. 11 L-shaped non-convex domain. The numerical results from mixed FEM (left) and TIGA (right, created by the PS method) converge to different solutions; the results from mixed FEM are 100% larger than those from TIGA

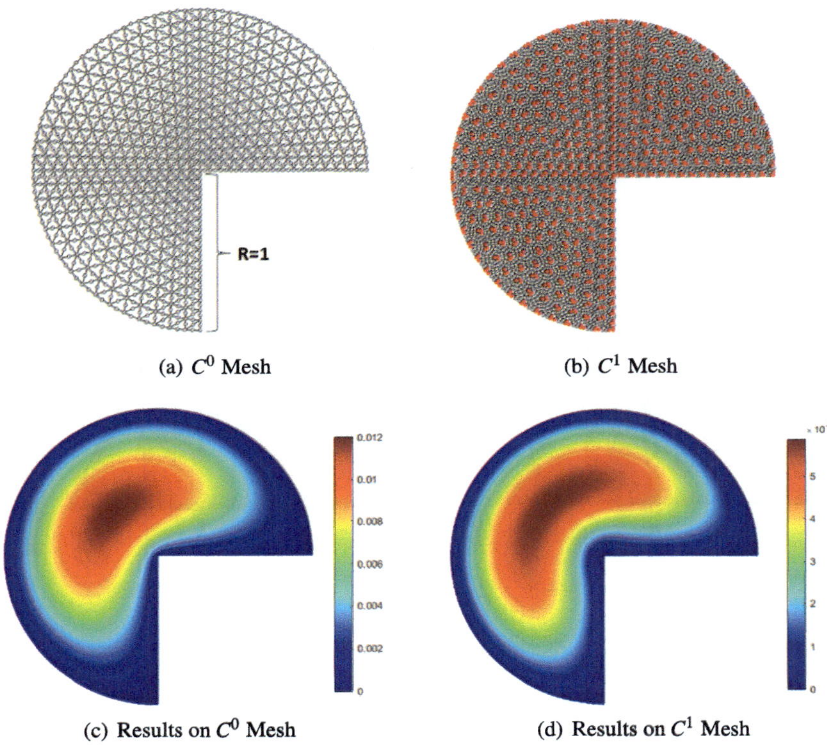

Fig. 12 Circular L-shaped non-convex domain. The numerical results from mixed FEM (left) and TIGA (right, created by the PS method) converge to different solutions; the results from mixed FEM are 100% larger than those from TIGA

result from mixed FEM is 100% larger than the result from C^1 TIGA. the same relative difference is observed in Fig. 12 (1275 free nodes in the C^1 mesh and 1617 nodes in the C^0 mesh). In Fig. 13, the maximum of the approximated u is 12×10^{-5} and 8×10^{-5} for mixed FEM (1617 nodes) and C^1 TIGA (1275 free nodes), respectively. Moreover, in the same case, we can observe a slight difference in the approximated distribution of u over the domain.

5 Conclusion

In this study we implemented C^1 TIGA and C^0 TIGA (mixed FEM) for the numerical solution of the von Karman equations. The error analysis of the presented methods demonstrate that we can obtain the best convergence rates for L^2, H^1 and H^2 using C^1 TIGA, using both quadratic and cubic elements.

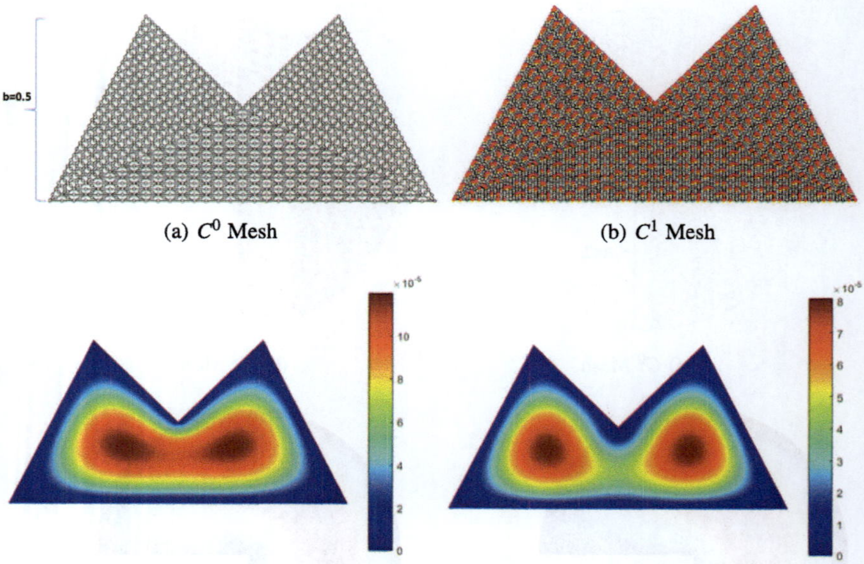

(a) C^0 Mesh (b) C^1 Mesh

(c) Results on C^0 Mesh (d) Results on C^1 Mesh

Fig. 13 M-shaped non-convex domain. The numerical results from mixed FEM (left) and TIGA (right, created by the PS method) converge to different solutions; the results from mixed FEM are 50% larger than those from TIGA

Regarding efficiency in the needed DOFs, TIGA is highly efficient. Mixed FEM is computationally expensive compared with C^1 TIGA for the same accuracy, e.g., for cubic elements, mixed formulations could cost 490% more DOFs for the same accuracy. This can also lead to convergence difficulties in nonlinear problems such as the von Karman equations for geometrically nonlinear plates.

Furthermore, results of the current work demonstrate that when the domain has a re-entrant corner, mixed FEM does not converge to the solution obtained from TIGA (C^1 elements). The difference between the two types of solutions is significant; the same observation can be found in the papers [7, 21].

This study shows that TIGA is a promising tool in investigating more engineering problems for which consideration of high accuracy, efficiency and nonlinearity are of particular interest.

Acknowledgements The authors would like to acknowledge the financial support from ARO grant W911NF-17-1-0020.

References

1. Adam, N., Le Tallec, P., Zarroug, M.: Multipatch isogeometric mortar methods for thick shells. Comput. Methods Appl. Mech. Eng. **372**, 113403 (2020)
2. Bartezzaghi, A., Dedè, L., Quarteroni, A.: Isogeometric analysis of high order partial differential equations on surfaces. Comput. Methods Appl. Mech. Eng. **295**, 446–469 (2015)
3. Benzaken, J., Evans, J.A., McCormick, S.F., Tamstorf, R.: Weak boundary condition enforcement for linear Kirchhoff-Love shells: Formulation, error analysis, and verification (2020). arXiv Preprint arXiv: 2007.01279
4. Brenner, S.C., Neilan, M., Reiser, A., Sung, L.Y.: A C^0 interior penalty method for a von Kármán plate. Numer. Math. **135**, 803–832 (2017)
5. Carstensen, C., Mallik, G., Nataraj, N.: A priori and a posteriori error control of discontinuous Galerkin finite element methods for the von Kármán equations. IMA J. Numer. Anal. **39**, 167–200 (2019)
6. Chen, H., Pani, A.K., Qiu, W.: A mixed finite element scheme for biharmonic equation with variable coefficient and von Karman equations (2020). arXiv Preprint arXiv:2005.11734
7. Gerasimov, T., Stylianou, A., Sweers, G.: Corners give problems when decoupling fourth order equations into second order systems. SIAM J. Numer. Anal. **50**, 1604–1623 (2012)
8. Herrema, A., Johnson, E., Proserpio, D., Wu, M.C.H., Kiendl, J., Hsu, M.C.: Penalty coupling of non-matching isogeometric Kirchhoff-Love shell patches with application to composite wind turbine blades. Comput. Methods Appl. Mech. Eng. **346**, 810–840 (2019)
9. Jaxon, N., Qian, X.: Isogeometric analysis on triangulations. Comput. Aided Des. **46**, 45–57 (2014)
10. Kapl, M., Sangalli, G., Takacs, T.: Isogeometric analysis with C^1 functions on planar, unstructured quadrilateral meshes. SMAI J. Comput. Math. **S5**, 67–86 (2019)
11. Kapl, M., Sangalli, G., Takacs, T.: A family of C^1 quadrilateral finite elements (2020). arXiv Preprint arXiv:2005.04251
12. Kiendl, J., Bazilevs, Y., Hsu, M.C., Wüchner, R., Bletzinger, K.U.: The bending strip method for isogeometric analysis of Kirchhoff-Love shell structures comprised of multiple patches. Comput. Methods Appl. Mech. Eng. **199**, 2403–2416 (2010)
13. Kiendl, J., Bletzinger, K.U., Linhard, J., Wüchner, R.: Isogeometric shell analysis with Kirchhoff-Love elements. Comput. Methods Appl. Mech. Eng. **198**, 3902–3914 (2009)
14. Lai, M.J., Schumaker, L.L.: Spline Functions on Triangulations, Encyclopedia of Mathematics and its Applications. Cambridge University Press, Cambridge (2007)
15. Liu, N., Jeffers, A.E.: A geometrically exact isogeometric Kirchhoff plate: Feature-preserving automatic meshing and C^1 rational triangular Bézier spline discretizations. Int. J. Numer. Meth. Eng. **115**, 395–409 (2018)
16. Maes, J., Bultheel, A.: C^1 hierarchical Riesz bases of Lagrange type on Powell-Sabin triangulations. J. Comput. Appl. Math. **196**, 1–19 (2006)
17. Mallik, G., Nataraj, N.: Conforming finite element methods for the von Kármán equations. Adv. Comput. Math. **42**, 1031–1054 (2016)
18. May, S., Vignollet, J., Borst, R.d.: Powell-Sabin B-splines and unstructured standard T-splines for the solution of the Kirchhoff-Love plate theory exploiting Beźier extraction. Int. J. Numer. Methods Eng. **107**, 205–233 (2016)
19. Michopoulos, J.G., Iliopoulos, A.P.: Modeling the multiphysics wrinkling instability of ionic polymer composite plates for artificial muscle applications. In: Proceedings of the ASME 2016 International Design Engineering Technical Conferences and Computers and Information in Engineering Conference. Volume 1A: 36th Computers and Information in Engineering Conference (2016)
20. Miyoshi, T.: A mixed finite element method for the solution of the von Kármán equations. Numer. Math. **26**, 255–269 (1976)
21. Nazarov, S.A., Sweers, G.: A hinged plate equation and iterated Dirichlet Laplace operator on domains with concave corners. J. Differ. Equ. **233**, 151–180 (2007)

22. Nelson, M.R., King, J.R., Jensen, O.E.: Buckling of a growing tissue and the emergence of two-dimensional patterns. Math. Biosci. **246**, 229–241 (2013)
23. Reinhart, L.: On the numerical analysis of the Von Karman equations: Mixed finite element approximation and continuation techniques. Numer. Math. **39**, 371–404 (1982)
24. Speleers, H., Dierckx, P., Vandewalle, S.: Multigrid methods with Powell-Sabin splines. IMA J. Numer. Anal. **28**, 888–908 (2008)
25. Speleers, H., Manni, C., Pelosi, F., Sampoli, M.L.: Isogeometric analysis with Powell-Sabin splines for advection-diffusion-reaction problems. Comput. Methods Appl. Mech. Eng. **221–222**, 132–148 (2012)
26. Toshniwal, D., Speleers, H., Hughes, T.J.R.: Smooth cubic spline spaces on unstructured quadrilateral meshes with particular emphasis on extraordinary points: Geometric design and isogeometric analysis considerations. Comput. Methods Appl. Mech. Eng. **327**, 411–458 (2017)
27. Wang, C., Xia, S., Wang, X., Qian, X.: Isogeometric shape optimization on triangulations. Comput. Methods Appl. Mech. Eng. **331**, 585–622 (2018)
28. Xia, S., Qian, X.: Isogeometric analysis with Bézier tetrahedra. Comput. Methods Appl. Mech. Eng. **316**, 782–816 (2017)
29. Xia, S., Wang, X., Qian, X.: Continuity and convergence in rational triangular Bézier spline based isogeometric analysis. Comput. Methods Appl. Mech. Eng. **297**, 292–324 (2015)
30. Yang, H.S., Dong, C.Y., Wu, Y.H.: Non-conforming interface coupling and symmetric iterative solution in isogeometric FE-BE analysis. Comput. Methods Appl. Mech. Eng. **373**, 113561 (2021)
31. Zareh, M., Qian, X.: A new plate formulation based on triangular isogeometric analysis. Proceedings of the ASME 2018 International Design Engineering Technical Conferences and Computers and Information in Engineering Conference. Volume 1A: 38th Computers and Information in Engineering Conference (2018)
32. Zareh, M., Qian, X.: Kirchhoff-Love shell formulation based on triangular isogeometric analysis. Comput. Methods Appl. Mech. Eng. **347**, 853–873 (2019)
33. Zhang, R., Qian, X.: Triangulation-based isogeometric analysis of the Cahn-Hilliard phase-field model. Comput. Methods Appl. Mech. Eng. **357**, 112569 (2019)

MIX
Papier aus verantwortungsvollen Quellen
Paper from responsible sources
FSC® C105338

Printed by Books on Demand, Germany